ADVANCES IN
X-RAY ANALYSIS
Volume 32

ADVANCES IN X-RAY ANALYSIS

Volume 32

Edited by

Charles S. Barrett

University of Denver
Denver, Colorado

John V. Gilfrich

Sachs/Freeman Associates
Washington, D.C.

Ron Jenkins

JCPDS–International Centre for Diffraction Data
Swarthmore, Pennsylvania

Ting C. Huang

IBM Almaden Research Center
San Jose, California

and

Paul K. Predecki

University of Denver
Denver, Colorado

Sponsored by
University of Denver Department of Engineering
and
JCPDS – International Centre for Diffraction Data

PLENUM PRESS • NEW YORK AND LONDON

The Library of Congress cataloged the first volume of this title as follows:

Conference on Application of X-ray Analysis.
Proceedings 6th–   1957–   [Denver]

v. illus. 24-28 cm. annual.
No proceedings published for the first 5 conferences.
Vols. for 1958–   called also: Advances in X-ray analysis, v. 2-
Proceedings for 1957 issued by the conference under an earlier name: Conference on Industrial Applications of X-ray Analysis. Other slight variations in name of conference.
Vol. for 1957 published by the University of Denver, Denver Research Institute, Metallurgy Division.
Vols. for 1958–   distributed by Plenum Press, New York.
Conferences sponsored by University of Denver, Denver Research Institute.
1. X-rays — Industrial applications — Congresses. I. Denver University. Denver Research Institute   II. Title: Advances in X-ray analysis.
TA406.5.C6   58-35928

ISBN 0-306-43236-6

Proceedings of the Thirty-seventh Annual Conference on Applications of X-Ray Analysis, held August 1–5, 1988, in Steamboat Springs, Colorado

© 1989 Plenum Press, New York
A Division of Plenum Publishing Corporation
233 Spring Street, New York, N.Y. 10013

Printed in the United States of America

The 37th Annual Denver Conference on Applications of X-Ray Analysis
was held August 1-5, 1988, at the Sheraton Steamboat Resort and Conference
Center, Steamboat Springs, Colorado. As usual, alternating with x-ray
diffraction, the emphasis this year was x-ray fluorescence, but as has been
the pattern for several occasions over the last few years, the Plenary
Session did not deal with that subject, specifically. In an attempt to
introduce the audience to one of the new developments in x-ray analysis,
the title of the session was "High Brilliance Sources/Applications," and
dealt exclusively with synchrotron radiation, a topic which has made a very
large impact on the x-ray community over the last decade. As the organizer
and co-chairman of the Plenary Session (with Paul Predecki), it is my
responsibility to report on that session here.

The Conference had the privilege of obtaining the services of some of
the preeminent practitioners of research using this remarkable x-ray
source; they presented the audience with unusually lucid descriptions of
the work which has been accomplished in the development and application of
the continuous, high intensity, tunable, polarized and collimated x-rays
available from no facility other than these specialized storage rings.

The opening lecture (and I use that term intentionally) was an
enthusiastic description of "What is Synchrotron Radiation?" by Professor
Boris Batterman of Cornell University and the Cornell High Energy
Synchrotron Source (CHESS). His many years of teaching and research were
obvious in the manner of his presentation, holding the attention of the
audience as he dealth with the sophisticated physics involved in the
production, extraction and manipulation of the x-ray beam for experimental
purposes. Dean Chapman, from the National Synchrotron Light Source (NSLS)
at Brookhaven, followed with a description of the general configuration of
the beamlines necessary to make use of the x-radiation. Such features as
shutters, vacuum, shielding and safety were discussed. Particular emphasis
was placed on the design of beamline optics, the monochromators and mirrors
required to provide focussed or collimated, monochromatic or continuous
spectrum beams, as well as the problems encountered due to the power
loading which those components must withstand because of the high intensity
of the radiation.

Following those two presentations, two talks were given describing
specific applications. John Newsam (with co-authors H. E. King, Jr., and
K. S. Liang), from Exxon Research and Engineering Company, narrated the
myriad ways in which x-ray diffraction could be conducted, using either
monochromatic or white radiation. Various single crystal experiments were
mentioned, including high resolution measurements, Laue techniques and
microcrystal analyses. Powder diffraction results addressed were high
resolution, energy-dispersion diffraction and anomalous scattering
measurements. I then had the privilege of presenting my thoughts on the

present status of x-ray fluorescence analysis using synchrotron radiation, polychromatic and monochromatic, as the excitation source. Research into the capabilities of this technique is being conducted around the world, and I took advantage of my personal knowledge of that effort to present a brief review of the accomplishments, including the use of both energy- and wavelength-dispersive detection.

I found the experience of organizing and participating in such a Plenary Session very rewarding. Even though I have been associated with work going on in the field for about seven years, I was surprised to discover how little I knew about the wide variety of research being conducted at the different facilities. It is even reasonable to assume, as John Newsam said, that the next few years will witness a great increase in the number of materials problems that are tackled using synchrotron radiation techniques.

John V. Gilfrich
Washington, D.C.
January 1989

PREFACE

This volume constitutes the proceedings of the 1988 Denver Conference on Applications of X-Ray Analysis and is the 32nd in the series. The conference was held August 1-5, 1988, at the Sheraton Steamboat Resort & Conference Center, Steamboat Springs, Colorado. The general chairmen were: J. V. Gilfrich, Sachs/Freeman Associates, NRL; and P. K. Predecki, University of Denver; with C. S. Barrett of the University of Denver as honorary chairman. The conference advisory committee this year consisted of: C. S. Barrett - University of Denver, J. V. Gilfrich - Sachs/Freeman Associates, NRL, R. Jenkins - International Centre for Diffraction Data, D. E. Leyden - Philip Morris USA, J. C. Russ - North Carolina State University, C. O. Ruud - The Pennsylvania State University, and P. K. Predecki - University of Denver. We take this opportunity to thank the advisory committee for their active participation, tireless efforts and able guidance which made this conference successful.

The conference plenary session was organized and chaired by J. V. Gilfrich, Sachs/Freeman Associates/NRL, Washington, D.C., and was entitled: "HIGH BRILLIANCE SOURCES/APPLICATIONS."

The invited papers on the program are listed below.

"What is Synchrotron Radiation?" B. W. Batterman

"Synchrotron Radiation Facilities," D. Chapman

"X-Ray Diffraction Using Synchrotron Radiation,"
    J. M. Newsam, H. E. King, Jr., and K. S. Liang

"XRF Using Synchrotron Radiation," J. V. Gilfrich

"X-Ray Diffraction Analysis of High Tc Superconducting Thin Films,"
    T. C. Huang, A. Segmüller, W. L. Lee, D. C. Bullock and R. Karimi

"Parallel Beam and Focusing Powder Diffractometry,"
    W. Parrish and M. Hart

"Applications of Dual-Energy X-Ray Computed Tomography to Structural Ceramics," W. A. Ellingson and M. W. Vannier

"X-Ray Microtomography,"
    H. W. Deckman, B. F. Flannery, J. H. Dunsmuir and K. D'Amico

"Advances in Elemental and Chemical State Imaging in Two and Three Dimensions Using Synchrotron Radiation,"
    J. H. Kinney, M. C. Nichols, Q. C. Johnson, U. Bonse, R. A. Saroyan and R. Nusshardt

In addition to the Plenary Session, the following Special Sessions were held:

- o X-RAY TOMOGRAPHY, IMAGING, AND TOPOGRAPHY, chaired by P. Engler, Standard Oil R&D; and N. Gurker, Technical University of Vienna, Austria.
- o APPLICATIONS OF DIGITIZED XRD PATTERNS, chaired by D. K. Smith, Penn State University; and R. Jenkins, International Centre for Diffraction Data.
- o X-RAY STRESS ANALYSIS, chaired by A. Krawitz, University of Missouri; and Y. Hirose, Kanazawa University, Japan.
- o HANDLING TECHNIQUES IN QUALITATIVE AND QUANTITATVE ANALYSIS, chaired by W. Schreiner, Philips Laboratories; and J. Nusinovici, Socabim, Paris, France.
- o THIN FILM APPLICATIONS OF XRD AND XRF, chaired by T. Huang, IBM Almaden Research Center; and J. C. Russ, North Carolina State University.
- o ON-LINE X-RAY ANALYSIS: XRF AND XRD, chaired by A. Harding, Colorado State University; and B. Cross, Kevex Corp.

Tutorial workshops on various XRF and XRD topics were held during the first two days of the conference. These are listed below with the names of the workshop organizers and instructors.

WF1 XRF QUANTITATION FOR BEGINNERS I; G. R. Lachance, Geological Survey of Canada (chair); and Michael Rokosz, Ford Motor Co.
WD1 RANDOM AND SYSTEMATIC ERRORS IN XRD; G. J. McCarthy, North Dakota State University (chair); T. G. Fawcett, Dow Chemical Co.; G. P. Hamill, GTE Laboratories; R. Jenkins, International Centre for Diffraction Data; and W. N. Schreiner, Philips Laboratories.
WF2 XRF QUANTITATION FOR BEGINNERS II; G. R. Lachance, Geological Survey of Canada (chair); and Michael Rokosz, Ford Motor Co.
WD2 USE/ABUSE OF STANDARDS FOR XRD; C. R. Hubbard, Oak Ridge National Lab (chair); D. Beard, Siemens Allis; W. Wong-Ng, National Bureau of Standards; G. J. McCarthy, North Dakota State University; and T. G. Fawcett, Dow Chemical Co.
WD3 MICRODENSITOMETER METHODS; Allan Brown, Studsvik Nuclear, Sweden (chair); and C. M. Foris, E.I. duPont de Nemours (chair).
WF3 NEW X-RAY CHARACTERISTIC LINE NOMENCLATURE; R. Jenkins, International Centre for Diffraction Data (chair).
WF4 XRF SAMPLE PREPARATION I; V. E. Buhrke, The Buhrke Co. (chair); Vlad Kocman, Domtar Inc., Canada; Robert Wilson, Westinghouse Hanford; Signa Fegley, Jessop Steel; Brad Wheeler, Link Analytical; Alyssa Malen, Construction Tech Labs Div. Portland Cement; and Fernand Claisse, Corporation Scientifique Claisse, Quebec.
WD4 DIFFRACTION PEAK BROADENING ANALYSIS; R. J. DeAngelis, University of Kentucky (chair); and Ashok G. Dhere, E. I. duPont., Chattanooga, TN.
WF5 XRF SAMPLE PREPARATION II; V. E. Buhrke, The Buhrke Co. (chair); Vlad Kocman, Domtar Inc., Canada; Robert Wilson, Westinghouse Hanford; Signa Fegley, Jessop Steel; Brad Wheeler, Link Analytical; Alyssa Malen, Construction Tech Labs Div. Portland Cement; and Fernand Claisse, Corporation Scientifique Claisse, Quebec.
WF6 INTERFACING XRF/XRD EQUIPMENT TO PC's; J. C. Russ, North Carolina State University (chair); and Ted Satterfield, The Nucleus, Oak Ridge, TN; D. C. Leepa, Brimrose Corp.; and Tom Baum, Kaufman, TX.
WD5 MONOCHROMATIZATION TECHNIQUES; R. Jenkins, International Centre for Diffraction Data (chair); Don Beard, Siemens Energy & Automation.

The total number registered for the conference was 413, over 260 of whom registered for one or more workshops. We are particularly indebted to the workshop organizers and instructors who gave unselfishly of their time and experience to make the workshops an outstanding part of the conference.

The conference dinner attendance was 185. A memorable evening of entertainment was provided by "The Powdermen," after the dinner.

On behalf of the organizing committee, I would like to sincerely thank the plenary session chairman, the invited speakers, the special session chairmen, the contributed session chairmen (J. P. Willis, G. R. Lachance, T. K. Smith, M. Rokosz, I. C. Noyan, G. M. Borgonovi, J. D. Zahrt and R. Rousseau), the poster session chairmen (S. Piorek, Y. Gohshi, C. S. Barrett and C. R. Hubbard) and the authors for their contributions. The exceptional efforts of all these people made the sessions a great success.

My special thanks to the Conference staff: Louise Carlson, Penny Eucker, John Getty, Lucien Hehn, Jim Ludlam, Brenda Ziegler, and to the Conference secretary, Lynne Bonno, all of whom worked long and unusual hours to make the Conference successful.

> Paul K. Predecki
> for the Organizing Committee
> January 1989

UNPUBLISHED PAPERS

The following papers were presented at the conference but are not published here for various reasons.

"Quantitative Determination of Mineral Phases in Sand by X-Ray Diffraction," R. M. Abu-Eid and I. Abdul-Rahman, Kuwait Institute for Scientific Research, Safat, Kuwait.
"Matrix Effect in X-Ray Fluorescence Analysis: A Practical Approach," R. M. Abu-Eid and I. Abdul-Rahman, Kuwait Institute for Scientific Research, Safat, Kuwait.
"What is Synchrotron Radiation?" B. W. Batterman, Cornell University, Ithaca, NY.
"Enhancement of L_3 Subshell X-Ray Fluorescence Cross-Sections due to Coster-Kronig Transitions," C. Bhan and B. Singh, Haryana Agricultural University, Hisar, India; S. N. Chaturvedi and N. Nath, Kurukshetra University, India.
"Quantitative Multicomponent Analysis by X-Ray Diffraction Using the Rietveld Method," D. L. Bish, Los Alamos National Lab, NM; and S. A. Howard, University of Missouri, Rolla, MO.
"Comparison of a Solid State Si Detector to a Conventional Scintillation Detector-Monochromator System in X-Ray Diffraction Analysis of Geological Materials," D. L. Bish and S. J. Chipera, Los Alamos National Laboratory, NM.
"Development and Optimization of a Thin Film Attachment for a Standard D500 Diffractometer," N. Broll and M. Haase, Siemens AG, Karlsruhe, West Germany.

"Synchrotron Radiation Facilities," D. Chapman, Brookhaven National Lab, Upton, NY.

"An Evaluation of the Rietveld Technique for Quantitative Analysis by X-Ray Powder Diffraction," J. Cline, National Bureau of Standards, Gaithersburg, MD.

"The Measurement of the Residual Strain in Polytetrafluoroethylene and Polypropylene Pipe Liners Using X-Ray Diffraction," C. E. Crowder, Dow Chemical Company, Midland, MI.

"Quantitative Phase Analysis through Digitized XRD Patterns," E. P. Farley, SRI International, Menlo Park, CA.

"Changes in Cu K-alpha Emission from $YBa_2Cu_3O_{7-\delta}$ Compounds," S. Fukushima and Y. Gohshi, Univ. of Tokyo, Japan; S. Kohiki, Matsushita Technoresearch, Osaka, Japan; T. Wada, Matsushita Electric Industries, Osaka, Japan.

"FDPP - An Interactive Program for the Processing of Digitized Fiber Diffraction Patterns," K. H. Gardner and R. M. Hilmer, E. I. DuPont, Wilmington, DE.

"Digital X-Ray Imaging of the Crystallinity in Polymers," R. W. Green, M. F. Garbauskas and D. G. LeGrand, General Electric Co., Schenectady, NY.

"Computed Tomography Concepts in Micro XRF-Imaging," N. Gurker, M. Bavdaz, Technical University Vienna, Austria; A. Knochel and W. Peterson, University Hamburg, Germany

"XRD Methods for Thin Film Analysis," E. Houtman, Philips I&E Division, Almelo, The Netherlands

"Applications of X-Ray Thin Film Diffraction Method," M. Katayama and M. Shimizu, Kawasaki Steel Corporation, Japan.

"Thickness Monitoring of Thin Composite Films Using DQM and X-Ray Fluorescence Technique," D. K. Kaushik, Dayanand College, Hisar, India; C. Bhan, Haryana Agricultural Univ., Hisar, India; S. K. Chattopadhyaya and N. Nath, Kurukshetra Univ., India.

"Observation of Grain Growth during Secondary Recrystallization in Grain-Oriented Silicon Steel with White X-Ray Topography by Synchrotron Radiation," K. Kawasaki and M. Matsuo, Nippon Steel Corporation, Kawasaki, Japan.

"X-Ray Diffraction Pattern Fitting Analysis of Phases in High-Temperature Superconducting Copper Oxides," M. Kimura and M. Matsuo, Nippon Steel Corp., Kawasaki, Japan.

"Advances in Elemental and Chemical State Imaging in Two and Three Dimensions using Synchrotron Radiation," J. H. Kinney, Q. C. Johnson and R. A. Saroyan, Lawrence Livermore National Laboratory, Livermore, CA; M. C. Nichols, Sandia National Laboratories, Livermore, CA; U. Bonse and R. Nusshardt, University of Dortmund, W. Germany.

"Determination of Fillers in Paper by X-Ray Fluorescence Spectrometry using Pulverized Samples," V. Kocman and L. M. Foley, Domtar Inc., Quebec.

"Application of Low-Cost XRF Technology to On-Line Analysis," S. R. Little, ASOMA Instruments, Inc., Austin, TX

"Nondestructive Measurement of Layer Thicknesses in Double Heterostructures by X-Ray Diffraction," A. T. Macrander, S. Lau, K. Strege and S.N.G. Chu, AT&T Bell Laboratories, Murray Hill, NJ.

"X-Ray Microdiffraction using Dual Nickel-Coated Focusing Mirrors and a Two Dimensional Position Sensitive Detector," T. F. McNulty, R. A. Larsen and J. M. Quigley, Nicolet Instrument Corp., Madison, WI.

"Development of a High Resolution X-Ray Powder Diffractometer and its Evaluation," S. Munekawa, Rigaku Corporation, Akishima, Japan; and H. Toraya, Nagoya Institute of Technology, Tajimi, Japan.

"Flow Visualization in Heterogeneous Core Samples Using X-Ray Computed Tomography," K. Narayanan, H. A. Deans, and W. F. Massell, University of Houston, Houston, TX.

"A General Geometric Modelling Approach for the Monte Carlo Simulation of Tomographic Systems," T. H. Prettyman, R. P. Gardner and K. Verghese, North Carolina State University, Raleigh, NC.

"XRF Analysis of Tool Steel with Theoretical Alpha Correction," X. Ronghou, L. Jinsheng and L. Yun, Central Iron & Steel Research Institute, Beijing, China.

"A Simple and Accurate Technique for Thin-Film X-Ray Fluorescence Analysis of Geologic Samples," D. K. Smith, Lawrence Livermore National Lab, Livermore, CA; and L. H. Cohen, University of California, Riverside, CA.

"On-Line X-Ray Fluorescence Measurements," G. B. Ury, Amoco Research Center, Naperville, IL.

"A Critical Comparison of X-Ray Diffraction and Barkhausen Noise Measurement of Residual Stress in 4340M Steel," D. M. Walker, Boeing Commercial Airplane Company, Seattle WA.

"Resolution of Practical Problems in the Residual Stress Analysis of CP Titanium Welds," D. M. Walker, Boeing Commercial Airplane Company, Seattle, WA.

"Criteria for Selecting a WD-XRF Spectrometer," P. L. Warren and A. E. Smith, Wilton Materials Research Centre, Cleveland, England.

"Mathematical Correction Models in X-Ray Fluorescence Utilizing Regression Analysis," B. D. Wheeler, LINK Analytical, Redwood City, CA

"Compton (Incoherent) Scattering of Sample Spectral Lines in XRFS (1): Evidence for the Phenomenon and Resulting Analytical Errors," J. P. Willis, University of Cape Town, Rondebosch, South Africa.

"Compton (Incoherent) Scattering of Sample Spectral Lines in XRFS (2): Problems in the Accurate Determination of Net Spectral Line Intensities in Samples of Low Mass Absorption Coefficient," J. P. Willis, University of Cape Town, Rondebosch, South Africa.

ESTABLISHMENT OF THE BARRETT AWARD

The Barrett Award in X-Ray Diffraction was established by the conference advisory committee in 1986 to recognize outstanding contributions to the field of powder diffraction.

The award was named in honor of Charles S. Barrett for his many years of exceptional work in the field. The Barrett Award thus is the counterpart of the Birks Award in X-Ray Spectrometry which was established at the same time. (The first Birks Award was made to R. Jenkins of the International Centre for Diffraction Data and is described in Volume 30 of Advances in X-Ray Analysis.)

Both awards are made possible through the generosity of the companies and organizations who exhibit their products at the Denver X-Ray Conference.

Paul K. Predecki
January 1989

PRESENTATION OF THE 1988 BIRKS AWARD TO EUGENE P. BERTIN

"For his many years of service to the x-ray analytical community, particularly the authorship of the monumental text which has been used to teach many fledgling workers in the field and which is used as a reference by all workers in the field."

In photo, left to right:
    L. S. Birks, Columbia, SC
    Eugene P. Bertin, Harrison, NJ, Recipient of the 1988 Birks Award

PRESENTATION OF THE FIRST BARRETT AWARD TO WILLIAM PARRISH

"For advances he has made in x-ray diffraction using conventional
and synchrotron sources; advances both in diffractometer instrumentation,
in powder methods and in the interpretation of diffraction patterns."

In photo, left to right:
    William Parrish, IBM Almaden Research Center, San Jose, CA,
        Recipient of the first Barrett Award in X-Ray Diffraction
    Charles S. Barrett, University of Denver, Denver, CO

CONTENTS

IV.   TECHNIQUES AND XRF INSTRUMENTATION

V.   XRF APPLICATIONS

CONTENTS

VI.   ANALYSIS OF THIN FILMS BY XRD AND XRF

VII. X-RAY STRESS ANALYSIS

VIII.   APPLICATIONS OF DIGITIZED XRD PATTERNS

IX.   QUALITATIVE AND QUANTITATIVE PHASE ANALYSIS
      DIFFRACTION APPLICATIONS

CONTENTS

SYNCHROTRON RADIATION X-RAY FLUORESCENCE ANALYSIS

John V. Gilfrich*

Dynamics of Solids Branch
Naval Research Laboratory
Washington, DC 20375-5000

*Also at:
Sachs/Freeman Associates, Inc.
1401 McCormick Drive
Landover, MD 20785-5396

INTRODUCTION

The physical principles giving rise to synchrotron radiation(SR), the facilities necessary to make use of this source of radiation and the way in which it can be used for x-ray diffraction experiments have been described in other parts of this proceedings. The use of synchrotron radiation as an excitation source for x-ray fluorescence takes advantage of many of its unique properties to provide the potential for an improved analytical capability, beyond that which can be realized with more conventional laboratory x-ray sources. Workers around the world are studying this application (as well as a wide variety of others), to establish the potential of this technique. Table 1 lists some of the facilities where significant XRF effort is being expended. This list is not intended to be complete, but just to convey some idea of the magnitude of the research which is being carried out.

Some (but certainly not all) of what will be presented here will reflect work in which we, at NRL, have been involved. For example, a few years ago[1], we illustrated the comparison of the intensity at a SR experimental station with an x-ray tube, as shown in Figure 1. Note the vertical scale, where the intensity is per milliamp; the conventional x-ray tube may operate up to 50 mA, while some storage rings are designed to operate up to 500 mA, making the comparison even more dramatic. Bear in mind that, for excitation of a sample, an x-ray tube may provide many more steradians than the 0.5 for which this comparison is made (1 mrad horizontal by 0.5 mrad vertical), making the relationship between the two curves considerably less dramatic. The import of this comparison is related to the particular value of SR for the application to an x-ray microprobe, where the strict collimation can be a distinct advantage for the optics necessary to focus the beam to a small size.

Table 1. Some synchrotron radiation facilities studying XRF

ENGLAND	Daresbury	SRS
Workers from Warwick, Eindhoven and Amsterdam		
FRANCE	Orsay	LURE
Workers from Institute Curie and University P. & M. Curie		
GERMANY	Hamburg	HASYLAB
Workers from University of Hamburg		
JAPAN	Tsukuba	PHOTON FACTORY
Workers from University of Tokyo and Photon Factory		
USA	Stanford	SSRL
Workers from Naval Research Lab., Oak Ridge National Lab. and Lawrence Berkeley Lab.		
	Brookhaven	NSLS
Workers from NRL, NSLS, National Bureau of Standards, United States Geological Survey, many Universities, and others.		
USSR	Novosibirsk	VEPP-3, VEPP-4
Workers from Institute of Nuclear Physics		

EXPERIMENTAL CONFIGURATIONS

The geometry for the measurements made by NRL at both SSRL[2] and NSLS[3] is illustrated in Figure 2. Although the details might differ at other facilities, this represents the way energy dispersion measurements have been made by most workers. At SSRL[2], we used the continuum straight out of the beam line, the helium path being used to avoid both air absorption and air scattering. Other workers have used an air path or, in a few cases, vacuum. At NSLS[3], we excited the samples using monochromatic radiation provided by beam-line optics consisting of a collimating mirror, a double-crystal monochromator and a focussing mirror. Other arrangements for monochromatic excitation have used the continuum emerging from the beam line with a crystal monochromator located in the hutch.

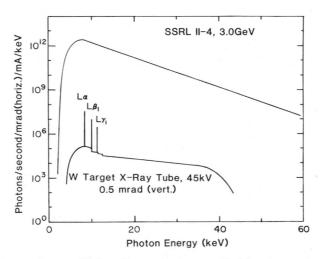

Fig. 1. Comparison of the SR spectrum available at a particular experimental station with that from a standard x-ray tube.

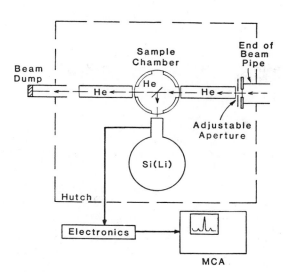

Fig. 2.   Experimental configuration for energy dispersion measurements.

Although the vast majority of SRXRF measurements have been made using energy dispersion, there have been some wavelength dispersion data taken. At both SSRL and NSLS, NRL used a small, homemade, crystal spectrometer, located in place of the Si(Li) detector in Figure 2, and enclosed in a helium-filled box.   And the workers from the University of Tokyo and the Photon Factory used a similar device to study the emission from very light elements (carbon, nitrogen and oxygen), excited by the quasi-monochromatic radiation produced by an undulator in the storage ring[4].   Several of the other facilities have suggested that the use of wavelength dispersion is a desirable option, but not many (if any) other reports of such use have been presented.

All of the experimental arrangements being utilized at the various SR laboratories have attempted to take advantage of the properties of this type of source which have the potential of demonstrating improvements in analytical capability.   Included among these are the very high intensity, as has been illustrated in Figure 1, the polarization of the beam in the plane of the electron orbit, the smooth continuum emitted from the tangent point, and the ability to monochromatize and/or focus the beam.   The high intensity, in fact, makes it a necessity to aperture the beam to a small size, usually, when making energy dispersion measurements, in order to avoid saturating the detector.   The polarization of the beam provides for a very significant reduction in background, by placing the detection system in the plane of the electron orbit; background is not completely eliminated because it is necessary to accept some vertical divergence, causing the beam to be somewhat less than 100% polarized.   The smooth continuum, devoid of any characteristic lines, avoids any interference such as might be experienced with x-ray tube excitation.   Using a monochromatic beam, produced by crystal optics, or a combination of crystals and mirrors, makes it possible to customize the excitation of specific elements, and to selectively excite particular elements to avoid interferences[5].   Focussing the beam to dimensions in the micrometer range provides the potential for an x-ray microprobe[6].

A SELECTION OF RESULTS

When the European Synchrotron Radiation Facility (ESRF) was first proposed about a decade ago, the Scientific Study included a section dealing with SRXRF, in which predictions were put forth concerning the energy dispersion detection limits which might be achieved using the operating parameters of HASYLAB at Hamburg, which was in operation at that time[7]. Figure 3 shows those predictions, compared to the initial measurements made by NRL at SSRL[1]. It can be seen that the predictions were not overly optimistic, particularly since the NRL measurements were made with only one-third of the storage ring current as that for which the predictions were made.

One of the first, if not <u>the</u> first, XRF experiment at a SR facility was performed by Sparks of Oak Ridge National Laboratory, who exposed some NBS SRM's to the continuum beam at SSRL[8]. Exciting the spectrum from NBS Orchard Leaves (SRM 1571), detection limits for Fe were determined to be 12 ppm and 5 ppm, with the storage ring at 1.83 and 3.1 GeV, respectively. But, the current in the ring was only 6 to 8 mA, about 10% of that used in its dedicated mode.

Some of the most definitive work has been performed at the Photon Factory (PF), by analysts from the University of Tokyo and the PF. Using thin samples of CHELEX 100 ion exchange resin, on which was deposited 20 ppm of Ca, Mn and Zn, they determined 100 second detection limits, for Mn, of 410 ppb for excitation by unfiltered continuum, 240 ppb for continuum filtered by 0.28 mm Al, and 70 ppb for 11 keV monochromatic excitation[5]. In this exercise they compared the SR results with conventional energy dispersion measurements, using a Ge fluorescer, which achieved a <u>600</u> second detection limit (again for Mn) of 2 pp<u>m</u>. These same workers have

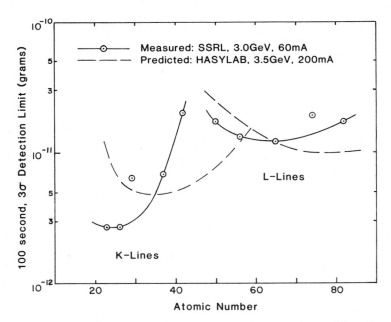

Fig. 3.   Measured detection limits (in grams) for thin films compared to predictions for HASYLAB.

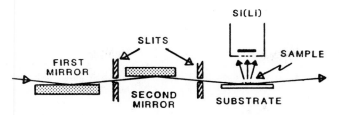

Figure 4.   The geometry for total reflection x-ray fluorescence analysis.

illustrated the way in which selective excitation can eliminate the interference between elements which have emission energies very close to one another.   By using a sample containing Ba and Ti in a concentration ratio of 20:1, selecting the excitation energy between the Ti K absorption edge (4.96 keV) and the Ba L_{III} edge (5.25 keV) prevented the Ba from being excited, and the Ti could be measured quantitatively, whereas, if the excitation energy was above 5.25 keV, the Ba L line overwhelmed the Ti.

The last application to be mentioned comes from Daresbury in England[9]. Using 15 keV primary radiation produced by a graphite monochromator with an energy resolution of 5% (dE/E), the concentration of arsenic at 10^{15} atoms/cm^2 in a Si semiconductor was measured with a HPGe detector.   The 100 second detection limit was determined to be 6 x 10^{13} atoms/cm^2.

LATEST DEVELOPMENTS

Focussing optics, to provide a microbeam at the sample, will make two things possible, primarily.   First, crystals and/or mirrors can produce a beam spot in the micrometer range, leading to an x-ray mocroprobe with spatial resolution comparable to an electron microprobe, but having the following advantages: no vacuum is required, preventing the deterioration of some kinds of samples; x-rays probe deeper into the sample than do electrons; and there is less damage (volatilization of components, redistribution of elements, changes in chemical bonding, and charring) to certain types of samples due to more diffuse energy deposition.   One of the primary efforts to produce such an x-ray microprobe is being implemented on beam line X-26 at NSLS[10], where applications will be attempted initially on geological and biomedical samples.   Second, for softer (lower energy) x-rays, Fresnel zone plates are an effective focussing device for an x-ray microscope, enabling the examination of living cells.   On the ten period mini-undulator at NSLS (beam line X-17T), soft x-ray images of zymogen granules, at a wavelength of 32 Å, have been shown with 750 Å spatial resolution[11].

The total reflection technique is a natural for SR, because of its inherent collimation.   The small divergence of the continuous beam can be caused to strike the sample at less than the critical angle, perhaps using a mirror (or mirrors) as a low pass filter.   A monochromatic beam can be used directly, or with focussing optics to confine the beam to the area of the sample on the totally reflecting substrate.   Figure 4 shows the geometry for this technique; the source may be the radiation from a conventional x-ray tube, rather severely collimated, or that emanating from the SR beam line.   Two mirrors are shown in the illustration, although a single mirror, or none at all, may sometimes be appropriate.   The workers from the PF have demonstrated a detection limit of 0.5 ppb for two microliters of a low concentration solution on a Si wafer[12].

One of the most practiced applications for SR is the study of the structure of materials by Extended X-Ray Absorption Fine Structure (EXAFS). Originally this technique involved the direct measurement of the absorption of the sample as the energy of the incident radiation was varied across the absorption edge of the element of interest.   The method was improved drastically (particularly for dilute or bulk samples) by the use of the fluorescence signal emitted by the element being studied, a direct measure of the photoelectric absorption.   Fluorescence EXAFS is, in fact, Appearance Potential X-Ray Fluorescence Analysis (APXRF)[13]; the magnitude of the increase in signal as the absorption edge is crossed, the absorption edge jump, is a measure of the number of atoms of the element of interest. Thus, it is possible to measure both the structure and elemental concentration in a sample in a single experiment.

CONCLUSION

This has been a necessarily brief introduction to SRXRF, which included an arbitrarily chosen set of examples of some of the work going on around the world.   It is impossible in a short review like this to include any more than a small fraction of the total effort; I hope that those researchers whose work has been neglected will understand that it is not because it is unappreciated.   In closing, it seems worthwhile to issue a caveat.   It seems unlikely that this technique will ever be used for routine analysis in any laboratory, but then who would have suspected, a few years ago, that we would see superconductors with critical temperatures above liquid nitrogen in our lifetime.

REFERENCES

1. J.V. Gilfrich, E.F. Skelton, D.J. Nagel, A.W. Webb, S.B.Qadri, and J.P. Kirkland, X-Ray Fluorescence Analysis using Synchrotron Radiation, in "Advances in X-Ray Analysis" Vol. 26, C.R. Hubbard, C.S. Barrett, P.K. Predecki and D.E. Leyden, eds., Plenum Publ. Co., New York (1983).
2. J.V. Gilfrich, E.F. Skelton, S.B. Qadri, J.P.Kirkland, and D.J. Nagel, Synchrotron Radiation X-Ray Fluorescence Analysis, Anal. Chem. 55:187 (1983).
3. J.V. Gilfrich, Naval Research Laboratory, unpublished data (1987).
4. A. Iida, Y. Gohshi, and H. Maezawa, Application of Synchrotron Radiation Excited X-Ray Fluorescence Analysis to Micro and Trace Element Determination, in "Advances in X-ray Analysis" Vol. 29, C.S. Barrett, J.B. Cohen, J. Faber, Jr., R. Jenkins, D.E. Leyden, J.C. Russ and P.K. Predecki, eds., Plenum Publ. Co., New York (1986).
5. A. Iida, K. Sakurai, T. Matsushita, and Y. Gohshi, Energy Dispersive X-Ray Fluorescence Analysis with Synchrotron Radiation, Nucl. Instr. and Meth. in Phys. Res. 228:556 (1985).
6. G.E. Ice and C.J. Sparks, Jr., Focusing Optics for a Synchrotron X-Radiation Microprobe, Nucl. Instr. and Meth. in Phys. Res. 222:121 (1984).
7. European Science Foundation, "European Synchrotron Radiation Facility: Supplement I, The Scientific Case," Y. Farge and P.J. Duke, eds., Strasbourg (1979), pp. 78-82.
8. C.J. Sparks, Jr., X-Ray Fluorescence Microprobe for Chemical Analysis, in "Synchrotron Radiation Research," H. Winick and S. Doniach, eds., Plenum Publ. Co., New York (1980).

9. D.K. Bowen, S.T. Davies, and T. Ambridge, Quantitative Analysis of
    Arsenic-Implanted Layers in Silicon by Synchrotron-Radiation-Excited
    X-Ray Fluorescence, <u>J</u>. <u>Appl</u>. <u>Phys</u>. 58:260 (1985).

10. K.W. Jones, B.M. Gordon, A.L. Hanson B.M. Johnson, M. Meron, J.G.
    Pounds, J.V. Smith, M.L. Rivers, and S.R. Sutton, FY 1987 Research
    Activities on the X-26 Beam Port, <u>in</u> "National Synchrotron Light
    Source Annual Report 1987," BNL Report 52131, S. White-DePace, N.F.
    Gmur and W. Thomlinson, eds., Brookhaven National Laboratory, New
    York (1987).

11. H. Rarback, D. Shu, S.C. Feng, H. Ade, J. Kirz, I. McNulty, D.P. Kern,
    T.H.P. Chang, Y. Vladimirsky, N. Iskander, D. Attwood, K. McQuaid,
    and S. Rothman, Scanning X-Ray Microscope with 75-nm Resolution,
    <u>Rev</u>. <u>Sci</u>. <u>Instrum</u>. 59:52 (1988).

12. A. Iida and Y. Gohshi, Energy Dispersive X-Ray Fluorescence Analysis
    using Synchrotron Radiation, <u>in</u> "Advances in X-Ray Analysis," Vol.
    28, C.S. Barrett, P.K. Predecki and D.E. Leyden, eds., Plenum Publ.
    Co., New York (1985).

13. J.P. Kirkland, J.V. Gilfrich, and W.T. Elam, Appearance Potential X-
    Ray Fluorescence Analysis, <u>in</u> "Advances in X-Ray Analysis," Vol. 31,
    C.S. Barrett, J.V. Gilfrich, R. Jenkins, J.C. Russ, J.W. Richardson,
    Jr., and P.K. Predecki, eds., Plenum Publ. Co., New York (1988).

X-RAY DIFFRACTION USING SYNCHROTRON RADIATION -

A CATALYSIS PERSPECTIVE

J. M. Newsam, H. E. King, Jr. and K. S. Liang

Exxon Research and Engineering Company
Route 22 East, Annandale, NJ 08801, USA

ABSTRACT

Synchrotron X-radiation provides unique opportunities for diffraction experiments and, therefore, for extending our understanding of the structure - property interplay in catalyst systems. The present status of opportunities and applications of synchrotron X-ray diffraction techniques in the structural chemistry and catalysis science areas is overviewed, and illustrated by selected recent results.

INTRODUCTION

Most of the properties of solids that are exploited commercially derive from particular structural characteristics. A knowledge of the relevant details of structure therefore provides a basis on which to rationalize and to understand performance. The developing appreciation of the structure - property interplay that diffraction studies provide offers the promise of being able to recognize and design those features on an atomic scale which are most desirable for a particular application. A range of materials are of importance in hydrocarbon conversions and separations in the petroleum and petrochemical industries, spanning metal surfaces, metal dispersions on amorphous or crystalline supports, bulk mixed metal oxides and crystalline microporous materials such as zeolites. The need to investigate the structures of these various classes of materials has necessitated implementing a range of synchrotron X-ray diffraction capabilities. These areas, which are overviewed here, illustrate well the particular advantages afforded by the synchrotron X-ray source.

The properties of synchrotron X-rays and available facilities have been introduced and discussed elsewhere [1-4]. The synchrotron X-ray source provides four particular qualities. It is extremely intense, or, more appropriately stated extremely bright (the photon flux as photons s^{-1} eV^{-1} cm^{-2} at the source is extremely high, but, in addition, the beam is intrinsically tightly focussed, enabling optimal use of X-ray focusing and optics technology and providing effective flux gains for diffraction experiments of many orders of magnitude.) It is continuous over a broad range of wavelengths which depends on the synchrotron ring energy and the structure of the magnet sections producing the X-ray beam. It is highly polarized and it has time structure (a bunch separation of the order of 10^{-8} s). Each of these qualities affords particular opportunities for diffraction methods. However, although synchrotron X-radiation capabilities have been in use for well over a decade [1-4], accessibility to a broad community has, until relatively recently, been limited and the potential for synchrotron X-diffraction in the inorganic, materials and catalyst science areas has only begun to be realized to a widespread extent.

EXXON BEAM LINES AT NSLS

The layout of one of three Exxon beam lines on the X10 port at NSLS (X10A) is shown schematically in Figure 1. The instrumentation on X10 was designed for flexibility and automated control [5]. Most of the range of experiments mentioned below were performed on X10A which accepts 3.7 mrad of radiation from a bending magnet. The radiation is focussed by a cylindrically bent, water cooled Pt-coated quartz mirror at 12m, the 1:1 position, and monochromated by a pair of parallel flat Si (111) crystals (the first of which is water cooled – Figure 1). In addition to the diffraction studies mentioned here, a series of other experiments have also been turned-over in the X10A station, including diffuse X-ray scattering [6], high resolution small angle scattering [7], and microtomography [8]. The side station, X10B, which was brought on-line just before the long shutdown began in March 1987, will be used primarily for crystallography experiments. This line has a cylindrical Pt-coated quartz mirror at 13m, and a single monochromator crystal at 14 - 16m. The monochromator crystal (which diffracts the beam sideways, providing room for the hutch installation) is asymmetrically cut and triangular, and may be bent about the vertical axis normal to the beam to provide horizontal focussing at the sample position. The centre station, X10C, will be equipped for EXAFS measurements under controlled, reaction conditions, and for diffraction experiments that require precise wavelength tuneability, such as diffraction using anomalous scattering from powder samples.

SURFACE X-RAY DIFFRACTION

Crystallographic studies of surfaces have traditionally relied heavily on low energy electron diffraction (LEED) [9], although the complication of multiple scattering (that is implicitly associated with the high scattering cross-sections that afford surface sensitivity) makes quantitative treatments of scattered intensities problematic. The lower cross-sections for X-rays generally permit scattered intensities to be treated in the simpler kinematical approximation. The concomitantly large penetration depths, however, imply a corresponding general insensitivity to the structure of the surface region. Altering the geometry of the scattering experiment such that incident and scattered photons make grazing angles with the surface (so that the scattering vector is close to parallel to the surface rather than, as is usual, normal), enhances dramatically the relative contribution of the surface region to the measured scattered intensitites [10,11]. Using the photon fluxes available at the synchrotron, monolayer surface sensitivity can be attained [12].

The structures formed by lead on various copper surfaces are excellent model systems for study by grazing incidence X-ray scattering (GIXS). Lead does not diffuse to any

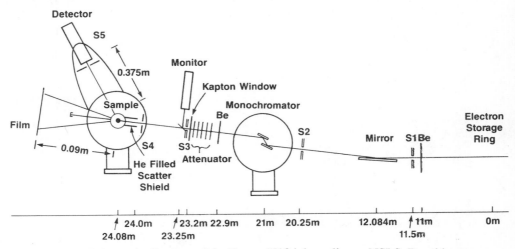

Figure 1. Schematic diagram of the Exxon X10A beamline at NSLS, Brookhaven.

appreciable extent into copper metal, and Pb monolayers can be easily prepared [13]. The system has been investigated in detail by conventional surface science techniques such as LEED, and the high Pb X-ray scattering power affords excellent count rates. LEED studies of monolayers of lead on the Cu (100) surface [14] revealed a $(5\sqrt{2} \times \sqrt{2})R45°$ structure with cmm diffraction symmetry and unit cell dimensions 18.074Å x 3.615Å. Detailed calculations of intensity versus electron energy based on different models indicated against a pseudo-hexagonal Pb coordination, but favored a c(2 x 2) antiphase domain structure [14]. GIXS studies of this system [15] yielded a set of 35 measured intensities that were corrected for Lorentz and sampling area factors, and used to compute a conventional Patterson synthesis. Interpretion of this map, together with subsequent least-squares optimization of structural parameters indicate rather complicated behavior, apparently with simultaneous near-equal occurrence of both pseudo-hexagonal and c(2 x 2) antiphase domain structures (both of which have c2mm symmetry). The observation of superlattice reflections close to b* suggests that the two structures exist in small domains with near-regular modulation along b, although detailed measurements of the superlattice reflection intensities are needed in order to evaluate the degree of coherence and perfection of this larger domain structure. The adsorbate - substrate and adsorbate - adsorbate interactions that are responsible for this unexpectedly complicated behavior are also being considered further.

In these GIXS studies of the Pb - Cu (100) system on X10A, the strongest measured intensities from the surface structure were over 1000 cps. The X-ray brightness of the bending magnet source is sufficient for studies of surface structures formed by atoms of low atomic number such as oxygen on copper [16], of surfactant monolayers on water [17], and of the growth and disappearance of surface steps on Cu(113) [18]. Further enhancements will permit studies of still more taxing problems, such as the structures of metal - hydrocarbon sorbate systems, and of oxide surface layers.

DIFFERENTIAL ANOMALOUS X-RAY SCATTERING

The continuous nature of the synchrotron source is exploited directly in EXAFS studies, in energy-dispersive powder diffraction and in single crystal Laue diffraction (mentioned further below). The ability to select a desired narrow wavelength band from the broad incident X-ray spectrum is the basis for studies that exploit anomalous scattering. This was early recognized as being of particular promise for tackling the phase problem in protein crystallography [19-22]. A recent study of $Zr_{0.81}Y_{0.19}O_{1.90}$ demonstrates how anomalous scattering effects can be used to provide separate information about different species occupying crystallographically equivalent sites [23]. Here we illustrate how this technique can be applied to study metal dispersions on crystalline or amorphous supports. Systems such as Pt - Al_2O_3, or Pt/Re - Al_2O_3 and related systems are the basis for commercial reforming catalysts.

Successive measurements were made on a sample of 5 wt % Pt on an η-Al_2O_3 support as prepared, after reduction in H_2 at 220°C for 60 min, and following re-exposure to air. The diffraction profiles $1 \le Q \le 11Å^{-1}$ ($Q = 4\pi\sin\theta/\lambda$) were in each case scanned using an intrinsic Ge detector ($\Delta E \sim 200eV$) at incident photon energies of 9 or 14eV, and 99 eV below the Pt L_{III} edge (~11,564 eV) [24]. The changes in intensity that accompany the ~3% change in the Pt atomic scattering factor are barely visible in the full patterns (Figure 2). The difference patterns of the low energy scan minus the scaled higher energy scan (that closer to the edge), however, are well defined and can be indexed completely on the basis of the face-centred cubic unit cell of Pt metal. The widths of the peaks indicate mean Pt particle sizes that range from ~ 5Å in the near-amorphous structure observed for the as-received and air-exposed materials, to ~ 20Å for the H_2-reduced sample.

The character of this system (that maximizes the Pt - Pt component in the Pt scattering) permits direct application of the differential anomalous X-ray scattering (DAXS) technique. For a system in which the target atomic species (Pt) is an integral, coherent part of the substrate structure, the DAXS pattern would contain both Pt - Pt and Pt - substrate information. Although interpretation then becomes more complex, such studies do afford insight into the Pt - substrate interaction.

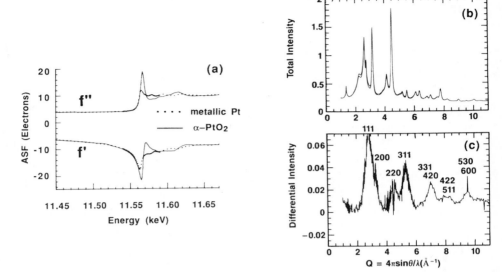

Figure 2. Real (f') and imaginary (f") components of the Pt scattering factor in the vicinity of the L_{III} absorption edge (a), and total (b) and DAXS (c) profiles for a sample of 5 wt% Pt on η-Al_2O_3.

DIFFRACTION FROM POLYCRYSTALLINE MATERIALS

As introduced above, in several heterogeneous catalyst systems reaction occurs or is initiated at external surfaces, which may represent only a small proportion of the total material. For other systems, the bulk structure is implicated in the properties of interest. Zeolites, for example, are crystalline aluminosilicates in which the active sites occur as part of the internal pore space [25]. Diffraction techniques that reveal bulk crystal structure details can thus yield direct data on the structural features of interest. Synthetic zeolites occur almost invariably as small crystallites, ≤ 5 μm, necessitating, conventionally, the use of powder diffraction techniques. The intensity, small source size and low divergence of the synchrotron X-ray source enable much higher resolution to be achieved for powder diffraction scans than is readily available in-house using focussing Bragg - Brentano geometry and a fine-focus X-ray tube source. In the high resolution experiment (Figure 3) a narrow wavelength band is extracted from the incident continuous X-ray spectrum by a single or double perfect crystal monochromator (Si (111) with $\Delta q \sim 7 \times 10^{-4}$ Å^{-1}, or Ge (111) with $\Delta q \sim 1.4 \times 10^{-3}$ Å^{-1} are commonly used). The diffracted beam acceptance angle can be defined by slits [26] or by an analyzer crystal [27,28]. Use of a single, adjustable slit permits the resolution to be continuously tuned, although precise slit adjustment can be difficult to reproduce, and, more importantly, the region of the sample that is viewed by the detector is also determined by the width of the slits. At very fine resolution, only a small region of the sample is viewed, limiting count rates and exacerbating problems associated with obtaining a true powder average. Use of a soller slit [29] circumvents this limitation, but prevents simple resolution adjustment and has, in practice, an upper resolution limit. A diffracted beam monochromator crystal acts, in essence, like a perfect soller slit, but with, in addition, energy selection which can help to reduce background due to sample fluorescence. For optimum signal intensities, the instrumental resolution can be matched to the crystallite size in the sample via choice of analyzer crystal.

High resolution synchrotron X-ray powder diffraction (PXD) experiments were first described only in 1983 [27,28], but already the wide potential of the technique has begun to be explored [30]. Higher instrumental resolution facilitates phase identification or analysis (Figure 4). Greater resolution effectively increases the amount of information that is

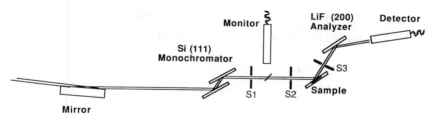

Figure 3. Schematic diffractometer configuration for high resolution powder diffraction as implemented on beamline X10A (Figure 1).

extractable from the PXD profile and hence enables more complicated structures to be refined by Rietveld analysis [31]. Several recent reports illustrate the power of this approach, including studies of $MnPO_4.H_2O$ (in which, remarkably, hydrogen atom positions were determined and refined) [33], and of a relatively high quality sample of zeolite ZSM-11 [34]. The latter refinement involved 54 atomic variables and represents the first improvement on the initial atomic positions determined on the basis solely of distance least squares in 1978 [35].

Higher resolution PXD data yield a much larger number of reflection intensities with uniquely determined values. Although such data are still considerably less precise than those obtained by measurements on a single crystal, they are still sufficient, in favorable circumstances, to permit *ab initio* structure solution. Examples that have already appeared

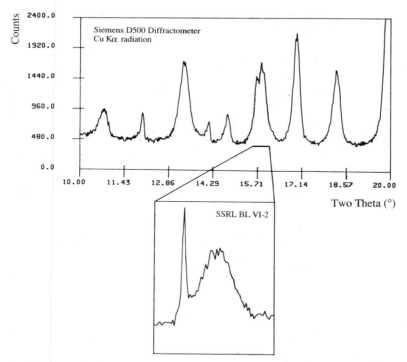

Figure 4. Sections of the powder diffraction profile of the product of a hydrothermal crystallization from a $Cs_2O \cdot SnO_2 \cdot SiO_2 \cdot H_2O$ gel composition [32]. The higher resolution synchrotron scan (inset) resolves the pattern into a mixture of both very sharp and broad peaks, indicating, in this case, the presence of at least two different phases.

include the structures of α-CrPO$_4$ [36] and the clathrasil Sigma-2 [37] that were both determined based on PXD data collected on beam line X13A at NSLS [38]. The PXD profile, $5 \le 2\theta \le 120°$, for sigma-2 was indexed, and then decomposed into individual intensities using the ALLHKL program [39]. Application of Direct Methods to the corrected structure factors revealed, in the best solution, positions for all four unique Si atoms, and for five out of the total of seven oxygen atoms. The structure was completed based on Fourier syntheses, and subsequent Rietveld refinement [37]. Although *ab initio* structure solutions based on synchrotron PXD data remain far from routine, these successful studies illustrate that higher instrumental resolution does significantly enhance the probability of a successful outcome by this approach.

The high resolution promised by a synchrotron powder diffractometer is frequently not realized in practice because of sample dependent contributions to the peak widths. Such effects and/or fine peak splittings indicative of lower than expected symmetry are not always welcome and can complicate the extraction of precise atomic scale models from the observed data. Index-dependent peak shapes and widths, however, convey detailed information about strain, stacking defects and disorder. As analysis methods develop further, the richness contained in the peak profiles themselves will more readily yield interesting information. In addition to experimental problems that can compromise an in-house powder diffraction experiment such as preferred orientation, the synchrotron experiment also has its own complexities. The rocking curve width of an individual crystallite is very narrow. To insure that a full orientational (powder) average is sampled, rotation of the specimen about an axis perpendicular to the scattering vector is desirable, and crystallite sizes should be optimally small. There is, thus, a relatively narrow optimum particle size window, between an upper bound determined by the powder averaging requirement and a lower bound dictated by the acceptable contribution of the finite particle size to the measured peak widths. In the ZSM-11 study cited above [34], for example, although a good quantitative fit of the entire PXD pattern was achieved, the mean particle size of ~0.4μm resulted in peak widths some 3 - 4x greater than the intrumental resolution. Realization of the instrumental resolution would have required crystallites $\ge$ ~ 1μm. As is discussed below, crystallites in this size range can now be examined singly by microcrystal diffraction techniques.

The continuous character of the synchrotron X-ray source permits measurements in energy-dispersive mode. Such a scan can be accomplished (using the arrangement illustrated in Figure 3) at a fixed Bragg angle by varying the incident wavelength. Alternatively, a white incident beam may be used, with a (fixed) energy-dispersive detector. The latter mode permits extended regions of the complete diffraction profile to be accumulated in a very short time frame. Further, the use of fixed incident and diffracted beam directions facilitates the use of sample environment control apparatus, such as high pressure cells. Energy dispersive diffraction studies of systems under extreme pressure are being conducted at most of the major synchrotron facilities [40]. The resolution afforded by today's energy dispersive detectors is of the order of 150 - 250 eV (compared to the Si (111) bandpass of $\Delta E/E \sim 10^{-4}$Å), permitting powder diffraction profiles to be obtained at modest resolution, with peak shapes that have been found, conveniently, to be gaussian [41]. Corrections need to be applied to the measured pattern for the distribution of incident intensity with wavelength, and for wavelength dependent scattering and absorption processes within the sample. Relatively simple systems can, however, be studied quantitatively [41].

SINGLE CRYSTAL DIFFRACTION EXPERIMENTS

Single crystal diffraction experiments can exploit each of the special features of synchrotron X-radiation. The source time structure has been exploited, for example, in diffraction studies of silicon during pulsed laser annealing [42]. The low source divergence provides high resolution also in the single crystal case, enabling studies of strain or mosaic [43], or, indeed, of how the mosaic character evolves with varying conditions (see below). The source brightness permits subtle incommensurate or superlattice structures to be examined [44]. The continuous character of the source, coupled with a high ring energy (such as the 5.5GeV facility at CHESS) can provide intense beams of X-rays of short wavelength, ≤ 0.5 Å. In this spectral range, the effects of multiple scattering which give rise to extinction effects are minimized and essentially extinction-free measurements can, therefore, be performed. In a feasibility study, > 7000 reflections were collected at 0.32Å at CHESS from a ~100μm crystal of hexamminechromium hexacyanochromate. Full structure refinement gave an excellent residual, R = 0.028, and yielded good X – X deformation density maps [45]. The polarized character of the source also permits exploitation of X-ray dichroism. The absorption cross sections and hence the anomalous scattering factors for an atom in a site symmetry lower than tetrahedral depend on the polarization of the incident light. Measurements on potassium tetrachloroplatinate(II) at close to the Pt L_{III} edge have demonstrated this effect [46].

The extreme brightness of the synchrotron radiation source facilitates studies of samples at high pressures in a diamond-anvil cell. Such samples are intrinsically small (10 nl – 0.1 pl) because the sample chamber is a small circular hole (~ 75 μm diameter x 20 μm) in a thin metal gasket between two flat diamond faces. Single crystal diffraction experiments on hydrogen at pressures of up to 26.5 GPa [47], and on sodium zeolite X at pressures of up to 3.2 GPa [48] have been performed. The objective of the latter experiments was to explore the causes of pressure induced broadening of the Bragg peaks from sodium zeolite X crystals in hydrostatic media that had been observed 'in-house' using a conventional sealed tube X-ray source. A water-saturated crystal (≈0.1 nl) mounted in a Merrill and Bassett type diamond anvil cell using 9:1 MEOH:H_2O as the pressurizing medium was examined on beamline X10A at NSLS with an incident X-ray photon energy of 12.4 keV. Unit cell dimensions, peak widths, and approximate relative intensities were measured at five pressures, $1.5 \leq P \leq 3.2$ GPa, encompassing the pressure range where peak broadening had been observed in the in-house experiments. A set of 12 reflections which covered a range of 2θ values and various directions in reciprocal space were measured. At all pressures the cubic unit cell was retained, and within the (limited) experimental precision the intensities were unchanged. Unlike the in-house experiments, the rocking curve widths initially decreased by ≈10% with increasing pressure. This, however, is a short-term effect. Measurements made in-house after annealing the crystal for several days at the highest pressure show a 300% increase in width over that at low-pressure. Thus in addition to the pressure-induced reduction in mosaiciy, the data provide information on the kinetics of the broadening phenomenon. The 18h elapsed time for the entire set of synchrotron experiments is clearly shorter than the halftime of the degradation in crystal quality effected by pressures above 2.0 GPa in this system.

The brightness of the synchrotron source permits measurements to be made on much smaller samples [49-55]. Initial microcrystal diffraction measurements at SSRL [49,50] demonstrated that crystallite volumes of the order of a few cubic microns ($1\mu m^3$ = 1fl or 10^{-15}l) were sufficient, given the brightness of today's synchrotron sources, to enable full sets of single crystal diffraction data to be obtained. More recent work has focussed on developing methods for crystallite selection and mounting, alignment definition, and full data

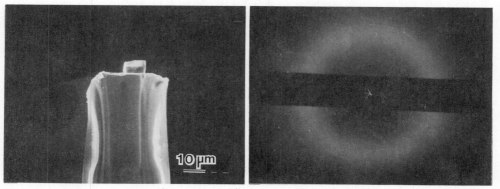

Figure 5. SEM micrograph of a 12μm zeolite X crystallite mounted on a 50μm diameter solid glass fibre (left), and a 120s rotation exposure recorded from a similar crystal on beamline X10A [53]. Sharp diffraction spots are clearly discerned. The diffuse halo represents scattering from the tip of the glass fibre mount.

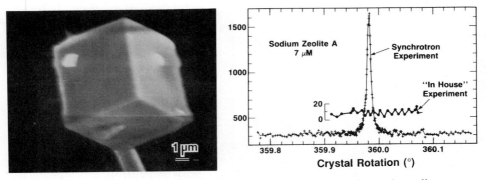

Figure 6. SEM micrograph of a 6μm zeolite A crystallite mounted on a 1μm diameter tapered hollow glass fibre (left), and a single reflection profile, $(0\ 1\ 4)_{subcell}$, from a 7x7x7 μm³ zeolite A crystallite recorded on beamline X10A. For contrast, typical data for a similarly sized crystal recorded in-house are also shown.

set acquisition [53]. Particles ≥ 5 μm can now be relatively routinely selected and mounted with excellent reproducibility (Figures 5 and 6) [53]. Careful control of background has enabled traditional rotation photographs to be used in initial reflection location (Figure 5). Given the typically small crystallite rocking curve widths, an extremely fine mesh (with a correspondingly long search time) would be required for a random search for reflections in 2θ - χ - φ space. Once an adequate subset of reflections has been centered, unit cell and orientation matrix optimization can be performed in the usual way. Issues associated with data acquisition over several beam fills in single counter mode, however, remain. The incident beam is susceptible to instabilities deriving from slight changes in the orbit of the electrons stored in the ring, and from heat up and cool down of beam line components such as the mirror and first monochromator crystal. The missetting of the second monochromator crystal that arises when the lattice constant of the first suffers thermal expansion or contraction can be partially compensated in a feedback loop. The bias of a piezoelectric drive on the second crystal is optimized with respect to the signal recorded by an ionization chamber prior to or after the sample position. To counteract slight variations in wavelength or beam direction, data is best accumulated in ω-scan mode, using telescoping scan step widths (Figure 6) [45]. Acceptable residuals have been obtained in refinements of data collected from larger crystals using these procedures [45,53]. Similar results have, however, not yet been obtained for microcrystals and data accumulated at the synchrotron is, as yet, typically not of precision equal to that collected with a completely stable, in-house source.

Early on, when we were proposing microcrystal diffraction experiments, questions as to sample heating, gross radiation damage, or damage of the sample mount in the finely focussed very intense X-ray beam were raised. Experience has shown that many of these factors are not of major significance, although in zeolitic materials degradation has been observed to occur over a relatively long half-time (several weeks) after X-ray exposure at the synchrotron. Problems associated with the quality of materials that occur only as very small particles have also been noted [54,55]. Certainly, several materials do occur only in highly faulted or poorly crystalline form, although many of the systems that we have investigated to date [49,50,53] have yielded crystallites with acceptable scattering characteristics.

Single crystal Laue diffraction exploits, simultaneously, both the white character of the synchrotron X-ray spectral distribution, and the high source brightness. The method has certain experimental advantages. The various planes in the (stationary) crystal diffract those beams from the continuous incident spectrum for which Bragg's law is satisfied in the given orientation. The accessible reflections, which are typically recorded on film packs, are thus scanned as a function of wavelength. The proportion of the unique intensity data that is recorded in a single exposure depends of the instrumental geometry, and on the crystal symmetry. Successive rotations of the crystal and fresh exposures permit accumulation of a full data set. To place all of the measured intensities on the same absolute scale requires that the incident intensity distribution, and all relevant wavelength dependent intensity corrections be known with reasonable precision. Structure refinement [56] and structure solution [55,57] based on reflection intensities measured in Laue geometry have been demonstrated. The required exposure time is extremely short, enabling sufficient data for structure determination to be accumulated in a few minutes (circumventing several of the problems encountered with monochromatic radiation in single counter mode). Single Laue exposures from macromolecular structures have been recorded in a few seconds at SRS, Daresbury [58] and, very recently, from photons emitted from a single 150 picosecond bunch in the storage ring at CHESS [59]. Although exposing samples to the full white beam greatly increases the potential for radiation damage, and although Laue data present some difficulties to structure solution [55], the use of Laue diffraction using synchrotron X-rays clearly promises to be an powerful technique.

CONCLUSION

Using synchrotron X-radiation for diffraction experiments is less convenient that performing conventional measurements in-house. However, the high brightness, low divergence, white character, and polarized nature of synchrotron X-ray sources provide unique opportunities for diffraction experiments on catalyst systems, on surfaces and on ceramics. Many of the techniques outlined above have been developed to a stage where routine application is, or will shortly be, possible. Access to facilities for performing experiments such as those mentioned here is available to the general scientific community, and it is easy to foresee that the contribution of such techniques to our understanding of materials and structural chemistries will continue to grow at a rapid pace. Experiments exploiting synchrotron X-ray diffraction remain at a relatively early stage of development, however, and with the prospect of this boom in application, it is an exciting time to be involved in diffraction experiments at synchrotron X-ray facilities.

ACKNOWLEDGMENTS

It is a pleasure to acknowledge the insight and enthusiasm of the many individuals at Exxon Research and Engineering Company who have contributed to and supported our own synchrotron X-ray diffraction experiments, including P. M. Eisenberger, D. E. Moncton, R. C. Hewitt, M. G. Sansone, G. J. Hughes, M. A. Modrick, K. L. D'Amico, S. K. Sinha, C. R. Safinya, M. E. Leonowicz, G. B. Ansell, R. Abramowitz, S. E. Bennett and E. W. Corcoran. Thanks are also due to J. L. Pizzulli for the SEM micrographs reproduced here.

REFERENCES

1. A. A. Sokolov and I. M. Ternov "Synchrotron Radiation" Pergamon, Oxford. (1968)
2. C. Kunz ed. "Topics in Current Physics - Synchrotron Radiation; Techniques and Applications" Springer-Verlag, Heidelberg. (1979)
3. H. Winick and S. Doniach eds. "Synchrotron Radiation Research" Plenum Press, New York. (1980)
4. H. Winick, Synchrotron Radiation, Scientific American 257 88-101 (1987).
5. R. C. Hewitt, M. Sansone, K. L. D'Amico, K. S. Liang and D. E. Moncton, Design, construction and use of the Exxon beam lines X10A, X10B and X10C at NSLS, unpublished (1988).
6. C. R. Safinya, D. Roux, G. S. Smith, S. K. Sinha, P. Dimon, N. A. Clark and A. M. Bellocq, Steric interactions in a model multimembrane system: a synchrotron X-ray study, Phys. Rev. Lett. 57 2718-2721(1986).
7. P. Dimon, S. K. Sinha, D. A. Weitz, C. R. Safinya, G. S. Smith, W. A. Varady and H. M. Lindsay, Structure of aggregated gold colloids, Phys. Rev. Lett. 57 595-598 (1986).
8. B. P. Flannery, H. W. Deckman, W. G. Roberge and K. L. D'Amico, Three dimensional X-ray microtomography, Science 237 1439-1444 (1987).
9. M. A. Van Hove and S. Y. Tong, "Surface Crystallography by LEED" Spinger-Verlag, Berlin. (1979).

10. W. C. Marra, P. Eisenberger and A. Y. Cho, X-ray total external reflection Bragg diffraction: a structural study of the GaAs-Al interface, J. Appl. Phys. **50** 6927-6933 (1979).
11. P. H. Fuoss, K. S. Liang and P. Eisenberger, in: "Synchrotron Radiation Research: Advances in Surface Science" R. Z. Bachrach ed. Plenum, New York. in press (1988).
12. P. Eisenberger and W. C. Marra, X-ray diffraction study of the Ge(100) reconstructed surface, Phys. Rev. Lett. **46** 1081-1084 (1981).
13. J. Henrion and G. E. Rhead, LEED studies of the first stages of deposition and melting of lead on low index faces of copper, Surf. Sci. **29** 20-36 (1972).
14. W. Hoesler and W. Moritz, LEED analysis of a dense lead monolayer on copper (100), Surf. Sci. **175** 63-77 (1986).
15. K. S. Liang et al., Structure and melting of lead overlayers on copper (100), in preparation (1988); Bull. Am. Phys. Soc. **32**(3) p. 452 (1987).
16. K. S. Liang, P. H. Fuoss, G. J. Hughes and P. Eisenberger, in "The Structure of Surfaces" M. A. Van Hove and S. Y. Tong eds. Springer-Verlag, Berlin pp. 246-250 (1985).
17. S. G. Wolf, L. Leiserowitz, M. Lahav, M. Deutsch, K. Kjaer and J. Als-Nielsen, Elucidation of the two-dimensional structure of an α-amino acid surfactant monolayer on water using synchrotron X-ray diffraction, Nature **328** 63-66 (1987).
18. K. S. Liang, E. B. Sirota, K. L. D'Amico, G. J. Hughes and S. K. Sinha, Roughening transition of a stepped Cu(113) surface: a synchrotron X-ray scattering study, Phys. Rev. Lett. **59** 2447-2450 (1988).
19. G. Rosenbaum, K. C. Holmes and J. Witz, Synchrotron radiation as a source for X-ray diffraction, Nature **230** 434-437 (1971)
20. J. C. Phillips, A. Wlodawer, M. M. Yevitz and K. O. Hodgson, Applications of synchrotron radiation to protein crystallography - preliminary results, Proc. Natl. Acad. Sci. USA **73** 128-132 (1976).
21. J. C. Phillips and K. O. Hodgson, Single-crystal X-ray diffraction and anomalous scattering using synchrotron radiation, in: "Synchrotron Radiation Research", H. Winick and S. Doniach eds., Plenum Press, New York. pp. 565-605 (1980).
22. J. R. Helliwell, Synchrotron X-radiation protein crystallography: instrumentation, methods and applications, Rep. Progr. Phys. Vol. **47** 1403-1497 (1984).
23. L. M. Moroney, P. Thompson and D. E. Cox, ADPD: a new approach to shared site problems in crystallography, J. Appl. Cryst. **21** 206-208 (1988).
24. K. S. Liang, S. S. Laderman and J. H. Sinfelt, Structural study of small catalytic particles using differential anomalous X-ray scattering, J. Chem. Phys. **86** 2352-2355 (1987).
25. J. M. Newsam, The zeolite cage structure, Science **231** 1093-1099 (1986).
26. W. Parrish, M. Hart and T. C. Huang, Synchrotron X-ray polycrystalline diffractometry, J. Appl. Cryst. **19** 92-100 (1986).
27. D. E. Cox, J. B. Hastings, W. Thomlinson and C. T. Prewitt, Application of synchrotron radiation to high-resolution powder diffraction and Rietveld refienement, Nucl. Instrum. Methods **208**, 573-578 (1983).
28. J. B. Hastings, W. Thomlinson and D. E. Cox, Synchrotron X-ray powder diffraction, J. Appl. Cryst. **17** 85-95 (1984).
29. W. Parrish and M. Hart , Advantages of synchrotron radiation for polycrystalline diffractometry, Zeit. Kristallogr. **179** 161-173 (1988).
30. C. R. A. Catlow ed.,"High Resolution Powder Diffraction" Materials Science Forum, Trans Tech Publications, Switzerland. Vol. **9** (1986).
31. H. M. Rietveld, A profile refinement method for nuclear and magnetic structures, J. Appl. Cryst. **2** 65-71 (1969).
32. J. M. Newsam, K. S. Liang, G. J. Hughes, High resolution powder X-ray diffraction, unpublished (1987).
33. P. Lightfoot, A. K. Cheetham and A. W. Sleight, Structure of $MnPO_4.H_2O$ by synchrotron X-ray powder diffraction, Inorg. Chem. **26** 3544-3547 (1987).
34. B. H. Toby, M. M. Eddy, C. A. Fyfe, G. T. Kokotailo, H. Strobl and D. E. Cox, A high resolution NMR and synchrotron X-ray powder diffraction study of zeolite ZSM-11, J. Mater. Res. **3** 563-569 (1988).

35. G. T. Kokotailo, P. Chu, S. L. Lawton and W. M. Meier, Synthesis and structure of synthetic zeolite ZSM-11, Nature 275 119-120 (1978).

36. J. P. Attfield, A. W. Sleight and A. K. Cheetham, Structure determination of α-$CrPO_4$ from powder synchrotron X-ray data, Nature 322 620-622 (1986).

37. L. McKusker, The ab initio structure determination of sigma-2 (a new clathrasil phase) from synchrotron powder diffraction data, J. Appl. Cryst. 21 305-310 (1988).

38. D. E. Cox, J. B. Hastings, L. P. Cardoso and L. W. Finger, Synchrotron X-ray powder diffraction at X13A; a dedicated powder diffractometer at the national synchrotron light source, in: "High Resolution Powder Diffraction" C. R. A. Catlow ed. Materials Science Forum, Trans Tech Publications, Switzerland. Vol. 9 pp. 1-20 (1986).

39. G. S. Pawley, Unit cell refinement from powder diffraction scans, J. Appl. Cryst. 14 357-361 (1981).

40. E. F. Skelton, High-pressure research with synchrotron radiation, Physics Today 37 44-52 (1984).

41. B. Buras, L. Gerward, A. M. Glazer, M. Hidaka and J. S. Olsen, Quantitative structural studies by means of the energy-dispersive method with X-rays from a storage ring, J. Appl. Cryst. 12 531-536 (1979).

42. B. C. Larson, C. W. White, T. S. Noggle and D. M. Mills, Synchrotron X-ray diffraction study of silicon during pulsed lase annealing, Phys. Rev. Lett. 48 337-340 (1982).

43. A. Kvick, Applications of synchrotron X-rays to chemical crystallography, in: "Chemical Crystallography with Pulsed Neutrons and Synchrotron X-Rays" M. A. Carrondo and G. A. Jeffrey eds. Nato Advanced Study Institute Series C, Vol. 221, D. Reidel, Dordrecht, Holland. pp. 187-203 (1988)

44. D. E. Moncton, K. L. D'Amico, J. Bohr, J. Als-Nielsen, R. M. Fleming, J. P. Remeika and D. Vaknin, Scattering studies of La_2CuO_4 single crystals: charge density modulations, unpublished (1987).

45. F. S. Nielsen, P. Lee and P. Coppens, Crystallography at 0.3Å: Single crystal study of $Cr(NH_3)_6Cr(CN)_6$ at the Cornell high-energy synchrotron source, Acta Cryst. B42 359-364 (1986).

46. D. H. Templeton and L. K. Templeton, X-ray dichroism and anomalous scattering of potassium tetrachloroplatinate(II), Acta Cryst. A41 365-371 (1985).

47. H. K. Mao, A. P. Jephcoat, R. J. Hemley, L. W. Finger, C. S. Zha, R. M. Hazen and D. E. Cox, Synchrotron X-ray diffraction measurements of single-crystal hydrogen to 26.5 gigapascals, Science 239 1131-1134 (1988).

48. H. E. King and J. M. Newsam, Studies of sodium zeolite X under hydrostatic pressures up to 3.5GPa, in preparation (1988).

49. P. Eisenberger, J. M. Newsam, M. E. Leonowicz and D. E. W. Vaughan, Synchrotron X-ray diffraction from a 800μm³ zeolite microcrystal, Nature 309 45-47 (1984).

50. J. M. Newsam and D. E. W. Vaughan, The impact of new diffraction techniques in zeolite structural chemistry, in: "ZEOLITES: Synthesis, Structure, Technology and Application" B. Drzaj, S. Hocevar and S. Pejovnik eds. Stud. Surf. Sci. Cat. 24, Elsevier, Holland. pp 239-248 (1985).

51. R. Bachmann, H. Kohler, H. Schulz, H-P. Weber, V. Kupcik, M. Wendschuh-Josties, A. Wolf and R. Wulf, Structure analysis of a CaF_2 single crystal with an edge length of only 6μm: an experiment using synchrotron radiation, Angew. Chem. Int. Ed. 22 1011-1012 (1983).

52. R. Bachmann, H. Kohler, H. Schulz, H-P. Weber, Structure investigation of a 6μm CaF_2 crystal with synchrotron radiation, Acta Cryst. A41 35-40 (1985).

53. J. M. Newsam, H. E. King and M. A. Modrick, Microcrystal diffraction techniques, in preparation (1988).

54. S. J. Andrews, M. Z. Papiz, R. McMeeking, A. J. Blake, B. M. Lowe, K. R. Franklin, J. R. Helliwell and M. M. Harding, Piperazine silicate (EU 19): the structure of a very small crystal determined with synchrotron radiation, Acta Cryst. B44 73-77 (1988).

55. M. M. Harding,The use of synchrotron radiation for Laue diffraction and for the study of very small crystals, in: "Chemical Crystallography with Pulsed Neutrons and Synchrotron X-Rays", M. A. Carrondo and G. A. Jeffrey eds. Nato Advanced Study Institute Series C, Vol. **221**, D. Reidel, Dordrecht Holland. pp. 537-561 (1988)
56. I. G. Wood, P. Thompson and J. C. Matthewman, A crystal structure refinement from Laue photographs taken with synchrotron radiation, Acta Cryst. B**39** 543-547 (1983).
57. J. A. Clucas, M. M. Harding and S. J. Maginn, Crystal structure determination of $Rh_6(CO)_{14}(dppm)$ using intensity data from synchrotron radiation Laue diffraction photographs, J. Chem. Soc. Chem. Commun. 185-187 (1988).
58. J. Hajdu, P. A. Machin, J. W. Campbell, T. J. Greenhough, I. J. Clifton, S. Zurek, S. Glover, L. N. Johnson and M. Elder, Millisecond X-ray diffraction and the first electron density map from Laue photographs of a protein crystal, Nature **329** 178-181 (1987).
59. K. Moffat, W. Schildkamp, D. Bilderback and M. Szebenyi, unpublished (1988).

ON-LINE X-RAY FLUORESCENCE SPECTROMETER FOR COATING THICKNESS MEASUREMENTS

N. Matsuura and T. Arai

Rigaku Industrial Corporation
Takatsuki, Osaka, Japan

INTRODUCTION

The principles and techniques for performing coating thickness measurements in the laboratory[1,2,3] and on-line[5,6,7] using nuclear radiation have been established for tin or zinc coated steels. Recently, additional engineering efforts have been made toward the development of new coating substances consisting of complex layered materials. For example, zinc-iron alloy metal has higher corrosion resistance than pure zinc-coated steel sheets and evaluations have been made of beneficial characteristics in automobile production processes, such as easy welding, good paintability, high efficiency for press plastics forming, etc.[8]

For on-line measurements with an X-ray spectrometer, each application requires its own independent and specific features and measurement performances depending on whether the coated materials are thick or thin, multi-component or multilayered, etc.[8] For thick coatings ($\sim$50 μm), e.g., galvanized iron, the measuring system consists of radioactive isotope (^{241}Am) excitation, non-dispersive optics for Zn-K alpha X-rays and an ionization chamber detector with high counting rate and good linearity. For the thin coatings ($\sim$10 μm), e.g., tin, zinc or zinc-iron, a combination of X-ray tube excitation, crystal monochromators, regular X-ray detectors and measurement of X-rays from substrate or coating materials can be employed. For thickness measurements of magnetic materials on plastic films, Fe-K alpha X-rays emitted from the coated layer, consisting of organic binder and fine iron oxide powder, are measured using a low-load X-ray tube, non-dispersive optics and a proportional counter with pulse height selection.

In the design and development of on-line instruments, one of the most important engineering tasks is the reduction of measuring errors caused by movement of the sample sheet such as low and high frequency vertical vibration, sheet inclination and uneven or crumpled sheets.

An X-ray sensing head, incorporated in the production line, is surrounded by a hostile atmosphere. Thus, the instruments incorporate counter-measure devices for excluding undesirable fumes and gases. For measurement of complex materials deposited on a substrate, rapid and real time conversion of X-ray intensity to coating thickness and composition should be carried out with the aid of a fast-response computer system.

This paper will report on design strategy, instrumentation, analysis of error factors and practical applications of the on-line X-ray spectrometers.

INSTRUMENT

(1) Outline of Instruments

The instrument consists of an X-ray sensing head installed in the movable traverse; field control boxes, including high voltage generators for an X-ray tube and a control panel box, are set in the production control room. A block diagram of the on-line X-ray instrument is shown in Fig. (1). A movable traverse, which is designed to minimize the measuring errors arising from the moving samples, is equipped with a standard sample turret and incorporated in the production line of the coating steel sheet or magnetic tape finishing process. Because the sheet sample width is from 50 cm to 200 cm, the deviation of coating thickness in the cross direction is measured by scanning with the movable traverse as shown in Fig. 1.

The atmosphere surrounding the sensing head is ghastly, namely, high and changeable temperature, high humidity, dusty acidified air, inflammable vapors, vibration of the floor, large fluctuation in the electric power line and electromagnetic wave interference. In general, control of the instruments and the display of measured results can be done in the specially prepared control room which is located 5 to 50 m away from the traverses.

In an on-line measurement, a one-second measured intensity corresponds to the accumulated intensity of a one to ten meter length of a running sample. Therefore, a total system with fast response is accomplished using high intensity measurements (for a reduction of X-ray statistical error) high count-rate measurements with good linearity and a fast-response computer system for the conversion of measured intensities to analytical results. Its performance is critical in controlling the range of quality of material in the production process.

(2) Instrument and X-Ray Measurement

Instruments and measuring conditions are shown in Table (1). There are many different kinds of samples to be measured, namely, magnetic tape, zinc electro-plating, alloy coating, zinc hot-dip coating, tin electro-plating and

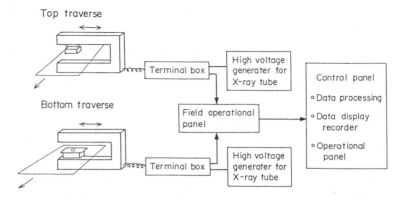

Fig. (1)   Block diagram of on-line X-ray instrument

Table (1)   Instruments and measuring conditions

Instrument	Excitation	Optics and detectors	Time response sec	Measuring X-rays
Magnetic tape	X-ray tube low power	Filter and proportional or scintillation counter	2 ~ 4	Fe-Kα
Zinc electro-plating thin zinc coating	X-ray tube 3 KW	LiF bent crystal and proportional counter	0.3~ 1.	X-rays from substrate and coating material
alloy coating $\left(\begin{array}{c}Zn\text{-}Fe\\Zn\text{-}Ni\end{array}\right)$	ibid	ibid		ibid
Zinc hot dip coating thick Zn coating	X-ray tube low power	Filter and iron chamber	0.5~ 4	Fe-Kα
Zinc hot dip coating	^{241}Am 60 KV	Proportional counter	4	Zn-Kα
Tin electro-plating	X-ray tube low power	Filter and iron chamber	0.5~ 1.	Fe-Kα
Tin free steel (Cr + Cr oxide)	X-ray tube low power	Filter and proportional counter	2	Cr-Kα

tin-free steel.   Fluorescent X-rays can be excited with low power X-ray tubes, high power X-ray tubes and a radioisotope, ^{241}Am, which emits 60 keV y-rays.

Optical systems use a nondispersive method and a wavelength dispersive method.   The former uses a scintillation or proportional counter, or an ionization chamber with a specified filter.   The latter uses crystal monochromators for complex sample measurement.   The response times of the various instruments are also shown.

The principle X-ray measurements for coating thickness consist of the intensity measurements of the fluorescent X-rays from the coating and substrate and the ratio of the zinc and iron X-ray intensities.   In the case of the X-ray ratio method, an improvement in the measuring accuracy can be expected in addition to the elimination of pass line variations.

(3)   Analysis of On-Line Measuring Errors

Measuring errors may be classified into two groups, stationary errors and dynamic errors.   The stationary errors consist of an X-ray statistical error depending upon measured intensity, drift compensation for the long term use, analytical accuracy based on the difference between X-ray and chemical analysis values and microscopic and macroscopic homogeneity of measuring samples related to the sampling procedures and the quality of products.

The dynamic errors may be classified into two groups; dynamic precision and dynamic accuracy.   Dynamic precision errors are identical to instrumental errors and are influenced by a moving sample, temperature change of sample and environmental condition and traverse scanning.   The dynamic accuracy, named the consolidated error, is defined as the synthetic error of dynamic precision and homogeneity of measuring materials and is related to the quality of products against the production volume.   In the practical applications shown in the next section, various error factors and their numerical values will be presented.

X-ray ratio method   X-ray shadow combination method

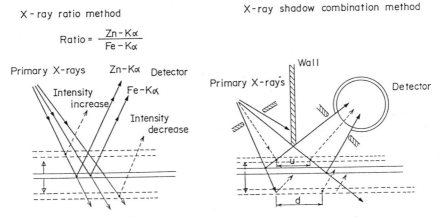

Fig. (2)   Elimination of pass line variation

(4)   Elimination of Pass Line Variation

It is necessary to provide a mechanism which eliminates the variation of X-ray intensity generated from a running sample. In Fig. (2), two methods are shown, one is an X-ray fluorescence method and the other is a shadow combination method. The former is the intensity ratio method using Zn-K alpha and Fe-K alpha X-rays. When X-ray optics are geometrically the same to both X-rays, the error in X-ray intensity originating from up and down vibration or motion of measuring samples may be cancelled. The latter utilizes the aberration effect which occurs between the irradiating area and the detectable area when sample variation occurs.

APPLICATIONS

(1)   Measurement of Magnetic Tape

The structure of magnetic tape for audio or video tapes is shown in Fig. (3). The purpose of the X-ray measurement is to obtain the thickness of the magnetic layer which is a mixture of a fine powder of iron oxide and organic binder. The regular sample speed is about 100 m/min. The mixing ratio of the binder to γ-Fe_2O_3 is about one to three by weight ratio.

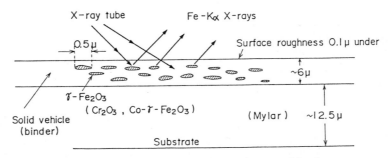

Fig. (3)   Sample structure of magnetic tape

Table (2)   On-line measurement errors of magnetic tape

Sample : Magnetic tape film thickness 6.0 μm

Sample speed : max. 150 m/min

Method : Fe-Kα measurement

	Measuring error	Standard deviation
Stationary		μm
① X-ray statistical error	±0.08 (2σ)	0.04
② Drift-long term stability	±0.08 (8h)	0.027
③ Calibration curve error	~±0.2 (2σ)	~0.1
Dynamic		
④ Pass line variation distance	+0 / -0.03 / ±1mm	0.005
⑤ Pass line variation angle along the line	+0.03 / -0 / ±0.5°	0.005
⑥ Pass line variation angle across the line	±0.08 / ±0.5°	0.027
⑦ Mechanical reproducibility of traverse scanning	±0.02	0.007
⑧ Integral temperature stabilization	±0.05 / ±5°C	0.017

All-round composit error at the condition of
±1.0mm, ±0.5°, ±0.5°, ±5°C and 8 hour's operation

$$\left[①^2 + ②^2 + ④^2 + ⑤^2 + ⑥^2 + ⑦^2 + ⑧^2 \right]^{1/2} \longrightarrow 0.06$$

In Table (2) the error components in measuring the thickness of a magnetic layer of 6 microns are summarized. The error factors are listed on the left. The range of variation is shown in the middle. The standard deviation of the errors is shown on the right. The all-round composite error of this on-line instrument was estimated using a formula of propagation of errors and is shown at the bottom. For a 6 micron sample it is 0.06 microns and the coefficient of variation is one percent.

(2) Measurement of Zinc Electro-Plating Steel

This measuring method for the zinc electro-plated steel involves taking the intensity ratio of Zn-K alpha and Fe-K alpha X-rays, and is effective in eliminating the errors arising from the pass line variations and improves the static and dynamic accuracies because of higher sensitivity to the thickness variation of the coated material.

The experimental results of the ratio method under static measurement conditions are shown in the upper part of Fig. (4), and X-Ray and chemical analysis values are compared in the lower part. The analytical accuracy is 0.4 g/m². The on-line measurement errors of zinc electro-plated steel are shown in Table (3). The coating thickness of this sample is 20 g/m² which is approximately 2.8 microns. The regular sample sheet speed is about 200 m/min.

The pass line variation of the No. 4 error factor is the vertical amplitude variation of a sample sheet. The pass line variation of No. 5 is the inclined angle of a sample sheet along the sample moving direction. The pass line variation of No. 6 is the inclined angle of a sample sheet at a right angle to the sample running direction. The all-round composite error

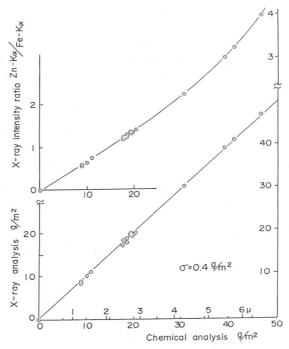

Fig. (4)   Calibration curves of zinc electro-plating steel

Table (3)   On-line measurement errors of zinc
             electro-plating steel

Sample : EGL 20 g/m² ~ 2.8 μm   Method : Zn-Kα/Fe-Kα
Sample speed : max 250 m/min

Error factors	Measuring error	Standard deviation
		g/m²
Stationary		
① X-ray statistical error	± 0.12 (2σ)	0.06
② Drift — long term stability	± 0.2 (8h)	0.07
③ Calibration curve error	10~41 g/m² ± 0.8 (2σ)	0.4
Dynamic		
④ Pass line variation distance	± 0.15 / ± 1 mm	0.05
⑤ Pass line variation angle along the line	± 0.3 / ± 0.5°	0.10
⑥ Pass line variation angle across the line	± 0.3 / ± 0.5°	0.10
⑦ Mechanical reproducibility of traverse scanning	± 0.1	0.03
⑧ Integral temperature stabilization	± 0.15 / ± 5°C	0.05

All-round composit error

$$\left\{ ①^2 + ②^2 + ④^2 + ⑤^2 + ⑥^2 + ⑦^2 + ⑧^2 \right\}^{1/2} \longrightarrow \frac{0.19}{20 \text{ g/m}^2}$$

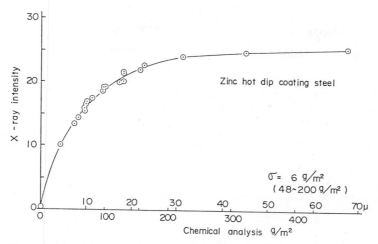

Fig. (5)   Relationship of Zn-K X-ray intensity against
            chemically determined values

Table (4)   On-line measurement errors of zinc
             hot dip coating steel

Sample : CGL 50 and 200 g/m^2 , 7 and 28 μm
Sample speed : max 200 m/min   Method : Zn-K_α

Error factors	Measuring error		Standard deviation	
			g/m^2	
Stationary	50	200	50	200
① X-ray statistical error	±0.2	±0.8 (2σ)	0.1	0.4
② Drift - long term stability	±0.1	±0.7 (8h)	0.033	0.23
③ Calibration curve error		±12 (2σ) 48~200	6.0	6.0
Dynamic				
④ Pass line variation distance	±0.2	±1.2/±1 mm	0.07	0.4
⑤ Pass line variation angle along the line	±0.2	±1.0/±1°	0.07	0.33
⑥ Pass line variation angle across the line	±0.2	±1.0/±0.5°	0.07	0.33
⑦ Mechanical reproducibility of traverse scanning	±0.1	±0.8	0.033	0.27
⑧ Integral temperature stabilization	±0.1	±0.8/±5℃	0.07	0.5

All-round composit error

$$\{①^2 + ②^2 + ④^2 + ⑤^2 + ⑥^2 + ⑦^2 + ⑧^2\}^{\frac{1}{2}} \longrightarrow \quad 0.18/50 \quad 1.0/200$$

is 0.19 g/m^2 for the 20 g/m^2 coating thickness. The coefficient of variation is about one percent and is half of the calibration curve error No. 3.

(3) Measurement of Zinc Hot Dip Coating Steel

The experimental relation of Zn-K X-ray intensity against chemically determined values for the thicker zinc coating steel sheet is shown in Fig.(5). For samples with 30 microns or greater thickness, the intensity of Zn-K X-rays is almost constant. The analytical accuracy of the stationary measurement is 6 g/m^2 for the range of 48 to 200 g/m^2 samples. The numerical errors of 50 g/m^2 and 200 g/m^2 of zinc coating steel sheet are shown in Table (4).

The mechanical reproducibility of traverse scanning is No. 7 in Table 4. It is more important to the measurement of thicker coated samples. The X-ray intensity was measured with the sample stationary and only the traverse was scanned.

The values of the integral temperature stabilization, No. 8, arise from the change of atmospheric temperature which is mainly caused by the variation of the absorption attenuation of air.

For the thicker sample measurements, a large variation in coating thickness has frequently been found and thus influences the analytical accuracy. The large difference between the calibration curve error of No. 3 and the all-round composite error is worthy of attention.

(4) Measurement of Tin Free Steel (TFS)

The structure of tin free steel is shown in Fig. (6). This coating material consists of two layers which are metallic chromium and hydrated chromium oxide. The total amount of the chromium coating on the steel sheet is measured by this method.

The comparison between chemical and X-ray analysis is shown in Fig. (7). The coefficient of variation of analytical accuracy is 4 percent against 100 mg/m^2 and 1 percent against 400 mg/m^2, even though a double layered coating material was measured.

The on-line measurement error of tin free steel for 100 mg/m^2 is shown in Table (5). Note that the standard deviation of the X-ray statistical error, No. 1, the calibration curve error, No. 3, and the all-round composite error are almost identical. There is an additional error factor, which is the influence of impurity chromium in the steel.

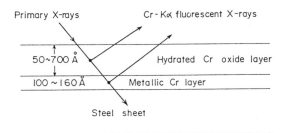

Fig. (6)   Sample structures of tin free steel (TFS)

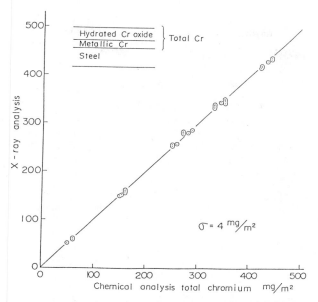

Fig. (7)   Comparison between chemical and X-ray
analysis of tin free steel (TFS)

Table (5)   On-line measurement errors of tin free steel (TFS)

Sample : TFS 100 mg/m²
Sample speed : Max. 450 m/min
Method : Cr-Kα measurement

	Measuring error	Standard deviation
Stationary		
① X-ray statistical error	7.6 (2σ)	3.8 mg/m²
② Drift-longstability	2.0 (8hs)	0.7
③ Calibration curve error	8.0 (2σ) 0~450	4.0
Dynamic		
④ Pass line variation distance	+0.0 / -1.1/±1.0mm	0.24
⑤ Pass line variation angle along the line	+0.0 / -0.5/±1°	0.17
⑥ Pass line variation angle across the line	+2.0 / -2.0/±1°	0.7
⑦ Mechanical reproducibility of traverse scanning	2.0	0.7
⑧ Integral temperature stabilization	±2.0/±5°C	0.7

All-round composite error

$$\left\{ ①^2 + ②^2 + ④^2 + ⑤^2 + ⑥^2 + ⑦^2 + ⑧^2 \right\}^{1/2} \longrightarrow 4.1/100 \, mg/m^2$$

CONCLUSION

Essential engineering items are summarized below for development and practical use of on-line X-ray fluorescence spectrometers for monitoring mass-production materials:

1. Instrumental reliability for continuous and long term use.
2. Counter measure against moving samples.
3. Counter measure against unusual environmental conditions.
4. Evaluation of all-round composite errors, and analysis of error factors.
5. Studies of sample characteristics related to a consolidated error.
6. Improvement of control efficiency for products with the aid of faster response measuring and computer systems.

With the callaboration of users and instrument manufacturers, on-line X-ray fluorescence spectrometers can be optimized for real-life requirements by making improvements in the above.

REFERENCES

1) Frideman and Birks, RSI Vol. 17, No. 3 (1946), P.99-101

2) Liebhafsky etc., X-Rays, Electrons and Analytical Chemistry Published by Wiley-Interscience (1972), P.281-327

3) Tsumura and Oshiba, Tetsu-to-Hagane, Vol. 63, No. 7 (1977), P.1170-1176 (in Japanese)

4) Shanfield and Bertin, Advances in X-Ray Analysis, Vol. 21 (1978), P.93-104

5) Cass and Kelly, Norelco Reporter, Vol. 10, No. 2 (1963) P.49-56 and 69 and Dunne idde Vol. 10, No. 2 (1963) P.59-60 and 70

6) Bertin and Waterman, Proceedings of the First International Tin Plate Conference, London, England. International Tin Research Institute Publication No. 530 (1976), P.363-371

7) Donhoffer and Beswick, Nuclear Techniques in the Basic Metal Industries, IAEA Vienna (1973), P.299-317

8) Tetsu-to-Hagane (Journal of the Iron and Steel Institute of Japan), Special Issue on Surface Treatment for Steel, Vol. 66, No. 7 June (1980) (in Japanese) and Special Issue on Metal Finishing, Vol. 72, No. 8, June (1986) (in Japanese)

PROCESS CONTROL APPLICATIONS OF THE PELTIER

COOLED SI(LI) DETECTOR BASED EDXRF SPECTROMETER

Anthony R. Harding
Tracor Xray, Inc.
Mountain View, CA

ABSTRACT

The demand for on-line analyzers capable of compositional determinations in petroleum and chemical process streams has increased dramatically in recent years[1,2]. Total control of production plant processes and resources requires the analysis of feed, intermediate, and product materials. This paper will describe a rugged, on-stream energy dispersive X-ray fluorescence (EDXRF) analyzer configured for the continuous determination of composition in solid and liquid samples. The detector in the X-ray sensor is a lithium-drifted silicon crystal which is thermoelectrically cooled (Peltier effect) to achieve operational temperatures. This approach to detector cooling offers advantages over traditional cryogenic liquid cooling when EDXRF is used in the process control environment.

Three applications of the thermoelectrically-cooled detector-based EDXRF spectrometer will be presented here. The first is the analysis of a catalyst solution to monitor depletion of the active species. Second, two components of a plating bath solution will be determined simultaneously in a flowing sample stream. In the third application, the spectrometer will be oriented in a downward looking configuration to monitor inorganic constituents in absorbent samples as they are transported by conveyor belt to the X-ray measurement area.

INSTRUMENTATION

The Tracor Xray Spectrace 7000 (Tracor Xray, Inc., Mountain View, CA) continuous EDXRF analyzer system consists of three, separate, interacting modules as shown in Figure 1. When the spectrometer is configured for continuous analysis of liquid streams, the X-ray sensor module is equipped with a flowcell. The flowcell entrains the stream with a thin polymeric window which is supported in a vertical plane. The X-ray optics face the flowcell window with 90° geometry between the X-ray detector and the X-ray tube relative to the flowcell window. The X-ray components consist of a 50 kV, 0.35 mA Rh anode X-ray tube, two selectable primary radiation filters, and the Si(Li) detector. The flowcell, which fits into the X-ray

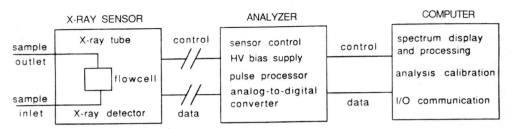

Figure 1. Block diagram of the on-stream EDXRF analyzer system.

sensor lid, is easily removed and can be fabricated from stainless steels or machinable plastics depending on the nature of the sample stream. The flowcell window is 0.65" in diameter. The thickness and composition is selected considering the nature of the sample stream and the X-ray emission line energies to be analyzed. An aluminum plate equipped with a second polymer window protects the X-ray optics from any flowcell window leakage.

The x-ray sensor and analyzer in Figure 1 can be separated by as much as 1000 ft. when equipped with a long distance transmission option. The analyzer module outputs control signals to, and accepts data from, the X-ray sensor module. The analyzer module houses the control board and the electronics necessary for pulse processing. An IBM PC/AT or compatible controls the X-ray excitation conditions automatically via the analyzer and accepts spectral and system status data. The software running on the PC provides continuous operation via a C-language, menu-driven program that provides system control as well as fundamental parameters and empirical matrix correction data treament.

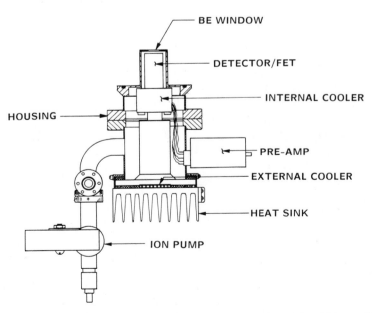

Figure 2.   Schematic diagram of the thermoelectrically cooled
            Si(Li) detector.

The heart of the EDXRF system used here is the thermoelectrically (Peltier effect) cooled Si(Li) X-ray detector (ECD). Coupling Peltier effect cooling elements to lithium-drifted silicon crystals has been described elsewhere[3]. The commercially available detector design is shown schematically in Figure 2. A multi-stage thermoelectric cooling unit is in thermal contact with a 20mm^2 Si(Li) crystal. The cooling apparatus reduces the crystal temperature to -85 C for detector operation. The hot side of the cooling unit is in thermal contact with a heat sink which is cooled by forced ambient air flow. Typical resolution of the ECD is 185 eV (Mn K-alpha, 5.90 keV).

Incorporation of the ECD into the EDXRF spectrometer system has advantages in process control applications when compared with liquid nitrogen cooled detectors. Since liquid nitrogen is not used, filling is not necessary, routine detector maintenance is eliminated, and the absence of cryogenic liquids makes the work environment safer. Another advantage of the ECD is the range of X-ray optic configurations attainable with the standard detector design. Since there is no liquid level to be concerned with, the detector can be oriented in any position in space. This option will be explored when applying the EDXRF spectrometer to the continuous analysis of solid samples on a conveyer belt.

EXPERIMENTAL

Catalyst standard solutions were prepared by dissolving the nitrate salts of Al, Co, and Zn into nitric acid. The standard compositions ranged from 0.53 to 3.9 g/mL, 2.0 to 9.0 g/mL, and 2.1 to 6.6 g/mL for Al, Co, and Zn, respectively. The standards were introduced into a multi-port presentation cell and analyzed in a static mode. Earlier work has shown no significant variation between static and flow through measurements using this system[4].

Two sets of spectral acquisition conditions were employed for the analysis of catalyst solutions. For the determination of Al, the analysis conditions were as follows: 6 kV tube voltage, 0.35 mA tube current, no primary radiation filter, and a 1000 second counting time. The X-ray optics were under a 1.5 SCFH He flush atmosphere. For the determination of Co and Zn, the excitation parameters were 30 kV tube voltage, 0.35 mA tube current, 0.05 mm Rh filter with a 60 second spectrum acquisition time.

Four plating solution standards were prepared using sodium phosphate monobasic and chromium nitrate dissolved in distilled-deionized water and nitric acid. Compositions of the standard solutions were 3900 ppm to 5500 ppm P and 4400 ppm to 6100 ppm Cr. The X-ray spectral acquisition parameters were 15 kV tube voltage, 0.35 mA tube current, no filter and a 200 second spectrum acquisition time. A helium flush was also used in this application.

Absorbent samples with large (3 mm), irregularly shaped particles were conveyed for analysis using the configuration illustrated in Figure 3. The chamber door of the Spectrace 7000 was removed and the protection plate window was positioned face down 2mm above the packed powder surface. absorbent powder was loaded into the hopper and continuously transported to the analysis area at a rate of 0.5 lbs/min. Spectrum acquisition conditions were 25 kV tube voltage, 0.35 mA tube current, cellulose primary radiation filter, and a counting time of 100 seconds. A helium atmosphere was used in

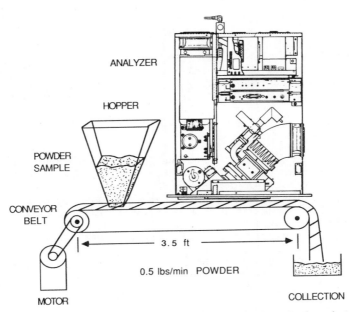

Figure 3. Equipment configuration for the analysis of absorbent samples.

this application also. Elements of interest for absorbent product
comparison were K, Ca, Ti, and Fe.

RESULTS

        Calibration curves were obtained for the analyte elements in catalyst
solutions. Wide concentration variations of the measurable elements in
these solutions result in matrix effects, as observed by the lack of
correlation when a linear plot is made of measured emission intensity versus
concentration. Matrix effects were treated using the empirical Lucas-Tooth
and Price model. Figure 4 is a plot of Al intensity corrected for the Co

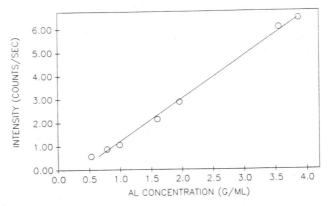

Figure 4. Aluminum calibration curve determined using seven
                    standard solutions.

Table 1. Computed calibration curve parameters for
three analytes in catalyst standard solutions

ELEMENT	SLOPE	INTERCEPT	CORRELATION COEFFICIENT
Al	1.83 ± 0.05	−0.62 ± 0.11	0.998
Zn	706.7 ± 6.0	−51.9 ± 25.2	0.999
Co	593.6 ± 5.2	−140.5 ± 26.9	0.999

Table 2. Results of the analysis of a synthetic plating bath
unknown using P and Cr calibration curves

ANALYTE	GIVEN (PPM)	DETERMINED AT 100 ML/MIN (PPM)	RELATIVE ERROR
P	4714	4683 ± 1.6% (RSD)	0.6 %
CR	5150	5194 ± 0.5% (RSD)	0.8 %

and Zn effects versus Al concentration for seven standards. The lower limit
of detection for Al is 0.32 g/mL. Calibration curve data for all three
analytes corrected for the concentration variation of the concomitant
elements are listed in Table 1. A synthetic unknown (3.10 g/mL Al and 2.35
g/mL of each Co and Zn) was analyzed and the determined concentrations were
3.15 g/mL, 2.37 g/mL, and 2.37 g/mL for the analytes Al, Co, and Zn,
respectively. Measurement precision for Al at the 3.90 g/mL level was found
to be 3.8% relative standard deviation (RSD). In the case of Co and Zn,
both at the 2.58 g/mL level, a precision of 0.45% was found for each metal.

Calibration curves for P and Cr were determined in the plating bath
standard solutions using a linear relationship of measured emission line
intensity versus analyte concentration. The linear regression parameters
for phosphorous are: slope = 2.33E-2 ± 0.10 E-2 cps/ppm, intercept = −12.06
± 5.01 cps, and a correlation coefficient of 0.982. The calibration curve
for Cr has the following linear regression parameters: slope = 4.19E-1
± 8.22E-3 cps/ppm, intercept = −459.07 ± 44.92 cps, and a correlation
coefficient of 0.996. A synthetic unknown was analyzed while flowing
through the flowcell at 100 mL/min and the results are listed in Table 2.
Precision of the measurement was determined using five replicate spectra of
the unknown and those data are listed in Table 2. The lack of fit for P
(cc= 0.982) may be caused by instability of the phosphate component in the
standards or by an interelement effect caused by chromium. A larger suite
of standards would be required to determine if an interelement effect was
present in this system.

An acquired spectrum of absorbent product 1 is shown in Figure 5.
Resolution provided by the ECD is sufficient to separate adjacent peaks
found in the spectrum. Since the absorbent samples are in an air atmosphere
an argon peak is observed in the spectrum. To perform a reliable comparison
of products, good precision is required for this analysis. Table 3 lists
the precision obtained for net emission peak counts. Note that all values
are above 10% RSD. Imprecision is caused by packing density variations in
large, irregularly shaped particles. To account for particle size effects,
peak counts were ratioed to a bremsstrahlung scatter region in the spectrum.
As shown in Table 3, the measurement to measurement precision improves so
that peak intensity differences between products 1 and 2 are statistically
significant.

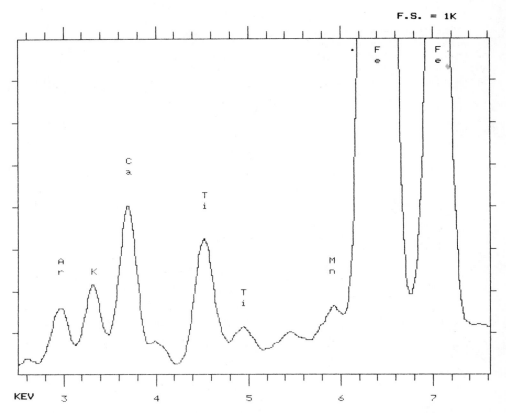

Figure 5. Spectrum acquired for absorbent product 1.

Table 3. Results of the comparison of two absorbent products

PRODUCT 1

ANALYTE	P/BREM (X100)	RSD (%)	PEAK RSD (%)
FE	606.91	1.3	14.6
K	9.70	4.8	11.2
CA	22.11	0.8	15.4
TI	17.20	4.1	13.6

PRODUCT 2

ANALYTE	P/BREM (X100)	RSD (%)
FE	695.02	1.3
K	17.43	4.0
CA	32.46	4.1
TI	20.30	5.7

CONCLUSIONS

The three analytical applications discussed here illustrate the usefulness of EDXRF spectrometry in the process control environment using the thermoelectrically cooled lithium-drifted silicon detector. The ECD based spectrometer offers numerous advantages over cryogenic liquid cooled detectors. Flexible configurations are available with the ECD since there are no cold finger orientation limitations. The absence of the liquid nitrogen means that no routine detector maintenance is required. Additionally, the absence of cryogenic liquids from the production floor facilitates a safe working environment.

REFERENCES

1.   J.B. Callis, D.L. Illman, B.R. Kowalski, Process Analytical Chemistry, Anal. Chem. , 59 (9) (1987).

2.   J.T.Y. Yeh, On-Line Compositional Analyzers, Chem, Eng. 93 (2), 55 (1986).

3.   N. W. Madden, G.H. Hanepen, B.C. Clark, "A Low Power High Resolution Thermoelectrically cooled Si(Li) Spectrometer", IEEE Trans. Nuc. Sci., 33 (1), 303 (1986).

4.   A.R. Harding, J.D. Leland, D.E. Leyden, On-Stream Determination of Trace Elements Using Energy Dispersive X-ray Fluorescence Analysis, FACSS, 583, Detroit, MI , Oct. 4-9,1987.

APPLICATION OF FUNDAMENTAL PARAMETER SOFTWARE

TO ON-LINE XRF ANALYSIS

D.J. Leland and D.E. Leyden

Department of Chemistry
Colorado State University
Fort Collins, CO 80523

A.R. Harding

Tracor Xray, Inc.
Mountain View, CA 94043

INTRODUCTION

X-ray fluorescence analysis (XRF) is an analytical method
which has been adapted with considerable success to on-line
industrial process analysis with various degrees of
sophistication. Process analysis XRF systems range from
relatively simple units utilizing radioisotope sources with
non-dispersive analyzers to complex wavelength dispersive
systems in a central location receiving samples from a number
of process streams. The advantages of on-line process
analytical instrumentation for quality control, regulatory
compliance and safety considerations are well documented.[1,2]
Advances in the development of low maintenance
thermoelectrically cooled Si(Li) detectors have made energy
dispersive X-ray fluorescence analysis (EDXRF) even more
amendable to on-line process analysis. EDXRF is an important
method of on-line instrumentation because of its ability to
simultaneously detect many elements. The sensitivity of the
measurement will depend on the element of interest and
acquisition parameters used. It is the goal of any process
monitoring program to provide accurate and timely information
regarding the composition of a process stream in a reliable,
economic manner.

A consideration fundamental to any XRF analysis is that
of standardization. This becomes increasingly important in an
on-line application where conventional laboratory methods for
standardization may not be acceptable because of time
considerations and the experience of personnel responsible for
the day to day operation of the instrumentation. There are
several different approaches to standardization in XRF. The
method selected will depend in part upon the type of sample to

39

be analyzed. When working with a relatively simple system
with a limited concentration range, a linear or perhaps a
quadratic fit to the intensity data of the standards may be
adequate. However, for more complicated samples where matrix
or interelement effects are present, a linear calibration
curve may not be valid and the analyst should consider using
an empirical model such as an intensity or concentration
correction. While empirical models have been shown to be quite
useful for a number of applications, they do require a large
set of standards of similar composition to the unknowns, which
generally makes them impractical for on-line applications.

     The use of fundamental parameters software for on-line
determinations is attractive for several reasons. First,
there are a number of fundamental parameter programs available
to the user. Secondly, the relatively low cost of
microcomputers makes it possible to dedicate them to an on-
line analyzer. Lastly, fundamental parameters has fewer
restrictions on the number of standards required for an
analysis. This can simplify the standardization protocol for
maintaining an on-line XRF system, and permits greater
flexibility in dealing with different types of materials.

     This paper will describe two applications of the use of
fundamental parameters for on-line analysis. The first
example, determination of lead and bromine in leaded gasoline
is an example where accuracy as well as precision are
important from a regulatory perspective. The second example
is the determination of additives in lubricating oils.

EXPERIMENTAL

     A Tracor 7000 X-ray spectrometer (Tracor Xray Inc,
Mountain View, CA) was used equipped with a rhodium anode X-
ray tube and a thermoelectrically cooled Si(Li) detector with
a resolution of 190 eV at 5.9 keV. The primary filters used
were a thick (0.127 mm) and thin (0.05 mm) Rh filters. The
window material for the protective base plate and flow cell
was 0.2 mil polycarbonate. The acquisition parameters used to
determine lead and bromine in gasoline and establish
calibration curves were as follows: 40 kV, 0.35 mA, thick Rh
filter and 250 s livetime. Lead standards were prepared by
dilution of 5000 ppm Conostan lead standard (Conoco, Ponca
City, OK) in isooctane. Bromine standards were prepared by
diluting dibromoethane in isooctane.

     Two sets of acquisition parameters were selected for the
determination of additives in lubricating oils. Acquisition
parameters used to measure phosphorus, sulfur and calcium
were as follows: 10 kV, 0.35 mA, no primary filter and 500 s
livetime. The second set of acquisition parameters used to
determine zinc, and copper when present were: 30 kV, 0.15 mA,
thin Rh filter and 250 s livetime. The two sets of
acquisition parameters were run sequentially and automatically
for each sample. A helium flush was used to purge the area
behind the base plate of the spectrometer sample area to
improve sensitivity for low atomic number elements. The
helium flush was run at a flow rate of 1 SCFH.

Table 1.   Calibration Curves for Pb and Br

Pb Conc. (ppm)	Pb Intensity (cps)	Br Conc. (ppm)	Br Intensity (cps)
3.73	0.798±0.332	2.44	1.366±0.265
18.7	2.928±0.065	8.14	4.218±0.477
37.3	6.700±0.098	15.0	7.744±0.633
187	28.84 ±0.28	30.0	14.31 ±0.22
373	65.98 ±0.31	50.0	25.03 ±0.63
		70.0	34.61 ±0.37
Slope (cps/ppm)	0.1488		0.4933
Intercept (ppm)	0.0178		0.0998
R	0.9997		0.9995

RESULTS

In the period between July 1985 to January 1986, the U.S. Environmental Protection Agency lowered the maximum permissible levels of lead in leaded gasoline from a level of 1.1 g/gal to 0.1 g/gal(about 37 ppm).[3]   This reduction was accompanied by the requirement for enhanced monitoring of the lead content in gasoline.   To investigate the potential of a flow cell approach for the determine of lead and bromine leaded, gasolines were purchased from seven different local gasoline stations.   Samples were taken directly from the pump. Determinations of lead and bromine were done using both a linear calibration curve and by fundamental parameters utilizing a single standard.   Table 1 shows the concentration and intensity data used to prepare the linear calibration curve.   Table 2 shows the results obtained from the seven gasoline samples.   The precision values represent ±1 standard deviation of three replicate measurements.   A paired t-test shows that there is no statistical difference in the results of determinations using the linear calibration curve and the fundamental parameters approach.   A set of NBS reference fuels was analyzed to compare measured lead values with the certified value for each reference material.   The results are

Table 2.   Determination of lead and bromine in gasoline

| Sample | Pb (ppm) | | Br (ppm) | |
	Linear	Fund. Para.	Linear	Fund. Para.
1	23.3±1.4	23.1±1.5	5.1±0.7	5.2±0.7
2	27.6±1.6	27.4±1.5	7.8±0.5	7.8±0.4
3	47.3±2.4	47.7±2.4	12.2±1.2	12.3±1.2
4	34.0±1.5	33.7±1.5	8.9±0.3	8.9±0.3
5	51.5±1.5	51.3±1.5	19.8±0.9	19.9±0.9
6	48.3±2.3	48.2±2.5	16.0±0.5	16.1±0.5
7	33.1±2.9	32.9±3.0	10.8±0.6	10.8±0.6

Table 3. Results for lead in NBS Certified Reference Fuel
1636a (ppm)

Sample	Linear	Fund. Para.	Certified Value
1	11.2±2.2	10.7±2.1	11.2±0.2
2	18.5±1.2	17.6±1.1	18.8±0.2
3	24.6±1.1	24.6±1.1	25.1±0.2

given in Table 3. The fundamental parameters approach as well
as the linear calibration curve gave results in good agreement
with the certified values.

The second example to be described is the determination
of additives in lubricating oils. These additives improve
performance through the use of antioxidants, detergents and
organometallic compounds which improve the lubricating
properties of the oil. The samples investigated were eight
motor oils of various viscosity obtained from three different
producers. The elements found by energy dispersive X-ray
spectrometer were phosphorus, sulfur, calcium, zinc. Copper
was found in three of the samples. The samples were analyzed
in an independent laboratory for comparison with XRF results.

Phosphorus and sulfur were determined by wet chemical
analysis, and calcium, zinc and copper were determined by
atomic absorption. The results of the comparison are given in
Table 4; the independent laboratory values are given in
parentheses. Agreement between the data are generally good,
although the wet chemical methods were somewhat less sensitive
than XRF. It must be pointed out that sample number eight was
arbitrarily chosen as the single standard for fundamental
parameter calculations, so a comparison for this sample is not
valid.

Although the gasoline and oil were analyzed in a flow
cell mode, an oil sample was placed in the flow cell and
counted repetitively for a period of 12 hours using both
static and flow acquisition conditions to evaluate long term

Table 4. Results of the determination of elements present in
lubricating oil additives (%)

Sample	P	S	Ca	Zn
1	0.15 (0.13)	0.36 (0.42)	0.05 (0.06)	0.09 (0.10)
2	0.18 (0.13)	0.57 (0.59)	0.09 (0.09)	0.09 (0.11)
3	0.17 (0.13)	0.49 (0.58)	0.09 (0.12)	0.09 (0.11)
4	0.17 (0.12)	0.55 (0.52)	0.09 (0.08)	0.08 (0.10)
5	0.14 (0.17)	0.31 (0.33)	0.01 (0.01)	0.09 (0.08)
6	0.16 (0.12)	0.36 (0.37)	0.02 (0.02)	0.10 (0.11)
7	0.17 (0.17)	0.42 (0.47)	0.16 (0.18)	0.09 (0.12)
8	0.19 (0.19)	0.51 (0.52)	0.11 (0.11)	0.10 (0.11)

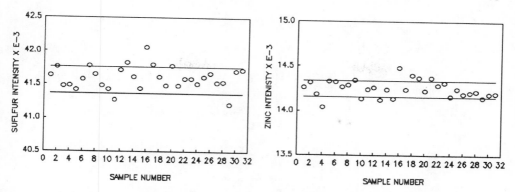

Fig. 1.   Zinc and sulfur X-ray intensities taken
under static conditions over a 12 hour
period.

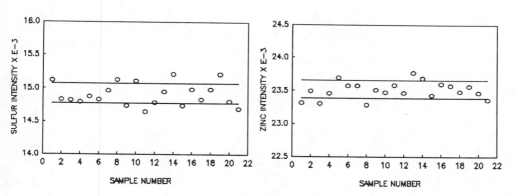

Fig. 2.   Zinc and sulfur X-ray intensities taken
under flow conditions over a 9 hour period.

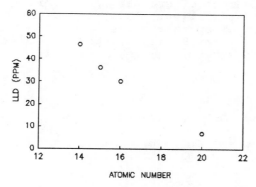

Fig. 3.   Lower limits of detection for lower atomic
number elements in oil.

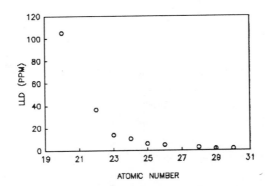

Fig. 4.   Lower limits of detection for middle atomic
           number elements in oil.

precision.   Figure 1 shows the intensities for zinc and sulfur
plotted versus time.   The   solid lines represent ±1 sigma
from the mean value.

    Figure 2 shows data from a second oil sample counted under
the same conditions as the static sample for a period 9 hours,
but circulated at a flow rate of 100 mL/min thorough the flow
cell.   Again, the solid lines represent ± 1 sigma from the mean.
The long term precision for both static and flow modes were
comparable.

    Figures 3 and 4 show lower limits of detection for
acquisition parameters used for the oil samples.   Detection
limits were calculated using the three sigma definition.
Standards were prepared by weighing and diluting 5000 ppm
Conostan standard in to Conostan 75 base oil.

CONCLUSION

    The use of fundamental parameter programs can be valuable
in reducing the number of standards required and time needed
for on-line standardization.   This results in a simplified
protocol required of operators responsible for the instrument
and allows greater flexibility in dealing with process streams
of different composition.

REFERENCES

1.   T. Hirschfeld, J.B. Callis, B.R. Kowalski, Science, 26:312
     (1984).
2.   J.T.Y. Yeh, Chem. Eng. News, 98(2):55 (1986).
3.   Chem. & Eng. News, 63(10):8 (1985).

ON-STREAM XRF MEASURING SYSTEM FOR ORE SLURRY ANALYSIS

B. Holyńska, M. Lankosz, J. Ostachowicz

Institute of Physics and Nuclear Techniques
Academy of Mining and Metallurgy, al. Mickiewicza 30
Krakow, Poland

T. Wesołowski, J. Zalewski

"Boleslav" Mining and Smelting Works, Bukowno, Poland

A modern metal-ore flotation process requires a continuous instrumental measurement of the metal content in main flotation streams. X-ray fluorescence technique is most often used for on-stream analysis of metal-ore slurries.[1] Especially the radioisotope energy-dispersion XRF (REDXRF) method provides the possibility of using measuring probes immersed in the slurry at the point of analysis. The main advantage of this technique is the elimination of the necessity to transport the slurry to a control laboratory which houses the X-ray spectrometer.

The REDXRF measuring system has been developed by us for on-stream determination of Fe, Zn and Pb in zinc-lead ore slurries.[2] In the XRF method it is necessary to eliminate the influence of variation in the chemical composition of the solid fraction, and the water content on the analysis. This is obtained by additional measurements of backscattered primary radiation and the intensities of the characteristic X-rays from the main interfering elements. For the determination of zinc, lead and iron in zinc-lead ore slurries a simple calibration equation is used:[2]

$$C_i = a_{Oi} + a_{ii} I_i + \sum a_{ij} I_j + a_{ibs} I_{bs} \qquad (1)$$

where $a_{Oi}, a_{ii}, a_{ij}, a_{ibs}$ are the constants to be determined from regression analysis. C_i is the concentration of the determined element. I_i, I_j, I_{bs} are the relative values of the measured intensities of characteristic X-rays of the determined i-th element, interfering j-th element and backscattered radiation, respectively.

The system consists essentially of X-ray fluorescence immersion probes, the electronic unit and the minicomputer with its peripherals. The probes are mounted in the plant in a flor-cell fed by a by-line from the process streams. The flow cell consists of two vessels connected by siphon. Entrainment of air in the analysis zone is minimized by having baffle plates and de-aerator. The flow cell is equipped with a sampler, allowing samples to be taken for calibration and control analysis. The XRF probe consists of a

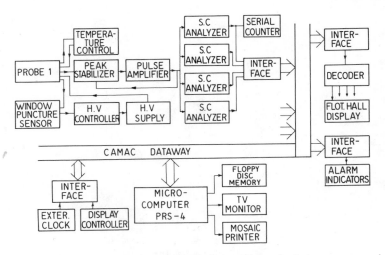

Fig. 1. A scheme of an electronic unit and data processing system.

rigid, water-proof cover with the measuring Melinex window, an argon-filled proportional counter (energy resolution-17% at the energy of 6.4 keV) and a radioisotope source. Usually a Cd-109 radioisotope is used with an activity of 75-185 MBq, depending on the flotation stream. The probe is additionally equipped with a heater for stabilizing the temperature inside the probe and a window puncture sensor.

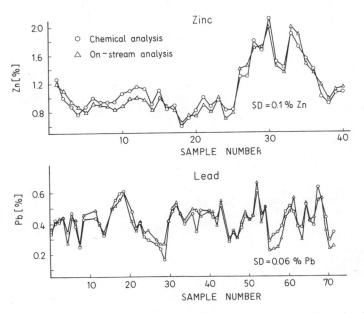

Fig. 2. Comparison of on-stream and chemical determination of Pb in tailings and Zn in feed.

In Fig. 1 the scheme of the electronic unit and data processing system is shown. The electronic unit consists of a set of four-channel spectrometers equipped with peak stabilizers. Selection of the energy channels is made in a differential (window) mode corresponding to the energies of the characteristic X-ray lines of the determined and the interfering elements as well as to the energy of backscattered radiation. The data processing system provides the number of counts every 100 s in particular energy channels and calculates the concentration of the determined elements in slurry solids. The mean values of Fe, Zn and Pb concentrations are printed at fixed time intervals (15 min) and are also displayed in the flotation room. Also, the computer activates an alarm when the electronic equipment or the technological process is malfunctioning. Two on-stream analysis systems have been installed in mineral processing plants for permanent work. Fig. 2 shows a one-month comparison of the results of on-stream and chemical determinations of lead in tailings and zinc in feed. Accuracy of on-stream analysis of zinc-lead ore slurries is within 3-15% relative, depending on metal concentration.

The results obtained so far confirm a high degree of applicability of the on-stream REDXRF measuring system. The introduction of this system to mineral processing plants leads to better control of the flotation process, which results in economic benefits through increased recovery of valuable minerals, reduced consumption of reagents and reduced labor requirements for sampling and analysis.

References

1.   J. S. Watt, On-Stream Analysis of Metalliferous Ore Slurries, Int. J. Appl. Radiat. Isot. 34: 309 (1983).
2.   B. Holyńska, M. Lankosz, J. Ostachowicz and K. Wolski, On-Stream XRF Measuring System for Ore Slurry Analysis and Particle-size Control, Int. J. Appl. Radiat. 36: 369 (1985).

APPLICATIONS OF ON-LINE XRF AND XRD ANALYSIS TECHNIQUES TO INDUSTRIAL PROCESS CONTROL

Matti Hietala
Outokumpu Oy, Espoo, Finland

Dennis J. Kalnicky
Princeton Gamma-Tech, Princeton, New Jersey

INTRODUCTION

Temperature, pressure, and flow measurements are considered standard for process control purposes. It is vital that they be made on-line in real-time and not manually in the laboratory. Chemical assays should be done as fast and continuous as temperature measurements in order to be useful for process control. Until recently, this has not been the case because the assay methods have been difficult to automate.

The introduction of minicomputers in the late sixties made it possible to construct on-line analyzers which incorporated extensive data manipulation capabilities to convert original measured values into useful forms and to control the operations of the analyzer. Current developments in microprocessors, electronic components, and detectors have made it possible to design compact, smart analyzers for on-line uses.

X-ray fluorescence (XRF) instruments have been used in laboratories as elemental analyzers since the fifties and were first used in commercial on-line analyzers in the sixties. The largest on-line XRF user group has been mineral concentrator and metallurgical plants.

X-ray diffraction (XRD) analysis, recently introduced by Outokumpu(1), represents the latest development in on-line X-ray technology. Their approach was to modify a field-proven on-line XRF analyzer to accomodate on-line XRD while others have unsuccesfully attempted to modify laboratory analyzers. This on-line XRD analyzer is capable of both qualitative and quantitative analysis of minerals in slurry samples and, unlike on-line XRF, has no light-element limitations.

Below, we describe practical aspects of on-line XRF and XRD analysis in process control applications.

TABLE 1. Benefits On-line Measurements

-'S of Off-line	+'S of On-line
- Inadequate process control	+ Reliable process control
- Infrequent measurements	+ Better understanding of the process
- Labor intensive	+ Material and labor savings
- Questions about reliability	+ Increased throughputs
- Automation not feasible	+ Closed-loop control
- Not a tool for Stastical Process Control(SPC)	+ Improved product quality, minimize rejects

ON-LINE VS. OFF-LINE ANALYSIS

When comparing on-line vs. off-line measurements from the viewpoint of process control, a number of significant benefits become clear. Table 1 summarizes differences for the important factors in this comparison. It is evident that off-line measurements are not adaquate for process control and are of marginal use in stabilizing product quality.

Most processes inevitably have periodic fluctuations and disturbances. On-line analysis provides immediate information about the state of the process which can be used to control it. This concept is illustrated in Figure 1 which shows the variation over time of an analysis in a process point.

Because of the long intervals between consecutive measurements and the time delay in conventional analyzing methods, the operator is constantly lagging behind the process and

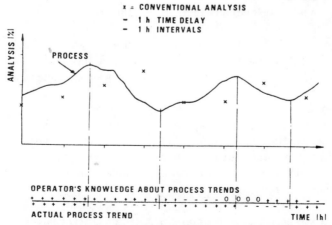

FIGURE 1. Information about the process from conventional routine analysis

does not have a clear understanding of its current state even though the analysis itself is nearly free of error. Moreover, the operator cannot know the result of the next analysis before he gets it, and often has a delayed opinion of the actual trend of the process. The process could actually be going down when the operator gets the information that it is going up, hence, his control action will often be made in the wrong direction.

On-line analytical information can improve the situation considerably. The operator has a much better understanding of the actual process state even though the on-line analysis is somewhat less accurate than a laboratory analysis. Because of this understanding, the operator is able to respond quickly to correct for changes in the process.

Constant information about ongoing changes and disturbances in the process gives the operator the ability to stabilize it as close as possible to the specification. This leads to material savings and a more satisfied customer who purchases the product. Figure 2 illustrates the concept of process stabilization where a sudden disturbance causes the impurities analysis to climb above the specification limit. Prompt information about this event allows corrective action to be taken immediately; with less frequent analysis the disturbance is detected much later which leads to increased production costs and poor product quality.

When process control is based on laboratory analysis a heavy burden is placed on laboratory personnel. A large number of routine analyses have to be done, usually on a shift work schedule. When an on-line analyzer is installed, the number of routine laboratory analyses is reduced and the remaining work can generally be done in the day shift only. On the other hand the important task of periodically checking the operation of the analyzer and improving its calibration will be added to the responsibilities of the laboratory.

By using continuous on-line analytical information the operator is able to study the process carefully and thus achieve material and labor savings. Necessary process modifications can be made when they are needed. Furthermore, the process control strategy can be designed and tested with use of the on-line analytical information.

SAMPLE PRESENTATION FOR ON-LINE XRF ANALYSIS

The depth from which XRF radiation in the sample can be observed by the detector is often small and depends on the energy of the X-ray line being measured as well as the absorption properties of the sample for that line. Typical depths range from a few microns

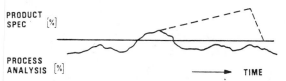

FIGURE 2. *Stabilizing product quality*

to a few millimeters for common materials measured with on-line analyzers. This means that a thin layer of sample next to the flow cell window must be representative of the process to be measured. If the sample is homogeneous and has low viscosity, this will be easily achieved. However, if it is inhomogeneous, of high viscosity, or is attacking the measuring cell window, the sampling system and presentation will require much attention.

There are two principal ways to install an on-line analyzer, either immerse the probe directly in the process or use a fast loop and sample cell concept. Outokumpu has employed the fast loop and sample cell concept(Figure 3) for better representation of inhomogeneous samples and for safety aspects in chemical processes. Compared to the immersible probe approach, the fast loop concept provides better control of flow conditions and sample cell window cleaning. Furthermore, calibration samples can be taken from the actual measured process stream.

In cases where several samples are to be multiplexed to the analyzer, the fast loop can be activated just before the start of each measurement. During intermediate periods the lines may be filled with water or other suitable cleaning liquid. The process pressure is often enough to keep the fast loop circulating; sometimes extra pumps or pneumatic transport will be required to circulate the sample to the analyzer.

Outokumpu has developed a number of sampling systems for homogeneous liquids, slurry samples, dry powder, and granulate materials. Also, liquid-solid separation systems (filtering and sedimentation principle) have been designed to extract clear solution from slurry samples.

An often misunderstood factor in on-line X-ray analysis is the effect of sample cell window absorption for the radiation being measured. Figure 4 shows transmission curves for several

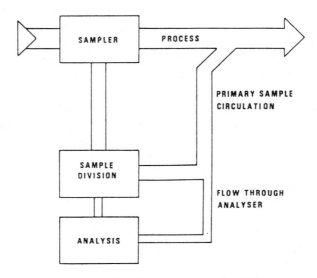

FIGURE 3. Sampling —fast loop concept

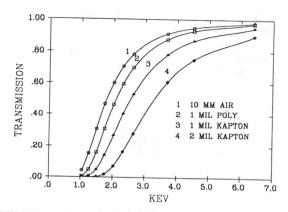

FIGURE 4. Cell window and air path absorption effects

window materials and air as a function of energy. It is evident that the light element limitation is principally due to window effects with the practical lower limit being phosphorus(P). In many applications the chemical inertness and temperature resistance of Kapton is required for reliable sample presentation to the analyzer. For a typical system with a 1/2 mil (13 micron) Kapton cell window and another 1/2 mil Kapton flow guard window, only 27% of P radiation is transmitted through the windows. Figure 4 also shows that for close coupled geometry, even 10 mm of air path transmits twice the P radiation compared to 1 mil Kapton.

STRUCTURE OF FIELD PROVEN ON-LINE XRF AND XRD ANALYZERS

On-line analyzers must be designed for much harsher conditions than laboratory systems. The process environment can have high temperature and humidity variations, dusty conditions, aggressive gases such as H_2S or SO_2, excessive vibration, and high electrical fields. In addition, the analyzer may periodically be washed with water or accidentally get a shower of process liquid. This is especially true in the mineral and metallurgical industry; in the chemical industry, explosion proof housings present additional sampling considerations.

A proven design principle is to select high quality components that are not sensitive to vibration, temperature and humidity changes, corrosive atmospheres, or low quality mains supplies. Outokumpu has paid much attention to these considerations in developing and manufacturing good quality detectors and crystal spectrometers for on-line XRF and XRD analyzers. Ruggedness and excellant performance can be achieved simultaneously when X-ray systems are designed for on-line or portable use. Detection limits are frequently as good as those for laboratory analyzers.

In the following section, application examples for three different types of on-line XRF/ XRD analyzers are presented: 1) the Courier 20(XRF), 2) the Courier 30(XRF), and 3) the Courier 40(XRF and XRD) analyzers. The Courier 20 is a source-excited, solid-state

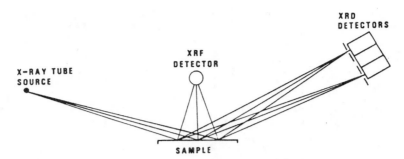

FIGURE 5. Courier 40 measurement geometry

detector Energy-Dispersive XRF (EDXRF) system with two flow cells each of which may be multiplexed. The Courier 30 is a tube-excited, proportional counter detector Wavelength-Dispersive XRF (WDXRF) system with a single multiplexed flow cell. The Courier 40 combined XRF/XRD analyzer is based on diffraction tube excitation of the sample with seperate proportional counters to detect diffracted beams (at fixed angles) originating from particles rotating in the turbulant flow of the sample cell. As a by-product, XRF radiation from the sample may be measured by energy dispersive methods using one or two proportional counters. The geometry of the Courier 40 is illustrated in Figure 5 and the Courier family of analyzers are described in more detail in references 2-5.

ON-LINE XRF AND XRD APPLICATION EXAMPLES

Courier 30 in a Hydrometallurgical Application

Outokumpu has used a Courier 30 at its Kokkola cobalt refinery since 1981 to analyze Fe, Cu, Co, Ni, and Mo from five different points in a solution purification process. The environment is harsh with H_2S fumes and the sample is acidic and at 80° C. The sample must be filtered before measurement because only clear solution assays are important for process control purposes. Metal contents in solution vary from 10 mg/l to 35 g/l with good analyses achieved over the entire range. Typical calibration results for Co analysis in waste solution are summarized below:

Concentration range	0.5 - 204 mg/l
Average	70 mg/l
Correlation coefficient	0.999
Absolute error(1-sigma)	2.6 mg/l
Relative error	3.7 %
Variables in correction model	Co, Cu
Number of samples	10

The details of this application are described elsewhere(6).

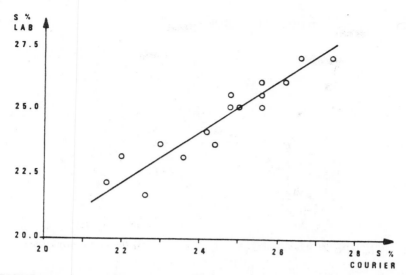

FIGURE 6. On-line sulphur calibration results for Harjavalta smelter feed

Courier 20 for Inhomogeneous Powders

At the Harjavalta Copper flash smelter, the combustion properties of the feed going to the smelter furnace must be measured continuously. These analyses are used to calculate the setpoint for the oxygen feed to control the furnace. Combined with other measurements, the Courier 20 analyses are used to stabilize furnace operation.

The sample is taken from the conveyor that transfers the feed mixture to the flash furnace burners. The primary sample is further divided with a secondary sampler and fed to the measuring cell of the analyzer. The tight geometry of the Courier 20 enables measurement of light elements, such as sulphur, without the need for helium or vacuum. A multi-source technique is used for simultaneous measurement of S, Fe, Cu, Zn, Pb and other elements.

Results to date have demonstrated the viability of the analysis technique. Figure 6 shows sulphur calibration results with the Courier 20 for a highly variable, multi-source concentrate feed mixture. The details of this application are described in reference 7.

Apatite Ore Analysis with On-line XRD

Many minerals are difficult to measure using on-line XRF because they contain P and lighter elements; the X-ray penetration is too weak. By using XRD the minerals can be measured.

The most important phosphate minerals are fluorapatite, $Ca_5(PO_4)_3F$, (in igneous rocks) and francolite, a carbonate fluorapatite (in sedimentary rocks). The most intense XRD lines of several apatite minerals are very close to that of fluorapatite which is an advantage

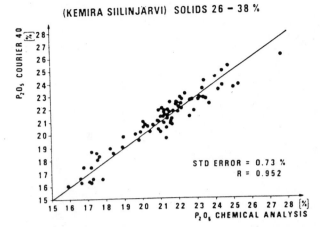

FIGURE 7. *Apatite rougher concentrate results with Courier 40
on-line XRD/ XRF analyzer*

for on-line XRD analysis. Figure 7 shows calibration results achieved by the Courier 40 when it was installed in an apatite concentrator plant. The details of the apatite application are given in reference 8.

On-site plant tests of the Courier 40 for apatite and talc analyses have shown results to be accurate enough for good process control(5).

CONCLUSIONS

On-line XRF analyzers have been successfully used for about twenty years in mineral processing applications, mainly in metallic ore beneficiation plants to measure metal concentrations of solids in water slurries. The number of applications in other industries (eg., chemicals) is expected to increase in the future. The introduction of on-line XRD analysis by Outokumpu represents the latest development in X-ray technology and is well suited to continuous monitoring of mineral concentrations in process environments.

The advantages of on-line analysis are clear. It provides: 1) immediate information on the state of the process with no time lag, 2) stabilization of product quality with minimum material give-away, 3) reliable measurements with proper sample representation using instruments designed for industrial environments, 4) automated analysis capabilities with little or no operator intervention, and 5) an interface to process control to implement closed-loop control and statistical process control(SPC).

REFERENCES

1. "On-stream X-ray Diffraction Analyzer for Mineral Concentrators," H. Sipila and J. Koskinen, Denver X-ray Conference, August, 1986; ADVANCES IN X-RAY ANALYSIS, Vol. 30, pp. 367-372(1987).

2. "Application and Economic Aspects of WDXRF and EDXRF Techniques in Industry," P. Rautala, M. Hietala, and H. Sipila, Proceedings of the Advisory Group Meeting on Practical Aspects of Energy Dispersive X-ray Fluorescence Analysis, Organized by the International Atomic energy Agency, Vienna, 29 May - 2 June, 1978.

3. "New Energy and Wavelength Dispersive On-stream XRF Probes," M. Hietala and E. Kiuru, Third IFAC Symposium on Automation in Mining, Mineral and Metal Processing, Montreal, August, 1980.

4. "Automation of the assays in the KCl-production by X-ray methods," B. J. Golovkov and H. J. Sipila, IFAC Symposium on automation in Mining, Mineral and Metal Processing, Tokyo, August, 1986.

5. "On-site Tests of a new XRD/XRF On-line Process Analyzer," A. Ahonen et al., Denver X-ray Conference, August, 1988.

6. "Experiences of a new on-stream X-ray analyzer in a metal refinery," K. Saarhelo, International Federation of Automatic Control—4th IFAC symposium, Helsinki, Pergamon Press, Oxford, 1983.

7. "Application Announcement: automatic Analysis of Flash Furnace Feed using COURIER 20 On-stream Analyzer," Outokumpu Electronics publication, Espoo, Finland, 1988.

8. "On-line Analysis in Industrial Mineral Applications," K. Saarhelo, 8th Industrial Minerals International Congress, Boston, April, 1988.

ON-SITE TESTS OF A NEW XRD/XRF ON-LINE PROCESS ANALYZER

A.Ahonen, C.v.Alfthan, M.Hirvonen, J.Ollikainen,
M.Rintamäki, K.Saloheimo, and P.Virtanen

Outokumpu Oy, Electronics Division
P.O. Box 85, 02201 Espoo, Finland

INTRODUCTION

X-ray fluorescence (XRF) has been successfully used for more than 20 years as an on-line analysis method for controlling various industrial processes where the concentrations of different elements must be known in the process streams. In many industrial processes, however, the knowledge of the concentrations of different minerals, rather than those of the elements, is of prime importance. Examples of such processes are numerous; plants concentrating apatite for fertilizer and detergent manufacturers, paper filler producers manufacturing kaoline, titanium dioxide (rutile) and talc, potash concentrators and different crystallization processes.

For this very branch of process industry Outokumpu has developed the Courier 40 analyzer discussed in this paper. The analyzer is based on the combination of X-ray diffraction (XRD) and XRF. The idea of using on-line XRD to determine the concentrations of relevant minerals in a process stream is not new; the first attempts[1,2] to employ XRD on-line were made in England as early as in the 1960's and more recently in South Africa[3]. Obviously the Courier 40 analyzer by Outokumpu, however, is the first commercially available on-line XRD-instrument especially designed for the industrial environment.

This paper reports on the results from the first on-site tests of the new XRD/XRF-analyzer. First, the measuring geometry, construction, and the application range of the analyzer are described. Then the results from a test-series on an apatite concentrator are discussed. Finally, the results from a similar test set-up in a talc concentrator are described.

MEASURING GEOMETRY AND CONSTRUCTION OF
THE COURIER 40 XRD/XRF-ANALYZER

The measuring geometry of the analyzer in its present form
is shown in Figure 1. X-rays emitted by a small 30 kV/200 W
side-window tube with a spot size of roughly 0.6 mm x 0.6 mm
are first collimated into a narrow well-defined beam by means
of a tubular collimator made of nickel. So far, solely X-ray
tubes with a copper anode producing a strong K_α-line at 1.54Å
have been used. A nickel filter in front of the window of the
X-ray tube has been employed to eliminate the Cu K_β-line and
to reduce the intensity in the bremsstrahlung regime. The
collimated X-ray beam irradiates the sample through a thin
plastic sample cell window. The length and the width of the
elliptical irradiated area are 14 mm and 3.5 mm, respectively.
The incident angle of the radiation, the angle between the cell
window and the X-ray beam, is 15°. So far, only slurry samples
have been studied. The sample flows through the measuring cell
at a rate of roughly 20 l/min. To obtain a representative
sample, the flow conditions in the cell have been carefully
optimized. The flow in the vicinity of the cell window is
highly turbulent with a rather slow translatory flow velocity.
Typically, a 25 μm Kapton window can be used in a mineral
slurry measurement for 48 h without rupture.

The diffracted X-rays travel through two sets of fixed
analyzing slits with a typical width of 0.6 mm. The diffracted
intensity is then measured by means of a xenon-filled proportional counter equipped with five anode wires in separate
compartments. The analyzing slits are custom made according to
the application. X-ray lithography employing diffracted radiation from a powder sample is used to correctly mark the
positions of the analyzing slits on the slit block before
machining. Typically, the error in the effective slit positions
can be kept below $\delta(2\theta) \approx 0.03°$.

The actual construction of the front end of the analyzer
is shown in Figure 2. The main goals in designing the analyzer

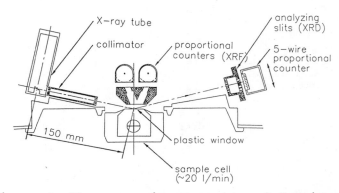

Figure 1. The measuring geometry of Courier 40
          XRD/XRF-analyzer.

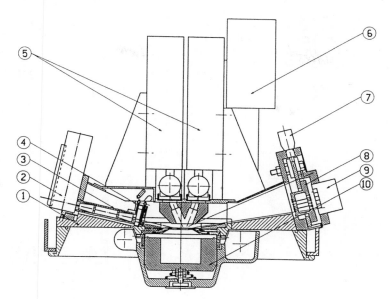

Figure 2. Construction of the front end of the analyzer.
1 Analyzer body, machined out of a single
metal block, 2 X-ray tube in its holder,
3 Collimator, 4 Shutter assembly, 5 Preamplifiers and high voltage sources for XRF
proportional counters, 6 Preamplifiers
and the common HV-supply for the 5-wire
proportional counter, 7 Caliper screw for
fine adjusting the overall position of the
analyzing slits 8 and the XRD-detector 9,
10 The flow cell with the plastic window
support ring in front.

have been the mechanical stability of the construction and high
degree of reproducibility in mounting the X-ray tube, window
support ring and the analyzing slits. For fine tuning of the
position of the analyzing slits a separate caliper screw
adjustment can be employed.

    The applicable range of the Bragg angle 2θ is from 22°
to 45°, and the maximum angular spread $\Delta(2\theta)$ of the diffracted
beams simultaneously covered is 9.5°. For the time being, the
minimum applicable separation between two neighbouring analyzing slits corresponds to a difference of 0.5° in the Bragg
angles. Examples of Bragg angles of some minerals, relevant to
this work, are given in Table 1.

    The actual Courier 40 analyzer with a five-stream sample
multiplexer for slurry applications is schematically illustrated in Figure 3.

    The sample multiplexer selects one of the five sample
streams at a time and feeds it through the flow cell of the
analyzer at a rate of 20...30 l/min. A separate calibration

Table 1. Examples of the Bragg angles for some relevant minerals. The data correspond to the most intense diffraction signal at the wavelength of the Cu Kα-line, 1.542 Å.

Mineral		2θ
Quartz	SiO_2	26.6°
Talc	$Mg_3Si_4O_{10}(OH)_2$	28.6°
Fluorapatite	$Ca_5F(PO_4)_3$	31.8°
Magnesite	$MgCO_3$	32.6°

sampler can be used to extract a representative sample for subsequent laboratory analysis out of the stream simultaneously being analyzed by Courier 40.

RESULTS FROM ON-SITE TESTS OF
THE COURIER 40 ANALYZER

So far, the new XRD/XRF on-line analyzer has been tested on site in two mineral concentrators under real industrial conditions. One of the test sites was an apatite concentrator producing raw material for fertilizer industry. The other test series was carried out in a talc concentrator producing high purity talc for use as a filler material in paper mills.

Traditionally both processes have to rely on rather complicated time consuming laboratory analysis techniques for controlling the process and the quality of the end product. Due

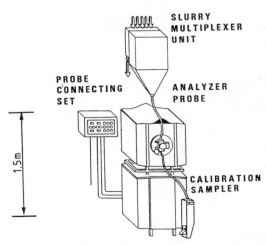

Figure 3. A schematic illustration of the actual Courier 40 analyzer.

Table 2. The typical solid fractions and P_2O_5 concentrations in the three process streams studied in this work.

Stream	Solid fraction (% by weight)	P_2O_5 concentration in solids (% by weight)
Rougher concentrate	26 ... 38	21
Concentrate	36 ... 47	36
Tailings	27 ... 35	1

to its low atomic number the concentration of phosphorus in a mineral slurry cannot be determined on-line by e.g. traditional XRF-methods. On the other hand, the determination of the talc concentration to a good accuracy from an industrial sample is quite difficult even in a well equipped industrial laboratory and great care is traditionally taken to avoid systematic errors.

Both apatite and talc have distinct crystal structures giving rise to easily resolved diffraction signals (cf. Table 1). Therefore, the new XRD/XRF on-line analyzer was considered to offer significant new potential in controlling both of the example processes.

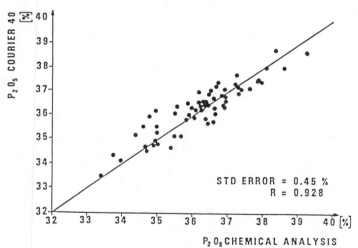

Figure 4. P_2O_5 concentration in the final concentrate, Courier 40 assay vs. laboratory analysis. 62 samples were collected over four weeks. The model is a linear combination of the apatite diffraction intensity, the diffraction background, the XRF-intensity of iron impurities, and the intensity of Compton-scattered CuKα-radiation. Correlation coefficient R = 0.93.

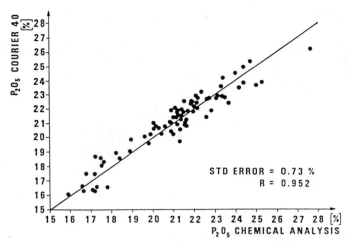

Figure 5. P_2O_5 concentration in the rougher concentrate,
Courier 40 assay vs. laboratory analysis. 83 samples
collected over four weeks. Type of model equivalent
to that used for the data in Fig.4. Correlation
coefficient R = 0.95.

On-site tests in an apatite concentrator

During the tests in the apatite concentrator, the apatite
content, hereafter given as the concentration of phosphoric
pentoxide (P_2O_5), was followed in three different process
streams containing rougher concentrate, final concentrate, and
tailings. Sample streams from all these three process locations
in addition to a fourth line containing pure water were connected to a multiplexer of the type shown in Figure 3. The water
sample was used as a background reference. Typical values of
solid fractions and P_2O_5 concentrations in the three process
streams are listed in Table 2.

For the calibration of the analyzer, calibration samples
were collected during six weeks employing the calibration
sampler shown in Fig.3. The laboratory analyses of these
samples were then compared to the concentrations calculated
from the corresponding XRD and XRF intensities using several
empirical models and regression analysis. The model giving the
best agreement with the laboratory analyses was then selected.
The P_2O_5 concentrations calculated from final models are shown
as functions of the corresponding laboratory data in Figures 4,
5, and 6 for the final concentrate, the rougher concentrate,
and the tailings, respectively. The laboratory data are based
on routine analysis methods the accuracy of which was not
studied in the course of this work.

A summary of the data in Figures 4, 5, and 6 is given in
Table 3, where the correlation coefficients between the Courier
40 assays and the corresponding laboratory analyses, and the
relative standard errors in P_2O_5-concentrations are listed for
different process streams.

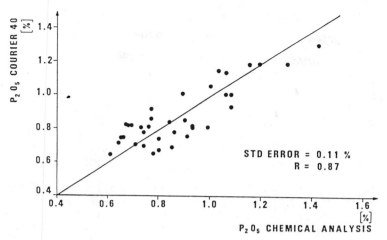

Figure 6. P_2O_5 concentration in the tailings, Courier 40 assay
            vs. laboratory analysis. 36 samples collected over
            six weeks. The model is a linear combination of the
            apatite diffraction intensity, the diffraction
            background, and the XRF-intensity of iron impuri-
            ties. Correlation coefficient R = 0.87.

    The accuracy of the calibration for both the concentrate
and the rougher concentrate is quite good, and similar to the
accuracies typically obtained by XRF-techniques in metal
concentrators. The less pleasing accuracy in tailings simply
reflects the fact that the apatite concentration, 0.9 % on the
average, approaches the detection limit of the analyzer.

    For any process analyzer, long-term stability is of essen-
tial importance. The stability of Courier 40 analyzer was
tested in the apatite concentrator by simply following the
output of the device and comparing it to routine laboratory
analysis. In the typical comparison of Figure 7 the difference
between the two analyses stays well below one per cent unit for
most of the time.

Table 3. Correlation coefficients between Courier 40 P_2O_5
          assays and laboratory analyses in different process
          streams and the corresponding relative standard
          deviations of P_2O_5 concentrations.

Stream	Correlation coefficient	Rel. standard error
Concentrate	0.93	1.2 %
Rougher concentrate	0.95	3.4 %
Tailings	0.87	12 %

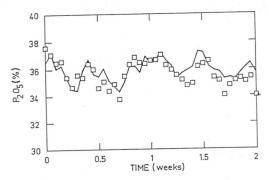

Figure 7. The concentration of P_2O_5 in the final
           concentrate as given by Courier 40, solid
           curve, and the laboratory analysis, open
           boxes, during a period of two weeks.

On-site tests in a talc concentrator

During the on-site tests in a talc concentrator the concentration of the main product, talc, and that of the main
impurity, magnesite, were measured in four process locations
employing a sampling system similar to the one used in the
apatite concentrator. The sampled process streams with their
typical solid fractions and compositions are listed in Table 4.

Calibration samples were collected over a period of six
weeks from all the four process streams, and models for concentration calculations were developed using the methods
already discussed in conjunction with the tests in the apatite
concentrator.

The measured talc concentrations are shown as functions of
the corresponding laboratory data in Figures 8 and 9 for the

Table 4. Solid fractions and approximate compositions
          of the four process streams studied in this
          work.

Stream	Solid fraction (% by weight)	Composition (% of solids)	
		talc	magnesite
Feed (ore)	10 ... 20	50	40
Rougher concentrate	15 ... 35	85	10
Final Concentrate	10 ... 20	97	2
Tailings	7 ... 15	20	67

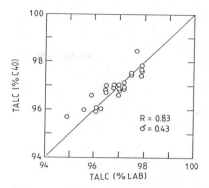

Figure 8.   The talc concentration in the final concent-
            rate, Courier 40 assay vs. laboratory analy-
            sis. 28 samples collected over six weeks. The
            model relies solely on the measurements of
            the diffraction background and the XRD
            intensity of magnesite, the main impurity.
            Correlation coefficient R = 0.83.

final concentrate and the tailings, respectively. Like in the
case of the apatite studies, the laboratory data are based on
routine analysis, the accuracy of which was not controlled
during this work.

     A summary of the data in Figures 8 and 9 is given in Table
5, which also lists the corresponding data for the feed and the
rougher concentrate.

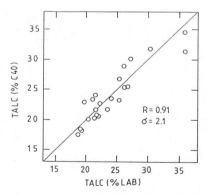

Figure 9.   The talc concentration in the tailings,
            Courier 40 assay vs. laboratory analysis.
            24 samples collected over six weeks. The
            model is a linear combination of the diffrac-
            tion intensity of talc, the XRF-intensity of
            calcium, and the intensity of the Compton-
            scattered CuK_α-radiation. Correlation
            coefficient R = 0.91.

Table 5. Correlation coefficients between Courier 40 talc
          assays and laboratory analyses in different process
          streams and the corresponding relative standard
          errors of the talc concentrations.

Stream	Correlation coefficient	Rel.standard error
Feed	0.65	3.3 %
Rougher concentrate	0.90	1.4 %
Concentrate	0.83	0.45 %
Tailings	0.91	9 %

     The accuracy of the Courier 40 assays as reflected by the
data in Table 5 was quite sufficient for process control. The
lowest correlation was obtained for the calibration of the feed
stream. The less satisfactory result for the feed is due to the
narrow concentration range of the calibration samples and the
large variation in the mineralogy of the ore.

     The long-term stability of the analyzer is demonstrated
in Figure 10, where daily averages of the talc concentration
in the final concentrate both as measured by Courier 40 and as
analyzed in the laboratory are shown over a period of six
weeks.

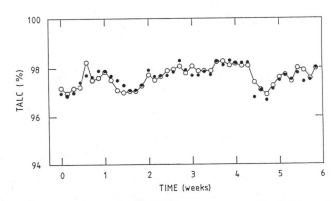

Figure 10. The daily averages of the concentration of
            talc in the final concentrate measured by
            Courier 40, open circles, and the corres-
            ponding laboratory analysis, filled circles,
            as functions of time during a period of six
            weeks.

CONCEPTS OF INFLUENCE COEFFICIENTS IN XRF ANALYSIS AND CALIBRATION

Richard M. Rousseau

Geological Survey of Canada
601 Booth St., Ottawa, Ont., CANADA, K1A 0E8

ABSTRACT

It will be shown that there are different types of influence coefficients to correct for matrix effects in XRF analysis: the empirical coefficients calculated from regression analysis and valid for the standards on hand only; the theoretical binary coefficients that are essentially constant in a given concentration range; then the theoretical multi-element coefficients calculated from one given composition and valid for one specimen only.

With theoretical coefficients, however, a special calibration process is needed, because theory cannot take into account all the instrumental parameters. In the proposed calibration method it will be shown that the slope of the calibration curves is the link to correlate the theoretical formalism to the experimental reality; that the standards used do not have to be similar to the unknowns and that a few only are required; that the calibration curves can be extrapolated by a factor of two or three; that pure analytes are not required.

INTRODUCTION

In X-Ray Fluorescence (XRF) analysis, the influence coefficient approach assumes that measured intensities can be corrected for matrix effects (absorption and enhancement) by means of numerical coefficients that depend on the matrix composition. These "multi-element coefficients" can be calculated from theory, using the Fundamental Parameters equations[1,2,3], or can be determined empirically by regression analysis[4]. While the empirical coefficients depend specifically on the composition of the standards used, a set of theoretical coefficients calculated from a specimen composition is "unique" because each specimen composition is "unique". Any other specimen containing the same elements but in different proportions, or any other set of standards, will generate different sets of influence coefficients.

This last property of influence coefficients has caused problems to the XRF analysts and is the reason why approximate methods have proliferated[2,4,5,6] between the years 1960 and 1980. Those methods were based on the hypothesis that the total matrix effect on the analyte i is equal to the sum of the effects of each element j of the matrix, calculated independently of each other. With these methods it was then possible to calculate a set of "binary coefficients" valid for a given composition range rather than for a given specimen. In other words, with binary coefficients it is assumed that the coefficient a_{ij} is a CONSTANT for a given range of C_i and C_j rather than being a VARIABLE dependant on the whole matrix composition. It will be shown that some of these methods are still valid as long as they are used with theoretical binary coefficients[5,7,8].

At the beginning of the 80's, new powerful quantitative methods, based on the "Fundamental-Parameters" approach and/or on the new concepts of theoretical coefficients, have been proposed mainly by J.W. Criss[9], R.M. Rousseau[3,10] and N. Broll[11]. The last two papers propose methods for calculating theoretical influence coefficients as a function of the composition of the matrix. These coefficients are called "multi-element coefficients" because they depend on the full composition of the matrix. For each specimen, an estimate of the matrix composition must be calculated first in order to calculate the multi-element coefficients and finally to calculate from them a more accurate composition of the specimen. This approach will be discussed in the present paper.

The two types of theoretical influence coefficients, binary and multi-element, are powerful tools to correct for matrix effects as long as a link exists for adapting the theory to the experimental reality, considering that no theory accounts for all experimental variables. It will be shown that an appropriate calibration process corrects for the imperfections of the theory and has a few other major qualities that facilitate the work of the analyst.

BINARY COEFFICIENTS

Lachance and Traill[5] were among the first to propose influence coefficients as a method of correction for matrix effects in XRF analysis. One of the most important ideas proposed in their paper is that the so-called "Alpha coefficients", can be calculated from theory. The proposed theoretical expression to calculate these is

$$\alpha_{ij} = \frac{\mu_j^*}{\mu_i^*} - 1 \qquad (1)$$

where $\qquad \mu_j^* = \mu_j(\lambda_k) \csc \phi' + \mu_j(\lambda_i) \csc \phi'' \qquad (2)$

$\mu_j(\lambda_k)$: mass absorption coefficient of j for the wavelength λ_k of the incident spectrum

$\mu_i(\lambda_i)$: mass absorption coefficient of j for the characteristic wavelength λ_i of i

ϕ', ϕ'' : incidence and take-off angles of X-ray beam

Equation (1) is an excellent start but it is valid for one given wavelength (λ_k) only of the incident spectrum. It also assumes incorrectly that enhancement behaves as a negative absorption and that α_{ij} is independent of the elements other than i and j.

To overcome these limitations, many approximate methods to correct for the matrix effects and to convert measured intensities (I_i) into concentrations (C_i) have been proposed. Among the proposed algorithms, three are more generally favoured for their accuracy and their solid theoretical foundation. They also use the modern concepts of theoretical binary coefficients. They are:

First, the Lachance-Traill algorithm[5]

$$C_i = R_i \left(1 + \Sigma_j \ a_{ij} \ C_j \right) \tag{3}$$

where R_i is the ratio of the measured net intensity I_i to the measured net intensity of the pure analyte i. The binary coefficient a_{ij} is calculated using equation (16) of reference (3) for the special case of a binary standard having a composition (C_{im}, C_{jm}), where C_{im} is the mid-value of the calibration range of the analyte i and where

$$C_{jm} = 1 - C_{im} \tag{4}$$

This approach assumes that the coefficient a_{ij} is a CONSTANT (it is an approximation!) when applied to samples with a limited concentration range (0-10%), for example, oxides in rock samples in fused disks. A theoretical mean relative error of 0.02% on the calculated concentrations[12] is expected.

Secondly, the Claisse-Quintin (CQ) algorithm[7]

$$C_i = R_i \left[1 + \Sigma_j (a_{ij} + a_{ijj} \ C_M)C_j + \Sigma_j \Sigma_k \ a_{ijk} \ C_j C_k \right] \tag{5}$$

where the matrix concentration

$$C_M = C_j + C_k + \ldots \tag{6}$$

and where the "crossed" coefficients a_{ijk} are included to compensate for the fact that the total interelement correction cannot be strictly represented by a sum of binary matrix effects. The coefficients are calculated by equations (4,5,6) of reference (10). Contrary to Wadleigh's claim[13], the latter equations do not contain errors when they are compared to those of Rousseau and Claisse[14], they are only improved and more accurate.

The influence coefficients of the CQ algorithm are considered as constant when applied to samples with a medium concentration range (0-40%), for example, oxides in cement samples in pressed pellets. In this case, the method introduces a theoretical mean relative error of 0.04% on the calculated concentrations[12].

Thirdly, the Lachance algorithm[8]

$$C_i = R_i \left\{ 1 + \Sigma_j \left[a_1 + \frac{a_2 \ C_M}{1 + a_3 (1 - C_M)} \right]C_j + \Sigma_j \Sigma_k \ a_{ijk} \ C_j C_k \right\} \tag{7}$$

where the binary coefficients a_1, a_2, and a_3 are calculated by means of equations (A6) of reference (8) and the crossed coefficients a_{ijk} by the equation (19). This algorithm can be applied to samples with a large concentration range (0-100%), for example, elements in alloys. In this case, the method introduces a theoretical mean relative error of 0.3% in the calculated concentrations[12].

An experimental verification of these three algorithms done by Pella et al.[15], confirms the expected theoretical accuracy. Consequently, the XRF analyst should consider the theoretical binary coefficient approach within these three algorithms as a good alternative to the fundamental parameters approach, specially when the matrix effects or the variations in composition are small.

However, the calculation of theoretical coefficients cannot be done without computer programs. The public domain program, NBSGSC[16], performs all the XRF analysis calculations on a DIGITAL PDP-11 minicomputer. The commercial version of this program[17] and a few others such as CiLT, CiLAC and CiROU[18], running on an IBM-PC, are also available.

EMPIRICAL COEFFICIENTS

Coefficients for matrix effect corrections can be obtained by multiple regression methods using measured intensities and standard compositions, which cover the analyte concentration ranges of interest. These correction coefficients are called empirical coefficients. The Rasberry and Heinrich model[4] is a good example of this approach. The success of empirical coefficients depends on the availability of standards that closely match the compositions of the unknowns. As the number of analytes increases, so does the number of standards required.

The preparation and measurement of an adequate number of reliable standards could be long and tedious and can easily yield empirical coefficients that are often not accurate and have no physical meaning. This statistical approach is sensitive to any error in the data and the calculated coefficients can be applied ONLY to unknowns of composition similar to those of the standards used.

An easier and much more practical solution to these problems is to compute theoretical coefficients. They can be calculated for any combination of elements and any experimental conditions, within of a few seconds! Use of empirical coefficients should then be discontinued to avoid potential problems and to minimize inaccurate results.

MULTI-ELEMENT COEFFICIENTS

A better approach than the empirical coefficients is the fundamental parameters method first proposed by Criss and Birks[2]. This method has been optimised by Criss[9] and can be applied through the computer program XRF-11 also developed by him. More details on the method and on the program performance are given in reference (19).

Alternately, new powerful quantitative methods, based on the fundamental parameters approach and on the new concepts of theoretical multi-element coefficients, have been proposed[3,10,20]. In particular, Rousseau[3], using the Sherman's equations[1], deduced new theoretical expressions for multi-element influence coefficients in which the correction for both absorption and enhancement effects are clearly and accurately defined. They are used in the fundamental algorithm

$$C_i = R_i \; \frac{1 + \Sigma_j \; \alpha_{ij} \; C_j}{1 + \Sigma_j \; \rho_{ij} \; C_j} \tag{8}$$

where α_{ij} and ρ_{ij} are the multi-element coefficients correcting for absorption and enhancement, respectively. These are calculated by means of equations (9) and (10) in reference (3). The fundamental algorithm can be applied to any type of samples with any composition. The method introduces a theoretical mean relative error of 0.07% only[12].

An experimental verification of this method done by Rousseau and Bouchard[21] on different types of alloys confirms the accuracy and versatility of the method. To test this method the computer program CiROU V3.0 was developed by the Geological Survey of Canada and is in the public domain. It runs on DIGITAL PDP-11 minicomputers under the RSX-11M operating system. A more user-friendly version of this program is available from Corporation Scientifique Claisse[18]. It runs on IBM-PC/XT/AT, PS/2 or compatible computers under the PC/MS-DOS operating system.

CALIBRATION

Although in many cases the analysis of samples with limited composition range may allow the use of linear calibration curves (Intensity versus Concentration), it is usually more desirable to work with a general purpose calibration process which is applicable to a larger variety of matrix types over wider concentration ranges.

With theoretical coefficients, however, a special calibration process should be used, because it is well known that theory cannot account for all the instrumental parameters. The proposed calibration process is the following:

The relative intensity R_i is defined as the ratio of the net intensity of the analyte i to the net intensity $I_{(i)}$ of the pure analyte i :

$$R_i = \frac{I_i}{I_{(i)}} \tag{9}$$

This equation can be written in the form of a straight line equation,

$$\frac{I_i}{I_{im}} = \frac{I_{(i)}}{I_{im}} R_i \tag{10}$$

where both side of the equation have been divided by I_{im}, the net intensity of the analyte i in any monitor. The relative intensity can be calculated by the expression

$$R_i = \frac{C_i}{[\ 1 + \ ...\ ]} \tag{11}$$

from any of the inverted algorithm (3), (5), (7) or (8) chosen by the analyst. The R_i intensities are calculated for all the standards on hand and their I_i and I_{im} intensities are measured. Then, if the MEASURED relative intensities I_i/I_{im} (Y-axis) are plotted as a function of the CALCULATED relative intensities R_i (X-axis), it gives a straight line of slope equal to $I_{(i)}/I_{im}$[10,21].

This calibration approach has the following clear advantages:

1. Since the curves compare two similar objects, calculated and measured relative intensities, they are independent of the composition of the standards used. Consequently, the standards do not have to be similar to the unknowns and such curves become universal calibration curves.

2. The calibration slope adjusts the theory to the experimental reality. If the pure analyte is used as monitor, the slope value is close to 1.0 but not necessarily equals to 1.0. The deviation from 1.0 come from the imperfection of the theory and the observed slope is the factor required to correct for that.

3. Pure analytes are not required, but their intensities are directly calculated from the slope of the calibration curves. If a few are available, they are handled as any other standard.

4. If a single point is off the calibration curve by more than the expected error, it clearly stands out and facilitates tracing eventual errors. In other words, the method is a "fail-safe" process.

5. Each analyte of a specimen can be calibrated independently of all other analytes. In this manner, if the calibration of one analyte has to be done again, this does not affect the calibration of the other analytes. Furthermore, the standards do not need to contain all the analytes of the specimen. Each calibration curve can be drawn from as few as one standard, which for example can be the pure analyte.

6. The calibration curves can be extrapolated by a factor of two or three, thus protecting the analyst from errors when the sample concentrations exceed the calibration range.

APPLICATION TO ANALYSIS

Once the slope m_i of the calibration curve of the analyte i is determined, the concentrations C_i of the unknowns are calculated from a combination of equations (10) and (11)

$$C_i = \frac{1}{m_i} \frac{I_i}{I_{im}} [1 + \ldots] \tag{12}$$

where the term in brackets is the matrix effect correction in the algorithm selected by the analyst.

CONCLUSION

Since there is no longer any need for empirical coefficients, only two types of influence coefficients remain. Firstly the theoretical binary coefficients that are constant in a given concentration range of C_i and are independent of the matrix composition. They are used in theoretically valid algorithms, such as the Lachance-Traill, Claisse-Quintin and Lachance algorithms. Secondly, the theoretical multi-element coefficients calculated from an estimate of the composition of each sample are then used in the fundamental algorithm to obtain a more accurate composition of the sample.

It is expected that the present explanations will eliminate the pseudo-ambiguities or pseudo-errors raised by some authors[13,22,23].

Without computer programs, theoretical coefficients can not be calculated. Some are already available but higher quality ones will be written in
the future, making empirical and approximate methods less and less popular,
specially when microcomputers become still more effective and less costly.

The proposed calibration process offers many advantages compared to the
traditional calibration of intensities versus concentrations; it increases
the accuracy of XRF analysis and facilitates the work of the analyst.

REFERENCES

1. J. Sherman, Spectrochim. Acta 7, 283, (1955).
2. J.W. Criss and L.S. Birks, Anal. Chem., 40, 1080, (1968).
3. R.M. Rousseau, X-Ray Spectrometry, 13, 115, (1984).
4. S.D. Rasberry and K.F.J. Heinrich, Anal. Chem., 46, 81, (1974).
5. G.R. Lachance and R.J. Traill, Can. Spectrosc., 11, 43, (1966).
6. R. Jenkins, Advances in X-Ray Analysis, 22, 281, (1979).
7. F. Claisse and M. Quintin, Can. Spectrosc., 12, 129, (1967).
8. R.M. Rousseau, X-Ray Spectrometry, 16, 103, (1987).
9. J.W. Criss, Advances in X-Ray Analysis, 23, 93, (1980).
10. R.M. Rousseau, X-Ray Spectrometry, 13, 121, (1984).
11. N. Broll, X-Ray Spectrometry, 15, 271, (1986).
12. R.M. Rousseau, G.S.C. Internal Report, October 1981.
13. K.R. Wadleigh, X-Ray Spectrometry, 16, 41, (1987).
14. R. Rousseau and F. Claisse, X-Ray Spectrometry, 3, 31, (1974).
15. P.A. Pella, G.Y. Tao and G. Lachance, X-Ray Spectrometry,
    15, 251, (1986).
16. G.Y. Tao, P.A. Pella and R.M. Rousseau, NBSGSC - A FORTRAN
    Program for Quantitative X-Ray Spectrometric Analysis.
    NBS Technical Note 1213, National Bureau of Standards,
    Washington, DC (1985).
17. RAINIER SOFTWARE, 2414 S.E. Mud Mt. Rd., Enumclaw, WA 98022.
18. CORPORATION SCIENTIFIQUE CLAISSE INC., 2522 chemin Sainte-Foy,
    Sainte-Foy, Quebec, Canada, G1V 1T5.
19. D.B. Bilbrey, G.R. Bogart, D.E. Leyden and A.R. Harding,
    X-Ray Spectrometry, 17, 63, (1988).
20. N. Broll and R. Tertian, X-Ray Spectrometry, 12, 30, (1983).
21. R.M. Rousseau and M. Bouchard, X-Ray Spectrometry, 15, 207, (1986).
22. R. Tertian, X-Ray Spectrometry, 15, 177, (1986).
23. G.R. Lachance, Advances in X-Ray Analysis, 31, 471, (1988).

PAINLESS XRF ANALYSIS USING NEW GENERATION COMPUTER PROGRAMS

Richard M. Rousseau

Geological Survey of Canada
601 Booth St., Ottawa, Ont., CANADA, K1A 0E8

ABSTRACT

A new computer program, named CiROU, has been written to determine the chemical composition of homogeneous samples analyzed by X-ray fluorescence spectrometry (XRF). It uses the fundamental algorithm to convert measured intensities into concentrations and to correct for matrix effects and runs on any IBM-PC/XT/AT, PS/2 or compatible computer.

The program is very easy to use mainly because it is a step-by-step guide to the routine process of analyzing specimens by XRF and because a user friendly interface makes up a main menu bar containing six major options, which provide several other options inside a pull-down window system.

The program is applicable to specimens of any composition with the utmost accuracy that can be obtained from the fundamental-parameters method. It's unique calibration procedure still adds to the accuracy by eliminating the bias between theory and reality.

The program is a complete "off-line" program allowing to perform all the steps of an XRF analysis: calculation of theoretical influence coefficients, calibration, reading of measured intensities, calculation of net intensities and concentrations of unknowns, print-out of analytical reports and data transmission to or from an external computer.

INTRODUCTION

For the last 30 years, X-Ray Fluorescence (XRF) analysis has undergone a spectacular evolution from all points of view. Today, WDS spectrometers supplied by the four major companies (ARL, PHILIPS, RIGAKU, SIEMENS) are completely computer controlled, making them very flexible. High tension generators are more compact, more stable and more powerful. Dual anode x-ray tubes are now available. New reliable sample changers can automatically read up to 72 specimens in any order. The new generation of spectrometers is very sensitive, particularly for the determination of trace ele-

ments, affording better detection limits. They can detect ultra light elements (B, C, F, etc.) mainly because of the new multilayer synthetic crystals. To operate these new spectrometers, cheaper more powerful computers
are now available. Control and qualitative software distributed with these
are of a high quality.

Sample preparation is still a problem but is improving every year.
Quantitative methods to convert measured intensities into concentrations
are no longer lacking as shown in the former paper[1]. Algorithms have been
proposed that effectively cover a much greater composition range with
greater accuracy. New calibration methodology to adapt theory to the real
world of XRF analysis, as explained in the former paper, as well as the
availability of standards of increasingly well defined and diverse compositions give greater precision and accuracy.

But aside from this impressive list of features, there is one place
where improvement can still take place: it is the software used for quantitative analysis. That is why the XRF analyst who developed the methods
for XRF analysis[2,3,4,5] decided to write, in collaboration with other analysts, the matching computer program. Three years ago, a team of XRF analysts then started to develop a computer program which was exclusively designed around XRF analysis and written by focusing on the point of view of
the XRF analyst. The program was not written by programmers, but by XRF analysts from their day to day experience, wishing to supply a tool which is a
step-by-step guide to the "nuts and bolts" process of analyzing specimens
by XRF and taking advantage of the most recent developments regarding mathematical models and methods of XRF analysis. The final product must be adapted to the needs of XRF analysts and be a professional product, not an
"amateur" one.

It is the purpose of this paper to share the experience thus obtained
and to inform the XRF analysis community of the potential of a new generation of XRF programs to analyze any kind of material.

OBJECTIVES OF THE PROGRAM

With the high technology of new spectrometers, comprehensive software
packages must now be available which can be interfaced with virtually any
XRF spectrometer via a mini or micro computer and by using the measured
intensities they can apply the most advanced procedures to suit the analytical requirements of different users.

In order to make the program accessible to a as large as possible number of users, it must run on the most popular computer, the least expensive
but at the same time it must also offer high speed calculations and high
accuracy. Therefore, the program runs on any IBM-PC, -XT, -AT, PS/2 or compatible computer with 640 KB of memory, a 360 KB floppy disk drive, a hard
disk drive and a math coprocessor (8087/80287/80387). A PC/MS-DOS operating system is also needed. If the spectrometer is controlled by an IBM-non-
compatible computer, a communication package, like PROCOMM V2.42[6], can be
used to transfer the measured intensity files.

The fundamental-parameters method associated with the theoretical
multi-element influence coefficient concept is certainly the most powerful and accurate method known today to convert measured intensities into
concentrations. And it seems that it will become the must popular one in
the near future, especially with computer performance increasing all the

time with decreasing cost. Empirical and approximate methods are becoming
obsolete. That is why the program uses the fundamental algorithm as developed by Rousseau[2] to calculate the sample compositions. The calibration
procedure used is the one proposed in the former paper.

Which features should an XRF analyst expect to find in a modern program
to make his or her work easier and to produce accurate results ? Most importantly, the program must be easy to use, with plenty of features designed to
avoid human errors. The program should be accessible to any non-expert by
being a step-by-step guide to routine analyses by XRF and by automatically
taking care of all complex aspects of the theory. To really make the program
easy to use, it should be written for the analyst having little or no experience with the fundamental-parameters approach. Any new user should be pro-
ductive after a few hours of training only. The documentation should not be
really necessary because the number of commands to learn and remember are
minimized. Keystrokes should be few and intuitive.

The program should make the analyst's work more accurate, more produc-
tive, and more enjoyable. It should be a highly flexible program that also
automatically performs many of the more trivial and repetitive tasks. All
of the tedious calculations, and parameter entries should be eliminated to
increase the efficiency. Each data entry should be saved by the program in
order to minimize the number of keystrokes. The time of the XRF analyst is
then conserved for the more demanding parts of the analysis, not wasted on
repetitive entries of information that the computer already has in memory.
You should not have to enter or remember a single code or parameter. Keyboard entries should basically be limited to filenames. With so much of the
work done for you, how to use the program should be much faster and easier
to learn.

MAIN MENU

All of the above objectives of the program were reached mainly because
of the user interface utilized. Older user interfaces employed a menu system, where the user was continually presented with lists of choices requiring that he or she type-in a letter or number to make a selection. Such me-
nus promote human error entries. Furthermore, the result of such a choice
was often simply another menu, then another one, and so on. In the end, it
was easy to lose the searched place in the menu structure, and to forget
how to get from one menu to another.

The pull-down menu concept is a simple improvement: a menu only appears
when the user points to the menu title (usually one of a number of choices
across the top of the screen). The menu "pulls down" over a portion of the
existing screen and allows the user to make a choice from the menu items.
After the choice is made, the menu disappears, restoring the initial screen.

A "pop-up" menu is similar, but appears automatically when certain
choices are made. For example, if a user selects the "Parameter File" option
from a pull-down menu, a pop-up menu appears under the first option, allowing to select a sub-menu choice. Once the choice is made, the pop-up menu
disappears. Each menu appears in a "window", a separate area of the screen,
surrounded by a border. These windows appear and disappear as needed. Each
window supplies a list of options, which can be selected by moving up and
down an highlight bar with the arrow-keys.

Thus, the ease of use of the program is mainly due to its menu with windowed input screens where most of the time the arrow-keys or the Enter-key only are used to dialogue with the computer. Any input is automatically checked for proper form and length. The menu is a dBASE-III-PLUS work-alike. It is made up of a main menu bar containing six major modules: ALPHA, CALI-BRATION, INTENSITY, SAMPLE, REPORT and UTILITIES. Each module provides seve-ral other options inside a pull-down window system. To dialogue with the computer is then like navigating through pop-up windows, which appear and disappear as fast as you can press the arrow-keys. Furthermore, each option provides context sensitive on-line HELP windows. To select an option, press the F1-key and a full screen of information on the option appears. Press the Esc-key and you are back to the initial screen. Each pop-up window has a different colour, which is not essential but makes the analyst's work easier.

SOME OTHER FEATURES OF THE PROGRAM

The program is used to determine the chemical composition of homoge-neous samples analyzed by X-Ray Fluorescence Spectrometry. It uses the fun-damental algorithm to convert measured intensities into concentrations and to correct for inter-element effects. It can be used to analyze any kind of samples with large or limited range of composition.

To carry out the various operations of the program, the user need not be familiar with the fundamental-parameters method or the formulas that are used; they are obtained automatically from data files as required. To run the program, the user should only supply the measured intensities from a few standard reference materials, their compositions, measured intensi-ties of unknowns and concentrations of unmeasured components, if any.

The program provides a wide range of facilities, which allow you:

* To create, update or print a file of all fundamental parameters required to analyze a new series of analytes.

* To calculate and print any X-ray tube output spectrum using the accurate and efficient mathematical model proposed by Pella et al.[7].

* To Calculate and print the theoretical influence coefficients in the Claisse-Quintin (CQ) algorithm for any kind of systems. For the user, calculations are as easy as entering the name of the analyte. The pro-gram takes care of all complex aspects of the theory and automatically supplies all fundamental parameters: mass absorption coefficients, flu-orescence yields, jump ratios, and line and edge wavelengths. Mass ab-sorption coefficients are calculated from a judicious mix of the Heinrich[8] and of the Leroux & Thinh[9] data.

* To enter, correct or print the composition of calibration standards.

* To calculate Alpha and Rho coefficients in the fundamental algorithm for any standard. The theoretical relative intensity of each analyte is also calculated.

* To perform a unique calibration process with high-quality graphical out-put. It matches the theory with the reality by referring to standards. A few only, two or three, are required for the calibration of a given analyte and these do not have to be similar to the unknowns. The pure elements are not required.

* To determine a series of working conditions appropriated to your laboratory. For example the program asks you if you want to enter intensities by hand or from a file, to enter net or gross intensities, to enter
  sample names with the present program or not, etc.

* To transform measured intensity files into working files of net intensities in a format compatible with the program, after having subtracted the background, any blank and corrected for any interference on
  each measured peak. It is also possible to enter intensities by hand.

* To create a master file to store all the calculated concentrations of
  unknown samples, which can be read, printed, updated or deleted.

* To convert the measured net intensities into concentrations using the
  CQ algorithm to get a first estimate of the sample composition. Then,
  the final composition is calculated with the fundamental algorithm for
  up to 24 analytes. The program allows you to enter the concentrations
  of unmeasured components and to correct for them.

* To print final results in a report of analysis, in a format selected
  by the analyst, ready for distribution.

* To provide data transmission to or from an external computer.

* To use some utility options for access to the data files, to edit them
  or to execute other related duties.

    In short, the program is a complete "off-line" package to perform XRF
analysis. It is also universal, flexible, easy to use, accurate, applicable
to specimens of any composition that are homogeneous with a flat and polished surface.

WORKING PHILOSOPHY

    The basic philosophy of the program is to enter a minimum of informa-
tion when you use it and every data entry must be saved for further use.
The program also allows easy editing of all these data. The program must
also supply all possible answers for any given question. To achieve all
these goals the following approach is used.

    When you execute an option, you dialogue with the system by answering
to a series of questions. And no matter what the question, all the possible
operator inputs are always shown between two square brackets ([ ]), followed by a colon, a default and then the symbol ' := ', after which the cur-
sor stops and waits for your answer, as in the following example:

```
X-ray tube anode [ Sc, Cr, Mo, Rh, Ag, W, Au ] ...:      Rh :=
X-ray tube voltage [ 10 - 100 ] KV ..............:       40 := 60
Anode angle [ 10.0 - 90.0 ] Degrees .............:       18.0 :=
Thickness of Be window [ 0.125 - 0.500 ] Mm .....:       0.150 :=
Last wavelength of spectrum [ 2.5 - 8.0 ] A .....:       7.09 :=
Tube filter [ Y/N ] ? ...........................:       No :=
Write X-ray tube spectrum to disk [ Y/N ] ? .....:       Yes :=

Is there any error [ Y/N ] ? ....................:       No :=
```

You can accept the default simply by pressing the Enter-key or enter a new answer, which may be either numbers or strings of alphanumeric characters and then press the Enter-key. In the above example, only the voltage is updated from 40 to 60 KV. For all the other questions, the proposed default was accepted as a new answer. Most of the time, the defaults are the last series of data entered. For example, if you enter "YES" to the last question, the same series of questions will be displayed again and all defaults become the last data just entered. Note that after any series of questions, the above last question is always displayed, giving the operator the opportunity to correct any wrong answer.

CONCLUSION

A new program has been described for the analysis of any kind of material by XRF. By so doing, an example of a modern program is presented. It is not claimed that this program is neither the best nor the only solution to the XRF analysis problem. There may exist a more elegant solution. However, the producers of some commercial products would be wise to inspire themselves from this work in order to make their programs much more easy to use and accurate, in other words, to make a reality the idea "painless XRF analysis".

Today, quantitative methods are no longer a problem and now software of high-quality is available. The last step now depends on the user himself, it is up to him to use it or not, to make pressure on companies producing WDS spectrometers to supply better quantitative software products. In the end the technique must benefit from it by becoming really easy and enjoyable to use and accurate.

Any analyst who has any suggestions to improve the XRF programs in general, or requires additional information, or wishes to test the present program, should write to the author.

REFERENCES

1. R.M. Rousseau, Concepts of Influence Coefficients in XRF Analysis and Calibration, (Also published in the present book).
2. R.M. Rousseau, X-Ray Spectrometry, 13, 115, (1984).
3. R.M. Rousseau, X-Ray Spectrometry, 13, 121, (1984).
4. R.M. Rousseau and M. Bouchard, X-Ray Spectrometry, 15, 207, (1986).
5. R.M. Rousseau, X-Ray Spectrometry, 16, 103, (1987).
6. Datastorm Technologies, Inc., P.O. Box 1471, Columbia, MO 65205, U.S.A.
7. P.A. Pella, L.Y. Feng and J.A. Small, X-Ray Spectrometry, 14, 125, (1985).
8. K.F.J. Heinrich, The Electron Microprobe, p. 296. Wiley, New York (1966).
9. T.P. Thinh and J. Leroux, X-Ray Spectrometry, 8, 85, (1979).

INTENSITY AND DISTRIBUTION OF BACKGROUND X-RAYS
IN A WAVELENGTH-DISPERSIVE SPECTROMETER. II. APPLICATIONS

K. Omote and T. Arai

Rigaku Industrial Corporation
Takatsuki, Osaka, Japan

INTRODUCTION

In the spectroscopic analysis of minor and trace elements by fluorescent X-rays, many improvements in the analytical performance of trace element measurements have been made. For the analysis of trace elements, the background intensity governs the analytical accuracy and the lowest detection limit in a sample. A comparison is made between experimental and theoretically calculated background X-ray intensities in a previous paper[1]. It is based on the formula for scattered X-ray intensity, from the estimation of Thomson and Compton scattered X-rays. Also, the asymmetrical peak profiles at the base of the giant intensity peak are discussed and are clearly shown in the skirt part of K beta X-rays, e.g., Ni-K beta or Fe-K beta X-rays. The purpose of this report is to investigate the intensity of background X-rays, using glass beads and powder samples of iron oxide and quartz, based on the previous fundamental studies and the overlapping correction procedure for cobalt determination in low-alloy and stainless steel.

CALCULATIONS FOR THEORETICAL INTENSITY FOR BACKGROUND X-RAYS

In order to predict the magnitude of background intensity on the basis of theoretical calculations of Thomson and Compton scattered X-rays, the simplified formula Eq. (1) is derived. Thomson X-rays are defined here as rays from crystal phases[1].

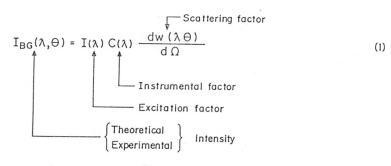

$$I_{BG}(\lambda,\theta) = I(\lambda) \, C(\lambda) \, \frac{dw(\lambda\theta)}{d\Omega} \tag{1}$$

The scattering factor in Eq. (1) is calculated with respect to individual X-rays, and consists of scattering terms in the numerator and an absorption term in the denominator, shown in Eq. (2).

$$\frac{dw}{d\Omega} = \frac{\sum n_i \, d\bar{\sigma}_i(\lambda,\theta)}{P \sum W_I \left(\frac{\mu}{\rho}\right)_i \left(\frac{1}{\sin\phi} + \frac{1}{\sin\psi}\right)} \tag{2}$$

It was assumed that the product of I and C was a constant for each X-ray and that the product of I, C and the computed scattering factor for the glass bead sample of Fe_2O_3 was the same as the measured intensity. Using this constant, the theoretical intensities of other samples were calculated and compared with the experimental values.

EXPERIMENTAL

Experimental results were measured by a RIGAKU 3070 sequential spectro-meter equipped with a Rh-target X-ray tube (OEG-75). The glass bead samples were prepared by a standard method for ISO-iron ore analysis. The weight ratio of sample to borax is one to ten, which influences the background intensity.

APPLICATIONS

(1) Measurement of Powder Samples of SiO_2 and Fe_2O_3

The background intensity distribution of powder samples of SiO_2 and Fe_2O_3 is shown in Fig. 1.

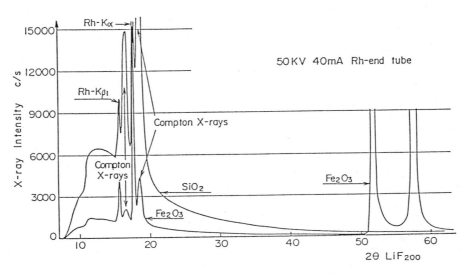

Fig. 1   Background intensity distribution of
           powder samples of SiO_2 and Fe_2O_3

When trace elements are analyzed in iron ores, powder pellet samples
should be used because of the high sensitivity. The background intensity of
the SiO_2 sample between 22 to 50 degrees 2θ is three times higher than that
of Fe_2O_3 samples. On the lower-angle side of the iron K absorption edge, the
background intensity from Fe_2O_3 is lower than that from SiO_2 because of the
high absorption of iron. On the higher-angle side of the Fe-K alpha X-rays,
the intensity from SiO_2 samples is about half that from Fe_2O_3 samples because
of the low absorption of iron.

(2) Measurement of Glass Bead Samples of SiO_2 and Fe_2O_3

The background intensity distributions of the qualitative low-angle and
high-angle side scans of glass bead samples of SiO_2 and Fe_2O_3 are shown in
Figs. 2 and 3. The background intensity of Fe_2O_3 is half the intensity of
SiO_2 and the difference for glass bead samples is smaller than that for
powder samples.

At the 2θ angle position of the iron K absorption edge shown in Fig. 3
the intensity of the SiO_2 sample is approximately twice that of the Fe_2O_3
sample. Over a range of 65 degrees 2θ , the background intensities of SiO_2
and Fe_2O_3 are almost the same owing to the canceling of the mass absorption
coefficient and the sum of the Thomson and Compton scattering terms in the
scattering factor in Eq. (2).

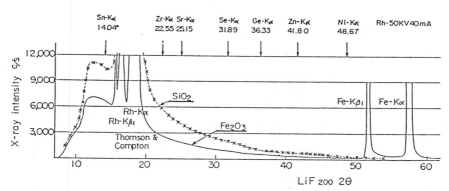

Fig. 2   Background intensity distribution of glass bead
samples of SiO_2 and Fe_2O_3 - low angle side

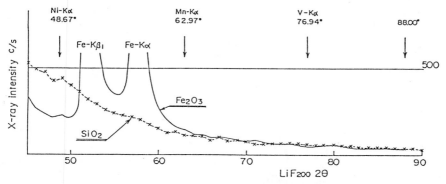

Fig. 3   Background intensity distribution of glass bead
samples of SiO_2 and Fe_2O_3 - high angle side

(3) Comparison Between Experimental and Theoretical Background Intensities
      of SiO_2, Fe_2O_3 and $CaO \cdot SiO_2$

In order to demonstrate that the calculated intensity can be converted
to the value to be measured, a comparison between experimentally and
theoretically calculated intensities was made among powder samples of SiO_2
and Fe_2O_3 and glass bead samples of SiO_2, Fe_2O_3, and a mixture of SiO_2 and
CaO. The comprison between measured and calculated values of background
intensity is shown in Table (1).

It is shown that on the whole, the calculated intensity which is the
product of the previously mentioned constant and the scattering factors for
each X-ray wavelength is in good agreement with the measured background
value. From the standpoint of the principle of X-ray scattering phenomena,
the following can be said. The Thomson X-rays scattered from powder samples
are diffracted from crystal grains. X-rays scattered from glass bead samples
are diffracted from an amorphous material. The Compton X-rays are
independent of the substance condition. Therefore, because the calculated
intensity is given by considering the phenomena between incident X-rays and a
scattering atom, a good agreement between measured and calculated intensity
values means that the measured intensity of background X-rays is proportional
to the scattering factor and the product of I and C given in Eq. (1).
Consideration should also be given to the relationship between an incident
X-ray distribution, the reflection efficiency of an analyzing crystal, etc.,
for developing wider applications.

Table 1   Comparison between experimental (Ex) and calculated (Th)
          values of background intensity of SiO_2, Fe_2O_3 and $CaO \cdot SiO_2$

			Powders and ISO Glass beads							Unit Kcps
BG	Sn-Kα	Zr-Kα	Sr-Kα	Se-Kα	Ge-Kα	Zn-Kα	Ni-Kα	Mn-Kα	V-Kα	——
0.4921Å	0.7873	0.8767	1.1061	1.2554	1.4364	1.6592	2.1031	2.5048	2.7970	
LiF(200)14.04°	22.55	25.15	31.89	36.33	41.80	48.67	62.97	76.94	88.00	
Glass bead samples										
Fe_2O_3 Ex	6.68	2.59	1.93	1.00	0.572	0.356	0.226	0.135	0.054	0.034
Th	6.68	2.59	1.93	1.00	0.572	0.356	0.224	0.135	0.054	0.034
CaO Ex	9.98	4.55	3.26	1.66	0.923	0.535	0.346	0.100	0.045	0.032
SiO_2 Th	9.23	4.37	3.25	1.63	0.912	0.553	0.347	0.102	0.041	0.025
SiO_2 Ex	10.9	5.75	4.09	2.02	1.13	0.689	0.423	0.124	0.053	0.039
Th	10.2	5.36	4.01	2.00	1.11	0.672	0.420	0.124	0.048	0.030
Powder samples										
Fe_2O_3 Ex	1.31	0.602	0.451	0.258	0.170	0.123	0.075	0.209	0.083	0.039
Th	1.45	0.612	0.480	0.284	0.173	0.111	0.061	0.145	0.063	0.040
SiO_2 Ex	6.34	2.81	2.02	1.11	0.698	0.408	0.236	0.099	0.033	0.022
Th	7.79	2.74	2.09	1.16	0.680	0.427	0.269	0.077	0.030	0.018

Using experimental and theoretical intensity of Fe_2O_3 beads, I x C values were determined
and theoretical intensity of SiO_2 and $SiO_2 \cdot CaO$, Fe_2O_3 were calculated

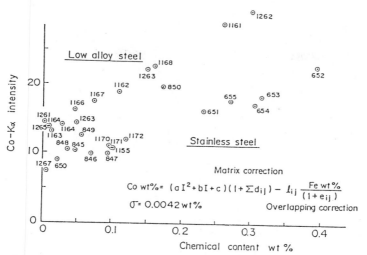

Fig. 4 Relationship of Co-Kα X-ray intensity and chemical
content of cobalt in low-alloy and stainless steel

(4) Studies of Overlapping Corrections for the Cobalt Determination in Steel

For the purpose of determining the cobalt content in low-alloy and
stainless steels, the relationship between the intensity of the Co-K alpha
X-rays and the cobalt content has been investigated and is shown in Fig. (4).

The measured intensities from low-alloy steel samples are higher than
those from stainless steel samples. With the lower cobalt content, the
background intensity of stainless steel samples is approximately half of that
of low-alloy steel samples. In order to determine the cobalt content in the
samples, matrix corrections for Co-K alpha X-rays and overlapping corrections
are applied, as shown in Fig. (4). In the overlapping correction term, the
iron content divided by the matrix correction factor for Fe-K beta X-rays is
proportional to the intensity of Fe-K beta X-rays. The accuracy, i.e., the
difference between chemically determined values and X-ray measured values was
0.0042 wt%.

CONCLUSION

It is shown that using the theoretically calculated intensities the
background intensity can be predicted on the basis of the Thomson and Compton
scattering formula for a single atom model. Comparing experimental values
with theoretically calculated values for glass bead and powder samples of
SiO_2 and Fe_2O_3, good agreement can be found between them.

In the trace element analysis, the influence of the high-angle side
tailing (long wavelength side) at the base of giant intensity peaks can be
eliminated with the optimized correction method.

REFERENCES

1) Omote and Arai: Advances in X-Ray Analysis, Vol. 31 (1988), P 507-514.

What Can Data Analysis Do For

X-Ray Microfluorescence Analysis?

John D. Zahrt

Applied Theoretical Physics Division
Los Alamos National Laboratory

INTRODUCTION

In 1985 Nichols and Ryon [1] first demonstrated their x-ray microfluorescence analysis (XRMF) system. By 1986 Nichols et al. [2] Boehme [3] and Gurker [4] provided us with spectacular photographs of x-ray images of geological materials, wire grids, and semiconductor chip carriers. During the delivery of the paper by Nichols et al. [2] the present author realized that a higher degree of spatial resolution could be accomplished by analysis of the raw data. A first step toward this end was accomplished in that same year [5] when it was demonstrated that the known incident beam intensity profile, $I_o(x)$ (also called the kernel, herein), the unknown analyte concentration $c(x)$, and the measured detector signal $s(x)$ are related by

$$A \int_{-\infty}^{\infty} I(x-y)c(y)dy = s(x) \quad , \tag{1}$$

where $I(x) = \exp(\frac{-x^2}{\sigma^2})$ and A is a constant of proportionality. It was also demonstrated that the map of the signal is broader and smoother than the map of the analyte concentration.

The equation above is categorized as a Fredholm integral equation of the first kind (IFK) with the kernel $I(x-y)$. These equations are in general particularly nasty because a) they do not admit to a unique solution[1] and b) small errors in $s(x)$ can cause large errors in the resolution of $c(y)$ (ill-posedness). Zahrt [6] recently reviewed some classical resolutions of Eq. (1) and supplied a new method of resolution. Philip [7] has proved that the Gaussian kernel used in Eq. (1) makes this equation the most ill-posed of any IFK.

We report here results from three other established numerical methods of resolution, one of which is quite rapid, (and could be used in real time) that do quite well on a tough test case.

---

[1] Because of the non-uniqueness we will henceforth speak of resolving (instead of solving) Eq. (1).

In summary, in XRMF (as well as in other x ray experiments) usually one does not measure (an intensity profile) what one desires to know (a concentration profile). If the concentration gradients are small and if the beam profile is as narrow as the information desired, there is nothing more to do. But if the concentration gradients are large and/or the beam profile is wide compared to the information desired, deconvolution of Eq. (1) can provide a spatial resolution many times that of the raw data.

RESOLUTIONS

We can rewrite Eq. (1) as

$$I * c = s \; , \tag{2}$$

where $*$ symbolizes the convolution operator. In mathematical jargon, s is the Weierstrass transform of c given that I is Gaussian as above. A table of Weierstrass transforms is given in [6]. In that table it is indicated that the sine ax and cos ax are eigenfunctions of the Gaussian kernel with eigenvalues of exp $\left(-\frac{a^2}{4}\right)$. Here we wish to present three numerical methods for the resolution of Eq. (1) and provide what is considered to be a critical test case. The data set used was essentially the same for all three methods. A 1 % white noise was added to the signals generated by the exact concentration profile.

Singular Value Decomposition (SVD)

Equation (1) can be discretized and written in matrix form as (leaving off the sub zero from the $I's$)

$$
\begin{pmatrix}
I_{11} & I_{12} & \cdots & I_{1N} \\
I_{21} & I_{22} & \cdots & I_{2N} \\
\vdots & & & \\
I_{M1} & I_{M2} & & I_{MN}
\end{pmatrix}
\begin{pmatrix}
c_1 \\
c_2 \\
\vdots \\
c_N
\end{pmatrix}
=
\begin{pmatrix}
s_1 \\
s_2 \\
\vdots \\
s_M
\end{pmatrix} , \tag{3}
$$

where we take M signal measurements and desire to know the concentration at N points. For the rest of this paper, $M = 40$ and $N = 100$.

If we form II^T and $I^T I$ and find the eigenvalues and eigenvectors of these square matrices we have

$$II^T u = \sigma^2 u \tag{4a}$$

$$I^T I v = \sigma^2 v \; . \tag{4b}$$

The first M singular values, σ_i, of Eq. (4b) are identical to the M singular (eigen) values, σ_i, of Eq. (4b). All singular values are

$$0 \le \sigma_i \le 1 \quad i = 1, 2, \cdots N. \tag{5}$$

In fact, the σ_i are ordered such that $\sigma_i \to 0$ as $i \to N$. It can then be shown that [8]

$$c = \sum_{j=i}^{N} \frac{(s, u_j)}{\sigma_j} v_j \quad , \tag{6}$$

where (,) indicates the dot product of two vectors. However, since $\sigma_j \to 0$ as $j \to N$, the terms for large j may blow up if (s, u_j) doesn't go to zero faster than σ_j. That is the case here for the Gaussian kernel. Figure 1 shows that the singular values rapidly approach zero. The computation of the u_i, v_i and σ_i, is essentially an N^3 problem. But these are only kernel (intensity profile) dependent and independent of the particular data set. They can be computed once and stored.

If the graph of σ_i vs. i shows a slowly declining plateau followed by a steep jump, the usual practice is to truncate the series in Eq. (6) just above this jump. Here the σ_i vs. i graph is very smooth [it goes as $\exp(-ai^2)$: compare to eigenvalues of sine and cosine at the beginning of this section] and it is not immediately obvious where to truncate. Faber [9] has given some criteria for such a decision. We use $i = 16$. Figure 2 shows the Gaussian kernel centered at 4.875 and the true $c(y)$ for our test cases. Figure 3 shows the SVD resolution for $c(y)$ and the signal $s(x)$. The kernel has a FWHM of about 1.6 units, and the signal a FWHM of about 3 units yet we are able to resolve structure at least at the 0.6 unit level. It is, however, quite apparent that the resolution gives regions of negative concentration although this might be removed by postfiltering.

To emphasize the difficulties of working with smooth kernels and to indicate how the experiment might be modified to provide even greater spatial resolution, Fig. 4 shows the resolution of the problem using a half Gaussian kernel. This could be obtained by intercepting half of the incident beam with some absorber. The map of c is much sharper and all 40 singular functions were used in the reconstruction.

Maximum Entropy (ME)

Instead of resolving a deduction problem, one may solve the inductive problem of finding the most probable distribution of $c(y)$ such that

$$\chi^2 = \sum_{i=i}^{m} (\sum_{j=1}^{N} I_{ij} c_j - s_i)^2 / \sigma_i^2 = K \quad .$$

Here σ_i is the standard deviation of the ith measurement. A unique solution to this problem is essentially brought about by the use of prior information that $c(y) > 0$ for all y and that $\chi^2 = K$. We thus maximize the functional

$$- \sum_{i=i}^{N} p_i \ln p_i + \mu C + \lambda K \quad , \tag{7}$$

where $p_i = c_i / C, C = \sum_{i=i}^{N} c_i$ and K is given above. μ and λ are Lagrangian multipliers. Figure 5 gives the maximum entropy solution given the same kernel and true concentration displayed in Fig. 2. The solution is superior to the SVD resolution in that it is everywhere positive.

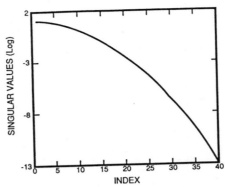

Fig. 1. The magnitude of the singular values of a Gaussian kernel as a function of number. Note the decrease of 14 orders of magnitude between σ_1 and σ_{40}.

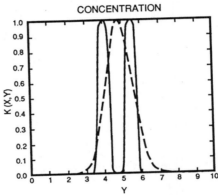

Fig. 2. The true concentration (solid) and Gaussian kernel (dash) centered at x = 4.875.

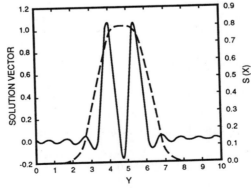

Fig. 3. The SVD resolution of the concentration (solid) using 16 singular functions and the signal (dash) obtained from the convolution of the curves in Fig. 2; 1% white noise was added to this signal before the deconvolutions.

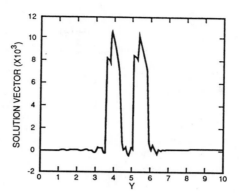

Fig. 4. A SVD resolution of the concentration using 40 singular functions for the half Gaussian kernel $(K(x, y) = 0$ if $y < x)$.

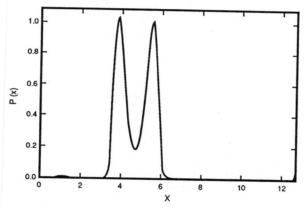

Fig. 5. The maximum entropy resolution of the concentration.

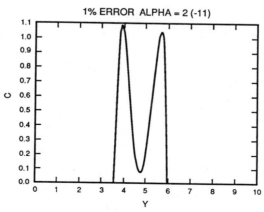

Fig. 6. The regularization resolution. Note that $\alpha = 2 \cdot 10^{-11}$.

Regularization

In the regularization algorithm we seek to minimize the functional

$$\Phi(c) = \frac{1}{2}\{\|Ic - s\|^2 + \alpha\|c\|^2\}$$

with respect to c. We let $\|\cdot\|^2$ denote the L^2 vector norm. This problem has a unique solution for a given α and a given data vector s. We would like to choose α such that the square error between the computed and true concentrations is a minimum, but this step is impossible because the true concentration is unknown. We thus choose to minimize another function of α designed to transfer our lack of knowledge about c to some statement about the magnitude of errors in the data vector s. The interested reader should read Butler, Reeds, and Dawson [10] for the details and further references. Groetsch's monograph [11] is also quite useful. Triay and Rundberg [12] lay out the computational algorithm very clearly. Figure 6 shows the regularization resolution of the problem.

CONCLUSIONS

The integral equation relationship between concentration and signal for x-ray microfluorescence has been displayed in Eq. (1). Available data indicate that the kernel (intensity) should be a Gaussian function. In fact, the Gaussian kernel was used throughout this paper, but the methods discussed herein (as opposed to those discussed in [6]) will work for any empirical intensity profile (kernel). The singular value decomposition code is faster than the maximum entropy or regularization codes. The singular vectors and singular values need be computed only once and can be stored to resolve any number of problems with the given intensity profile. However, the maximum entropy and regularization approaches are attractive in that their solution vectors are positive everywhere. All three of these standard methods of dealing with Fredholm integral equations of the first kind [Eq. (1)] can provide spatial resolution many times better than the signal profile itself in worst case scenarios. Any thing the experimenter can do to roughen the intensity profile will greatly increase the spatial resolution.

ACKNOWLEDGEMENTS

I would like to thank Devinder Sivia for performing the maximum entropy calculations and Ines Triay for performing the regularization calculations.

References

[1] Nichols, M. C., and Ryon, R. W., "An X-Ray Micro-Fluorescence Analysis System With Diffraction Capabilities," *Adv. in X-Ray Anal.* **29**, 423-426 (1986).

[2] Nichols, Monte C., et al., "Parameters Affecting X-Ray Microfluorescence Analysis," *Adv. in X-Ray Anal.* **30**, 45-51 (1987).

[3] Boehme, Dale R., "X-Ray Microfluorescence of Geologic Materials," *Adv. in X-Ray Anal.* **30**, 39-44 (1987).

[4] Gurker, N., "Imaging Techniques for X-Ray Fluorescence and X-Ray Diffraction," *Adv. in X-Ray Anal.* **30**, 53-65 (1987).

[5] Zahrt, John D., "High Spatial Resolution in X-Ray Fluorescence," *Adv. in X-Ray Anal.* **30**, 77-83 (1987).

[6] Zahrt, John D. and Carter, Natalie, "The Weierstrass Transform in X-Ray Analysis," Accepted for publication in *Lecture Notes in Pure and Applied Mathematics.*

[7] Philip, J., "The Most Ill-Posed Non-Negative Kernels in Discrete Deconvolution", *Invers. Prob.*, **3**, 309-328 (1987).

[8] Faber, V, and Wing, G. Milton, "The Abel Integral Equation," LA-11-16-MS, Los Alamos National Laboratory, 1987.

[9] Faber, V., personal communication, 1988.

[10] Butler, K., Reeds, K. and Dawson, S., "Estimating Solutions of First Kind Integral Eqns. with Nonnegative Constraints and Optical Smoothing," *Siam J. Numer. Anal.* **18**, 381-397 (1981).

[11] Groetsch, C., *The Theory of Tikhonov Regularization for Fredholm equations of the First Kind,* Pitman Pub., Boston, 1984.

[12] Triay, I., and Rundberg, R., "Determination of Selectivity Coefficient Distributions by Deconvolution of Ion-Exchange Isotherms," *J. Phys. Chem.* **91** 5269-5274 (1987).

THE DETERMINATION OF RARE EARTH ELEMENTS IN GEOLOGICAL SAMPLES BY XRF USING THE PROPORTIONAL FACTOR METHOD

Chen Yuanpan

Research Institute of Geology for Mineral Resources
China National Nonferrous Metals Industry Corporation
Sanlidian Guilin, China

ABSTRACT

This paper describes the principle, method and application of the proportional factor method for the determination of rare earth elements (REE) in geological samples by XRF. The analytical results of this method are compared with REE recommended values in geological standards, and good agreement was observed. Five of the REE (Pr, Tb, Ho, Tm and Lu) have also been determined by interpolation from chondrite-normalized plots, and more satisfactory results were obtained.

INTRODUCTION

Because of the continued increasing development of industrial applications of REE yearly, it is important to determine the location of REE in the earth's crust, the nature of the mineral types, and how the minerals are formed. In addition, since REE are sensitive indicators of the magmatic process, the study of the geochemical character of REE in rocks not only can help one differentiate between rock types, but also determines the origin of the rock and mineral beds (1-3). Therefore, measurement of REE in geological samples has significant implications.

In 1961, Turanskaya (4) determined REE in minerals using the so-called correction factor method. Vainstein and Borovsky (5) called it an internal factor method and transition factor method, respectively. These methods have only been used to analyze minerals of high REE content. In recent years geochemists often wish to know REE concentrations in the sub-ppm range.

A survey of the recent literature (6-11) shows that most methods for the determination of REE at low concentrations in geological samples consist of chemical preconcentration using external standards for calibration. This paper describes the analysis of REE by chemical preconcentration on a thin film using a suitable internal reference element normally found in the analyte specimen and applied to the determination of REE in a variety of mineral and rock samples. Experiments conducted in our laboratory over a period of several years have shown that this approach has several advantages,

among which are: (A) Samples having wide concentrations of REE from 0.01%
and above can be analyzed, (B) the procedure is simple, convenient, requires
only a small amount of standard, and does not require that standard samples
be routinely measured, and (C) matrix effects are small enough to ignore, and
instrumental drift can be eliminated by using Nd or Er as a variable internal
reference element.

This paper describes the principle, method and application of the
proportional factor method for the determination of REE. Results of this
method have been compared to data from recommended values of geochemical
standards and chondrite-normalized plots in the literature from which five
elements (Pr, Tb, Ho, Tm and Lu) have been determined by interpolation.

THEORETICAL

When the specimen thickness is less than the "critical thickness," the
absorption-enhancement effect can be ignored, and the spectral line intensity
(I_i) of the analyte element is directly proportional to concentration (C_i),
where

$$I_i = K_i \times C_i \tag{1}$$

If we select a suitable REE (Nd or Er) as an internal reference element j, we
may derive the following equation:

$$I_i/I_j = K_{ij} \times C_i/C_j \tag{2}$$

where $K_{ij} = K_i/K_j$, I_j and C_j represent the spectral line intensity and
concentration of the reference element respectively. Standards are prepared
in which the concentration of both analyte and reference element are equal;
therefore, $K_{ij} = I_i/I_j$. These ratios of REE in standards are referred to as
K values. The intensity ratio of the same spectral lines in the specimen is
referred to as the A value, i.e. $A_{ij} = I_i'/I_j'$. From Equation (2) we can
obtain

$$A_{ij}/K_{ij} = C_i'/C_j' \tag{3}$$

$$\sum \frac{A_{ij}}{K_{ij}} = \sum \frac{C_i'}{C_j'} \tag{4}$$

The concentration C_i' of individual REE in the specimen can now be derived
from (3) and (4), where

$$C_i' \% = \frac{A_{ij}/K_{ij}}{\sum (A_{ij}/K_{ij})} \times 100\% \tag{5}$$

From Equation (5) we can see that the relative content of individual REE in
the specimen can be calculated by two proportional factors K_{ij} and A_{ij}. If
we preconcentrate the REE by a chemical method and determine the total REE,
e.g., S, then the percent C_i'' of the individual REE in mineral and rocks can
be calculated from the equation

$$C_i'' \% = C_i' \% \times S = \frac{A_{1j}/K_{1j}}{\sum (A_{1j}/K_{1j})} \times S \times 100\% \tag{6}$$

If the total is unknown, then after the calculation of C_i' according to Equation (5), the absolute content C_i'' of a single REE (Nd or Er) in the specimen is determined using an external standard method. Thus, the total REE in the specimen can be calculated by Equation (6).

EXPERIMENTAL METHOD

1. Sample Preparation

For samples in which the total REE is higher than 0.2%, from 0.5000-1.000g sample is weighed. The total REE content is then determined by an oxalate gravimetric method. From 2-10mg of the obtained rare earth oxide is transferred into a 50ml beaker and then added to distilled water. The suspension is poured rapidly onto a special filter and filtered through a 40mm diameter filter paper (12). After air-drying, 10 drops of 1% celluloid acetone solution is used to fix the sample to the filter paper.

For samples in which total REE content is lower than 0.2%, the total REE are preconcentrated by PMBP-n-butyl acetate extraction. Chromium is added as an internal standard into a 5% formic acid solution of the back extract. The solution is transferred to a 6μ Mylar film made with 0.2% celuloid acetone solution, resulting in a 30mm diameter thin film sample ready for XRF measurement (13).

2. Working Conditions

The X-ray intensity of REE is measured using the following instrumental conditions:

W or Rh X-ray tube at 40kV-20mA; crystal LiF(200); scintillation detector. All analyte lines were $L\alpha$ except $YK\alpha$, $CrK\alpha$, $PrL\beta_1$ and $HoL\beta_1$.

3. Determination of K Values

Several sets of standards were prepared in which the content of the individual REE are equal. The net intensity of the analyte line is obtained after correcting the spectral line overlap. The K values of each REE are then calculated using Nd or Er as a variable internal reference. The K value represents an average of at least ten determinations. Experiments show that the K values are not affected by sample size and RSD are better than 6%.

4. Calculation of Analytical Results

A monazite sample is selected as an example, calculated results are shown in Table 1. The total (39.7%) of REE is determined by an oxalate gravimetric method.

COMPARISON OF XRF RESULTS WITH CHONDRITE-NORMALIZED PLOTS

Based on the proposed representative REE abundances, Nakamura et al. (14) have obtained plots of normalized curves of REE in samples based on the

Table 1. XRF results for a monazite sample

Line	I_i' (cps)	A_{ij}	K_{ij}	A_{ij}/K_{ij}	Oxide	$c_i'\%$	$c_i''\%$
LaLα	4649	0.664	0.519	1.28	La$_2$O$_3$	18.8	7.48
CeLα	12457	1.78	0.707	2.52	CeO$_2$	37.0	14.7
PrLβ_1	1413	0.202	0.701	0.288	Pr$_6$O$_{11}$	4.23	1.68
NdLα	7001	1.00	1.00	1.00	Nd$_2$O$_3$	14.7	5.84
SmLα	2411	0.344	1.38	0.249	Sm$_2$O$_3$	3.66	1.45
EuLα	71	0.0101	1.66	0.0061	Eu$_2$O$_3$	0.09	0.036
GdLα	2252	0.322	1.03	0.167	Gd$_2$O$_3$	2.45	0.97
TbLα	123	0.0176	2.11	0.0083	Tb$_4$O$_7$	0.12	0.047
DyLα	1398	0.200	2.31	0.0867	Dy$_2$O$_3$	1.28	0.51
HoLβ_1	142	0.0203	1.01	0.0201	Ho$_2$O$_3$	0.30	0.12
ErLα	811	0.116	2.65	0.0438	Er$_2$O$_3$	0.64	0.25
TmLα	29	0.00414	1.77	0.0023	Tm$_2$O$_3$	0.034	0.01
YbLα	431	0.00616	1.83	0.0337	Yb$_2$O$_3$	0.495	0.20
LuLα	137	0.0196	1.94	0.0101	Lu$_2$O$_3$	0.15	0.06
Y Kα	33820	4.83	11.4	0.424	Y$_2$O$_3$	6.24	2.48
ThLα	21730	3.10	4.70	0.660	ThO$_2$	9.71	3.86
Sum				6.799		100%	39.7%

average of ten ordinary chondrites. From such curves an estimation of the concentration of REE can be obtained.

Since the concentration of REE in geological samples obey a certain variable rule, namely, that simple plots (ppm REE versus atomic number) show striking differences in abundance between successive lanthanides having odd or even atomic numbers, chondrite-normalized curves are useful for comparison of the analytical results of REE in rocks and soils. According to the degree of smoothness of the curve (sometimes Eu and Ce show a certain anomaly), we can estimate whether the analytical results are reliable or not. With the exception of Eu and Ce, as the other elements show an obvious anomalous behavior, one reason for this may be due to inaccuracy in the X-ray line-overlap correction.

The analytical method proposed here has been tested using four GSD standards. We consider our method to be reliable because of the good agreement obtained with the GSD standards. On the other hand, chondrite-normalized curves of REE in these GSD samples are smooth. From Figure 1, except for Eu in GSD-2 where some negative anomaly observed, the other standards compare well with the international standard G-2. From Table 2, we can see a comparison of recommended and analyzed values. In general, good agreement was observed for many elements, but differences for Ho, Tm, Lu, Tb and Pr data were large. These elements were also determined by interpolation from chondrite-normalized plots for comparison.

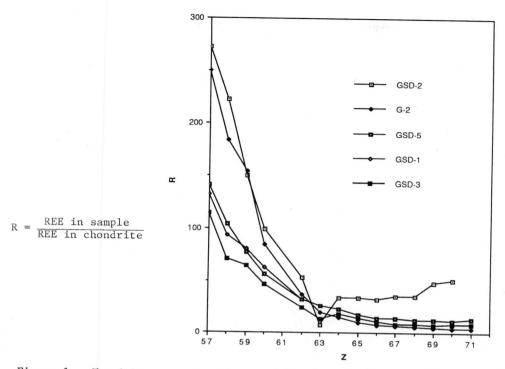

$$R = \frac{REE \text{ in sample}}{REE \text{ in chondrite}}$$

Figure 1. Chondrite-normalized curves of REE in GSDs and G-2 standards

QUANTITATIVE ANALYSIS BY INTERPOLATION

As indicated above, the REE distribution abundance in the earth's crust follow a certain rule, namely, that the content of elements with even atomic numbers is higher than the content of elemens with odd atomic numbers. Obviously the analytical accuracy of higher concentration is much better. Since the experimental XRF values for the REE at low levels (i.e., 0.1-1 ppm) have large associated errors, we suggest that elements such as Pr, Tb, Ho, Tm, Lu, and others be more reliably determined by interpolation from chondrite- normalized curves from the literature. For example, assume

$$R = G_X/C_X = f(Z) \tag{7}$$

then $G_X = R \times C_X$ (8)

where G_X is the abundance of each REE in geological samples, C_X is the abundance of the corresponding element in the chondrite, Z is the atomic number, and R is ratio. According to the measured values of REE, R is calculated by equation (7). We then plotted these R values against Z. From the obtained curves, we can find the corresponding R values of Pr, Tb, Ho, Tm, Lu, etc., elements by interpolation, and according to equation (8) calculate the results for these elements. We have calculated the REE in many standards and samples in a variety of rocks and soils using these equations and the results are very satisfactory. It is obvious from a comparison of values in Table 2 that the accuracy of interpolation is better than the XRF

Table 2.   Comparison of results (ppm)

E	GSD-1 A	B	RE%	C	RE%	GSD-2 A	B	RE%	C	RE%	G-2 (16)
La	43	48.1	12			90	97.9	8.8			86
Ce	81	96.2	19			192	168	-12			159
Pr	10	12.1	21	9.3	-7.0	18.6	21.0	13	17.6	-5.4	19
Nd	39	41.9	7.4			62	60.2	-2.9			53
Sm	7.2	9.0	25			10.8	11.7	8.1			7.2
Eu	1.8	2.4	33			0.49	0.6	22			1.41
Gd	6.1	7.1	16			9.5	10.1	6.3			4.1
Tb	0.86	1.25	45	0.99	15	1.8	2.25	25	1.77	-1.7	0.48
Dy	4.4	5.35	22			11	11.6	5.4			0.343
Ho	0.91	2.95	224	1.01	11	2.9	3.1	6.9	2.65	-8.6	0.078
Er	2.3	2.65	15			8.0	8.1	1.2			0.225
Tm	0.42	1.1	162	0.44	4.8	1.55	1.5	-3.2	1.43	-7.7	0.034
Yb	2.36	3.15	33			11	10.7	-2.7			0.220
Lu	0.45	0.15	-67	0.44	-2.2	1.6	1.8	12	1.70	6.2	0.034
Y	22.5	23.6	4.9			67	63.2	-5.7			18

E	GSD-3 A	B	RE%	C	RE%	GSD-5 A	B	RE%	C	RE%	REE in chondrite(15)
La	39	43.6	12			46	44.6	-3.0			0.328
Ce	64	64.7	1.0			89	89.5	0.6			0.865
Pr	8.3	7.35	-11	7.0	-16	9.6	9.95	3.6	9.29	-3.2	0.123
Nd	30	27.6	-8.0			35	30.8	-12			0.630
Sm	5.3	6.5	23			6.6	6.8	3.0			0.203
Eu	1.3	1.29	-1.3			1.4	1.25	-11			0.077
Gd	4.7	5.75	22			6.4	5.55	-13			0.276
Tb	0.70	0.85	21	0.83	18	0.90	1.25	39	0.94	4.4	0.052
Dy	4.0	4.05	1.2			5.0	4.1	-18			0.343
Ho	0.90	1.65	83	0.86	-4.4	1.1	2.65	141	1.09	-0.9	0.078
Er	2.4	2.25	-6.2			3.1	2.6	-16			0.225
Tm	0.43	0.4	-7.5	0.37	-14	0.48	0.7	46	0.42	-12	0.034
Yb	2.6	3.1	15			2.9	2.75	-5.2			0.220
Lu	0.39	0.35	-10	0.37	-5.1	0.46	0.3	-35	0.42	-8.7	0.034
Y	22	21.6	-1.8			26	21.2	-18			18

A-Recommended values (17)   B-XRF results (13)

C-Interpolation results   S-Samples   R-Results   E-Elements

values. At a 0.x ppm level, relative errors using interpolation are less
than 20%, while those from the XRF method used here are high from 150-224%.
In the determination of Ho by the XRF method, the error is high because in
preconcentrating REE by the extraction procedure, Ni was not completely
separated and $NiK\alpha$ interferes seriously with $HoL\beta_1$. Therefore, using the
interpolation method, we can obtain more accurate results.

ACKNOWLEDGEMENTS

     The author is greatly indebted to Dr. P. A. Pella for his kind help to
modify extensively this paper and for very useful discussions.

REFERENCES

(1) L. N. Kogarko, et al., Geokhimia 5 639-651 (1984)
(2) Zhao Zhenhua, et al., Geochimica 1 26-35 (1981)
(3) Wen Qizhong, Yu Suhua, Gu Xiongfei and Lei Jianquan, Geochimica 2
     151-161 (1981)
(4) N. V. Turanskaya, Method for determination and analysis of rare
     elements Moscow 143-144 (1961)
(5) E. E. Vainstein, Methods for Quantitative X-Ray Spectroscopic
     Analysis Moscow 176-177 (1956)
(6) V. P. Bellary, S. S. Deshpande, R. M. Dixitand and A. V. Sankaran,
     F. Z. Anal. Chem. 309 380-382 (1981)
(7) I. Roellandts, Anal. Chem. 54(4) 676-680 (1981)
(8) A. N. Sial, M. C. H. Figueiredo and L. E. Long, Chem. Geol. 31(3)
     271-283 (1981)
(9) P. Mitrapoulos, Chem. Geol. 35(3/4) 265-280 (1982)
(10) Feng Lianyuan, et al., Geochimica 1 35-46 (1982)
(11) An Qingxiang, Acta Ptrologica Minerallogica et Analytica 3(2)
     162-165 (1984)
(12) Chen Yuanpan, Analytical Chemistry (in China) 1 61-64 (1981)
(13) Sun Pinhui and Chen Yuanpan, Spectroscopy and Spectral Analysis
     5(4) 47-52 (1985)
(14) N. Nakamura, Geochim. Cosmochim. Acta 38(5) 757-775 (1974)
(15) P. J. Potts, O. W. Thorpe and J. S. Watson, Chem. Geol. 34(3/4)
     331-352 (1981)
(16) E. S. Gladney and C. E. Burns, Geostandards Newsletter 7(1)
     11-12 (1983)
(17) Xuejing Xie, Mingcai Yan, Lianzhong Li and Huijun Shen, Geostandards
     Newsletter Vol. IX No.1 83-159 (1985)

HOW TO USE THE FEATURES OF TOTAL REFLECTION OF X-RAYS

FOR ENERGY DISPERSIVE XRF

H. Schwenke, W. Berneike, J. Knoth, and U. Weisbrod

Institute of Physics, GKSS Research Centre Geesthacht
P.O.Box 1160, 2054 Geesthacht, Federal Republic of Germany

ABSTRACT

The total reflection of X-rays is mainly determined by three parameters, that is the critical angle, the reflectivity and the penetration
depth. For X-ray fluorescence analysis the respective characteristic
features can be exploited in two rather different fields of application.
In the analysis of trace elements in samples placed as thin films on
optical flats, detection limits as low as 2 pg or 0.05 ppb, respectively,
have been obtained. In addition, a penetration depth in the nanometer
regime renders Total Reflection XRF an inherently sensitive method for
the elemental analysis of surfaces. This paper outlines the main physical
and constructional parameters for instrumental design and quantitation in
both branches of TXRF.

INTRODUCTION

That variant of energy dispersive XRF, which works at grazing incidence of the primary radiation and profits from the effect of total reflection of the incident beam has been known since 1971 from the work of
Yoneda and Horiuchi[1]. Since that time, many papers have been published
documenting developments in Total Reflection XRF (TXRF) in trace element
analysis (A. Prange et al.[2]), and its recent entry into the circle of
instruments for surface analysis (P. Eichinger et al.[3]). The character of
both branches of TXRF is set by the total reflection effect, however, the
features of total reflection are exploited in a quite different manner.
This will be substantiated later on after a brief introduction into the
behaviour of X-rays impinging upon an interface in grazing incidence.

THE MAIN FEATURES OF THE TOTAL REFLECTION EFFECT

A description of the total reflection phenomenon within the approxi-
mations of the classical dispersion theory has been proven to be sufficient for the design and operation of TXRF-instruments.

The description is based on the fact that the real part of the index of refraction

$$n = 1 - \delta - i\beta \tag{1}$$

is smaller than unity, whereas the refractive index of vacuum (and, for practical purposes, air) is $n = 1$.

β describes the attenuation of X-rays in matter. It provides the damping which is an essential feature of the grazing incidence arrangement and is given by:

$$\beta = \frac{\mu\lambda}{4\pi}$$

μ = linear absorption coefficient
λ = wavelength of the x-radiation.

The variation of δ with X-ray energy in the approximation of the classical dispersion theory for regions far from any absorption edge of the material is given by:

$$\delta = n_e \, e^2 \, \lambda^2 / \, 2 \, \pi \, m \, c^2$$

with
n_e = electron density
e = elementary charge
m = mass of the electron
c = velocity of light.

The critical angle for total reflection is defined from Snell's law as:

$$\cos \phi_c = 1 - \delta$$

where ϕ_c is the glancing angle of incidence for which the angle inside the reflector is just zero.

In view of the physical importance of the imaginary part of the refractive index (eq. 1), which is not considered in the definition of ϕ_c, the critical angle appears to be a more formal term than is really significant. It remains, however, a useful tool to establish the range of angles, where in a particular situation total reflection occurs.

ϕ_c can be calculated from the properties of the reflecting material (atomic number Z and weight A, density ρ) and the energy (E) of the incident X-rays by:

$$\phi_c = \text{const} \cdot \sqrt{\frac{Z \cdot \rho}{A}} \cdot \frac{1}{E} \tag{2}$$

where const = 99.1 [(min. of arc) $\cdot$ kev $\cdot$ $(g/cm^3)^{-1/2}$].

For quartz and Mo K_α radiation, $\phi_c = 5.9$ min. of arc or 1.73 mrad, respectively.

The most expressive feature of total reflection is naturally the reflectivity, which is defined as the ratio of the reflected intensity I to the incident intensity I_0.

With the aid of the critical angle, the variation of the reflectivity R with ϕ is given by the Fresnel equation:

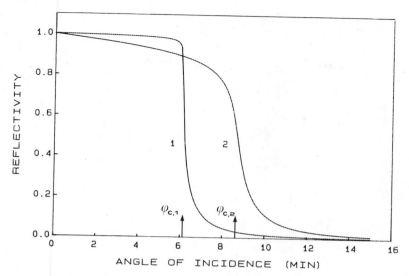

Fig. 1: Reflectivity for X-rays of 17.5 keV for Si; ρ = 2.3 g/cm³
(curve 1) and GaAs; ρ = 5.3 g/cm³ (curve 2).

$$R = \frac{[\sqrt{2}\, X-(((X^2-1)^2+Y^2)^{1/2}+(X^2-1))^{1/2}]^2+((X^2-1)^2+Y^2)^{1/2}-(X^2-1)}{[\sqrt{2}\, X+(((X^2-1)^2+Y^2)^{1/2}+(X^2-1))^{1/2}]^2+((X^2-1)^2+Y^2)^{1/2}-(X^2-1)} \quad (3)$$

with $\quad X = \dfrac{\phi}{\phi_c} \quad$ and $\quad Y = \dfrac{\beta}{\delta}$

The variation of the reflectivity with the angle of incidence is displayed in fig. 1 for two materials with different absorption characteristics. The curves represent the typical shapes for materials with low (curve 1) and high damping (curve 2).

From the Fresnel equation (eq. 3) one can obtain an expression for the depth of penetration of the X-rays in the reflection case (L.G. Parrat[4]). It is given in the form of the equation below.

$$z_p = \frac{\lambda}{4\pi \sqrt{\delta}\, [((X^2-1)^2 + Y^2)^{1/2} - (X^2-1)]^{1/2}} \quad (4)$$

where z_p is the distance normal to the interface at which the intensity of the refracted beam is reduced to 1/e. Fig. 2 shows the penetration depth for the example of gallium arsenide as a function of the angle of incidence using a copper (curve a) or a molybdenum anode (curve b), respectively.

It can be seen from fig. 2 that the penetration depth for very small glancing angles is independent of the energy of the primary beam.

The very slight penetration (some nm) at angles less than critical makes X-ray reflection inherently a sensitive method for exploring thin

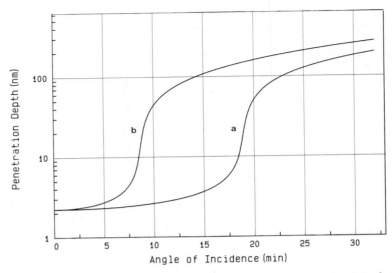

Fig. 2:   Penetration depth for GaAs at an incident energy of 8 keV (a)
           and 17.5 keV (b)

surface layers. This task, however, should be tackled with care, in order
to avoid a maze of instrumental problems and difficulties in the calibration and quantitative procedures.

ANALYSIS OF FILM-LIKE SAMPLES ON A SUBSTRATE

        In spite of the need for an additional sample preparation step, the
analysis of a sample placed on a reflecting surface turns out to be
basically (from a physical point of view) simpler and easier to survey
than the analysis of atoms embedded in a surface layer. Therefore, both
branches of TXRF ought to be based on the employment of standard solutions deposited on quartz or silicon wafer surfaces for instrumental
development, adjustment and calibration.

        The detection limit (DL) in XRF depends on two factors according to
the relation (eq. 5)

$$DL \sim 1/S \cdot \sqrt{B} \qquad\qquad (5)$$

where S is the sensitivity of the instrument (counts/quantity) and B the
background. For matrix-free samples placed on a flat substrate, the TXRF
technique provides an exceptionally good peak to background ratio, because the energy transfer through the interface, which is responsible for
scattered and fluorescence radiation contributed from the substrate, is
extremely hampered (low background). The primary intensity, however,
available for the excitation of the sample is not reduced by the total
reflection of the incident beam but approximately doubled (high peak).

        In order to quantify the effects occuring, the energy transfer
through the interface is given by

$$I_O \cdot \phi \cdot (1 - R) \tag{6}$$

This means that the primary intensity I_O is reduced twice, once by the low incident angle and once more by the fact that only the unreflected portion of the incident radiation must be considered for inelastic processes.

In contrast to eq. (6), the intensity of the radiation which excites the sample is given by:

$$I_O \cdot (1 + R) \tag{7}$$

The expression eq. (7) describes the fact that a film-like sample on a reflecting surface is hit by the incident as well as by the reflected X-rays. The course of eq. (7) as a function of the angle of incidence compared with measured data is shown in fig. 3.

Whereas under total reflection conditions the peak to background ratio appears to be in a way a present of nature, an adequate sensitivity (according to relation 5) must be acquired by constructional measures. Fig. 4 shows the design of a commercially available instrument (Rich. Seifert & Co[5]) in the latest stage of development. An additional mirror is inserted into the path of the primary X-rays which acts as a low pass filter. Therewith beneficial use has been made of a further feature of X-ray total reflection, the cut-off effect at higher energies (fig. 5), which is a consequence of the energy dependence of the reflectivity according to eq. 3 and eq. 2.

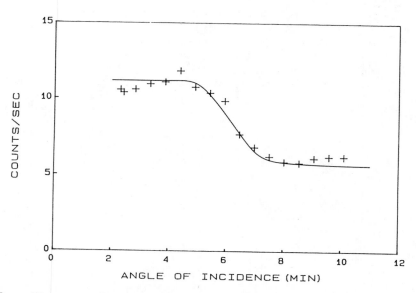

Fig. 3:   Fluorescence intensity of a film-like sample on a quartz surface vs. angle of incidence. E = 17.5 keV.
(+ measured values, — calculated curve)

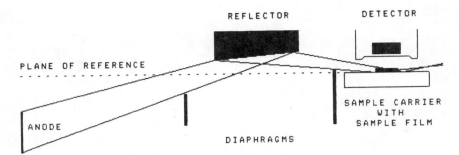

Fig. 4:   Arrangement with mirror acting as low-pass filter of the prima-
           ry radiation

        Figs. 6a and b and fig. 7 demonstrate the excellent detection per-
formance obtained with the latest TXRF-instrumentation for the analysis
of film-like samples (Rich. Seifert & Co[5]).

SURFACE ANALYSIS

        Applying the same formalism as above the primary intensity in a near
surface layer at a distance z normal to the interface is given by:

$$I(z) = I_0 \cdot (1 - R(\phi)) \cdot K(\phi) \cdot e^{-\frac{z}{z_p}} \tag{8}$$

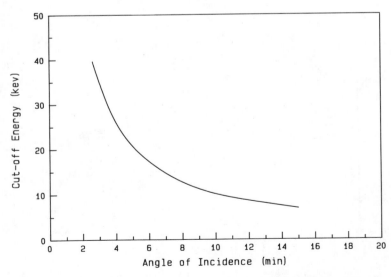

Fig. 5:   Cut-off energy as a function of the angle of incidence using a
           quartz mirror

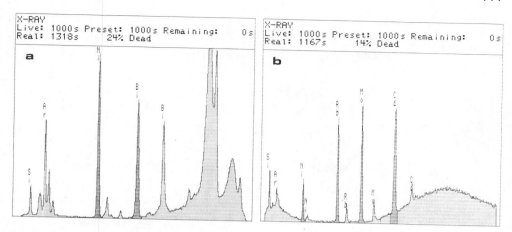

Fig.6a,b: Spectra of a 1 ng Ni, Bi sample using a Mo tube (a) and a 1 ng
            Ni, Rb, Mo, Cd sample using a W tube (b).
            Resulting detection limits: a) Ni 1.95, Bi 3.25 pg;
            b) Ni 18.3, Rb 5.8, Mo 5.1, Cd 10.4 pg

where

I_o = intensity of the primary beam above the interface
R = reflectivity
ϕ = angle of incidence
z_p = value of z at which the intensity of the refracted beam is
         reduced to 1/e.

The term (1-R) illustrates the fact that only that portion of the
primary radiation, not reflected at the interface, is available for in-

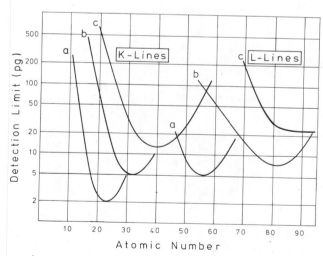

Fig. 7:   Limits of detection using a Mo tube (curve b) or a W tube
           (curve a: W-L excitation, curve c: 50 kV bremsstrahlung), re-
           spectively

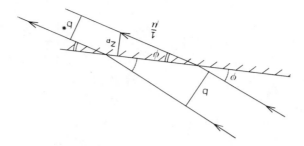

Fig. 8:   Refraction of X-rays for grazing incidence

elastic processes in the surface, e.g. fluorescence. The factor K, the so-called compression factor, describes the change in the density of the flux lines of the refracted field compared to the incoming beam. Changes in the field strength arise in the total reflection case, because the flux lines of the refracted beam are concentrated, or sometimes, even diluted, in the near surface layer, compared to normal incidence. Finally, the exponential function describes the decline of the intensity as a function of the distance from the interface.

Simple geometric considerations lead to eq. 9 for the compression factor K. The distances b and b* in fig. 8 between the incident and refracted rays, respectively, describe the radiation densities on both sides of the interface. K is defined as:

$$K = b/b*$$

For small angles and a range of the refracted rays of $1/\mu$ inside a material with linear absorption coefficient μ, it follows that:

$$b/b* = \phi/\psi \quad \text{and} \quad \psi = \mu \cdot z_p \quad \text{or:} \quad K = \phi/(\mu \cdot z_p) \qquad (9)$$

The determination of homogeneously distributed atoms in a near surface layer is an important special case in surface analysis e.g. for the determination of impurity atoms in silicon wafer surfaces. The fluorescence radiation I_F for this case is described by:

$$I_F = C_i \cdot A_v \int_0^H I(z) \, dz \qquad (10)$$

C_i = instrumental calibration factor for element i
A_v = concentration of impurity atoms
H = thickness of the near surface layer.

Inserting eq. (8) in eq. (10), one obtains after integration:

$$I_F = (C_i \cdot I_o) \cdot A_s \cdot (1-R) \cdot K \cdot (z_p/H) \cdot (1 - e^{-\frac{H}{z_p}}) \qquad (11)$$

with

$$A_s = A_v \cdot H \quad \text{for the atoms per unit area.}$$

Equation (11) can be written as:

$$A_s = \frac{I_F}{(C_i \cdot I_o) \cdot f} \tag{12}$$

with

$$f = (1-R) \cdot \frac{\phi}{\mu \cdot H} \cdot (1 - e^{-\frac{H}{z_p}}) \tag{13}$$

where I_F is the measured value.

The function f, we call it the form function, depends essentially on the layer thickness and the angle of incidence.

For calibration, the product $(C_i \cdot I_o)$ has to be determined by inspection of a film-like standard solution placed on a quartz carrier sample with known concentration A_s'. In this case the measured fluorescence intensity I_F' is given by (see eq. 7):

$$I_F' = (C_i \cdot I_o) \cdot A_s' (1+R) \tag{14}$$

At low angles of incidence, where R is nearly unity, the product $(C_i \cdot I_o)$ can easily be determined by writing eq. (14) as

$$(C_i \cdot I_o) = \frac{I_F'}{A_s' \cdot 2} \tag{14a}$$

After calibration of the instrument via eq. (14a) all values on the right hand side of eq. (12) are known in principle.

Not as yet considered, however, is in eq. 13 the angular spread of the primary beam. Due to the low energy transfer into the surface (see

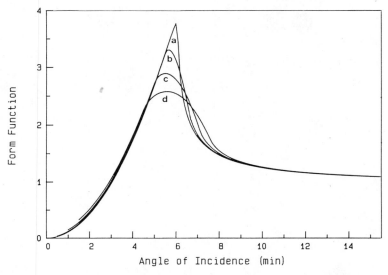

Fig. 9:   Fluorescence intensity of a near-surface layer with various beam divergences. (a=no div., b=1 min, c=2 min, d=3 min)

eqs. 6 and 8) one of the main requirements on the design of a TXRF surface analyzer is a short distance between tube anode and sample spot, in
order to avoid solid angle losses. Therefore, a substantial angular divergence has to be accepted. The divergence has been calculated by a
Monte Carlo simulation on the basis of data, which describe a particular,
commerically available instrument (Atomika, München[6]). The calculations
result in the introduction of an effective form function f*, which re-
places f in eq. 12. For practical purposes f* should be implemented in
the quantitation software which belongs to the TXRF instrumentation.
The effect of the angular spread on the fluorescence signal from a near
surface layer of 2 nm is displayed in fig. 9.

Using the instrument already mentioned (Atomika, München[6]) TXRF
analyses of metallic impurities in the surface of silicon have been shown
to agree with those from RBS-measurements (V. Penka[7]).

CONCLUSIONS

By exploiting the specific features of total reflection, both branches of TXRF surpass by far the respective capabilities of normal energy
dispersive XRF. In trace element analysis extremely low quantities can be
detected, while in surface analysis ultra thin surface layers are accessible. The advantages of EDXRF such as multielement performance, simplicity and partly nondestructiveness are preserved. The technique, however,
makes high demands on the instrumentation, both with regard to the mechanical design as well as to the control of additional parameters.

REFERENCES

1. Y. Yoneda, T. Horiuchi, Rev. Sci. Instr., 42: 1069 (1971).
2. A. Prange, J. Knoth, R.P. Stößel, H. Böddeker, and K. Kramer,
       Anal. Chim. Acta, 195: 275 (1987).
3. P. Eichinger, H.J. Rath, and H. Schwenke, Semiconductor Fabrica-
       tion: Technology and Metrology, ASTM STP 990,
       (D.C. Gupta, ed.) American Society for Testing and Materials,
       1988
4. L.G. Parrat, Phys. Rev., 95: 359 (1954).
5. Rich. Seifert & Co., Bogenstraße 41, D-2070 Ahrensburg, FRG.
6. Atomika Technische Physik GmbH, Postfach 450135, D-8000 München,
       FRG.
7. V. Penka, W. Hub, Spec. Chim. Acta B, in print.

APPLICATIONS OF A LABORATORY X-RAY

MICROPROBE TO MATERIALS ANALYSIS

D. A. Carpenter, M. A. Taylor, C. E. Holcombe

Oak Ridge Y-12 Plant*
Oak Ridge, Tennessee 37831-8084

INTRODUCTION

A laboratory-based X-ray microprobe, composed of a high-brilliance microfocus X-ray tube, coupled with a small glass capillary, has been developed for materials applications. Because of total external reflectance of X rays from the smooth inside bore of the glass capillary, the microprobe has a high sensitivity as well as a high spatial resolution. The use of X rays to excite elemental fluorescence offers the advantages of good peak-to-background, the ability to operate in air, and minimal specimen preparation.[1] In addition, the development of laboratory-based instrumentation has been of interest recently because of greater accessibility when compared with synchrotron X-ray microprobes.[2,3,4]

The most serious limitation to the use of X-rays in a microfluorescence on microprobe application is the lack of intensity. X-rays collimated through a pinhole aperture are subjected to $1/r^2$ losses. On the other hand, X-rays propagated through glass capillaries are subjected to reflection losses, which are significantly less than the $1/r^2$ losses. Previous work in this lab has shown that, compared with aperture collimation, Pyrex-glass capillaries provided intensity gains of as much as 180-fold for a 10 μm capillary that was 119 mm long and positioned 5 mm from the 30 μm focal spot.[5] The capillary tubes used in those tests ranged from 10 to 100 μm, inside diameter, and 6 mm outside diameter. The glass tubes were inserted directly into the X-ray tube through an o-ring fitting which replaced the X-ray tube window. The position and focus of the focal spot were adjusted by electromagnetic coils. Additional tests showed that within a few millimeters of the end of the capillary, the beam profile was uniform and was not the image of the foreshortened focal spot. The study also showed that the spectrum from a tungsten anode was distorted because of the energy selectivity of total external reflection. Lower energy radiation was favored over higher energy radiation.

---

* Operated for the U.S. DEPARTMENT OF ENERGY by Martin Marietta Energy Systems, Inc., under contract DE-AC04-84OR21400.

In this paper, some additional beam characterization experiments will be discussed. Also, applications to several types of materials will be demonstrated.

EXPERIMENTAL

The microprobe is based upon a radiographic microfocus X-ray tube with a Pyrex-glass capillary inserted into the tube and aligned as described above. The specimen surface is positioned normal to the X-ray beam. A Si(Li) detector is located in the horizontal plane, approximately 45° from the specimen surface. The specimen is translated in the plane normal to the beam with ball-screw-driven micropositioning tables capable of 0.1 μm resolution. Signals from the detector are processed through an amplifier, analogue-to-digital converter, and multichannel analyzer. Counts in selected regions of interest are routed through single channel analyzers and collected with individual counters. Scanning is "on-the-fly" in the x direction and step-wise in the y direction. The counting sequence is initialized by a timer using a predetermined number of pulses from the stepper-motor controller. The data are illustrated in either grey level or false color scales with an Amiga 1000 computer.

RESULTS AND DISCUSSION

The spatial resolution of the microprobe is primarily dependent upon the width of the beam profile. As pointed out by Hirsch,[6] the divergence of the beam would be expected to be the same as the divergence at the anode end of the capillary. For a 5 mm anode-to-capillary distance, the divergence should be relatively large. However, the profile width can be controlled by minimizing the capillary-to-specimen distance, h. By grinding a 10° bevel on the specimen end of the capillary, h can be reduced to as low as 200 μm without significant intensity loss due to blockage of the signal by the capillary. Fig. 1 shows scans in the horizontal direction, at h = 200 and

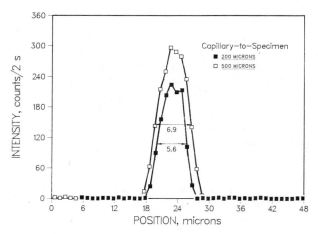

Fig. 1.   Horizontal scans using a 10-μm capillary of a 1.5 μm nickel wire at two capillary-to-specimen distances. The fwhm values, in microns, are shown on the plots.

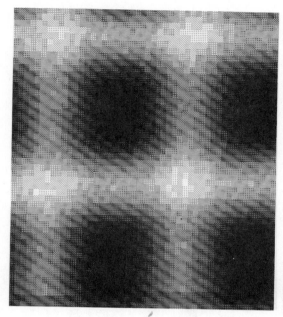

Fig. 2.   Nickel kα image of a 1000-mesh
          nickel grid made using a 10 μm
          capillary at **h** = 500 μm. Grid
          spacing = 25 μm, element
          size = 8 μm. (Conditions:
          Tungsten anode at 30 kV, 0.70 mA,
          step size = 1 μm/pixel, count
          time = 1 s/pixel, 10 μm capillary.

500 μm, of a 1.5 μm nickel wire past the beam emerging from a 10 μm
capillary. Thus, a spatial resolution of between 5 and 6 μm should be
possible with a 10 μm capillary.

As an indication of the uniformity and resolution of the beam, Fig. 2
shows a two-dimensional scan of a 1000-mesh nickel grid using a 10 μm
capillary.

The detection sensitivity is demonstrated in Table 1, where the count
rates for electron-microprobe standards are listed.

APPLICATIONS

The laboratory X-ray microprobe has been used to analyze a variety of
materials. The following examples illustrate applications to metals,
ceramics, and geological materials.

TUNGSTEN-NICKEL-IRON ALLOYS

Dilute alloys of nickel and iron in tungsten may be sintered at
relatively low temperatures due to the formation of low melting BCC solid
solution of nickel, iron, and a small amount of tungsten which resides

Table 1. Detection Sensitivity with Pure Metal Standards

Element	Count Rate, c/s
Ni	1280
Cu	1125
Mn	850
Ge	711
Se	668

Note: Anode: W
       Load: 30 kV, 0.70 mA
       Capillary size: 10 μm
       Capillary length: 119 mm

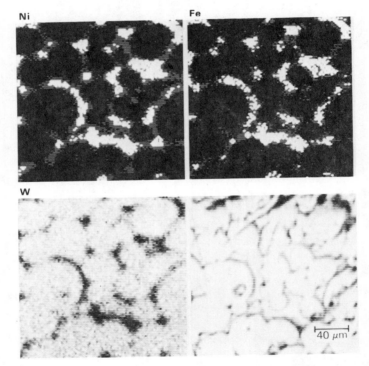

Fig. 3. Elemental images of a sintered W-5 wt % Ni-
         2.5 wt % Fe alloy. Lower right image is a
         digitized photomicrograph of the analyzed area.
         (Conditions: Mo anode, 50 kV, 0.30 mA, step
         size = 2 μm/pixel, count time = 0.5 s/pixel,
         10 μm capillary.)

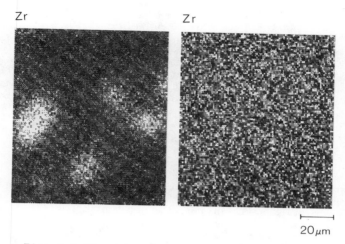

Fig. 4. Zirconium distributions in commercial
         (left) and in Y-12 (right) yttria-2
         wt % zirconia ceramics. (Conditions:
         Mo anode, 50 kV, 0.30 mA, step size =
         1 μm/pixel, count time = 1 s/pixel.)

in the grain boundaries. The matrix consists almost entirely of tungsten.
For a W-5 wt % Ni-2.5 wt % Fe alloy, Fig. 3 shows images of nickel and iron
Kα and tungsten Lα fluorescence, along with a digitized micrograph of the
analyzed area. The microstructure of the sintered condition is clearly
illustrated.

YTTRIA-2 WT % ZIRCONIA

Yttria-2 wt % zirconia ceramics fabricated at the Y-12 Plant have shown
greater thermal shock resistance than those available commercially. The

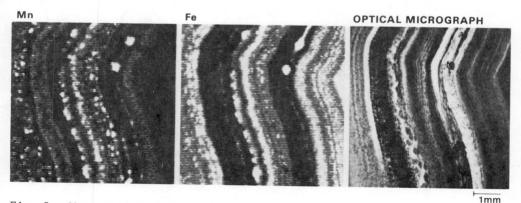

Fig. 5. Manganese (left) and iron (right) distributions in a 25-million-
         year-old Brazillian banded agate. Specimen area analyzed is shown
         in the micrograph to the right of the elemental distributions.
         (Conditions: W anode, 30 kV, 0.70 mA, step size = 50 μm/pixel,
         count time = 2 s/pixel, 100 μm capillary.)

Y-12 ceramics were prepared from slurries of yttria mixed with aqueous zirconyl nitrate and then sintered in a microwave oven.

The commercial ceramic was prepared by mechanically mixing the dry powders and then sintering in a conventional manner. In Fig. 4, the commercial ceramic is represented by the heterogeneous Zr distribution on the left, while the Y-12 ceramic on the right has a homogeneous distribution, at least within the resolution of the microprobe. The thermal shock resistance is believed to be related to the Zr distribution.

BRAZILLIAN BANDED AGATE

Banded agates contain color bands due to the presence of transition metal elements which dissolved in the quartz structure over a long time period. Fig. 5 depicts the iron and manganese distributions in a 25-million-year-old Brazillian banded agate containing approximately 0.75 wt % Fe and 0.35 wt % Mn. The large area scan was obtained with a 100 μm capillary and demonstrates the versatility of the probe. The data also illustrates some advantages over the use of electron beams. For example, beam penetration enabled access to a larger specimen volume and resulted in a higher count rate. In addition, no special coating was required to prevent charging.

REFERENCES

1. C. J. Sparks, Jr., "X-Ray Fluorescence Microprobe for Chemical Analysis," in: Synchrotron Radiation Research, H. Winich and S. Doniach, Eds., Plenum Press, New York, 1980.
2. M. C. Nichols, D. R. Boehm, R. W. Ryon, D. Wherry, B. Cross, G. Aden, "Parameters Affecting X-Ray Microfluorescence (XRMP) Analysis," in: Advances in X-Ray Analysis, vol. 30, C. S. Barrett et al., Eds., Plenum Press, New York, 1987.
3. D. Wherry and B. Cross, "XRF, Microbeam Analysis, and Digital Imageing Combined into a Powerful New Technique," Analyst, 37:8 (1986).
4. R. D. Giaque, A. C. Thompson, J. H. Underwood, Y. Wu, K. W. Jones, and M. L. Rivers, "Measusrements of Femtogram Quantities of Trace Elements Using an X-Ray Microprobe," Anal. Chem., 60:855 (1988).
5. D. A. Carpenter, "An Improved Laboratory X-Ray Source for Microfluorecence Analysis," submitted for publication.
6. P. B. Hirsch, "X-Ray Microbeam Techniques," in: X-Ray Diffraction by Polycrystalline Materials," H. S. Peiser et al., Eds., Chapman and Hall, London, 1960.

DEVELOPMENT OF INSTRUMENT CONTROL SOFTWARE FOR THE SRS/300 SPECTROMETER

ON A VAX/730 COMPUTER RUNNING THE VMS OPERATING SYSTEM

M. J. Rokosz and B. E. Artz

Research Staff, Ford Motor Company
20000 Rotunda Drive
Dearborn, MI 48121

I. INTRODUCTION

It is not practical to connect each controllable function of a WDXRF (Wavelength Dispersive X-Ray Fluorescence) spectrometer separately to the laboratory computer. Therefore, a spectrometer controller is used to interface with the hardware control actuators which operate the selectable or settable components of the spectrometer system. The basic function of a controller is to receive sequences of orders from the computer and transmit the data which has been collected. More advanced controllers, such as the one used in the Siemens SRS-300 spectrometer, may perform many additional functions. In addition to handling spectrometer-to-lab computer communication, the SMP (Siemens MicroProcessor controller) is capable of monitoring the state of the instrument, resolving conflicts in external or internal requests for action, and terminating spectrometer action which can be seen as hazardous or damaging to the instrument. The SMP is also used for maintaining a library of up to 80 parameter sets for the acquisition of data on defined "elements", allowing manual manipulation of all spectrometer settings and functions through a standard keyboard input, presenting current information in video terminal displays relating to spectrometer operation, contents of various storage areas, self diagnostics, managing the ratemeter scans of samples, and performing complex data acquisitions on samples using stored parameters, auxiliary conditions, and user-specified procedures. Many of the features described here are very advantageous and enhance the interaction between a laboratory computer and such a controller; however, the increased complexity of the SMP can also result in a substantial loss in flexibility by arbitrarily fixing the data-collection sequence. The intent of the current work was to develop a simple method to manipulate the SRS-300 spectrometer by issuing individual commands for each spectrometer action directly from the laboratory computer thereby permitting more flexibility in defining data-collection sequences and procedures.

The Siemens data-collection software referred to in this paper is
the SPECTRA-300 package as implemented on Digital Inc. laboratory
computers. The newest data-collection software for the SRS-300
implemented on the IBM Inc. PC has not been tested.

II. DATA ACQUISITION METHODS

During the data-acquisition process the Siemens software sends the
entire analytical request (ie. all the element parameter sets which
must be measured on all samples) to the SMP. Simultaneously Siemens
software stores the existing element parameter sets, which are in SMP
memory, in a temporary file which is used to restore the contents of
the SMP after the analysis has been completed. The SMP then conducts
the analysis as specified and transmits data to the lab computer in
variable size packets. The interchange of information is handled
through "Mailbox" routines on the VAX computer. Various SPECTRA-300
routines are given access to this information which is stored in a
temporary file.

This approach works well when routine or repetitive analysis enables
the spectroscopist to predetermine the acquisition parameters and the
data collection sequence. For our applications this was a major
drawback since the microprocessor did not have the capability to
optimize the data acquisition but had to collect data on all samples
for a particular element in the same way. Direct control of the data
acquisition by the computer, which has the capability to perform
optimization calculations, was not possible.

Two other potential problems with the data-collection procedure were
noted. If data collection is terminated by a failure in either the
software or hardware, the data which has already been collected can be
lost because of the complex communication procedures utilized between
the SMP and the software routines in the laboratory computer.
Information stored on disk by the data-collection software is recorded
in a binary format which is not ASCII-character based. Retrieval of
this information for review or editing requires the use of the SPECTRA-
300 software. We found this feature too restrictive for our
applications since we often needed to "adapt" the form of the data for
input to various analytical packages.

The following data-acquisition philosophy (1) has been implemented
in the software that we developed: The spectrometer controller
maintains the spectrometer instrument by monitoring the state of the
instrument, receiving commands from the laboratory computer, processing
commands, resolving conflicts between actuator actions, and
transmitting data to the laboratory computer. The laboratory computer
generates commands based on the scope and objective of the data
collection task, processes incoming data for storeage or further
optimization of data acquisition. The laboratory computer also
recognizes command completion and error state messages from the
spectrometer controller. Data-collection software should not
transgress any of the conflict conditions that are unresolvable by the
spectrometer controller and should allocate sufficient time for the
completion of commands which do not issue a completion code.

III. MICROPROCESSOR/CONTROLLER REQUIREMENTS

The SRS-300 was equipped with the version #2 microprocessor chip with revised spectrometer operating codes. Two major changes were implemented by Siemens in this version. The spectrometer did not return to sample position zero and atmospheric pressure at "job" completion and a series of commands were implemented for external (laboratory computer) control of individual parameters (hardware actuators) on the spectrometer. These changes made it possible to implement the control of the spectrometer by the laboratory computer using two different approaches. One could implement the standard Siemens "job" concept which would require the assembly of a 64 character (byte) buffer that specifies all of the operating parameters for each data-acquisition cycle, or one could use the individual commands to control each step in a data-acquisition cycle. The former option would use microprocessor logic to resolve timing problems and conflicts between actuator commands but would also require more complex, event-flag, controlled computer software to coordinate data transmission. The latter option requires more careful software design

CHARACTER STRING	ACTION REQUESTED
04	Computer mode on, spectrometer in external control
05	Computer mode off, spectrometer in local control
0120000	Start count rate measurement
00	Stop (everything)
13	Clear the data buffer
1501xx*	Sample changer to position xx (00-10)
1502xxxxxx	Spectrometer angle (xxx.xxx)
1503_____x	Crystal (1-6)[1]
1504x	Diffraction order (1, 2)
1505x	Detector (1-flow, 2-scint, or 3-both)
1506xx	Scint lower level discriminator (.xx [volt])
1507xxx	Scint upper level discriminator (x.xx [volts])
1508xx	Flow lower level discriminator (.xx [volt])
1509xxx	Flow upper level discriminator (x.xx [volts])
1510x	Soller slit (C-coarse, F-fine)
1511xxxxx	Preset maximum time to acquire [sec]
1512xxxxxxxx	Preset maximum counts to acquire
1513_x	Aperture mask (1-3)[1]
1514___x	Mask (1-4)[1]
1515xx	Tube voltage [kv]
1516xx	Tube current [ma]
1517_x	Spectrometer mode (V-vacuum, A-air, H-helium)
1518x	Sample rotation (Y-yes, N-no)
1519__x	Flow detector gas (1, 2)[1]

* The following instructions are all versions of Command #15 which allows specifying the settable functions of the spectrometer.

x   Indicates an ASCII character

_   Indicates a blank space.

[1] Arbitrary value. Actual crystal, mask, gas, or filter is determined by which one is installed for a specific positioner value.

Fig. 1. Siemens microprocessor commands.

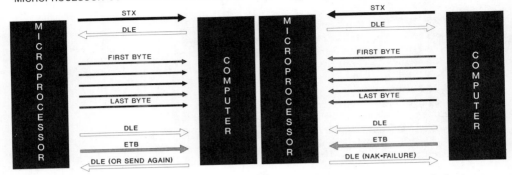

Fig. 2. Transmission of command to microprocessor.

Fig. 3. Transmission of data to computer.

to avoid conflicts in actuator requests but could utilize simple "sleep" routines with predetermined idle times for suspension of the computer software pending the completion of a spectrometer action. The second option was ultimately implemented.

Figure 1 contains the list of the microprocessor commands used to control the spectrometer. These commands include 19 instructions that relate specifically to spectrometer actions and a series of instructions used to initiate, monitor, and terminate communication with the computer. All commands are sent to the microprocessor as ASCII characters. Figure 2 illustrates the transmission of a command of arbitrary byte length to the microprocessor. Count rate data stored in the microprocessor output buffer are also transferred to the computer as an ASCII character string. Figure 3 illustrates this process. The ASCII characters STX, DLE, NAK, and ETB (octal codes 002, 020, 025, and 027) are used for handshaking purposes when transferring ASCII character (byte buffers). For information regarding the data format (start bit, character length, parity, stop bits, and baud rate) for asynchronous serial transfer of the character strings, refer to the Siemens Microprocessor Manual (2).

IV. SPECTROMETER CONTROL ROUTINES

The driver program is a collection of Fortran subroutines. Two subroutines, DCH (the Device Communication Handler is a separate MACRO subroutine) and CCA (the Command Character Assembler), are used to handle the transmission and reception of character strings by the VAX computer. DCH issues the VAX-VMS QIO (communication request) to the terminal device used for communication with the SMP for reading or writing of character strings. CCA assembles the character strings for receipt or transmission and imposes the SMP protocol for the mutual acknowledgement of the SMP and the VAX, definition of character string length, initiation of string exchange, and completion of string exchange as shown in Figures 2 and 3.

Some of the driver subroutines are called by the data-collection programs for selection of instrument parameters or for calculation of pertinent data, others are called by subroutines within the driver for error checking, calculations, or communication functions. Most of these subroutines contain timeouts, calls to the subroutine SLEEP, which is a utility subroutine included in the driver. SLEEP suspends the data acquisition task for an amount of time specified by the calling routine while the SRS counts or is changing instrument parameters. A list of the other subroutines and a brief explanation of each follows.

INIT   Initialize all instrument parameters and store these values in their corresponding buffers. Assign an I/O channel for communication between the VAX and SRS.

ERR   Error - checking subroutine called by all subroutines requesting action or information from SMP. It retains the name of the calling subroutine and the code for the action requested and types the subroutine name, action code and error code when an error occurs. Err allows 3 attempts by the COUNT routine in case the counting process has been interrupted by a temporary equipment failure.

DONE   Reset the spectrometer settings to the values used during idle state either upon completion of data acquisition or termination of data acquisition by the ERR subroutine. DONE relinquishes the I/O channel assigned by INIT.

POS   Set the goniometer angle of the spectrometer to a specified value. Check whether appropriate detector will be used at this angle. Check if an angle change will occur across the region where flip out of scintillation detector occurs and furnish additional timeout time for this to be accomplished.

FLT   Select appropriate beam filter or open-beam position.

CHANGE   Advance sample changer by specified number of positions.

XTAL   Select the specified crystal.

DET   Select the appropriate detector.

COL   Select the specified collimator.

COUNT   Cause the SRS to count for fixed counts or fixed time. Determine the time necessary to count for fixed counts. COUNT checks for a counter buffer overflow. If an overflow occurred the routine reacquires the counts using a reduced counting time.

GETPOS   Obtain the current spectrometer angle position from the BUFANG buffer.

VOL   Select the tube voltage to be used and check if the current-voltage limits of tube will be exceeded.

CUR   Select the tube current to be used and check if the current-voltage limits of tube will be exceeded.

MAS   Select the internal mask to be used.

MOD   Select the spectrometer atmosphere to be used.

ROT   Select if sample rotation is to be used.

GAS   Select the detector gas to be used.

LOAD   Permit loading of additional samples into the spectrometer
           during a run by moving the sample changer to position 0.

DETANG   This routine determines the proper order for selecting a new
           detector and spectrometer angle so that a conflict is not
           generated.

PKCALC   Using the background and peak positions, background and peak
           counts, and the background and peak counting times this
           routine determines the interpolated background level, net
           count rate and various peak statistics.

SCL, SCU   Set the upper and lower scintillation detector
           discriminators.

FCL, FCU   Set the upper and lower flow-counter discriminators.

V. IMPLEMENTATION FOR DATA COLLECTION

By customizing the initialization subroutine (INIT), some of the
subroutines in the driver need not be called in the data-acquisition
programs. Since all of the parameters are set to an initial value in
INIT, any given parameter can be set to the desired value and left the
same for all data acquisitions. Data-acquisition programs can also be
designed to allow the spectroscopist the choice of retaining the preset
value of a parameter or changing it for one data-acquisition session.
For most situations, however, the values for parameter settings are
stored in a parameter-data file which contains the predetermined
settings for the acquisition of count-rate data for a particular
analyte x-ray.

Two serious difficulties exist in the implementation of this
computer control philosophy. As currently implemented, the SMP does
not respond when a spectrometer parameter has been changed. There is
also no procedure to query the SMP for the value of a given
spectrometer parameter. This makes the computer essentially "blind" to
what actually happens in the spectrometer. In order to circumvent
these problems, all of the subroutines except SLEEP, PKCALC, DCH, and
CCA access a common buffer storage area named SRS. This area contains
the current values of the spectrometer parameters. It is essential
that all the values be initialized and transmitted to the spectrometer,
and then kept current throughout the program operation. In addition,
wait times (calls to SLEEP) have been determined for each parameter
change that represent the maximum worst-case time for that operation.
These times should not be shortened. Clearly, this is not the optimum
way to communicate with the SMP and results in significantly longer
data-collection times than are necessary.

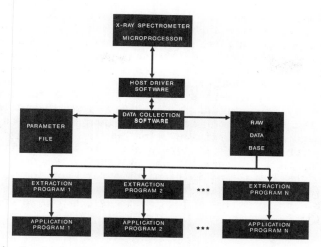

Fig. 4. Typical data acquisition/analysis software package.

Figure 4 shows a flow diagram of a typical data-collection data-analysis software package which we have implemented for the VAX/730-SRS-300. Arrows indicate the direction of flow for data or control commands. A strict functional hierarchy is observed in the organization of the software. The only programs which contain x-ray instrument-dependent code are the programs which communicate directly with the spectrometer. Data-collection programs only contain code that determines the order and manner of data collection and storage. Data analysis programs read stored data files, apply the correction equations, and store report files. This hierarchy insures that future changes in equipment (either of the computer or x-ray spectrometer), data collection, or analysis philosophy can be implemented in a straightforward way by modifying only those programs that are necessary. For a more complete discussion of this approach see Rokosz et al. (1). Figure 5 shows the portion of Figure 4 that illustrates the host driver software-microprocessor link.

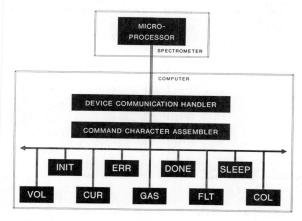

Fig. 5. Microprocessor-host driver link.

An excellent example of the optimization of a data-acquisition parameter performed on the laboratory computer is the selection of peak and background counting times for an analyte line. Most manufacturers of x-ray spectrometers have implemented data - collection software that preselects the counting times for each element. In some cases the same times are used for peak and background counting. Preselection of the counting times for an element should not be used if the quantity of the element varies among specimens. Counting times should be determined individually for each element in each specimen in order to minimize the total instrument data-acquisition time while providing for a given measurement error on each analyte peak. In order to limit the total data collection to a "reasonable" amount of time, an upper limit, T_{Max}, is placed on the total acquisition time for any individual element. The calculation of specific background and peak times is regulated by the calculations of time for an "optimum fixed - time counting experiment" (3).

$$\frac{T_p}{T_b} = \left(\frac{R_p}{R_b}\right)^{1/2} \qquad [1]$$

$$T_{total} = T_p + T_b \leq T_{Max} \qquad [2]$$

$$T_b = T_{low\ b} + T_{high\ b} \qquad [3]$$

$$\sigma = \frac{1}{T_{total}^{1/2}} \left(R_p^{\frac{1}{2}} + R_b^{\frac{1}{2}} \right) \qquad [4]$$

R_p, R_b, T_p, and T_b are the peak and background count rates and counting times, respectively, the background-counting time being split

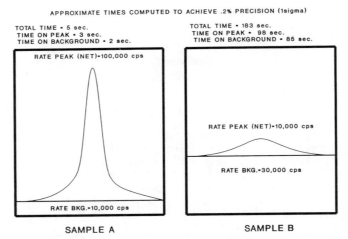

APPROXIMATE TIMES COMPUTED TO ACHIEVE .2% PRECISION (1sigma)

TOTAL TIME = 5 sec.
TIME ON PEAK = 3 sec.
TIME ON BACKGROUND = 2 sec.

RATE PEAK (NET)=100,000 cps

RATE BKG.=10,000 cps

SAMPLE A

TOTAL TIME = 183 sec.
TIME ON PEAK = 98 sec.
TIME ON BACKGROUND = 85 sec.

RATE PEAK (NET)=10,000 cps

RATE BKG.=30,000 cps

SAMPLE B

Fig. 6. Comparison of counting times for two samples.

between the low background and high background counting times. Estimates of the peak and background counting rates (and therefore T_{total}) are obtained by counting for one second at each position. The optimum total counting time to achieve a certain σ (up to the set limit) and the correct partition of that time between peak and backgrounds is then calculated from Equations 1, 2, 3 and 4 above. The peak and background count rates are then measured for the predetermined times. This process is repeated for each element, in each sample, each time the sample is measured. Figure 6 shows peak and background count rates that might be measured for a given element on different samples. If the analyst wished to achieve a given measurement precision (eg. < .2 %, 1σ), then the appropriate counting times are 5 seconds for sample A and 183 seconds for sample B. Clearly, if all the samples had to be measured for 183 seconds to guarantee the expected precision, substantially more time would be required. The use of computer-determined counting times can significantly reduce total data acquisition time while insuring that a given level of measurement precision is observed.

VI. CONCLUSIONS

A modular instrument-control software package has been developed and implemented on a VAX-730 computer for the SRS-300 spectrometer which allows a great amount of flexibility in designing data-acquisition procedures for various analytical applications in x-ray spectrometry. Various data-collection programs that use this control package have been monitored during operation through the use of VMS utility monitoring routines. It was found that a minimal amount of CPU time is required for the operation of these routines, as a result of the method of implementation. After extensive use, no instrument-control errors have been observed. It is estimated that in excess of 10^7 spectrometer instructions have been processed. The spectrometer hardware has also proven to be very durable and precise.

REFERENCES

1 M. J. Rokosz, B. E. Artz, Adv. X-Ray Anal. 29, 477 (1986)
2 Siemens SRS-300 Spectrometer Microprocessor Manual, Siemens AG, Karsruhe, West Germany
3 E. P. Bertin, "Principles and Practice of X-Ray Spectrometric Analysis", Plenum Press, New York (1978)

INSTRUMENTATION AND APPLICATIONS FOR TOTAL REFLECTION X-RAY FLUORESCENCE SPECTROMETRY

Y. Tada, Y. Sako, K. Iwamoto, S. Gonsui and T. Arai

Rigaku Industrial Corporation
Takatsuki, Osaka, Japan

INTRODUCTION

X-ray fluorescence spectrometry has been used in a broad spectrum of applications. These include elemental analysis, both qualitative and quantitative, based on wavelength dispersive (WDXRF) or energy dispersive (EDXRF) methods. In these methods the detection limit of analyte elements is mainly in the one to ten ppm range in solid samples. Therefore, improvement of these limits is desirable for many useful applications. In this context it is essential to remember that the excitation efficiency for fluorescent X-rays is very low when compared with electron or proton excitation. In the case of WDXRF, the dominant factor is the low reflectivity from the analyzing crystal. In EDXRF, it is the poor signal-to-background ratio caused by low-energy resolution of solid state detectors combined with large-angle optics. In order to reduce the background X-rays caused by scattered primary radiation, Yoneda and a co-worker introduced the total reflection technique, which totally reflects the unabsorbed primary radiation in the sample and uses a solid state detector.[1] Since the feature of low intensity of the background spectrum allows the benefit of large solid-angle optics and removes the restrictions caused by high counting rates on a multichannel analyzer, the detection limit has been improved down to the ppb range.[2,3]

An instrument based upon this new method has been developed. The incident beam, shaped like a narrow strip of paper by use of a very narrow slit, is emitted from a high-brilliance source. It strikes the thin layered sample, on an optically flat glass plate or silicon wafer, at a very small grazing angle. A solid state detector is located in front of the sample plate. The flat plate is placed on a sample stage which is specially designed to maintain the reproducibility of X-ray measurements. This paper will describe the instrumentation and typical applications of trace-element analysis of semiconductor samples.

INSTRUMENT

The block diagram of the X-ray total-reflection spectrometer is shown in Fig. (1).

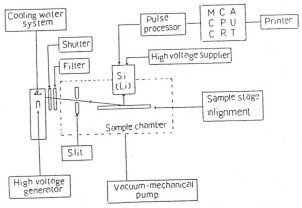

Fig. (1)   Block diagram of total reflection spectrometer

    This consists of a microfocus diffraction tube of tungsten (2kW),
molybdenum (2kW) and chromium (1.5kW) target elements, single slit optics, a
sample support plate holder with an accurate setting mechanism, a silicon
(Li) solid state detector located closely in front of an analyzing sample,
and electronic scaling devices including pulse amplifier, multichannel pulse-
height analyzer and display of measured results on a CRT and a printer.

    The irradiating X-ray beam taken from the target surface of an X-ray
tube with a four degree take-off angle is passed through a fine slit of 20
microns and has a shape like a flat paper sheet. Between the X-ray tube and
the slit, a specified filter is positioned to reduce the background X-ray
intensity. The X-ray beam with a grazing incident angle of 0.05 degrees hits
the flat support plate carrying a sample or a flat surface sample. The
irradiated area of the sample surface is approximately 1 cm^2. In order to
suppress air-scattered X-rays, which would contribute to background X-rays,
the X-ray beam path is evacuated with a mechanical pump.

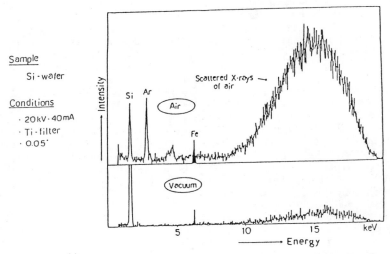

Fig. (2)   Comparison of vacuum path against air path

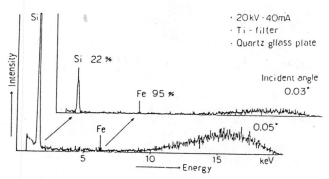

Fig. (3)   Effect of grazing incident angle

Characteristics of the Instrument and Analytical Performances

(1)   Comparison of a vacuum path versus an air path

Fig. (2) shows a comparison of measuring results between air and vacuum paths. In the air path measurement, the air- scattered background X-rays exhibit a broad peak which is the reflection of the energy distribution of the primary radiation. In the vacuum path measurement, the Ar-K alpha X-rays are missing; the background X-rays in the low-energy range and the intensity distribution of the broad peak generated from the silicon crystal wafer are much reduced.

(2)   Effect of grazing incident angle

A comparison of the intensity distributions between the grazing incident angles of 0.03 and 0.05 degrees is shown in Fig. (3). There is a reduction in the intensity of the Si-K alpha X-rays and of the broad X-ray distribution. Also, there is a small reduction in the Fe-K alpha X-rays.

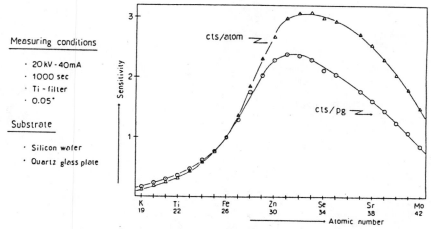

Fig. (4)   Relative sensitivity

(3) Relative sensitivity

The sensitivity was determined by the same method used for measurement of impurities in water. One drop of water (10 micro liters) containing 1 ppm of an analyte element is deposited on the support plate and dried. Over a range of one to ten ppb, an excitation condition without a primary beam filter can be used, but below this level the filter method should be applied in order to reduce the background X-ray intensity. When the Fe-K alpha X-rays were measured without a filter, their intensity was about 40 times higher than that with a titanium filter. The relative intensity of fluorescent X-rays against Fe-K alpha X-rays, using a titanium filter, is illustrated in Fig. (4) for a very thin layered aggregate of analyte elements on a sample support plate. In spite of the use of a titanium filter, which has a high absorption, a high relative sensitivity can be obtained.

(4) Instrumental detection limit

The titanium filter method was used to determine the detection limit because of the influence of the background X-rays. The background X-ray intensity is dominated by the total amount of the thin layered aggregate of analyte elements. Fig (5) shows the approximate detection limit for a very thin layer of sample on a substrate.

$$\text{Detection limit} = \frac{3 \times (\text{Background intensity of plain plate})^{1/2}}{\text{Sensitivity}} \qquad (1)$$

The limits were calculated using Eq. (1). In the low-energy region, higher limit values are obtained because of a low sensitivity. In the high-energy region, the integrated intensity of the primary radiation is not enough to excite the K alpha X-rays of analyte elements.

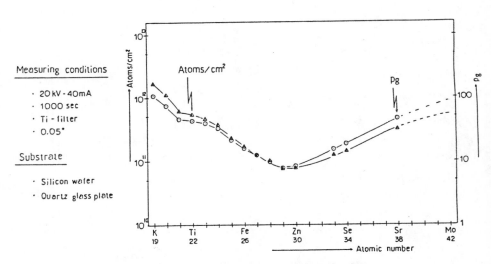

Fig. (5) Instrumental detection limits

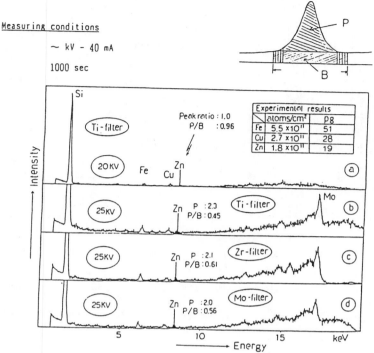

Fig. (6) Influence of filter effect and applied voltage
for X-ray tube

APPLICATIONS

(1) Detection of surface contamination on silicon wafer

It is very important to reduce the surface contamination on a silicon
single crystal wafer in the production process of semiconductor devices.
Fig. (6) shows the detection of surface contamination on a silicon wafer,
which was washed with extra-clean pure water, under various excitation
conditions using a Mo X-ray tube. The intensity of fluorescent X-rays and
peak to background intensity ratios were investigated by changing the applied
voltage and filter materials. For Zn, the peak-to-background intensity
ratio of the (a) case is the largest, but the Zn-K alpha peak intensities of
the (b) to (d) cases are approximately twice that of (a). From these results
we see that for the improvement of the Zn detection limit, the measurement
should be carried out using an applied voltage of 20kV and a titanium
filter. For higher atomic number elements a higher applied voltage becomes
more effective.

(2) Measurements of impurity in water

Fig. (7) shows the X-ray spectrum of NBS standard reference material
1643b (trace elements in water). Impurity elements in the 10 to 100 ppb
concentration range are detected.

The total amount of calcium (35), sodium (8), magnesium (15) and
potassium (3) in μg/ml governs the background intensity and its distribu-

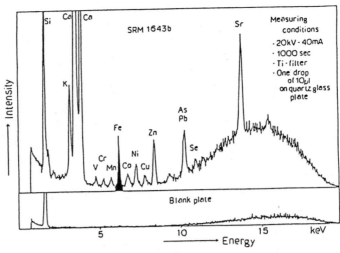

Fig. (7)   Detection of impurity elements in NBS water sample

tion.   In order to show the relationship between a thin sample amount on a
plate and the background intensity and distribution, measurements were made
by diluting the NBS water sample.   Fig. (8) shows the experimental results
between the dilution factor and peak intensity of fluorescent X-rays, and a
good correlation can be seen.

When the incident X-ray beam hits the thin layered material of an
analyzing sample on a plate, some part of incident X-rays is absorbed by the
thin material.   The transmitted X-rays are totally reflected by the support
plate.   Then some of the totally reflected X-rays are absorbed again by the

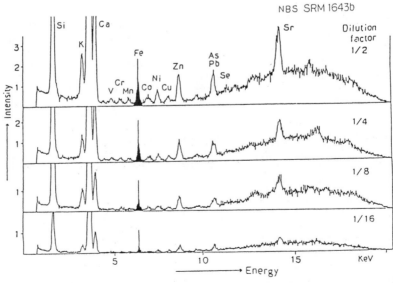

Fig. (8)   Effect of solid sample amounts in NBS water sample

Table (1)   X-ray analysis of NBS water sample

	NBS Certified value	X-ray results	σ [1]
V	45.2±0.4 ng/g [2]	42ng/g	6ng/g
C r	18.6±0.4	30	4
Mn	28 ±2	30	3
F e	99 ±8	114	4
C o	26 ±1	30	3
N i	49 ±3	39	2
C u	21.9±0.4	15	2
Z n	66 ±2	58	2
A s	(49) [3]	63	3
S e	9.7±0.5	10	3
S r	227 ±6	236	8
C a	(35000)	———	———

*1   σ : Calculated standard deviation
     based on X-ray statistical error
*2   ng/g = ppb
*3   ( ) : Not certified

thin material.   The X-rays absorbed by the thin material cause the fluorescent X-rays of the analyte elements and the background X-rays, consisting of
the Thomson and Compton scattered X-rays of the primary radiation to be
emitted as a result of the photoelectronic effect.   Both are dependent on the
total amount of the thin material.

Table (1) gives the numerical results for the impurity analysis of the
NBS water sample.   X-ray analytical results are calculated using the

Table (2)   X-ray analysis errors using NBS water sample

	Simple repeat 10 times			Setting reproducibility 10 times			Preparation reproducibility 5 samples		
	Ave (ng/g)	σ (ng/g)	CV (%)	Ave (ng/g)	σ (ng/g)	CV (%)	Ave (ng/g)	σ (ng/g)	CV (%)
V	44.6	3.8	9	45.5	5.5	12	42.2	6.3	15
C r	32.0	2.8	9	33.0	4.1	12	30.0	4.9	16
Mn	29.3	3.6	12	30.8	2.6	8	30.3	4.1	14
F e	113.3	5.6	5	113.0	4.7	4	114.8	6.4	6
C o	30.7	3.2	10	30.1	3.4	11	30.1	4.8	16
N i	38.5	1.6	4	38.9	1.9	5	39.1	4.2	11
C u	14.0	1.5	11	13.2	1.2	9	14.4	2.2	15
Z n	56.6	2.5	5	58.6	1.3	2	58.3	3.6	6
A s	64.0	2.0	3	63.6	2.7	4	62.7	4.8	8
S e	10.9	2.3	21	11.1	1.5	13	10.6	2.6	21
S r	242.6	6.0	3	226.8	6.3	3	231.7	14.6	6

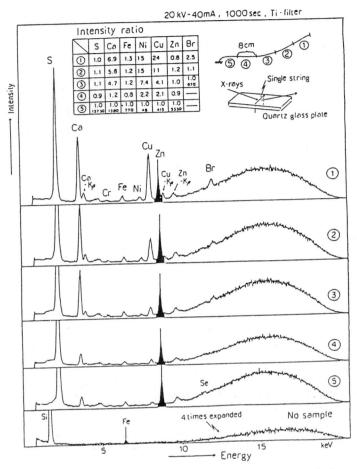

Fig. (9)   Quantitative analysis of a human hair

sensitivity gradient and the intensity of the measured element with background correction. Good agreement was obtained except for chromium. Table
(2) shows the practical errors associated with the X-ray analysis. Simple
statistical error for the X-ray intensity, repeat sample setting (replacement
positioning) and the sample preparation error are listed. The sample setting
error depends on the geometry of the grazing angle of the incident beam and
the sample support level. The sample preparation error depends on the
surface condition of the sample and the support plates. All three sources of
error are essentially identical.

(3) Quantitative analysis of a human hair

Fig. (9) shows the results of a quantitative analysis of a human hair.
A single hair of about 40 cm length was divided into five pieces. A piece of
the hair was positioned on the flat support plate. Because the diameter of a
hair is about 100 microns, the analyzed part is a small area of the surface.
The change in the concentration of calcium, nickel, copper and bromine can be
seen. The differences in the background X-ray intensity between sample and

no-sample plates are due to the different heights of the sample support plates.

CONCLUSION

The instrumental development of a total reflection X-ray fluorescent spectrometer and the studies of its applications have been made. The conclusions are as follows:

1) The elimination of background X-rays in an X-ray spectrum could be achieved successively using the total-reflection method. However, there are many factors which govern the background X-ray intensity.

2) The detection limit is approximately 10^{11} atoms/cm^2 or 10 pg/g. This is superior to that of regular WDXRF and EDXRF which have a range of 1-10 ppm.

3) The spectrometer should have a vacuum path for the primary radiation to reduce the intensity of air scattered background X-rays.

4) When a measurement of the lowest concentration level is made, the primary-beam filter method is recommended due to the reduction of background X-rays. At higher analyte element concentrations, the filter is not necessary.

5) A low applied voltage should be used in order to reduce the scattered X-rays of the primary radiation.

REFERENCES

1) Yoneda and Horiuchi, RSI 42 (1971) 1069.

2) Iida and Gohshi, Jpn J. Appl. Phys. 23 (1984) 1543.

3) Aiginger and Wobranschek, Adv. X-Ray Anal. 28 (1985) 7.

4) Michaelis, Knoth, Prange and Schwenke, Adv. X-Ray Anal. 28 (1985) 75.

5) Stosseland Pange, Anal. Chem. 57 (1985) 2880.

MICRO X-RAY FLUORESCENCE ANALYSIS WITH SYNCHROTRON RADIATION

Shinjiro Hayakawa, Atsuo Iida,* Sadao Aoki,**
and Yohichi Gohshi

Department of Industrial Chemistry, Faculty of Engineering
University of Tokyo, Hongo, Bunkyo-ku, Tokyo 113, Japan
*Photon Factory, National Laboratory for High Energy
Physics, Tsukuba-shi, Ibaraki 305, Japan
**Institute of Applied Physics, University of Tsukuba
Tsukuba-shi, Ibaraki 305, Japan

ABSTRACT

A synchrotron radiation X-ray micro analyzer(SRXMA) was developed
at Photon Factory in Japan. The present SRXMA combines a double
crystal monochromator and mirror optics and either a white or a
monochromatic microbeam can be used. Micro X-ray fluorescence analysis
was carried out, and a minimum detection limit of 1 ppm for Mn was
obtained for 100 sec measurement with the white beam. With
monochromatic beam excitation, micro X-ray spectroscopies are now
feasible.

The obtained beam size was 1.6 μm * 34 μm. The beam was blurred
in one direction by the scattered X-rays caused by the surface
irregularities of the focusing mirror. Improvements in the mirror
quality will ensure a beam spot of just a few microns with sufficient intensity.

INTRODUCTION

Micro X-ray fluorescence analysis(XRF) is nondestructive trace
element analysis with high spatial resolution, and some X-ray
microprobes have been developed with synchrotron radiation(SR)[1,2].
Generally various X-ray focusing optics have low angular acceptance of
divergent X-rays and are unable to achieve a higher photon flux
density with conventional sources. SR's low divergence and extreme
brightness are advantageous to micro XRF. Another important feature of
SR is its continuum energy distribution. The characteristics of SRXRF
with monochromatic beam have already been experimentally examined for
bulk analysis purpose[3]. The selective excitation of the element of
interest demonstrated the advantages of the energy tunability of SR.

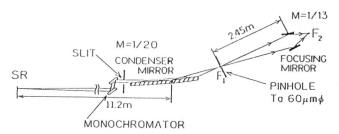

Fig.1   A schematic layout of the SRXMA optical system.

The combination of mirror optics and a monochromator makes possible an energy tunable synchrotron radiation X-ray micro analyzer(SRXMA), which will enable not only micro XRF but also various kinds of micro X-ray related spectroscopies to be carried out.

SRXMA SYSTEM

Fig. 1 shows the schematic representation of a SRXMA at Photon Factory. The design parameters and the initial experimental results were reported in previous papers[1,4,5]. Two Wolter type 1 mirrors (condenser mirror and focusing mirror) are used, and X-rays up to 10 keV are reflected and focused. A double crystal monochromator was installed upstream before the condenser mirror at BL-4A. Either a white beam or a monochromatic beam can be used as a light source. In the case of total reflection mirror optics, the focused beam position is independent from the energy of the incident beam. Furthermore the use of the pinhole at F1 removes the influence of the small displacement of the source caused by storage ring instability or misalignment of the double crystal monochromator.

Samples were placed at F2 and scanned with the stepping motors in x,y,z directions. An energy dispersive X-ray fluorescence(EDXRF) system with a Si(Li) detector was used to analyze the sample. The probe beam intensity was monitored by an ionization chamber before the sample.

EXCITATION MODE

The minimum detection limits(MDL) obtained with the white and the monochromatic beams for 100 sec measurement are summarized in Table 1. A Zn, Mn and Ca 100 ppm adsorbed chelate resign was measured as a reference sample. The sample was prepared to be 7 mg/cm^2. The intensity loss at the pinhole(60 μm φ) was about 30 %, but no pinhole was used at F1 to exclude the influence of the alignment. The beam size was estimated to be about 20 μm in diameter from the beam profile obtained with the edge scan described later.

The photon flux of the white beam was about two hundred times higher than that of the monochromatic beam. However, the differences

Table 1. Comparison of MDL for different excitation modes using a SRXMA. Collection time was 100 sec, and no source pinhole was used at F1.

Excitation mode	MDL in relative conc.		Irradiation area	MDL in abs. amount Mn
	Mn	Ca		
White	1 ppm	3 ppm	300 μm^2	0.02 pg
Monochromatic (7.5 keV)	8 ppm	29 ppm	300 μm^2	0.16 pg

of the MDL for Mn and Ca are only 8 times and about 10 times, respectively, between the white and the monochromatic beam excitation. The background caused by the scattered X-rays was reduced with the monochromatic beam and the energy of the incident beam was optimized for Mn.

The double crystal monochromator can be replaced with a layered synthetic microstructure(LSM) monochromator. With the use of LSM, higher S/B spectra can be obtained without any significant signal loss.

The other advantage of monochromatic beam excitation is the realization of X-ray spectroscopies in the micro region. Chemical state analysis using the shift of the absorption edge was studied with a conventional SRXRF system[6]. The position of the absorption edge depends on the chemical state and the distribution of each chemical state in a mixed system could be obtained separately[7]. To evaluate the intensity performance of the present system, a sample of mixed chemical states of iron(iron ore sinter[7]) was measured with a SRXMA. A count rate of about 800 cps was obtained above the absorption edge. With a collection time of 5 sec/point, the difference of the absorption edge position could be observed. Chemical state imaging with a SRXMA will be performed.

BEAM SIZE

To characterize the focusing mirror exclusively, the source pinhole(16 μm ϕ) at F1 was illuminated directly without using the condenser mirror. The focusing mirror has axial symmetrical shape and the beam spot should be isotropic when all of the mirror surface is used. However the mirror surface was not fully used with a pinhole of 16 μm because of the small divergence of X-rays at the mirror head.

To estimate the beam size, a knife edge was scanned horizontally and vertically across the beam spot, and the transmission was measured. The beam profile was obtained from the numerical derivative of the edge scan image. The beam size was 1.6 μm(V) * 34 μm (H) when

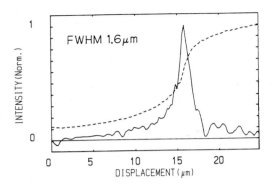

Fig. 2   The transmission(dotted line) obtained with the edge scan and
its derivative(solid line) in vertical direction, which represents the
beam profile.

the reflection plane was in a horizontal direction. In another case, a
beam size of 36 μm(V) * 8.6 μm(H) was obtained.   The beam size in the
reflection plane was blurred by the scattered X-rays caused by the
surface irregularities of the focusing mirror, but the beam size out
of the reflection plane was not affected by the scattered X-rays.
Fig. 2 shows the vertical beam profile obtained with the edge scan.
The beam size of 1.6 μm was almost the same as the size(1.2 μmφ)
calculated from the pinhole size(16 μm φ) and the magnification of the
focusing mirror(M=1/13). These results show both the high figure
accuracy and the inadequate surface smoothness of the focusing mirror.
Improvements in the surface smoothness of the mirror will permit a
beam size of a few microns in diameter.

ELEMENT MAPPINGS AND POINT ANALYSIS

A rock sample from W/Cu mineral deposits was analyzed with white
beam excitation. The sample contained several minerals such as
pyrrhotite($Fe_{1-x}S$), chalcopyrite($CuFeS_2$), scheelite($CaWO_4$), silicate
and sphalerite(ZnS). An area of 750 μm * 750 μm in the optical
micrograph (Fig. 3a taken with polarized light) was analyzed under
atmospheric conditions. The Fe, Cu, Ca and S images were obtained
simultaneously with a collection time of 4 sec/pixel.   The Fe
image(Fig. 3b) and the S image(Fig. 3d) correspond to the distribution
of the pyrrhotite and chalcopyrite. The Cu and the Ca images(Figs. 3
c,d) correspond to the chalcopyrite and the scheelite, respectively.
Silicate is easily identified in the area where the signals of Fe, Cu,
Ca and S are all weak.

The advantage of the scanning microprobe is its ability to
perform the trace element analysis in a micro spot on a sample. The
precise identification of the beam irradiated position is important in
this case. However, the resolution of the sample monitor system was
inferior compared with the beam size. Therefore point analysis was
performed after XRF imaging. The sample was translated to the position

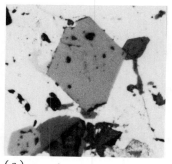

(a)

Fig. 3. Optical micrograph and XRF images of a rock sample which contains pyrrhotite, chalcopyrite, scheelite, silicate and sphalerite. Area of 750 μm * 750 μm was analyzed with 4 sec/pixel.
a) optical micrograph b) Fe Kα c) Cu Kα d) Ca Kα e) S Kα images.

(b)

(c)

(d)

(e)

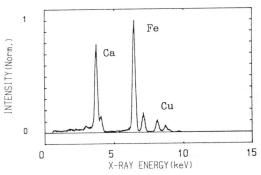

Fig. 4   An XRF spectrum of scheelite using SRXMA.

of interest which corresponded to the pixel of the XRF images. Fig. 4 shows a XRF spectrum of scheelite. A strong peak of Ca could be observed with a collection time of 200 sec. Easily understood from the 2-D images of Fe and Cu, the signals of Fe and Cu were weak at the position of the scheelite. However, their signals still remained in this spectrum. The contribution of background X-rays to point analysis should be evaluated and minimized to optimize the SRXMA.

To realize higher spatial resolution, improvements in the beam profile, more accurate determination of the focal plane and precise identification of the beam irradiation area are necessary.

ACKNOWLEDGMENT

We wish to thank the PF staff for helping us with experiments, Dr. Sakurai for his useful suggestions about the chemical state analysis and Dr. Sato for providing us with the rock sample. Our thanks also go to Nikon Co. for the focusing mirror and Nippon Steel Co. Ltd. for the iron ore sinter. This work was supported by the Grant-in-Aid for Scientific Research No.61234004 from the Ministry of Education Science and Culture of Japan.

REFERENCES

1. Y. Gohshi, S. Aoki, A. Iida, S. Hayakawa, H. Yamaji and K. Sakurai, A Scanning X-ray Microprobe with Synchrotron Radiation, Jpn. J. Appl. Phys. 26:L1260 (1987).
2. A. C. Thompson, J. H. Underwood, Y. Wu, R. D. Giauque, K. W. Jones and M. L. Rivers, Element Measurements with An X-ray Microprobe of Biological and Geological Samples with Femtogram Sensitivity, Nucl. Instr. and Methods A266:318 (1988).
3. A. Iida and Y. Gohshi, Energy Dispersive X-ray Fluorescence Analysis Using Synchrotron Radiation, in:"Advances In X-ray Analysis", Charles S. Barrett et al., ed. Plenum Press, NY, Vol. 29, (1986).

4. S. Aoki, Y. Gohshi and A. Iida, A 10 keV X-ray Microprobe with Grazing Incidence Mirrors, in: "X-ray Microscopy", P. C. Cheng and G. J. Jan ed. Springer-Verlag Berlin Heidelberg (1987).

5. Y. Gohshi, S. Aoki, A. Iida, S. Hayakawa, H. Yamaji and K. Sakurai, A Scanning X-ray Microprobe with Synchrotron Radiation, in: "Advances In X-ray Analysis", P. K. Predecki et al., ed. Plenum Press, NY, Vol. 31, (1988).

6. K. Sakurai, A. Iida and Y. Gohshi, Chemical State Analysis by X-ray Fluorescence Using Shifts of Iron K Absorption Edge, Anal. Sci. 4: 37(1988).

7. K. Sakurai, A. Iida, M. Takahashi and Y. Gohshi, Chemical State Mapping by X-Ray Fluorescence Using Absorption Edge Shifts, Jpn. J. Appl. Phys. in press.

X-RAY MICROPROBE STUDIES USING MULTILAYER FOCUSSING OPTICS

A.C. Thompson, J.H. Underwood, Y. Wu and R.D. Giauque

Lawrence Berkeley Laboratory
Berkeley, CA 94720

M.L. Rivers

University of Chicago
Chicago, IL 60637

R. Futernick

Fine Arts Museum
San Francisco, CA 94121

INTRODUCTION

The availability of intense x-rays from synchrotron radiation sources permits the elemental analysis of samples in new ways. An x-ray microprobe using these sources allows the analysis of much smaller samples with greatly improved elemental sensitivity.[1] In addition to the higher x-ray intensity obtained at synchrotron sources, the development of high efficiency x-ray reflectors using multilayer coated optical mirrors permits the achievement of spot sizes of less than 10 μm x 10 μm with enough x-ray intensity to simultaneously measure femtogram quantities of many elements in less than one minute. Since samples to be studied in an x-ray microprobe do not have to be placed in a vacuum, almost any sample can be conveniently analyzed. With an x-ray microprobe it is possible to obtain elemental distributions of elements in one, two or even three dimensions.

In an x-ray microprobe a beam of x-rays is either collimated or focussed to a fine spot which is then scanned over the specimen. The characteristic fluorescent x-rays excited in the specimen are then detected using an energy or wavelength dispersive detector. In our system we use a synchrotron radiation x-ray beam as the source of x-rays, a pair of multilayer mirrors to focus the x-rays, and a Si(Li) detector to measure the fluorescent x-rays. It allows us to simultaneously measure the concentration of elements from K to Zn with a sensitivity of better than 50 fg in 60 sec.

FOCUSSING X-RAYS WITH MULTILAYER MIRRORS

Multilayer mirrors make excellent x-ray optical components. The mirrors are "super-polished" to have a low scatter finish that has been measured by optical interferometric techniques to have a microroughness of around 2 Å RMS. A multilayer coating is placed on top of this substrate by depositing alternate layers of two elemental materials of very different atomic number (i.e. carbon and tungsten). These layers produce a periodic lattice which diffracts x-rays.

These mirrors have two important properties. X-rays can be focussed with them since the substrate can be properly figured. The curvature of the mirror substrate determines the focussing qualities of the mirror. The reflectivity of a multilayer reaches a maximum at an angle of incidence given by the Bragg relation:

$$n\lambda = 2(d_A + d_B)sin\theta = 2dsin\theta$$

where λ is the wavelength of the photons, d is the multilayer period (the sum of the thicknesses of the two component layers A and B) and θ is the glancing angle of incidence. The layers are deposited using a dual source sputtering system and are very thin. For the proper operation of the focusing system, the d spacings of the multilayer must be adjusted to reflect x-rays of given energy at selected angles. In this experiment the d spacings for the two mirrors were 58 Å and 87 Å respectively. The bandpass of our mirrors were measured to be 10% at 10 keV.

To achieve focussing in two directions a Kirkpatrick-Baez geometry was used to demagnify the x-ray source by a factor of over 200 in both directions. This geometry has several advantages for microprobe applications. Since only on-axis aberrations are important in a microprobe rather than aberrations over a large field of view, this geometry allows optics with a high demagnification with very little aberration. The multilayer coating on the mirrors allows them to be used at larger angles than the critical angle of total external reflection. This therefore allows improved solid angle for a given length of mirror. Because the multilayers limit the energy bandpass to 10%, the elemental sensitivity is better than a pinhole or glancing incidence mirror system since the background under the fluorescent x-ray peaks is much lower.

DESCRIPTION OF EXPERIMENT

Details of the initial operation of the microprobe have been presented elsewhere.[2-4] Figure 1 shows how one of the spherical mirrors is mounted with a flexure hinge and a sine bar arm to allow its angle to be varied. The mirror assembly has recently been redesigned and Figure 2 shows the new system. Six stepping motors are used to focus the mirrors. Two linear stages(not shown) permit the height and horizontal position of the pair of mirrors to be set. Another two linear stages allow the distance of each mirror from the sample to be adjusted. The last two stepping motors are used to set the angle of incidence of each mirror.

The results reported here were acquired during an experiment at the LBL/EXXON beamline of the Stanford Synchrotron Radiation Laboratory (SSRL) in November,1987. This beamline is powered by a 54 pole wiggler magnet that had a magnet field of 0.8 Tesla. During this run the electron storage ring operated at 3.0 GeV at a current of around 30 mA. On this beamline it is not possible to obtain a white radiation beam as we have done at the Brookhaven National Light Source (NSLS). Instead it was necessary to use the installed beamline monochromator which uses a pair a Si<111> crystals to produce a 10 keV beam with bandpass of less than 4 eV. This significantly reduced the x-ray flux that was produced at the sample. The beam flux was measured using a set of NBS thin glass Standard Reference Materials. A flux of 5 x 10^8 10 keV photons/sec was measured in a spot size 20 μm x 10 μm. Although this intensity was not as high as that obtained at NSLS (3 x 10^9 photons/sec at 10 keV), it was still possible to analyze a variety of interesting samples.

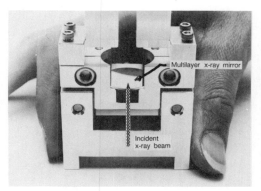

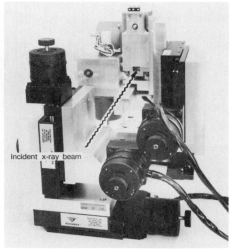

Figure 1. Picture of a multi-layer coated mirror with the mirror holding block.

Figure 2. Overview of the microprobe mirror assembly. Four motors are used to set the incidence angles and positions along the beam direction for the two mirrors to achieve proper focusing.

The beam spot size was measured by scanning a cross hair made from 8 μm diameter gold coated tungsten wire and measuring the transmitted radiation using an ion chamber. This allowed the position and angle of both mirrors to be optimized rapidly for best focus.

RESULTS

A variety of paper and ink samples obtained from the conservation department of the Fine Arts Museum in San Francisco were measured. A special mounting frame was built using plastic frames and small permanent magnets to hold these paper samples.

A series of stains or "fox" spots on different papers were scanned. These fox spots are a problem in art conservation since they grow as the document ages and can seriously degrade the appearance of the document. There are two types of fox spots. Some look like brown or grey circular stains with a diameter of about 1 mm. Others are more irregular and diffused and look like a mold growth. Fox spots of the first type on several different papers were scanned in the microprobe. In all spots scanned it was found that near the center of the stain there was a small area in which the x-ray microprobe measured relatively large quantities of iron. Figure 3 shows a one dimensional scan across one of these spots. A scan in the other direction was similar. From the figure it can be seen that the spot is less than 50 μm in diameter and that the highest Fe elemental density is 130 μg/cm^2. These spots, therefore, probably come from small iron particles that were imbedded in the paper when it was made. With time the iron has oxidized in the presence of moisture in the air. The oxide slowly diffuses through the paper result the coloring of the surroundings.

On the other hand, when fox spots of the other more diffuse kind were scanned, no increase in any of the elements from K to Zn was measured. These spots are therefore more likely of an organic rather than metallic origin.

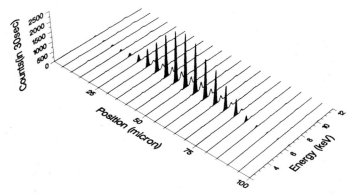

Figure 3. A scan across the "fox" spot shown very localized and quite concentrated elemental Fe distribution. The highest peak correspond to an elemental density of 130 $\mu g/cm^2$.

The second problem that was examined was whether the x-ray microprobe can measure the components of ink and separate them from the paper background. Small spot size is important for this experiment since the concentrations of ink relative to the paper is small. To study application of the microprobe to this type of problem, a scan across a line in a signature of an old document was taken. Figure 4 shows the energy spectra taken at each point across the line. There is clearly iron in the ink and it has an interesting distribution. It shows that the Fe concentration increases when the ink line is approached and that it has two maxima on either edge of the line. This is how the ink might be expected to concentrate as it dried.

CONCLUSIONS

The synchrotron radiation based x-ray microprobe using multilayer focussing optics provides a new analytical tool that can be used to study a wide variety of samples. The Kirkpatrick-Baez mirror geometry allows an x-ray beam with an adequate intensity to measure femtogram amounts of elements from K to Zn in less than 60 sec.

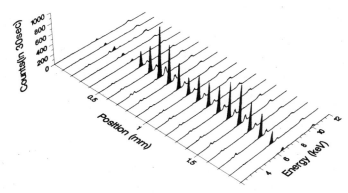

Figure 4. Scan across a dried ink line. It is interesting to notice the uneven distribution of Fe across the line. Upon drying, the liquid ink tend to get thicker at the outside edge and therefore result in a higher Fe concentration. The highest peak correspond to an Fe density of 60 $\mu g/cm^2$.

As an example of the application of this probe, fox spots and ink lines on paper samples were examined. Some of the fox spots were determined to have a small metallic particle near the center of the stain with a diameter of less than 50 μm. In other more diffused fox spots, no increase in elements from K to Zn was measured. A scan across an ink line showed that the microprobe has excellent sensitivity to inks containing elements from K to Zn.

There are many possible applications of this probe. One example would be the study of the "Vinland" map. There is a controversy about the authenticity of this document.[5,6] If titanium dioxide particles are present in the ink it would tend to indicate that the document is not very old. With the x-ray microprobe it is possible to study different points along the ink lines to see if titanium dioxide particles are present and if so what their sizes might be. Since the x-ray flux is not that high, there is no radiation damage to precious art documents that might be scanned with this probe.

It is also possible to reduce the spot size of this microprobe using improved multilayer mirrors. If spot sizes close to 1 μm x 1 μm can be achieved, there are a large number of biological studies that could be undertaken. Since the sample does not have to be in a vacuum, almost any sample can be scanned. The use of multilayer mirrors with laboratory x-ray sources is also being studied.

ACKNOWLEDGEMENTS

The authors wish to thank P. Batson and R. Delano for their help with the mechanical parts of this experiment. This work was supported by the Director, Office of Energy Research, Office of Basic Energy Sciences, Materials Sciences Division, of the U.S. Department of Energy under Contract DE-AC03-76SF00098. The experiment was carried out at the SSRL which is supported by the U.S. Department of Energy, Office of Basic Energy Sciences.

REFERENCES

1. C. J. Sparks Jr., X-ray Fluorescence Microprobe for Chemical Analysis, in: "Synchrotron Radiation Research", H. Winick and S. Doniach, ed., Plenum Press, New York (1980).
2. A.C. Thompson, J.H. Underwood, Y. Wu, R.D. Giauque, K.W. Jones and M.L. Rivers, Elemental Measurements With an X-Ray Microprobe of Biological and Geological Samples With Femtogram Sensitivity, Nucl. Instr. and Meth. A226:318 (1987).
3. J.H. Underwood, A.C. Thompson, Y. Wu and R.D. Giauque, X-Ray Microprobe Using Multilayer Mirrors, Nucl. Instr. and Meth. A226:296 (1987).
4. R.D. Giauque, A.C. Thompson, J.H. Underwood, Y. Wu, K.W. Jones and M.L. Rivers, Measurement of Femtogram Quantities of Trace Elements Using an X-Ray Microprobe, Anal. Chem., 60:885 (1988).
5. T. A. Cahill et al., The Vinland Map, Revisited: New Compositional Evidence on Its Inks and Parchment, Anal. Chem., 59:829 (1987).
6. W. C. McCrone, The Vinland Map, Anal. Chem. 60:1009 (1988).

RESOLUTION ENHANCEMENT FOR Cu Kα EMISSION OF Y-Ba-Cu-O COMPOUNDS

N. Saitoh

National Research Institute of Police Science, 6 Sanbancho
Chiyoda-ku, Tokyo 102, Japan

Y. Higashi, M. Minami, S. Fukushima, Y. Gohshi

Department of Industrial Chemistry, Faculty of Engineering
University of Tokyo, 7-3-1 Hongo, Bunkyo-ku, Tokyo 113, Japan

S. Kohiki

Matsushita Technoresearch Inc., Moriguchi, Osaka 570, Japan

T. Wada
Central Research Laboratories, Matsushita Electric Industries
Moriguchi, Osaka 570, Japan

INTRODUCTION

Since the discovery of the Y-Ba-Cu-O superconductors, their physical
properties have been investigated by various methods. The chemical state of
Cu in Y-Ba-Cu-O compounds is one of the greatest issues because the mechanism
of superconductivity in Y-Ba-Cu-O is not understood theoretically. We are
analyzing X-ray fluorescence spectra of Cu compounds including
superconductors, intending to analyze the chemical state of Cu in Y-Ba-Cu-O.
As for other 3d transition elements, structures due to unpaired electrons
appear clearly on the lower energy side of the Kα$_1$ line of the element.
However there are little differences observed among Cu Kα spectra of Cu
compounds even if they are measured by a high-resolution two-crystal
spectrometer (see Fig. 1). Although Cu is a member of 3d transition
elements, its Kα spectrum shows somewhat different behavior compared with
other 3d transition elements. This point is one subject we are interested
in. In order to clarify spectral details, we have studied the chemical state
of the Cu atom in compounds by processing Kα spectra numerically. Generally
speaking, measured spectra are deformed by various broadenings such as an
instrumental window function and a natural width. Therefore detailed
structures in spectral profiles are not observed clearly. A line width of
high-resolution X-ray fluorescence spectra is dominated by a natural width
which is due to the uncertainty principle. To clarify fine structures, it is
necessary to remove the natural width numerically. For such pruposes, there
are many methods proposed since van Cittert. As is well known, however, this
problem, called deconvolution, is ill-conditioned and it has no unique
solution mathematically. There are many inappropriate solutions which
include sidelobes or ringings. To avoid such solutions, nonlinear methods

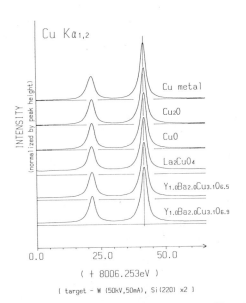

Fig. 1.   Cu Kα spectra for various Cu Compounds. These spectra
were measured with a high-resolution two-crystal spectrometer.

have been developed[1,2]. However it is not appropriate to use such nonlinear
results for quantitative analysis of spectral profiles. Therefore we
examined the behavior of several iterative deconvolution methods by using
simulations and then we applied the most appropriate method of these for
analyzing X-ray fluorescence spectra of Y-Ba-Cu-O superconductors.

EXAMINING DECONVOLUTION METHODS BY SIMULATIONS

In general, the relation between a true spectrum (T(x) and a measured
spectrum M(x) is expressed as a convolution integral as follows:

$$M(x)=\int_{-\infty}^{\infty} T(t)S(x-t)dt,$$

where S(x) represents a broadening function. Deconvolution aims to solve
this integral equation and obtain the true spectrum T(x). Van Cittert
proposed the following iterative method for this problem:

$$T_n(x)=T_{n-1}(x)-(T_{n-1}(x)*S(x)-M(x)),$$

where the symbol * means a convolution integral. By using this method, the
intensity of the noise component is amplified n+1 times after n iterations[3]
and ringing structures occur around a peak position. The latter is a serious
problem.

Jansson introduced a nonlinear acceleration coefficient for the
iterative method and indicated that such a nonlinear coefficient can suppress
ringing structures. Jansson's nonlinear iterative method is as follows:

$$T_n(x)=T_{n-1}(x)-g(T_{n-1}(x))(T_{n-1}(x)*S(x)-M(x)),$$

where $g(T_{n-1}(x))$ means a nonlinear acceleration coefficient and it is a function of an intermediate result $T_{n-1}(x)$. The van Cittert's method is equivalent to the case when g is constant and equals to 1 in the Jansson's method.

(1) Ringing Structures

In Fig. 2, deconvolution results obtained by using these two methods are shown. The original spectrum is a Lorentz function with its full width at half maximum (FWHM) being 20 channels and the peak intensity being 15000 counts. The total number of data channels is 512. When an original spectrum included noise, it was smoothed by using 4-th order spline functions[4]. The FWHM of the broadening function is 18 channels, that is, 90% of the original FWHM.

As is seen in Fig. 2, ringing structures are large and noise effects are conspicuous with van Cittert's method. But with Jansson's method, ringing structures are not observed and also noise effects are rather small. In Fig. 3, results for spectra with two peaks are shown. Larger differences

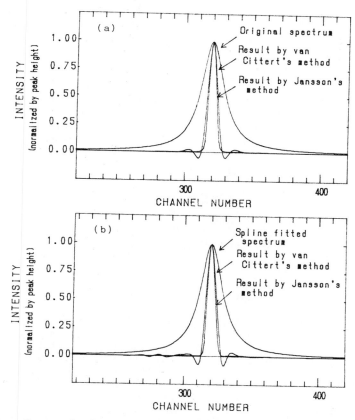

Fig. 2.   Deconvolved results obtained by the van Cittert's method and the Jansson's method for a synthetic spectrum which consists of one Lorentzian peak. (a) Noise-free case. (b) Noise-inclusive case. Noise-inclusive spectra are smoothed by using spline-functions fitting.

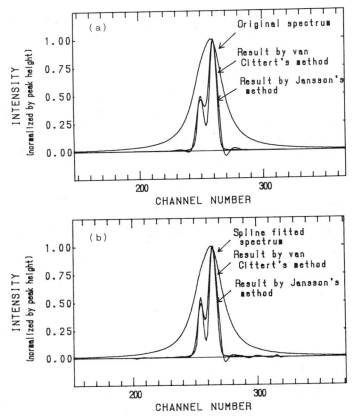

Fig. 3.   Deconvolved results for a synthetic spectrum which
includes two Lorentzian peaks. (a) Noise-free case.
(b) Noise-inclusive case.

between results of the two methods are observed. While by van Cittert's
method, there are ringing structures and two peaks are not resolved
sufficiently, Jansson's method gives highly resolved spectra almost without
generating ringing structures.

(2) Nonlinearity Problem and Convergence

Since the Jansson's method imposes a kind of non-negativity constraint,
there exists a nonlinearity problem. For example, when there are two or more
peaks, the relative intensity of peaks is not restored exactly. In Fig. 4,
peak relative intensities and the residual sum of squares (RSS) between
$T_n(x) * S(x)$ and $M(x)$ are plotted against iteration times. In this figure, the
main peak intensity of $T_n(x) * S(x)$ with respect to that of $M(x)$ is also
shown. The relative intensity of the subpeak is originally 0.5 with respect
to the main peak intensity. After deconvolution, it becomes almost equal to
0.5 at about 50 iterations and becomes larger as the iteration number
increases. Thus on using Jansson's method, it is important to determine the
iteration time. The main peak intensity after convolution reaches about 1.0
at about 30 iterations. But at that step, RSS has not converged
sufficiently. RSS converges at about 50 iterations. Main peak intensity
becomes constant at 1.0 at that step. The relative intensity of the subpeak

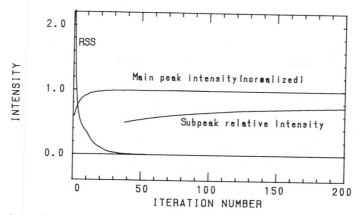

Fig. 4.   Variation of RSS, main peak intensity and subpeak
          relative intensity for a synthetic spectrum in Fig. 3.

almost reaches 0.5 at about 50 iterations. Calculating more iterations only
makes the relative intensity of the subpeak larger. In this work, we
performed 50 iterations for Jansson's method.

(3) Effects of Background

     When no discernible background exists, subpeaks are resolved clearly
even for 10 channel separation. But when the background becomes conspicuous,
many artificial peaks may be produced if there is no background subtraction.
When there exists a significant background, it must be subtracted from
spectra. After that, regarding the relative intensity of narrow peaks,
results similar to those obtained without backgrounds are obtained. But wide
peaks, which are regarded as backgrounds, cannot be restored exactly. In
Fig. 5, a result is shown for a case with three peaks, one of which is wide
and can be regarded as a background. The original intensity of the
background in Fig. 5 is half of the main peak intensity. But after Jansson's

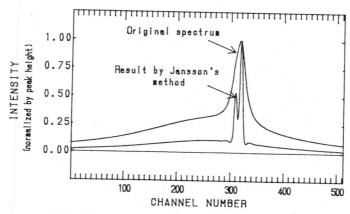

Fig. 5.   Deconvolved results for a synthetic spectrum with
          3 Lorentzian peaks, one of which is wide and can be
          regarded as a background.

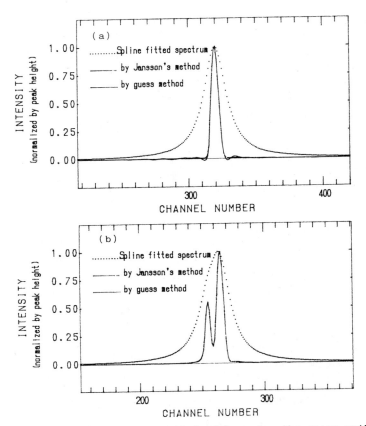

Fig. 6.   Deconvolved results obtained by using the guess method.
           (a) Spectra with only one Lorentzian peak. (b) Spectra
           including two Lorentzian peaks.

deconvolution, its relative intensity is about 0.1. However the relative
intensity of the narrow peak, which is also 0.5 originally, is restored
sufficiently. Therefore as for the peak intensity, Jansson's method does not
give exact results for wide and large peaks (backgrounds). Thus because of
nonlinearity, Jansson's method is not appropriate for analyzing spectral
profiles. In order to overcome this problem, we examined the pseudo
deconvolution method with guess spectra, which is first introduced by
Gohshi[3]. Gohshi's method is introduced at first to reduce ringing
structures. Its principle is as follows: in general, in calculating
successive correction terms by means of van Cittert's method, a measured
spectrum is used as an initial spectrum. That is, the large difference
between a measured spectrum and a true spectrum appears as ringings in a
deconvoluted spectrum. Therefore ringing structures will be reduced by using
a spectrum closer to a true profile as an initial spectrum. In our case, we
use a resultant spectrum obtained by means of Jansson's method as an initial
spectrum. Simulation results obtained with the guess method are shown in
Fig. 6. This figure shows that the guess method with the Jansson result
provides results similar to Jansson's method. By the guess method, linear
results are obtained in principle and at the same time ringing structures are
very small, as are those for Jansson's method. As a conclusion, we adopted
the guess method using a Jansson's result for strange peaks as an initial
spectrum.

RESOLUTION ENHANCEMENT OF Cu Kα SPECTRA

In Fig. 7, deconvolved results for Cu Kα spectra of seven Cu compounds
are shown. These seven compounds are shown in Table 1. There are 2 superconductors included. As the natural width is more than 2.0eV for the Cu Kα
line[5], a smearing width is set to be 2.1eV. Spectra were measured 3 times
for each compound and after deconvolution these 3 spectra were averaged to be
displayed in the figure. In order to evaluate the effect of noise, standard
deviation is also shown in Fig. 7. Table 1 gives the Kα$_1$ line observed
positions; there seems to be no clear peak shift with respect to the Cu Kα$_1$
line of Cu metal. For all compounds in Fig. 7, there is a structure in the
lower energy region of the Kα$_1$ line, profiles that differ somewhat from each
other. In Fig. 8, Kα$_1$ spectra for CuO, Cu$_2$O and Y$_{1.0}$Ba$_{2.0}$Cu$_{3.0}$O$_{6.6}$ are
shown. Comparing the CuO spectum with that of Cu$_2$O, it is seen that the
profiles in the low energy region of Kα$_1$ line are different. The peak
position of the structure for CuO seems to shift to lower energy than that of
Cu$_2$O. As for Y$_{1.0}$Ba$_{2.0}$Cu$_{3.0}$O$_{6.6}$, the profile of the structure seems to take
a position between the profiles of CuO and Cu$_2$O. But it is not clear which
profile is similar to that of Y-Ba-Cu-O. Since there is no spectra measured
for Cu 3+ compounds, it is not appropriate to discuss the valency state of Cu
atom in Y-Ba-Cu-O compounds. The only thing we can point out at the present
stage is that there is no clear and positive evidence that the chemical state
of Cu atoms in Y-Ba-Cu-O superconductors are different from 2+ or 1+.

SUMMARY

We have measured Cu Kα spectra of Cu compounds including Y-Ba-Cu-O high
temperature superconductors. There are very small differences observed in
the profiles, peak shifts and FWHM of measured spectra. Therefore we
performed deconvolution processing for each measured spectrum and found that
the profiles of the structure in the low energy region of the Cu Kα$_1$ line
differs and depends on the chemical state of the Cu atom. As for Y-Ba-Cu-O
superconductors, no clear similarity is observed between the profile of the
structure of Y-Ba-Cu-O and that of CuO or Cu$_2$O. Therefore it is not easy to

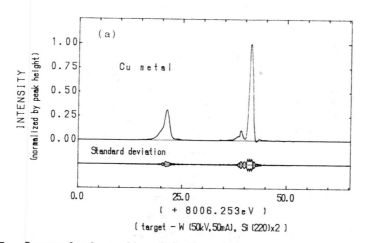

Fig. 7.   Deconvolved results of Cu Kα spectra for various Cu
          compounds. Each spectrum is an average of 3 measurements;
          standard deviation of each data point is depicted.

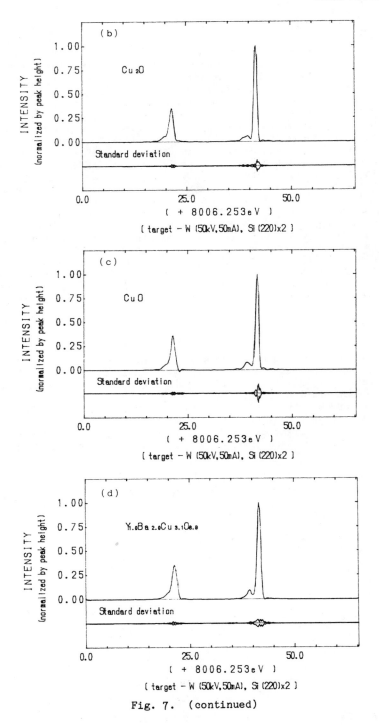

Fig. 7.   (continued)

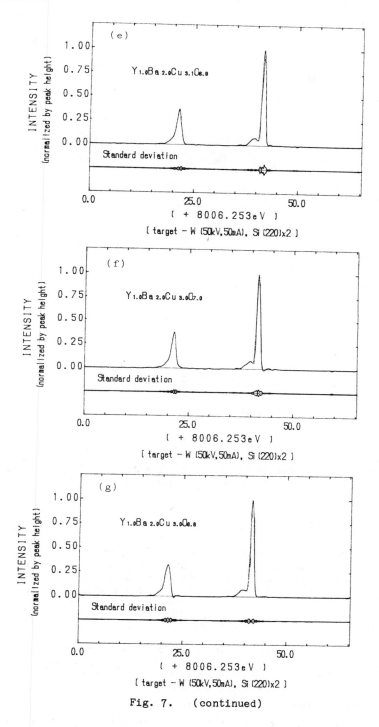

Fig. 7.   (continued)

Table 1.   Peak position and FWHM of Cu Kα$_1$ line for various compounds

	peak position	FWHM
Cu metal	8047.86eV	1.05eV
Cu$_2$O	8047.99eV	0.95eV
CuO	8048.01eV	0.94eV
Y$_{1.0}$Ba$_{2.0}$Cu$_{3.1}$O$_{6.9}$ (not single phase)	8047.94eV	1.00eV
Y$_{1.0}$Ba$_{2.0}$Cu$_{3.1}$O$_{6.9}$ (not single phase)	8047.96eV	1.04eV
Y$_{1.0}$Ba$_{2.0}$Cu$_{3.0}$O$_{7.0}$ (superconductor)	8047.98eV	1.02eV
Y$_{1.0}$Ba$_{2.0}$Cu$_{3.0}$O$_{6.6}$ (superconductor)	8047.99eV	0.99eV

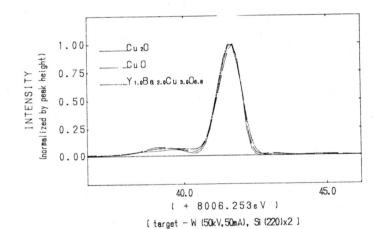

Fig. 8.   Comparison of Cu Kα$_1$ spectra for Cu$_2$O, CuO and
          Y$_{1.0}$Ba$_{2.0}$Cu$_{3.0}$O$_{6.6}$ after deconvolution.

indicate the Cu valency from the point of view of the profile of the
structure in the lower energy side of the Cu Kα$_1$ line. Since we did not
measure Cu Kα spectra of Cu3+ compounds, we cannot arrive at a definite
conclusion about the existence Cu3+ in Y-Ba-Cu-O superconductors.

REFERENCES

1.   P. A. Jansson, Modern Constrained Nonlinear Methods, in:
        "Deconvolution with Applications in Spectroscopy", Chapter 4,
        P. A. Jansson ed., Academic Press(1984)

2.   G. W. Halsey and W. E. Blass, Deconvolution Examples, in:
        "Deconvolution with Applications in Spectroscopy", Chapter 7,
        P. A. Jansson ed., Academic Press(1984)

3.   Y. Gohshi and J. Kashiwakura, Resolution Enhancement in X-ray
     Emission Spectroscopy, _Physica Fennica_, 9-S1:327(1974)

4.   N. Saitoh, A. Iida and Y. Gohshi, Data Processing in X-ray
     Fluorescence Spectroscopy-I.A smoothing method using B-splines,
     Spectrochimica Acta, 38B:1277(1983)

5.   O. Keski-Rahkonen and M. O. Krause, Total and Partial Atomic-Level
     Widths, _Atomic Data and Nuclear Data Tables_ 14:139(1974)

CHEMICAL STATE ANALYSIS BY X-RAY FLUORESCENCE

USING ABSORPTION EDGES SHIFTS

Kenji Sakurai*, Atsuo Iida[+] and Yohichi Gohshi

Department of Industrial Chemistry, University of Tokyo
7-3-1, Hongo, Bunkyo-ku, Tokyo 113, Japan

[+] Photon Factory, National Laboratory for High Energy Physics
1-1, Oho, Tsukuba, Ibaraki 305, Japan

INTRODUCTION

Recently synchrotron radiation (SR) sources have been extensively used for the study of materials science.[1,2] The high intensity of tunable monochromatic X-rays from SR facilitates many types of spectroscopic/ diffraction studies which have otherwise not been possible. Regarding the X-ray fluorescence technique, significant improvement of the minimum detection limit has been performed and has enabled trace element analysis in the order of tens of ppb or 10^{-12} g.[3-8] SR microanalyzers[3,9,10] and near surface analysis using grazing incidence geometry[11,12] are also attractive applications of synchrotron X-ray fluorescence technique. From a point of materials characterization, chemical state analysis is not less important than ordinary element analysis.

The selectively induced X-ray emission spectroscopy (SIXES)[13-16], which is based on the selective excitation of specific chemical species by tunable SR, has recently been realized. This technique uses the chemical shifts of the absorption edges, which reflect systematic changes of the binding energies of inner shell electrons corresponding to the chemical environment. High sensitivity is one of the most important advantages of SIXES. It is significant for practical analysis because sensitive methods for chemical state analysis are extremely limited.

In this paper, chemical state analysis by SIXES is described, with emphasis on the quantitative analysis of the mixed system and also on chemical state mapping.

\# Present address: Materials Characterization Division
National Research Institute for Metals
1-2-1, Sengen, Tsukuba, Ibaraki 305, Japan

PRINCIPLE OF SIXES

The energy of the absorption edge is the minimum energy necessary to excite the inner shell electron to the outer empty level, and shifts slightly with change of chemical environment. Since the fluorescent X-rays are emitted accompanied with the excitation of inner shell electrons, chemical shifts of the absorption edge lead to the differences in the threshold energy of the X-ray emission along the chemical species. This means that distinction of chemical states is possible with the measurement of X-ray fluorescence selectively emitted by tuning the incident energy. This is the reason why we call the present technique selectively induced X-ray emission spectroscopy (SIXES).

Chemical shifts of the absorption edges have been measured mainly by the transmission technique since the early days of X-ray spectroscopy.[17,18] The direct transmission method, however, has inherent limitations of the sensitivity and the sample conditions[18,19], consequently restricting its analytical applications. On the contrary, the absorption spectroscopy by X-ray fluorescence detection is expected to be suitable for the characterization of thin or dilute systems because of its high sensitivity. This procedure has become available since the advent of SR sources, and is employed mostly for EXAFS experiments.[20-23] The present study was designed to realize chemical state analysis with X-ray fluorescence by the observation of the systematic shifts of absorption edges.

With respect to chemical state analysis by SIXES, there are largely two types of experiments: (i) direct evaluation of the amount of chemical shift by measuring the energy dependence of the intensity of fluorescent X-rays, which gives the near-edge absorption spectrum, (ii) observation of the change or distribution of specific chemical species by measuring the dependence of fluorescent X-ray intensity on experimental parameters such as position of the sample, temperature, pressure or process of chemical reaction. In the present work, as examples for (i) and (ii), the quantitative analysis of the mixed system and chemical-state mapping are demonstrated, respectively.

EXPERIMENTAL

The experiment was carried out using SR at the Photon Factory (PF) on beam lines 4A and 6B. The apparatus used was the energy dispersive X-ray fluorescence system.[11] SR beams were monochromatized by a channel-cut Si(111) crystal monochromator. The intensity of $K\alpha$ fluorescent X-rays was measured by a Si(Li) detector as a function of the incident energy, which was scanned around the absorption edge. Intensities of incident and transmitted X-rays are measured by ionization chambers. Other experimental details were described elsewhere.

NEAR EDGE ABSORPTION SPECTRUM

Figure 1 shows the near edge absorption spectra of some 3d-transition metals and their oxides. The samples were prepared in the form of pellets mixed homogeneously with fine cellulose powder.[13] The first inflection point of each metal spectrum is set at zero. It is clearly seen that the absorption curve shifts to higher energy as the oxidation number increases. The energy resolution of the incident X-rays is less than 2 eV and is sufficient to reveal shifts of the absorption edge.

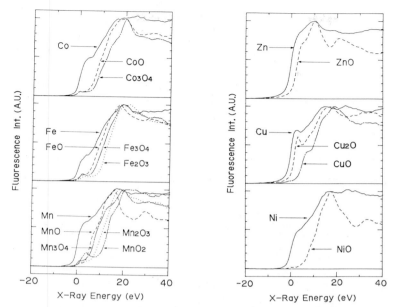

Fig. 1   Near edge absorption spectra of 3d transition metal
oxides obtained by X-ray fluorescence detection.

The absorption spectra shown in Fig. 1 include two discontinuities:
the initial absorption jump (lower energy side) corresponds to the transition
of the 1s-electron into an empty state of the 3d-4s band which has some
p-character, and the second jump (higher energy side) corresponds to that of
1s → 4p-level which is completely unoccupied.[18] The initial absorption jump,
which is a forbidden transition according to Laporte's rule, is much less
pronounced for the oxides than for the metal, because the atomic wave
functions are much more maintained in the oxides than in the solid metal, due
to the larger distance between neighboring iron atoms. These explanations
seem appropriate except for oxides of copper from Fig. 1. Pre-edge
structures of copper oxides are interpreted as a split of 1s → 4p transition
to 1s → 4pσ and 1s → 4pπ.[24] It is important to analyze such fine structures
and complicated shapes of the spectra. From an analytical point of view,
however, chemical shifts are most conspicuous and feasible features appear in
the experimental spectra as shown in Fig. 1.

A comparison between the present technique and the transmission
experiments is summarized in Table 1. Near-edge absorption spectra obtained
from both methods were in good agreement. However, transmission experiments,
which measure absorption directly, have limitations with respect to
sensitivity and sample conditions: application to the analysis of thin films
and dilute systems is difficult due to the restriction of the statistical
errors. The optimum thickness for maximum contrast of X-ray intensity is
usually of the order of μm. An uniform sample is also essential for the
transmission method; on the contrary, with the present technique, much higher
sensitivity is attainable. Trace systems in the order of ppm and thin films
of less than 0.1 monolayer can be analyzed by SIXES.[13,15] Furthermore, it is
far more versatile for sample conditions than is the direct absorption
technique. Though distortion of the spectra is observed for concentrated or

Table 1.   Comparison of Transmission and Fluorescence Methods

	Transmission Technique	Fluorescence Technique (SIXES)
Analytical Information	essentially the same	
Sensitivity	optimum thickness $\sim\mu m$	$\sim$ppm; less than ng
Inhomogeneous Sample	not suitable	suitable

thick samples, the position of the absorption edge can be determined using reference spectra.[14]

DETERMINATION OF THE CHEMICAL COMPOSITION OF MIXTURES

The determination of the chemical composition of a mixed system is one of the most interesting analytical applications of SIXES. Figure 2 shows the near-edge spectra from mixed samples containing FeO and Fe_2O_3 in various ratios. The absorption edge systematically shifts to the higher energy side as the concentration of Fe_2O_3 increases. The concentration of each oxidation state can be determined by the position of the absoprtion edge. The calibration curve, which is obtained experimentally using reference samples by plotting the edge energy as a function of the ratio of the components, was used for quantitative analysis. The precision of the present method was found to be about $\pm$ 6%.[13]

As an example of the practical application of SIXES, thin Co-O films which show perpendicular magnetic properties were analyzed.[15] According to the recent studies,[25,26] thin films of Co-O are expected to be composed of

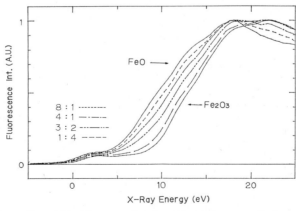

Fig. 2   Quantitative analysis of mixtures of iron oxides

Table 2   Preparation Conditions of Co-O Thin Films

	O content	Magnetism	Color
A		magnetic	metallic
B	increasing	magnetic	black
C	increasing	magnetic	black
D	critical pressure	non-magnetic	colorless transparent
E	↓	non-magnetic	colorless transparent

ferromagnetic cobalt metal and non-ferromagnetic CoO below a critical pressure. However, it is sometimes difficult to analyze these films by X-ray crystal structural analysis, because very fine particles of cobalt metal are surrounded by the layer of CoO in the Co-O film.[25,26]

The preparation conditions of Co-O thin films are listed in Table 2. They were prepared by vacuum evaporation with an electron beam gun and deposited on thin polyimide films. Oxygen gas was introduced during the cobalt evaporation, at a rate of approximately 1600 Å/sec, at various pressures ($4 \times 10^{-4} \sim 1 \times 10^{-3}$ Torr), and oxygen content was gradually increased for samples A through E. Sample D was prepared at the critical pressure, where saturation magnetization of the film became zero. The cobalt absolute mass of the films was about 200 $\mu g/cm_2$ (~ 2000 Å). Sample E was prepared by drastically lowering the cobalt evaporation rate, and contains an extremely high proportion of oxygen.

The experimental results for thin Co-O films are shown in Fig. 3. Absorption edges shift to the higher energy side as the oxygen pressure increases. These spectra were analyzed by direct comparison with the

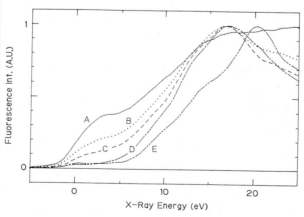

Fig. 3   The absorption spectrum for Co-O thin films. Partial pressure of oxygen was increased in the order A (solid line), B (dotted line), C (broken line), D (broken-and-dotted line), E (broken-and-double-dotted line).

reference spectra of cobalt oxides as shown in Fig. 1. As a result, curves
A, D and E are in good agreement with the spectra of cobalt metal, CoO and
Co_3O_4, respectively. Curves B and C correspond to the mixture of cobalt
metal and CoO. The results show a gradual change in the Co-O films from pure
cobalt metal into CoO through the mixed state of cobalt metal and CoO.
Furthermore Co-O films become Co_3O_4 above the critical oxygen pressure. The
ratios of cobalt metal to CoO in mixtures can also be determined; they were
about 1 to 1 and 3 to 7 for samples B and C, respectively.

Accordingly, it was demonstrated that SIXES can be applied to the
analysis of thin films even when there is no long range order suited for
using ordinary X-ray diffraction techniques. Capability of quantitative
treatment is its most important feature.

CHEMICAL STATE MAPPING

Chemical state mapping is another important analytical application of
SIXES.[16] Such a technique, especially with high sensitivity, has hardly been
developed yet in practice, in spite of increasing demands in scientific
studies. The absorption imaging technique,[27] which is promising for three
dimensional imaging in combination with X-ray computed tomography, is
severely restricted with respect to sensitivity.

At the lower energy side of the absorption edge (0 $\sim$ 10 eV in Fig.1)
excitation efficiency differs considerably among the chemical species. The
intensities of fluorescent X-rays of lower oxidation states are more intense
than those of higher oxidation states for the same quantity. This means that
information of specific chemical species can be separated from others, though
it is difficult to realize perfect selective exictation. On the contrary,
excitation efficiency is almost the same at the higher energy side of the
absorption edge (above 20 eV in Fig.1), in the same way as the ordinary
element analysis with fluorescent measurement.

The intensity of fluorescent X-rays from a specific chemical species
is proportional to the product of its concentration and its excitation
efficiency. The observed intensity at the incident excitation energy E_j can
be written as follows:

$$I(E_j) = A \cdot \sum_{i=1}^{n} F_i(E_j) \cdot C_i \qquad (1)$$

where C_i and $F_i(E_j)$ are the concentration of the chemical state i and the
excitation efficiency of i and E_j, respectively; n is the number of chemical
states contained in the sample and A is a constant. Using Eq.(1), C_1, C_2,
$\cdots$, C_n are determined when $I(E_1)$, $I(E_2)$, $\cdots$, $I(E_n)$ are given
experimentally. For example, in the case of a binary system, for the
determination of each concentration, it is necessary to measure the
fluorescent intensity at two energies.

For an experiment on chemical state mapping, a pellet sample having
separated regions of FeO and Fe_2O_3, was studied.[16] The intensity
distribution of Fe Kα was measured by point-by-point scanning with a beam
size of 250 μm x 500 μm. The measurements were performed at the lower
(8.1 eV in Fig.1) and higher energy sides (20.0 eV in Fig.1) of the
absorption edge with pixels of 100 x 20 points, and the collection time was
2 sec for each pixel. The sample was scanned from the FeO region through the
Fe_2O_3 region.

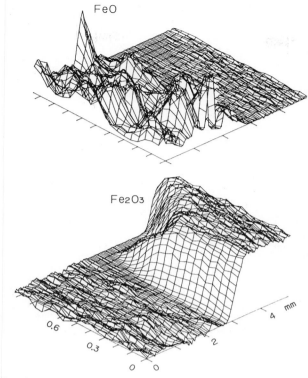

Fig. 4   Chemical state mapping of the iron oxide sample by SIXES.

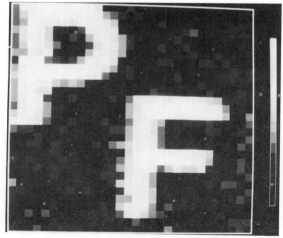

Fig. 5   Chemical state mapping (Cr image). The Cr_2O_3
image obtained was a reversal of the Cr image.

By substituting the experimental intensity distribution of Fe $K\alpha$ data in Eq.(1), the spatial distributions of FeO and Fe_2O_3 were determined separately as shown in Fig. 4. The results were in good agreement with the distributions obtained by an optical microscope. The concentration of FeO has many peaks because of an inhomogeneous distribution of relatively large size particles of a few hundreds μm in diamter, whereas the fine powders of Fe_2O_3 are distributed almost uniformly.

Another example is given in Fig. 5. The sample is a thin film of chromium on glass. Letters of 'PF' are chromium metal, and chromium oxide Cr_2O_3 is deposited around them. Figure 5 shows the mapping of chromium metal which is in good agreement with the optical image.

Though focusing optics was not used in the present work, it is very attractive for providing high resolution imaging with an SR microprobe[9,10] having a μm-order resolution. Microstructures of various materials will be analyzed more clearly by investigating the chemical state of the specific element. Application of the image reconstruction technique is also important.[16,28] This experiment is performed using a line beam with translational and rotational scannings of the sample, and is superior in getting a high counting rate of the signal as a result of a relatively larger irradiation area. Since the SR beam has an inherent line shape, high-resolution imaging with X-ray demagnification optics is feasible.[28] The present technique indicates possible new developments in extensive scientific fields, considering its nondestructive nature and versatility in regard to the requirements of measurement and the form of the sample.

ACKNOWLEDGEMENTS

The authors thank the PF staff for their help during the experiment. Our thanks also go to Mr. Nambu of Matsushita Electric Industrial Co., Ltd., for his valuable comments on the properties of thin Co-O films. This work was supported by Grant-in-Aid for Science Research No. 60470064 from the Ministry of Education, Science and Culture.

REFERENCES

1. H. Winick and S. Doniach Eds., "Synchrotron Radiation Research", Plenum, New York (1980)
2. E. E. Koch Ed., "Handbook on Synchrotron Radiation", North-Holland, Amsterdam (1983)
3. C. J. Sparks, Jr., X-Ray Fluorescence Microprobe for Chemical Analysis, in "Synchrotron Radiation Research", Ch. 14, H. Winick and S. Doniach Eds., Plenum, New York (1980)
4. J. V. Gilfrich, E. F. Skelton, D. J. Nagel, A. W. Webb, S. B. Qadri and J. P. Kirkland, X-Ray Fluorescence Analysis Using Synchrotron Radiation, Adv. in X-Ray Anal. 26:313 (1983)
5. R. D. Giauque, J. M. Jaklevic and A. C. Thompson, The Application of Tunable Monochromatic Synchrotron Radiation to the Quantitative Determination of Trace Elements, Adv. in X-Ray Anal. 28:53 (1985)
6. S. T. Davies, D. K. Bowen, M. Prins and A. J. J. Bos, Trace Element Analysis by Synchrotron Radiation Excited XRF, Adv. in X-Ray Anal. 27:557 (1984)

7. A. Iida, K. Sakurai, T. Matsushita and Y. Gohshi, Energy Dispersive X-Ray Fluorescence Analysis with Synchrotron Radiation, Nucl. Instr. and Methods 228:556 (1985)

8. K. Sakurai, A. Iida and Y. Gohshi, Analysis of Signal to Background Ratio in Synchrotron Radiation X-Ray Fluorescence, Anal. Sci. 4:3 (1988)

9. R. D. Giauque, A. C. Thompson, J. H. Underwood, Y. Wu, K. W. Jones, M. L. Rivers, Measurement of Femtogram Quantities of Trace Elements Using an X-ray Microprobe, Anal. Chem. 60:855 (1988)

10. Y. Gohshi, S. Aoki, A. Iida, S. Hayakawa, H. Yamaji and K. Sakurai, A Scanning X-Ray Fluorescence Microprobe with Synchrotron Radiation, Adv. in X-Ray Anal. 31:495 (1988)

11. A. Iida, K. Sakurai, and Y. Gohshi, Grazing Incidence X-Ray Fluorescence Analysis, Nucl. Instr. and Methods A246:736 (1986)

12. A. Iida, K. Sakurai, and Y. Gohshi, Near-Surface Analysis of Semiconductor Using Grazing Incidence X-Ray Fluorescence, Adv. in X-Ray Anal. 31:487 (1988)

13. K. Sakurai, A. Iida and Y. Gohshi, Chemical State Analysis by X-Ray Fluorescence Using Shifts Of Iron K Absorption Edge, Anal. Sci. 4:37 (1988)

14. K. Sakurai, A. Iida and Y. Gohshi, Chemical State Analysis of Trace Elements by X-Ray Fluorescence Using Absorption Edge Shifts, Adv. in X-Ray Chem. Anal. Japan 19:57 (1988) [in Japanese]

15. K. Sakurai, A. Iida and Y. Gohshi, Characterization of Co-O Thin Films by X-Ray Fluorescence Using Chemical Shifts of Absorption Edges, Jpn. J. Appl. Phys. 26:1937 (1987)

16. K. Sakurai, A. Iida, M. Takahashi and Y. Gohshi, Chemical State Mapping by X-Ray Fluorescence Using Absorption Edge Shifts, to be published in Jpn. J. Appl. Phys.

17 L. V. Azaroff and D. M. Pease, in "X-ray Spectrometry", Ch. 6, L. V. Azaroff, Ed., McGraw-Hill, New York (1974)

18. B. K. Agarwal, "X-ray Spectroscopy", Springer-Verlag, New York (1979)

19. M. E. Rose and M. M. Shapiro, Statistical Error in Absorption Experiments, Phys. Rev. 74:1853 (1948)

20. J. M. Jaklevic, J. A. Kirby, M. P. Klein, A. S. Robrtson, G. S. Brown and P. Eisenberger, Fluorescence Detection of EXAFS: Sensitivity Enhancement for Dilute Species and Thin Films, Sol. Stat. Commu. 23:679 (1977)

21. J. M. Jaklevic, J. A. Kirby, A. J. Ramponi, and A. C. Thompson, Chemical Characterization of Air Particulate Samples Using X-ray Absorption Spectroscopy, Environ. Sci. Technol. 14:437 (1980)

22. E. A. Stern, and S. M. Heald, in "Handbook on Synchrotron Radiation", Ch. 10, E. E. Koch Ed., North-Holland, Amsterdam (1983)

23. H. Oyanagi, T. Matsushita, H. Tanoue, T. Ishiguro and K. Kohra, Fluorescence-Detected X-Ray Absorption Spectroscopy Applied to Structural Characterization of Very Thin Films: Ion-Beam-Induced Modification of Thin Ni Layers on Si(100), Jpn. J. Appl. Phys. 24:610 (1985)

24. N. Kosugi, XANES and Molecular Orbital Picture: 3d Transition Metal Compounds, in "Core Level Spectroscopy in Condensed Systems", A. Kotani Ed., Springer-Verlag (1988)

25. K. Nakamura, N. Tani, M. Ishikawa, T. Yamada, Y. Ota and A. Itoh, A New Perpendicular Magnetic Film of Co-O by Evaporation, Jpn. J. Appl. Phys. 23:L397 (1984)

26. M. Ohkoshi, K. Tamari, S. Honda and T. Kusuda, Microstructure and exchange anisotrophy of Co-CoO films with perpendicular magnetization, J. Appl. Phys. 57:4034 (1985)

27. J. Kinney, Q. Johnson, M. C. Nichols. U. Bonse and R. Nusshardt,
    Elemental and Chemical-State Imaging Using Synchrotron Radiation,
    Appl. Opt. 25:4583 (1986)
28. A. Iida, M. Takahashi, K. Sakurai and Y. Gohshi, SR X-Ray
    Fluorescence Imaging by Image Reconstruction Technique, to be
    published in Rev. Sci. Instrum.

HIGH RESOLUTION X-RAY FLUORESCENCE Si Kβ SPECTRA: A POSSIBLE

NEW METHOD FOR THE DETERMINATION OF FREE SILICA IN AIRBORNE DUSTS

J Purton and D S Urch
Chemistry Department, Queen Mary College
Mile End Road, London E1 4NS, UK

N G West
Occupational Medicine and Hygiene Laboratories
Health and Safety Executive
403 Edgware Road, London NW2 6LN, UK

ABSTRACT

High resolution x-ray emission spectroscopy (HRXES) has been used to record the Si Kβ spectra of a variety of minerals. Distinct changes in peak profile can be related to mineral type. Representatives spectra were chosen and incorporated into a computer programme to allow the determination of free silica in binary and quaternary mixtures. The potential of HRXES for the analysis of airborne dust samples is discussed.

INTRODUCTION

Inhalation of α-quartz can cause fibrotic lung disease so that it is necessary to monitor the quantity of quartz within airborne dust samples. Present methods of analysis, including X-ray diffraction (XRD) and infra-red spectroscopy (IRS) (Pickard et al., 1985), are largely successful but are prone to interference from other minerals. Consequently an investigation was initiated to determine the viability of developing an analytical technique using high resolution X-ray emission spectroscopy. This technique has been suggested for the determination of quartz within coal dust samples (Hurley and White, 1973).

If an atom is bombarded with high energy particles (such as photons, electrons and protons) ionisation can occur creating a vacancy or hole within a core orbital. The atom may relax by the transfer of an electron from an orbital higher in energy to the vacancy and the emission of a photon whose energy is the difference in energy of the two orbitals. For light elements the electron transfer process is governed by the dipole selection rule (l=±1) allowing only p → s and s or d → p transitions.

X-ray emission lines can be classified into two groups:

a)   core orbital → core orbital transition
b)   valence orbital → core orbital transition

The second group has the greatest potential for providing chemical
information and exhibits a much wider variation in peak profile and
energy.   The Si $K\beta_{1,3}$ belongs to this group (Si3p → 1s transition)
and is suitably intense that it can easily be measured on a
conventional spectrometer.   It was considered that the Si Kβ had the
greatest potential for the development of this analytical technique.

EXPERIMENTAL

A commercial single, flat crystal spectrometer (Philips PW 1410)
was employed for the collection of the Si Kβ spectra using an ADP
(200) crystal for the dispersion of the X-rays.   The resolution of the
spectrometer has been enhanced using an extra collimator (Haycock and
Urch, 1982) to control vertical divergence.   Excitation was by means
of a Sc side window X-ray tube operating at 40mA and 50 kV.   Pulses
from the proportional counter were amplified by Harwell 2000 series
electronics before being fed via an interface into a Commodore Pet
Computer.   The computer allowed a repetitive step scanning technique
to be employed improving the signal to noise ratio.   Initially the
complete Si Kβ spectrum was recorded, taking measurements every 0.30°
2θ over a 127.0° - 132.5° 2θ interval.   However, when recording the
spectra of mixtures, measurements were taken at 0.02° 2θ intervals
throughout a 127.5° - 130.5° 2θ range.   This ignores the Si Kβ' peak
since it is a broad, low intensity feature and consequently has little
analytical potential (cf. Esmail et al., 1973).

Samples were prepared by thoroughly mixing the required
proportion of each mineral and pressing it into a disc using

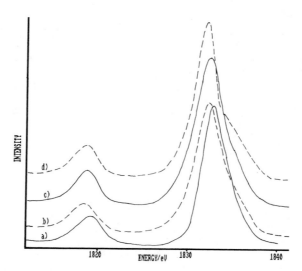

Fig. 1. Si Kβ spectra for (a) Zircon, (b) Kaolin, (c) Sodic feldspar
          and (d) Silica

polyethylene powder as a matrix.   The source of each mineral is
listed below:

Zircon   British Chemical Standard (Number 388)
Sodic feldspar   British Chemical Standard (Number 375)
Silica (pure precipitated) Hopkin & Williams Ltd
Kaolin (China Clay)   Hopkin & Williams Ltd

PRESENTATION OF SPECTRA

     The spectra obtained for the pure minerals phases are shown
in figure 1 and are comparable to those of Dodd and Glen (1969). The
spectra are composed of three dominant features.

1.   The main Si Kβ peak at 1832-1833 eV.
2.   The Si Kβ' peak at 1818-1819 eV.
3.   A high energy shoulder at approximately 1834 eV.

The spectra of silica, sodic feldspar and kaolin also have an
unresolved component to the low energy side of the main Si Kβ peak.

INTERPRETATION OF SPECTRA

     Urch (1970) has shown that the X-ray emission spectra of the SiO_4^{4-}
anion can be interpreted by considering the irreducible representations
of the valence electrons of the Si and O atoms:

Si (central)   4 O (ligands)

3s (1 orbital) a_1 2s   (4 orbitals) $a_1 + \underline{t}_2$
3p (3 orbitals) $\underline{t}_2$ 2p' (4 orbitals $\parallel$ Si-O
                                                bond) $a_1 + \underline{t}_2$
3d (5 orbitals) $e + t_2$ 2p   (8 orbitals $\perp$ Si-O
                                                bond $a + \underline{t}_2 + t_1$

Orbitals that are a_1 have the same symmetry properties as s orbitals
on the central silicon atom whilst those that are t_2 are of the same
symmetry type as the trio of silicon p orbitals. The advantage of
being able to classify the orbitals of silicon and oxygen in this way
is that only orbitals of the same symmetry type (ie irreducible representation) can interact. In this paper we are concerned with the Si Kβ
spectra and possible interactions of the Si 3p orbitals have been under-
lined.

     Core orbitals also need to be included in the overall picture of
electronic structure and are labelled in descending binding energy
according to their symmetry type. Thus the Si 1s orbital is $1a_1$,
being the most tightly bound, with the O 1s orbitals, which transform
as $2a_1$ and $1t_2$, less tightly bound. The 2s and 2p orbitals on the
silicon are then numbered $3a_1$ and $2t_2$ respectively and in the valence
band orbitals, that are mostly oxygen 2s in character, will be $4a_1$ and
$3t_2$ (the $4a_1$ orbital may also contain some Si 3s character and the $3t_2$
orbital some Si 3p character). The bonding sigma orbitals will be
generated by Si 3s and Si 3p interacting with O 2p' orbitals and will
be designated $5a_1$ and $4t_2$. The remaining occupied orbitals, $1e$, $1t_1$

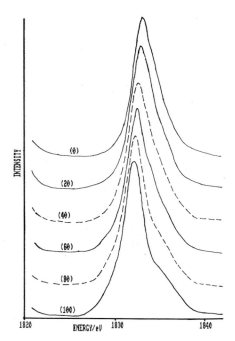

Fig. 2.   Si Kβ spectra for a series of binary mixtures containing
          zircon and silica.   The number in brackets indicates the
          mole percent of silica within the mixture.

and $5t_2$, are mostly oxygen 2p (not 2p') in character.   Because the
Si 3d orbitals also transform as e and t_2, Si 3d character may be
present in le and $5t_2$.

     The Si Kβ reflects the bonding of the Si 3p electrons which are
present in the $4t_2$ and $3t_2$ molecular orbitals.   The $4t_2$ orbital
contains more Si 3p character and thus forms the main Si Kβ peak
while the $3t_2$ orbital generates the Si Kβ' peak.   The high energy
shoulder to the main Si Kβ peak is probably due to Si 3p character
entering the $5t_2$ molecular orbital.   This interpretation is in
agreement with more sophisticated calculations of Tossel et al. (1973)
and Tossel (1975).   The feature to the low energy side of the Kβ can
be related to the reduction in symmetry from Td on the formation of
bridging oxygen bonds allowing mixing of the Si 3s and 3p orbitals.

     The relationship between the crystal structure and the Si Kβ
spectra has been discussed in a previous paper (Purton and Urch, 1988)
however it should be pointed that the spectra of zircon, sodic
feldspar and kaolin are typical of orthosilicates, framework silicates,
and sheet silicates respectively.   Although precipitated silica was
used in the mixtures of minerals the Si Kβ spectrum is identical with
that of quartz and cristobalite.   Wiech and Zurmaev (1985) have shown
that small differences do exist but they are beyond the resolution of
our spectrometer).

ANALYSIS OF MINERAL MIXTURES

A comparision of the silica spectrum in figure 1 with the spectra
of the other minerals shows that spectra of zircon and sodic feldspar
are significantly different and the kaolin spectrum is very similar.
The changes in peak profile can be used to determine the quantity of
silica within binary mixtures of minerals. Figure 2 shows the gradual
change of the Si Kβ spectrum for a series of binary mixtures containing
zircon and silica.

In order to best use the information in the standard spectra for
the analysis of multicomponent mixtures an algorithm was developed
based on the least squares principle. If a mixture consists of n
components $(x_1, x_2, \ldots x_n)$ and the X-ray emission intensity at wavelength λ for component x_k is $X_{k\lambda}$ then the predicted intensity (R_λ)
for a mixture would be

$$R_\lambda = \sum_1^n C_k \cdot X_{k\lambda},$$

where C_k is mole fraction of x_k. Let the observed intensity from the
mixture at this wavelength be R_λ', then by minimising

$$\sum_1^s (R_\lambda' - R_\lambda)^2$$

with respect to each value of C_k for the s wavelengths at which
readings were taken, it is possible to show that,

$$\sum_1^s R_\lambda' \cdot X_{j\lambda} = \sum_1^s (X_{j\lambda} \cdot \sum_1^n C_k \cdot X_{k\lambda}).$$

There will be n such equations, one for each of the n values of j.
This enables the n values of C_k to be calculated, by solving the
appropriate determinants.

In this work s was 17 measurements being made at 0.5 eV intervals
in the range 1830-1838 eV. It is reasonable to suppose that self
absorption would not vary significantly over the range of wavelengths

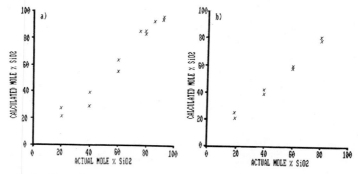

Fig. 3.   Calculated verses actual mole percentage silica in binary
          mixtures of (a) silica and sodic feldspar and (b) silica
          and zircon.

of interest so this factor should not affect peak profiles. For
binary mixtures (n=2) of silica and sodic feldspar and silica and
zircon the results are presented in figure 3.

The accuracy of the results varies according to the difference
between the mineral spectrum and that of silica. Thus for zircon-
silica mixtures, which show the greatest difference, the results
diverge little from the expected values. On the other hand the
spectra of silica and kaolin are too similar to allow any meaningful
analysis to be carried out.

Quaternary mixtures were also prepared, containing the above
minerals, and analysed in a similar manner. As expected the quantity
of zircon and feldspar could be determined with reasonable accuracy
but the programme was unable to distinguish between silica and kaolin.

CONCLUSION

For HRXES to be a useful method for analysis of dust samples a
library of mineral spectra would be necessary although this can be
reduced since many minerals (eg the feldspar family) have very similar
spectra. HRXES suffers less from interference than XRD and IRS; 2nd
order Ca Kα lies close to the Si Kβ line and can cause problems when
present in appreciable quantities. This problem and the small
differences between mineral spectra could be overcome by the use of a
higher resolution curved crystal spectrometer. The advantage of HRXES
is the speed of analysis requiring the determination of only 17
experimental data points.

ACKNOWLEDGEMENTS

One of the authors (J P) wishes to thank Science and Engineering
Research Council and the Health and Safety Executive for funding this
project. Thanks are also due to Miss Fatima Cader for her excellent
typing support.

REFERENCES

1. Dodd, C.G. and Glen, G.L, 1969, A survey of chemical bonding in
   silicate minerals by X-ray emission spectroscopy, Am. Mineral.,
   54: 1299.
2. Esmail, E.I., Nicholls, C.J. and Urch, D.S., 1973, The detection
   of light elements by X-ray emission spectroscopy with use of low
   energy satellite peaks, Analyst, 98: 725.
3. Haycock, D.E. and Urch, D.S., 1982, Resolution enhancement for a
   commercial X-ray fluorescence spectrometer, J.Phys. E:Sci.
   Instrum., 15: 40.
4. Hurley, R.G. and White, E.W., 1973, Rapid new soft x-ray technique
   for determination of quartz in coal mine dusts, Am. Ind. Hyg. Ass. J.,
   34: 229.
5. Pickard, K.J., Walker, R.F. and West, N.G. 1985, A comparison of
   X-ray diffraction and infra-red spectrophotometric methods for the
   analysis of α-quartz in airborne dusts, Ann. Occup. Hyg., 29: 149.

6. Purton, J.A. and Urch, D.S., (in press), High resolution Si KB
   X-ray spectra and crystal structure, Min. Mag.
7. Tossel, J.A., 1975, The electronic structures of silicon, aluminum,
   and magnesium in tetrahedral co-ordination with oxygen from SCF-Xα
   MO calculations, J. Am. Chem. Soc., 97: 4840.
8. Tossell, J.A., Vaughan, D.J. and Johnson, K.H., 1973,
   X-ray emission and UV spectra of SiO_2 calculated by the SCF Xα
   scattered wave method, Chem. Phys. Lett., 20: 329.
9. Urch, D.S., 1970, The origin and intensities of low energy satellite
   lines in X-ray emission spectra: A molecular orbital interpretation,
   J. Phys. C., 3: 1275.
10. Wiech, G and Kurmaev, E.Z., 1985, X-ray emission bands and electronic
    structure of crystalline and vitreous silica (SiO_2), J. Phys. C., 18:
    4393.

QUANTITATIVE ANALYSIS OF FLUORINE AND OXYGEN
BY X-RAY FLUORESCENCE SPECTROMETRY USING
A LAYERED STRUCTURE ANALYZER

Momoko Takemura and Hirobumi Ohmori

Toshiba R & D Center
Komukai Toshiba-cho, Saiwai-ku
Kawasaki, 210, Japan

INTRODUCTION

Recently, layered structure analyzers (called LSA for short) or layered synthetic microstructures (called LSM[1]) with d spacing of several tens of Å, have been developed for use as X-ray analyzing devices in wavelength dispersive spectrometers. The lower detection limit for light elements of atomic numbers lower than 13, such as aluminum, sodium, fluorine, oxygen, carbon and so on, has been greatly improved.

There have been several reports[2~5] published regarding LSA (or LSM) applications to light element analyses. Although they mostly emphasized better detection limits for light elements, LSA development has also caused precision improvement for light elements, and has made possible precise determination of very light elements possible. Accordingly, the authors applied the XRF method, using LSA to determine fluorine and oxygen.

INSTRUMENTATION

Table 1 shows instrumental conditions for fluorine and oxygen. A Toshiba AFV 202 spectrometer with a W/Si LSA was used for fluorine analysis and a PHILIPS PW 1404 spectrometer with an LSA, named PX-1 by PHILIPS, was used for oxygen analysis.

The W/Si LSA used for fluorine is made by Ovonic Synthetic Materials Co. LTD and the 2d spacing is 80 Å. Comparative measurements on magnesium, sodium and fluorine, using the LSA and crystal analyzers are shown in Table 2.

As shown in Table 2, when using LSA, peak intensities were multiplied 15 times for magnesium Kα, 25 times for sodium and 5 times for fluorine, compared with the values using crystals such as rubidium acid phthalate (RAP) or ammonium dihydrogen phosphate (ADP).

Table 1. Instrumental conditions for fluorine and
          oxygen measurements

	F	O
Analytical line	F Kα (λ=18.32A)	O Kα (λ=23.71A)
X-ray tube	Cr, 50kV-40mA	Sc, 45kV-75mA
Dispersing device	W/Si LSA*[1], 2d=80Å	PX1*[2]
Detector	P.C. (t1μm window)	P.C. (t1μm window)
Counting time	40s, 2times	40s, 2times
Measured area	D25mm	D25mm

*1 Ovonic Synthetic Materials Co., Inc.
*2 PHILIPS

Table 2. Peak intensity and FWHM using W/Si LSA
          compared to crystal analyzer*

Line	Wavelength (Å)	Specimen	Peak intensity ratio to crystal*	FWHM ratio to crystal
Mg Kα	9.89	metal(Mg 100%)	15.5	1.1
Na Kα	11.91	NaCl (Na 39%)	25.0	1.2
F Kα	18.32	Teflon (F 76%)	5.2	1.8

*ADP for Mg, RAP for Na and F

FLUORINE DETERMINATON

     The first application was fluorine determination of polytetra-
fluoroethylene-graphite mixtures for use in the fuel cell electrode.
Fluorine is a major constituent of polytetrafluoroethylene-graphite mixtures. The two methods mentioned below were used and the results were
compared.

     First three synthetic standards were prepared containing about 15 to
30% fluorine. The calibration curve for fluorine using them is shown in
solid line in Figure 1. The standard deviation of this calibration curve
was 1.2 %, and its linearity was satisfactory. Two samples were analyzed,
using this calibration. Results were shown in Table 3, with the results
determined by the second method.

     In the second method, polytetrafluoroethylene (PTFE for short),
namely Teflon, was used as the only standard. The calibration curve, made
from the PTFE measurement, and the zero point was shown in the dashed line
in Figure 1. The fluorine content for Teflon was 76%. This was a considerably different value from the samples. As FKα X-ray is strongly absorbed by carbon, which was another major constituent, correction to this absorption should be considered. As there is no enhancing element, it is
not necessary to consider enhancement.

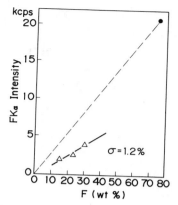

Figure 1.   Measured F Kα intensity versus F content
Δ: Measured intensity for
synthetic standards
●: Measured intensity for PTFE

The absorption correction was accomplished in the following way.   The
XRF intensity $I_{\lambda i}(\lambda_o)$ for wavelength λ_i, excited by a single wavelength λ_o
from a specimen of finite thickness t, is generally expressed as

$$I_{\lambda i}(\lambda_o) = \frac{K_i W_i}{A} [1-\exp(-A\rho t)] \tag{1}$$

where K_i is a constant including geometric factors, excitation probability
and primary X-ray intensity $I_{\lambda o}$, W_i is the weight fraction for an element
i, ρ is density for the specimen and A is a total mass absorption coefficient, expressed as

$$A = \frac{\Sigma\mu(\lambda_o)W_i}{\sin\psi_1} \quad \frac{\Sigma\mu(\lambda_j)W_i}{\sin\psi_2} \tag{2}$$

Equation (1) is transformed for thick samples (t→∞) to Eq. (3), then
further transformed to Eq. (4) for absorption correction.

$$I_{\lambda i}(\lambda_o) = \frac{K_i W_i}{A} \tag{3}$$

$$W_i = \frac{I_{\lambda i}(\lambda_o)}{K_i} \left(\frac{\Sigma\mu(\lambda_o)W_i}{\sin\psi_1} + \frac{\Sigma\mu(\lambda_j)W_i}{\sin\psi_2} \right) \tag{4}$$

Table 3.   Fluorine determination results

			(wt%)
Method Sample	(1)With synthetic standard series	(2)With single standard	Difference (1)-(2)
(A)	32.7	31.9	0.8
(B)	37.4	35.3	2.1

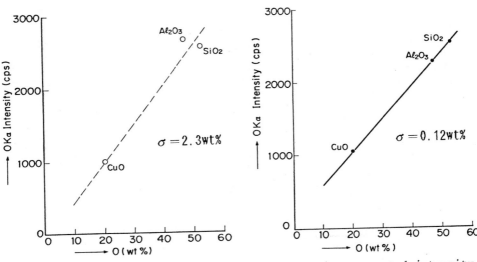

2-a Measured intensity of OKα
versus O contents

2-b Absortion corrected intensity
versus O contents

Figure 2. Effect of absorption correction on oxygen

Using Eq. (4) and the calibration curve made from the PTFE measurement and the zero point, the fluorine in the two samples were determined. Cr Kα from Cr tube was regarded as primary X-ray (λ_O). Because eq. (4) has W_i (the weight fraction of fluorine) on both sides, the fluorine contents were determined by iteration.

The analysis results of fluorine for samples A and B were shown in Table 3. The determined values for sample A by the two methods were in good agreement. However, the determined value difference for sample B was somewhat larger, compared with the standard deviation in the synthetic standards calibration. The reason for this difference can be explained as that the determined value 37.4 wt% or 35.3 wt% is larger than the maximum value of synthetic standards, so the value obtained by the calibration with synthetic standards should be regarded to be less reliable.

Thus, by LSA/XRF method, the major component fluorine could be successfully determined, using the one available standard.

OXYGEN DETERMINATION

The second application was oxygen determination for an Y-Ba-Cu oxide superconductor. As it is difficult to prepare synthetic standard series for oxide superconductors, the calibration curve was made using available oxides, such as silicon oxide, aluminum oxide and copper oxide. Absorption correction, using the total mass absorption coefficient, was carried out. Sc Kα from Sc tube was regarded as primary X-ray (λ_O). The results are in Figure 2. The standard deviation was 2.3 wt % without correction(Figure 2-a), but it decreased to 0.12 wt% after correction (Figure 2-b).

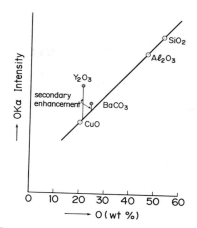

Figure 3. Absorption corrected OKα intensity versus O content

In this case, matrix enhancement could not be neglected. Figure 3 shows a plot for absorption corrected oxygen Kα intensity versus oxygen content for four oxides (SiO2, Al2O3, CuO, Y2O3) and barium carbonate. As shown before, silicon oxide, aluminum oxide and copper oxide are in good linearity, but, barium carbonate and yttrium oxide deviate upward. These deviations could be considered as secondary enhanced fluorescent X-ray.

Using these data, enhancement yield per weight % for barium was calculated. The same was done for yttrium. Then, the oxygen content was calculated, using the results.

Table 4 shows the oxygen analysis results for an oxide superconductor with a high Tc value. The compositions of yttrium, barium and copper were determined by wet analysis. In the top column, the oxygen atomic ratio X, estimated from the critical temperature value is given. In the middle column, the oxygen values given were determined by the LSA/XRF method in which only absorption correction was considered. This 8.0 value is considerably different from the value estimated from the Tc value. The bottom column shows the X value, determined by the LSA/XRF method, considering both absorption and enhancement corrections. This is in good agreement with the value estimated from the critical temperature.

Table 4. Oxygen determination results for $Y_{1.01}-Ba_{2.00}-Cu_{2.99}-Ox$

	X
Estimation from Tc value	6.7 ~ 6.8 (15.9wt%~16.2wt%)
Determination without enhancement correction	8.0 (19.2wt%)
Determination with enhancement correction	6.8 (16.2wt%)

CONCLUSION

The XRF method, using a layered structure analyzer as a dispersing device, was demonstrated to be effective and convenient for light element determination.

A major element fluorine could be determined by single standard calibration and absorption correction. The precision was 1 - 2 % at a 30 % level.

Oxygen in an yttrium, barium and copper oxide superconductor was determined by absorption corrected calibration, using available oxide standards, and the empirical correction of enhancement by barium and yttrium proved to be effective for accuracy improvement.

ACKNOWLEDGEMENT

The authors wish to thank Nihon Philips Corporation for contributing some of the XRF measurements.

REFERENCES

1. J. V. Gilfrich, D.J. Nagel and T.W. Barbee, Jr., Appl. Spectrosc., 36:58 (1982).
2. T.Arai, T. Shoji and R.W. Ryon, Adv. in X-Ray Anal., 28:137 (1985).
3. H. Kohno and T. Arai, Adv. X-Ray Chem. Anal. Jpn., 18:15 (1987).
4. J.A. Nicolosi, J.P. Groven and D. Merlo, Adv. in X-Ray Anal., 30:183 (1987).
5. M. Kunugi and M. Murata, Anal. Sci., 4:303 (1988).

THE HOMOGENEITY OF Fe, Sr and Zr IN SL-3/LAKE SEDIMENT STANDARD REFERENCE

MATERIAL BY RADIOISOTOPE INDUCED X-RAY EMISSION

J. J. LaBrecque and P. A. Rosales

Instituto Venezolano de Investigaciones Cientificas, IVIC
Apartado 21827, Caracas 1020A
Venezuela

INTRODUCTION

Following an international interlaboratory comparison study of SL-3
Lake Sediment sample, the results showed a large range of values for the iron
concentration (0.538 to 1.270% Fe) with an overall standard deviation of 18
percent from 38 accepted laboratory averages, see Figure 1 (1). Thus, it was

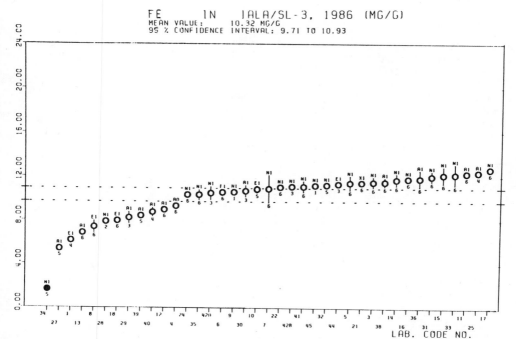

Figure 1. Results of iron concentrate in IAEA/SL-3 interlaboratory study
taken from reference (1). Note only about 1/3 of the values with
their error bars enter the 95% confidence interval.

191

decided to re-measure the iron homogeneity again in this material. The original homogeneity test was performed by neutron activation analysis and i was shown that, for 10 sub-samples from the bulk sample, the iron measurements did not differ significantly for sample sizes equal to or greater than 150 mg.

Preliminary measurements may need to be made to accept material for conformance with the specifications and to decide on such questions as premixing and subdivision into units of issue (e.g., bottling) prior to certification analysis. When a multicomponent/parameter standard reference material (SRM) is involved, this can be a major undertaking at this time. When possible, a quick and precise method is sought to evaluate homogeneity. In multiparameter materials, this may need to be judged on the basis of that of a limited number of typical constituents (2).

In this work, we have re-checked the homogeneity of iron in 24 bottles of SL-3 as well as strontium and zirconium by radioisotope-induced X-ray emission. Zirconium values from the international interlaboratory study were also widely distributed and thus were not certified in this material. Sample sizes from 90 mg to 4.5 grams with Cr_2O_3 added as the internal standard, as well as 20-50 mg portions without any sample preparation, were studied. Data will be presented to show that this material is indeed homogeneous for sample sizes >50 mg for iron, strontium and zirconium considering a 5% difference limit for 95% confidence limit using the student t-test. Finally, we shall discuss the advantage of using radioisotope X-ray fluorescence for homogeneity testing as compared with neutron activation, atomic absorption, etc.

EXPERIMENTAL

Sample Preparation

The sample material from each of the 24 different bottles were mixed and quartered separately. They were oven dried at 105°C overnight and then placed in a desicator. METHOD No. 1: 4.5 grams of sample were mixed and 0.5 grams of Cr_2O_3 as the internal standard in 20 ml plastic vials with 6 plastic balls employing a SPEX MILL/MIXER 8000 for 3 periods of 10 minutes each, with about 15 minutes between mixing to ensure the temperature of the mixture did not become too hot. The total mixture, about 5 grams, was then transferred to a Chemplex sample cup 1850 and sealed with 6 μMylar Film for measurement. The measurement time was 250 seconds with a Cd-109 annular source of about 4 mCi activity. The data analysis employed corrected peak areas (net peak areas). METHOD No. 2: was the same as Method No. 1 except only a portion of the mixture was measured. About 20-50 mg of the mixture was placed in the 1 cm aperture of a 2x6 cm piece of IBM card sealed with Scotch Magic tape. These samples were irradiated for 10 minutes (600 seconds). This thin-film technique has been reported in full elsewhere (3). METHOD No. 3: the same as Method No. 2 except gross peak areas were employed for the data analysis. METHOD No. 4: the same as Method No. 1 but smaller quantities. METHOD No. 5: no sample preparation; a typical spectrum is shown in Figure 2.

Instrumentation

The samples were excited by a weak Cd-109 annular radioisotope source (about 4 mCi). The secondary (characteristic) X-rays were collected by a high resolution Si(Li) detector with a thick Be window (about 0.150 mm). The data were collected and analyzed by an Apple IIe microprocessor with a Nucleus ADC/interface card. This system and its software have been described before (4).

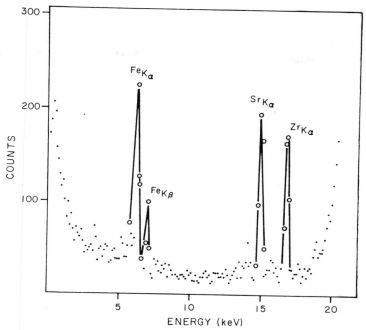

Figure 2.   A typical X-ray fluorescence spectrum of 20-50 mg of the original
SL-3 sample with no sample preparation (no internal standard);
Method No. 5.

The gross peaks areas were determined by summing all the counts within
the 0.6 FWHM window of the centroid of the K peaks.   The net peak areas were
calculated similarly by subtracting a similar energy region before and after
the peak of interest, or a nearby region where no characteristic X-rays were
present.   These operations were performed using a routine supplied as part of
an XRF software package by Dapple (Sunnyvale, California).

RESULTS AND DISCUSSION

Homogeneity within One Bottle

The homogeneity within one bottle of S1-1, SL-3 and Soil-7 was
determined for iron with Cr_2O_3 added as an internal standard employing Method
No. 1.   Table 1 presents the results of 5 independently prepared mixtures

Table 1. Homogeneity within one bottle for the Fe/Cr ratio
from 4.5 grams of sample plus 0.5 grams of Cr_2O_3

	Fe/Cr		
	SL-1	SL-3	Soil-7
replicates (n)	5	5	5
mean (x̄)	0.60	2.30	0.86
Standard deviation (SD)	0.02	0.06	0.03
Relative standard deviation (RSD)	3.2%	2.6%	3.5%

Table 2. Comparison of different sample preparation and sample measurement sizes for elemental ratios from 24 different bottles of IAEA/SL-3 Lake sediment

METHOD	Ratio	mean	Standard deviation	Relative standard deviation (%)	Students' t-test accepted m_0's values for $\alpha=0.05$
(1) 4.5 grams of sample plus 0.5 grams of Cr_2O_3;	Fe/Cr	0.602	0.036	6.0	>97%
	Sr/Cr	0.85	0.040	4.7	>98%
Total mixture measured; Net peak areas	Zr/Cr	0.91	0.056	6.2	>98%
(2) Same sample preparation as N° 1; but only measured 20–50 mg portions; Net peak areas	Fe/Cr	0.673	0.028	4.2	>98%
(3) Same as N° 2 but with Gross peak areas	Fe/Cr	0.706	0.025	3.5	>98%
(4) 90 mg of sample plus 10 mg of Cr_2O_3; measured 20–50 mg portions; net peak areas	Fe/Cr	0.640	0.09	14.1	>95%
(5) No sample preparation; measured 20–50 mg portions; Gross peak areas	Fe/Sr	1.34	0.105	7.8	>97%
	Fe/Zr	1.45	0.133	9.2	>96%
	Sr/Zr	1.08	0.073	6.8	>97%

show the relative standard deviations to be between 2.5-3.5%, thus the method is sufficiently precise for the homogeneity testing between the 24 different bottles.

Homogeneity between 24 Different Bottles

A comparison of different sample preparation and sample measurement sizes for various elemental ratios from 24 different bottles of IAEA/SL-3 Lake Sediment is given in Table 2. The characteristic X-ray spectrum from the 5 gram mixtures (Method No. 1) resulted in characteristic X-ray peak intensities about double of those when only 20-50 mg of the mixture is used as a thin-film method. (Note the second spectrum was for 600 seconds, not 250!) Thus, it is of little advantage to measure larger sample sizes which also produce more scatter. Also, one can note that the relative standard deviation is lower for the smaller thin-film samples, too.

It can concluded that when the peak intensities are small, <3000 counts, it seems that it is more precise to use gross peak areas rather than net peak areas as shown by the slight difference in the relative standard deviation of Method No. 2 and No. 3 in Table 2.

Method No. 4 shows a higher relative standard deviation. This can be assumed to be due to: a) the poor precision in weighing 10 mg of Cr_2O_3 and 90 mg of sample, and b) attempting to mix too small quantities to form a homogeneous mixture. But even so, it still is accepted by the student's t-test for m_0=95% of the mean and α=0.05. The authors attempted to not only use this method for the homogeneity test (a relative test) but also to quantify the amount of iron in SL-3 in a small sample size. Thus, it can be noted that applying the same method for determination of homogeneity and certification of an elemental concentration in a material is usually a compromise and possibly the amount of effort and time could be less if separately planned determinations were performed.

Method No. 5:   This method only requires that the selected sample in the range of 20-50 mg be placed in the sample holder, <u>it does not even have to be weighed</u>.   Thus, the sample preparation is only a minute or two.   But the limiting factor on how many samples can be analyzed depends upon the measurement time.   In this work it was 10 minutes, but the Cd-109 was only 4 mCi; in principle, a conventional 25 mCi source could produce similar peak areas in less than 2 minutes, or more precise peak areas in the same 10 minutes of measurement time.

CONCLUSIONS

It has been shown that radioisotope-induced X-ray emission is very suitable for homogeneity testing of geological materials, especially for iron when present as a minor component (>2%), and for strontium and zirconium when present in concentrations greater than 50 $\mu g/g$.   It has also been shown that using a small sample size (20-50 mg) without any sample preparation is at least as good as using an internal standard technique or other measurement technique that involves sample weight or geometric reproduction of small sample sizes.

Finally, when Method No. 5, without any sample preparation, is employed it is simple, rapid, economical, precise and allows small samples to be measured, as compared to atomic absorption and atomic emission, where the sample in most cases needs to undergo a long dissolution process to ensure complete decomposition.   With respect to neutron activation, various elements not only have a long irradiation time but a longer cooling time (weeks) before the samples can be counted.   In some cases, these techniques need to be employed for a variety of reasons including better detection limits.

REFERENCES

1. J.J. LaBrecque, A.N. Hanna, A.A. Abdad-Rassoul and R. Schelenz, IAEA/RL/
      143, July 1987, Vienna, Austria.
2. J.K. Taylor, "Standard Reference Materials: Handbook for SRM Users," NBS
      Special Publication 260-10, September 1988, p. 16
3. J.J. LaBrecque, Adv. in X-Ray Anal. 24(1981) 393-398
4. K. Borowski, I.L. Priess, J.J. LaBrecque and C. Pawley, Computer Enhanced
      Spectroscopy 1(1983) 99-104

QUANTITATIVE ANALYSIS OF ARSENIC ELEMENT IN A TRACE OF WATER

USING TOTAL REFLECTION X-RAY FLUORESCENCE SPECTROMETRY

T. Ninomiya, S. Nomura*, K. Taniguchi* and S. Ikeda**

Forensic Science Laboratory, Hyogo Prefectural Police
Headquarters, Hyogo, Japan
 * Osaka Electro-Communication University, Osaka, Japan
** Ryukoku University, Kyoto, Japan

INTRODUCTION

Various kinds of compounds containing arsenic have been used in the
world as rat poisons, agricultural chemicals and so on. In the field of
semiconductor materials, AsH_3 is used as a doping gas on silicon-wafer
substrates and GaAs is now also investigated as a wafer substrate instead of
Si wafer.

As for quantitative analysis of arsenic, atomic absorption spectrometry
(AAS) and inductively coupled plasma atomic-emission spectrometry (ICP-AES)
have often been used. In these methods, usually, complicated pretreatments
such as preconcentration and separation have been needed in order to obtain
reproducible values for arsenic. For instance, in AAS the procedure for
arsine gas evolution has been introduced. It is well-known that any nitric
acid remaining in the sample solution must be removed completely in order to
avoid its interference with the formation of arsine gas. Furthermore, both
methods need a considerable number of samples and the samples cannot be
recovered after the measurement in either method.

In practice, sample availability often dictates very small sample
masses, e.g., for forensic evidences, biological samples, archaeological or
extraterrestrial materials, valuable synthesized compounds and so on. Non-
destructive analysis is also ideal for these materials.

Total reflection X-ray fluorescence spectrometry (TRXRF) gives some of
the answers to the demands for the above valuable samples. This technique
has a favorable cost-benefit ratio and allows multielement determination even
in a micro-trace sample.

The use of TRXRF was first reported by Yoneda and Horiuchi[1]. The
authors have reported previously that TRXRF was quite useful to discriminate,
in forensic samples such as small chips of cloth[2], trace amounts of titanium
oxide pigments[3] and traces of plastic materials[4]. Iida et al. applied not
only conventional X-ray tube but also synchrotron radiation to TRXRF and
reported that this method was useful to analyze micro amounts of elements on

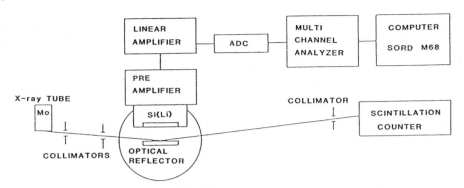

Fig. 1   Schematic diagram of TRXRF.

the surface[5]. As for quantitative analysis using TRXRF, Wobrauscheck and
Aiginger reported the determination of Mn and Cr in 5 μl of sample[6], Prange
and Knöchel reported multielemental determinations in sea water in which 300
ml of sample volume[7] was used, and Stössel and Prange reported trace elements
in rain water using 200 μl of sample[8].

In this paper, the authors have selected three kinds of sample volume,
that is, 0.5 μl, 1 μl and 10 μl and investigated factors influencing the
quantitative measurement for arsenic compared with ICP-AES.

EXPERIMENTAL

Figure 1 shows the schematic diagram of TRXRF. The arsenic samples
below 1000 ppm concentration were prepared by diluting the arsenic standard
solution (1000 ppm for arsenic, Wako Pure Chemicals Industries, for atomic
absorption spectrometry use) to 0.01, 0.02, 0.2, 1, 2, 5, 10, 20, 100, 500
ppm, respectively. The solutions of 5000 ppm and of 1 percent were prepared
by solvating directly arsenic trioxide into 5 percent sodium carbonate
solution. Optically flat Pyrex glasses (Furuuchi Chemicals Industries, grade
2676, 26 mm x 19 mm x 1 mm) were used as sample substrates. Before use, they
were cleaned with detergent, water and ethanol in the ultrasonic vessel,
successively.

The sample solutions were dropped on the center of the optically flat
glasses by micropipette (Eppendorf) and dried under vacuum conditions.

In order to improve the reproducibility of measurement, the Pyrex glass
surfaces were modified with the fluorocarbon polymer, RB-106 (Neos Company
Ltd.; it has been reported by the manufacturer that the surface tension of
the polymer should show 18.7 dyn/cm). The cleaned Pyrex glass was dipped
into RB-106, which was adjusted previously to be a 0.5 percent solution of
Freon 113, for 10 seconds and dried for several hours in atmosphere.

The measuring conditions are summarized in Table 1. The measuring time
was usually set to 200 seconds, while for samples with concentrations below
0.1 ppm, measuring time was set to 1000 seconds and the observed intensities
were converted to the 200 seconds unit. The sample solution was spotted
carefully at the center of the substrates and placed at the position facing
the center of the solid state detector (SSD). The angle of the incident

Table 1.   Measuring conditions

X-ray target	Mo
Exciting voltage	20kV
Tube current	30mA
Glancing angle	0.03°
Area of X-ray	10μm x 10mm
Detector	SSD(Si(Li))
Area of detector	10mm²
Distance between SSD and sample	7mm
Sample holder	Pyrex glass (26mm x 19mm x 1mm)
Sample volume	0.5, 1.0 or 10.0μl
Measuring time	200 or 1000sec

X-ray beam was monitored by the scintillation counter for setting samples normally under total reflection conditions.

Measurements by ICP-AES were performed using the Plasma Model 40 of Perkin Elmer Japan and the wavelength at 193 nm was selected for detection.

RESULTS AND DISCUSSIONS

In general, it is very important to understand the available dynamic range of the measurement for the proposed analytical techniques. The result is shown in Figure 2. The working curve gives a good linear relationship between the intensity of fluorescence X-rays and the concentration of arsenic in the range of 0.02 ppm to 1000 ppm. In cases of concentrations of 5000 ppm and 1 percent, intensities of the fluorescence X-rays deviate downward from

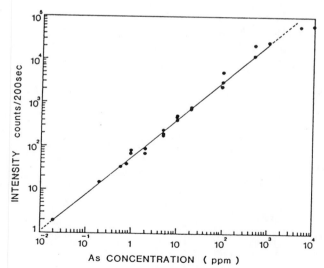

Fig.2.   Plots of the logarithm of the X-ray intensity versus the logarithm of arsenic concentration.(1μl sampling)

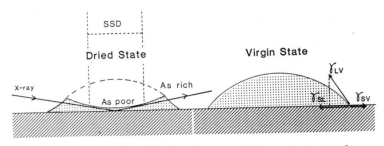

Fig. 4   Model of X-ray path expected an ill-mounted sample surface.
         (10 μl sampling)

        In cases of high concentration solution such as 1000 ppm, however, a
kind of crater shown in Figure 4 was usually observed by microscope after
vacuum evaporation, respectively.   In the left model of Figure 4, the edge
site of the crater was rich in arsenic content, while it was poor near the
center of the crater.   Therefore it can be supposed that the crater
phenomenon could exert a large effect on the reproducibility of the intensity
of fluorescence X-rays; see Figure 4.

        In order to improve the reproducibility for quantitative analysis of
arsenic and to overcome the crater phenomenon, the surface of the Pyrex glass
substrate was modified by the fluorocarbon polymer (the procedure is called
FC-modification as follows), so that the whole of the samples could be
accumulated near the center of the substrate.   The procedure of modification
with the fluorocarbon polymer is very simple as shown in the experimental
section.

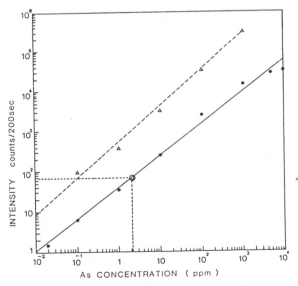

Fig. 5   Comparison of working curves for FC-TRXRF and ICP-AES.
         (1 μl sampling)

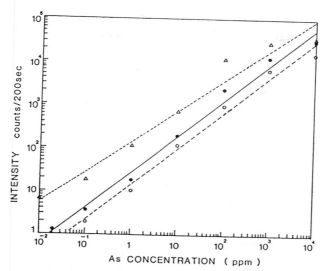

Fig.3.   Plots of the logarithm of the
X-ray intensity versus the logarithm of
arsenic concentration at 3 levels of
sampling: Δ,10μl; ●,1μl; O,0.5μl.

the curve. These deviations can be considered to be caused in part by
miscounting by the SSD because of the high concentration of arsenic.

Next, each aliquot of 0.5, 1 and 10 μl was chosen as a level of
sampling mass and the effect of sampling mass was examined on the intensity
of fluorescence X-rays in the range of 0.01 ppm to 1 percent concentration
for arsenic. The measurement was done on 3 samples for each concentration
level and the average value of the obtained intensity at each sampling mass
level has been plotted against concentration as shown in Figure 3.

In Figure 3, the intensity of fluorescence X-rays in the case of 1 μl
sampling is nearly 2 times as large as that in the case of 0.5 μl sampling
for each concentration. On the other hand, the intensity with 10 μl sampling
is at most 5 times as large as with 1 μl sampling at each concentration,
contrary to the theoretical expectation of being 10 times as large. This
must be due to the width of the sample exceeding the limited surface area
measurable by the SSD. The effective area of the SSD used in this experiment
is 10 mm so that the useful diameter of SSD is calculated to be nearly 3.6 mm.

Figure 4 shows the models for the sample in the case of 10 μl sampling.
The right model represents a virgin state immediately after spotting and the
left model represents a dried state. In this scheme, the effective diameter
of the SSD was 3.6 mm while the sample diameter was nearly 5 to 8 mm. The
sample diameter can be understood easily to be larger than that of the SSD as
shown in Figure 4. It can be concluded that the whole of the fluorescence
X-rays derived from dried samples in case of 10 μl sampling could not be
detected perfectly, even though a primary X-ray beam that could irradiate a
10 mm x 19 mm area could excite the whole area of dried samples. That is to
say, in the quantitative analysis of a trace of sample using TRXRF, the
reproducibility of X-ray fluorescence intensity would depend much on the
sample area in correlation with the effective area of SSD.

Table 2. Comparison of reproducibility for As by FC-TRXRF
and ICP-AES

	FC-TRXRF	ICP-AES
sample 1/ppm	1.9	2.7
sample 2/ppm	2.4	2.8
sample 3/ppm	2.0	3.0
average/ppm	2.1	2.8
CV/%	8.3	4.2

After sampling 1 μl solution for each concentration on the FC-modified
glass surface, each diameter of dried samples was nearly 1 mm. In other
words, the diameter in case of FC-modification was reduced so as to be nearly
one-third of that in case of non-modification. Also, the crater phenomenon
could scarcely be observed.

The effect of FC-modification on the reproducibility of measurement was
investigated with 1 μl sampling. Figure 5 shows the working curves for
FC-TRXRF and for ICP-AES comparatively. The measurement was done on 3
samples for each concentration and each averaged value was plotted in Fig. 5.
The intensities by FC-TRXRF were good correlation with that by ICP-AES. A
linear regression fit in case of FC-TRXRF yields lnY = 3.721 + 0.790 lnX, and
the relative coefficient is 0.919, while lnY = 6.225 + 0.893 lnX, and the
relative coefficient is 1.0 in case of ICP-AES.

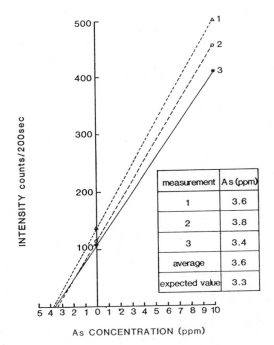

measurement	As (ppm)
1	3.6
2	3.8
3	3.4
average	3.6
expected value	3.3

Fig. 6 Arsenic determination by standard addition method using TRXRF.

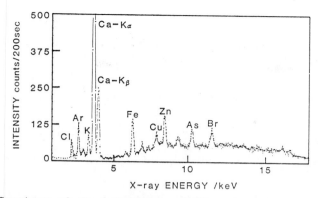

Fig. 7   Spectrum of the extracted solution from Hizikia fusiforme.
          (1 μl sampling)

Also a sample of unknown concentration for arsenic was measured with
the FC-TRXRF system for comparison with ICP-AES. The measurement was done on
3 samples (Table 2). The arsenic concentration with FC-TRXRF was 2.1 ppm in
average and that with ICP-AES was 2.8 ppm in average. Thus, the result by
FC-TRXRF coincides fairly well with that by ICP-AES. Although the value
obtained by FC-TRXRF deviates slightly from that by ICP-AES, the proposed
method is applicable to the non-destructive analysis of arsenic in forensic
samples.

Figure 6 shows the results of a trace of mineral water from the Japan
Alps Mountains to which an aliquot of arsenic solution had previously been
added. While the expected concentration for arsenic in the mineral water was
3.3 ppm, the value averaged after measurement of 3 samples using FC-TRXRF
system was 3.6 ppm. Hence, the value by FC-TRXRF agreed well with the
expected value.

As the last example, an extract solution from Hizikia fusiforme, a kind
of seaweed gathered in Wakayama prefecture, was analyzed for arsenic using
FC-TRXRF. The spectrum is shown in Figure 7. The peak of arsenic element is
detected clearly and the concentration for arsenic is calculated to be 1.8
ppm from the calibration curve shown in Figure 5.

CONCLUSIONS

In the quantitative analysis of elements in a sample using TRXRF, it
must be considered that the reproducibility of X-ray fluorescence intensity
would depend strongly on the sample area in correlation with the effective
area of the SSD. Also the crater phenomenon often observed after drying the
liquid sample could have a large influence on the reproducibility of the
measured value. The characteristics of FC-TRXRF can be summarized as follows
compared with that of ICP-AES. As a sample mass, 0.5 μl or 1 μl is
sufficient for FC-TRXRF, while ICP-AES needs about 0.1 ml to 1 ml. The
dynamic range is five orders of magnitude for FC-TRXRF; on the other hand it
is four orders of magnitude for ICP-AES. Minimum low limits are 20 pg for
FC-TRXRF against 10 ng for ICP-AES. Moreover, samples can be recovered in
FC-TRXRF because of non-destructiveness, but cannot be recovered in ICP-AES.
A critical value for FC-TRXRF is at most 11 percent in FC-TRXRF, and is
rather lower than that for ICP-AES.

REFERENCES

1. Y. Yoneda and T. Horiuchi, Optical Flats for Use in X-Ray Spectrochemical
    Microanalysis, Rev. Sci. Instrum., 42, 1069 (1971).

2. T. Ninomiya, S. Nomura and K. Taniguchi, Application of Total Reflection
    X-Ray Fluorescence Analysis to Forensic Model Samples, Memoirs of Osaka
    Electro-Communication University, 22, 51 (1986).

3. S. Nomura, T. Ninomiya and K. Taniguchi, Trace Elemental Analysis of
    Titanium Oxide Pigments using Total Reflection X-Ray Analysis,
    Advances in X-Ray Chemical Analysis Japan, 19, 217 (1988).

4. T. Ninomiya, S. Nomura and K. Taniguchi, Elemental Analysis of Trace
    Plastic Residuals using Total Reflection X-Ray Fluorescence Analysis,
    Advances in X-Ray Chemical Analysis Japan, 19, 227 (1988).

5. A. Iida, A. Yoshinaga, K. Sakurai and K. Gohshi, Synchrotron Radiation
    Excited X-Ray Fluorescence Analysis using Total Reflection of X-Rays,
    Anal. Chem., 58, 394 (1986).

6. P. Wobrauschek and H. Aiginger, Total-Reflection X-Ray Fluorescence
    Spectrometric Determination of Elements in Nanogram Amounts,
    Anal. Chem., 47, 852 (1975).

7. A. Prange and A. Knochel, Multi-Element Determination of Dissolved Heavy
    Metal Traces in Sea Waters by Total-Reflection X-Ray Fluorescence
    Spectrometry, Anal. Chim. Acta., 172, 79 (1985).

8. R. P. Stössel and A. Prange, Determination of Trace Elements in Rain Water
    by Total-Reflection X-Ray Fluorescence, Anal. Chem., 57, 2880 (1985).

IMPURITY ANALYSIS OF SILICON WAFERS

BY TOTAL REFLECTION X-RAY FLUORESCENCE ANALYSIS

Sigeaki Nomura, Kazuo Nishihagi,* and Kazuo Taniguchi

Osaka Electro-Communication University, Osaka, Japan
*Technos Corporation, Osaka, Japan

INTRODUCTION

High purity of silicon wafers is demanded by the high performance and highly integrated IC and LSI semiconductors. Impurities on, or in, the silicon wafer have a big influence on the characteristics of the semiconductor as a final product. Usually, these impurities are introduced by water during washing, by bad handling, or by reagents and processes.

Typical influences of these impurities are shown in Table 1. These elements are present at too small concentration to be detected by ordinary analytical methods (except for oxygen). Usually, XPS, AES, NAA, SIMS, ICP-AES, ICP-MS and AAS are used for trace analysis. XPS and AES are useful for surface analysis, but are seldom used for qualitative or quantitative analysis of the impurity analysis. NAA and SIMS have sufficient sensitivity for trace elements, but the former requires a nuclear reactor as an excitation source.

ICP-AES analysis is destructive and time for pretreatment of the sample is required. ICP-MS has higher sensitivity, but is very expensive and difficult to perform. Recently, total reflection x-ray fluorescence spectrometry (TRXRF) has been reported by many researchers as a trace analysis technique. This method has many advantages, for example, high sensitivity, non-destructiveness and rapidity; it requires no sample pretreatment.

In this paper, impurity analysis of silicon wafers, using TRXRF, is reported. A lower limit of detection (LLD) of the order of 10^{11} atoms/cm^2

Table 1. Influence of Impurity Elements on Silicon

Element	Influence
O	Single crystal of silicon trap included oxygen
Na, K	Instability of threshold voltage of MOS device
Fe, Ni, Cr, Cu	Increasing leak current of junction
Au, Ag, Pt	Decreasing life-time
As	Doping element - controls resistance
U	Introduces soft error by alpha-ray

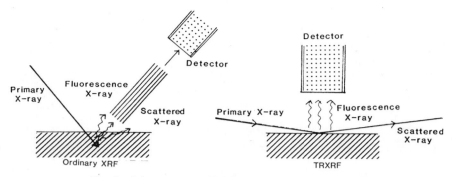

Fig.1. Comparison of TRXRF method with ordinary XRF

is achieved. This value is not sufficient at present for quality control, but could be used for on-line analysis in the near future by improving the excitation source and detector.

EXPERIMENTAL

    The comparison of ordinary x-ray fluorescence (XRF) with TRXRF is shown in Fig. 1. In ordinary XRF, primary radiation strikes the sample surface at a higher angle than in TRXRF. In XRF the fluorescent x-rays from the sample and the primary x-rays scattered by the sample and substrate are measured by the detector. The intensity of the signal from the sample must be above the background. Intensities of the fluorescent x-rays from trace elements are weak compared to the background. Therefore, regretfully, trace elements cannot be estimated using the ordinary XRF method.
    On the other hand, in TRXRF, the fluorescent x-rays are emitted from the sample, but the primary radiation is totally reflected by the optically flat sample holder, away from the detector. Therefore, TRXRF is expected to be useful for impurity analysis.
    The critical angle for total reflection is a function of the material having an optically flat surface and the energy of the primary x-ray beam, as given in eq. (1)

$$\theta_{critical} = \left(\frac{5.4 \times 10^{10} Z \cdot \rho \cdot \lambda^2}{A} \right)^{1/2} \tag{1}$$

where, Z, A, ρ and λ are atomic number, mass number, density and wavelength, respectively. The relationship between incident angle and critical angle is shown in Fig. 2. In the case of the incident angle being larger than the critical angle, the incident x-ray is refracted into the sample or the sample holder. In the case of the incident angle equal to the critical angle, the incident x-ray is neither refracted nor reflected. When the incident angle is smaller than the critical angle, the incident x-ray is totally reflected at that same angle.

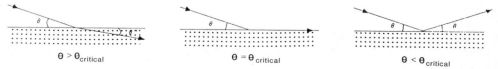

$\theta > \theta_{critical}$ $\theta = \theta_{critical}$ $\theta < \theta_{critical}$

Fig.2. Relation between incident angle and critical angle

The variations of the peak profile for the different incident angles is shown in Fig. 3. These spectra are obtained from 8 ng of copper (10 μL of an aqueous solution deposited on optically flat glass), excited by a Mo target at 20kV and 30mA. When the incident angle is 0.1°, the scattered primary x-ray is cut off at about 12 keV. As the incident angle decreases, the scattered intensity is reduced, and the fluorescent signal increased. At an angle less than the critical angle the primary radiation is virtually all reflected away from the detector, producing very low background. Also, the penetration depth of the primary x-rays is very low, only about 100 Å, TRXRF provides only surface information. In addition, the low background leads to optimum LLD.

Instrument

A block diagram of TRXRF is shown in Fig. 4. The fine-focused x-ray tube is used as the excitation source. The target is molybdenum, because of its few interference lines. The tube is operated at 30 kV and 40 mA; the beam size is 20 μm, defined by the collimators. The incident angle is 0.03°. A 30 mm^2 Si(Li) solid state detector is located 5 mm from the sample. The signal is processed to the personal computer via a linear amplifier and an A/D converter. A servo mechanism adjusts the sample position automatically to maintain the incidence angle at 0.03°, by controlling the Z axis and the theta angle, using a scintillation counter as a sensor. The reproducibility of this system is about 5%, sufficient for the impurity analysis.

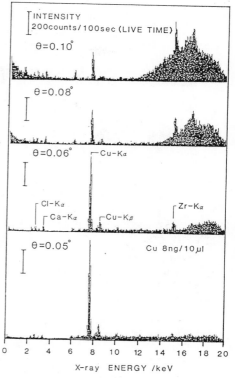

Fig.3. Variation of peak profile for diffrent incident angle

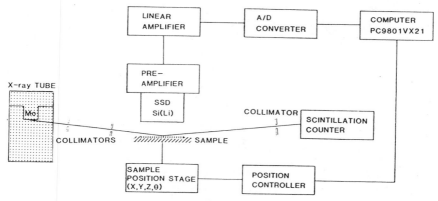

Fig.4. Block diagram of TRXRF

Working Curves

K, Ni, Cu, As and U were selected as typical trace elements. Working curves were produced by using samples made by a micro-drop method, according to the following steps:
1. Standard solutions of each element were prepared at concentration levels of 100 ppb, 1 ppm and 10 ppm, with pure water used as the blank.
2. 10 μL of each solution was placed on a virgin silicon wafer using a micropipette.
3. The sample is vacuum-dried.
4. The sample is placed in the instrument and analyzed by TRXRF.

Contaminated sample

A simulation of a contaminated silicon wafer was prepared by two different methods:
   The first method is called spin-contamination, and consists of the following steps:
1. A solution of each element is prepared at x ppm.
2. 6 mL of the solution is placed on the Si wafer and allowed to diffuse into the SiO_2 layer for 160 seconds.
3. The Si wafer is rotated at 2500 rpm for 20 seconds, removing the solution, and leaving the contaminant in the SiO_2 layer.
4. The sample is measured by TRXRF.
5. The SiO_2 layer is dissolved by HF.
6. The resultant solution is analyzed by AAS.
7. The impurity in the SiO_2 layer is calculated as atoms/cm^2.
8. The AAS result is compared to the TRXRF result.

The second method of contamination uses ion-implantation; the Si wafer doped with As ions is analyzed by TRXRF and NAA.

RESULTS AND DISCUSSION

The working curve for each element in the samples produced by the micro-drop method is generated by plotting the intensity versus the

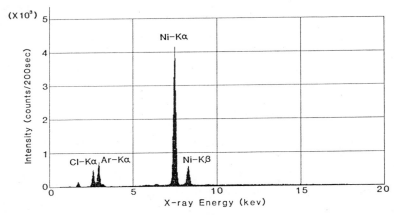

Fig.5. Peak profile of Ni (micro-drop:10ppm)

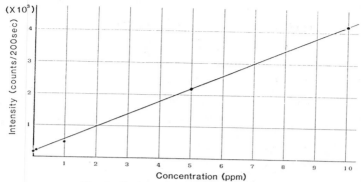

Fig.6. Working curve of Ni (micro-drop)

concentration of the contaminant.   The curves are linear over the
concentration range of 0 to 10 ppm. Fig. 5 shows the spectrum for the 10
ppm Ni sample.   Ni Kα and Ni Kβ are obvious, Ar is from the air, and the
Cl is the anion in the solution.   Figure 6 is the working curve for Ni.
     The lower limits of detection are calculated for each element using
the slope of the working curve and the background intensity, and are shown
in Table 2, after conversion to
atoms/cm^2.   This table represents
data from both the micro-drop method
and the spin-contamination method.
These values are not sufficient for
silicon   suppliers.   However,   we
assume that we shall be able to
improve those values by one order
of magnitude, by using a higher-
power excitation source, efficient
optics and efficient detector.

Table 2. LLD's on Si Wafers

Element	LLD (atoms/cm^2)
K	6.0×10^{11}
Cr	2.3×10^{11}
Ni	2.0×10^{11}
Cu	1.4×10^{11}
As	2.4×10^{11}
U	7.5×10^{11}

     The spectrum of a pure silicon wafer should show only peaks due to Ar
in the air and the Si. Figure 7 shows the spectrum of an actual Si wafer,
unacceptable because of the Fe peak; from the Fe intensity, it can be
estimated that the wafer contains about 10^{13} atoms/cm^2, close to the value

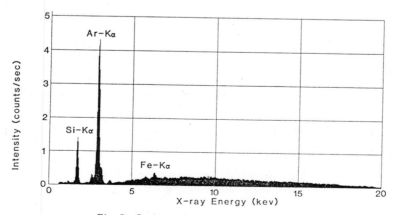

Fig.7. Peak profile of actual Si wafer

estimated by the manufacturer. The
working curve for Ni, prepared by spin-
contamination, is shown in Fig. 8. The
surface concentrations were measured by
AAS, and the x-ray intensity by TRXRF.
10^{12} atoms/cm^2 is equal to about 10
ppb, and 10^{14} atoms/cm^2 is equal to
about .1 ppm. These values are very
close to those realized using a working
curve by the micro-drop method. The
spectrum for the sample prepared by
using ion-implantation to dope the Si
wafer with As is shown in Fig. 9. The
doping time was 30 minutes with a
temperature of 1000° centigrade. The
As intensity of 5 counts per second
converts to about 1 x 10^{14} atoms/cm^2
impurity level, using the working curve
for As prepared by the drop method. In
this case also, the measured value of
As by TRXRF is very close to that
determined by NAA.

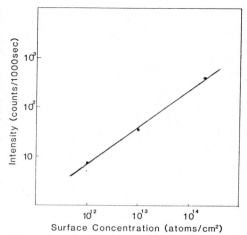

Fig.8. Working curve of Ni
(spin-contamination)

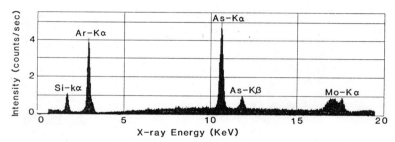

Fig.9. Peak profile of As (Doping:1.1x10^{14} atoms/cm)

CONCLUSION

TRXRF is suited for analyzing the impurities in silicon wafers, because
it is non-destructive, rapid, requires no sample preparation and is easy to
operate for surface analysis. However, the LLD's are not sufficient for
quality control; an improvement of an order of magnitude is necessary.
There are two ways to improve the LLD. The use of a rotating-anode x-ray
tube as an excitation source could improve the intensity by that order of
magnitude. And by monochromatization of the primary radiation, the
background can be reduced, further improving the LLD.

    The authors thank Professor S. Ikeda for his helpful suggestion, N.
Fujino, S. Sumita, H. Ohguro and H. Naganuma for their useful discussions,
and Inoue for his support.

SAMPLE TREATMENT FOR TXRF - REQUIREMENTS AND PROSPECTS

Andreas Prange and Heinrich Schwenke

Institut für Physik, GKSS-Forschungszentrum Geesthacht GmbH
P.O.Box 1160, D-2054 Geesthacht, Federal Republic of Germany

INTRODUCTION

Total-reflection X-ray fluorescence spectrometry, abbreviated as TXRF, is known for its high sensitivity down to the low pg-level or sub-ppb level, respectively, and its wide dynamic range of about three to four orders of magnitude (Yoneda and Horiuchi, 1971, Wobrauschek and Aiginger, 1980; Knoth and Schwenke, 1978 and 1980, Aiginger and Wobrauschek, 1985, Michaelis et al., 1985, Prange, 1987). Meanwhile several laboratories have purchased commercially available TXRF spectrometers and have started to report favourable about this technique. Applications have been reported from various disciplines: These are estuarine and marine water quality management and research, air pollution studies, mineralogical investigations, biology and medicine (Prange, 1987, Prange et al, 1985; Prange and Kremling, 1985, Prange et al., 1987, Stößel and Prange, 1985, Michaelis, 1986, Ketelsen and Knöchel, 1985, Leland et al., 1987, von Bohlen et al., 1987, Junge et al., 1983, Hentschke et al., 1985, Hentschke et al., 1985, Gerwinski and Goetz, 1987, von Bohlen et al., 1987). In spite of its close kinship to conventional EDXRF, TXRF is quite different with respect to operation and performance and provides complementary capabilities.

The present paper elucidates the special features of TXRF in view of the requirements for sample preparation for trace element analysis. Based on two examples of application a perspective on sample treatment will be presented which meets these special requirements.

REQUIREMENTS FOR SAMPLE PREPARATION

The measurement of thin film samples

For TXRF the angle of incidence of the exciting radiation is only a few minutes of arc, such that the radiation is totally reflected from a highly polished sample support. The total-reflection X-ray optics are shown in Figure 1. The sample is placed on the sample carrier as a thin film made from solutions or at least fine suspensions by evaporating the

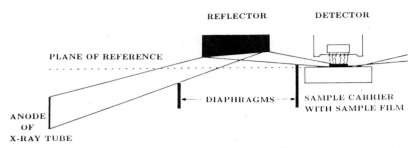

Fig. 1.   Schematic design of the total-reflection X-ray optics

solvent. The fluorescence radiation is doubled in intensity because of
excitation of the sample by both the incident and the reflected radiation. Because of the small angle of incidence, the primary beam has virtually no interaction with the sample support, thus leading to a drastic
reduction of scattered radiation which results in substantial improvements in peak to background ratios and detection limits. The reflector in
front of the sample carrier serves as a high-energy cut-off filter.
It ensures that only radiation fulfilling the total-reflection conditions
reaches the sample support, thus eliminating high-energy bremsstrahlung
by diffuse scattering and absorption. Direct primary radiation is retained by the diaphragms. For the excitation Mo- and W-anodes were used
(operating conditions: 50 kV and max. 38 mA). The fluorescence radiation
is detected at 90 degrees by a Si(Li)-detector. Finally, the compact construction of the total-reflection module provides for a high intensity of
the exciting radiation on the sample spot.

Internal standardization

Because of the use of thin films as measuring samples no corrections
for matrix absorption and enhancement effects of the fluorescence radiation are required. This means that a constant relationship of the fluorescence yield versus atomic number is valid as long as all instrumental
parameters are not changed.

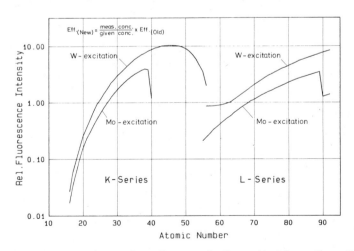

Fig. 2.   Calibration curves for Mo- and W-excitation for the K- and
          L-lines

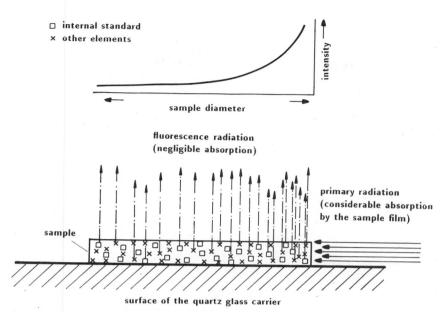

□ internal standard
× other elements

Fig. 3.   Schematically description of the excitation of a thin film
          sample

Figure 2 shows calibration curves for Mo- and W-excitation and for
the K- and L-lines, respectively. The slopes of the calibration curves do
not depend on the properties of the sample. Spiking the sample with a
well-defined quantity of an element which is not expected in the sample
and may serve as an internal standard, allows the quantitation of all
detected elements.

Due to the extremely low angle of incidence the primary radiation is
attenuated by the sample matrix along the pathway through the spot. Local
accumulations of one distinct element in the sample film would lead to a
systematic error. The conditions are shown in Figure 3. Therefore, an
important prerequisite for the precision and accuracy of the TXRF results
is a homogeneous distribution of the detected elements and the internal
standard within the sample spot.

Influence of the matrix

The interference-free detection limits of a few pg (e.g. 2 pg for
Ni, 1000 s counting time) are worsened - as is general with X-ray fluorescence methods - with increasing matrix content. The dependence of the
detection limits for TXRF on the matrix concentration is illustrated in
Figure 4.

It shows a linear correlation over several orders of magnitude for
the example of nickel in a sodiumchloride matrix. This demonstrates that
TXRF succeeded in achieving the fundamental limits of energy dispersive
XRF to a high degree. Apparative contributions to scattered radiation are
almost eliminated except for small residual effects perceptible from the
slope of the function at very low concentrations.

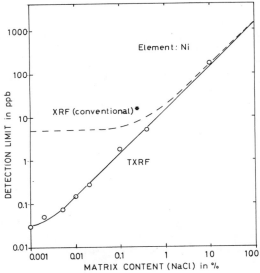

In contrast to TXRF, scattering from the instrument or sample support in conventional XRF introduces a background which dominates up to comparatively high concentrations. This is illustrated by the matrix-independent part of the broken line in Figure 4, based on data recently published for a conventional XRF instrument which was optimized for trace element analysis also using thin film samples (Rastegar et al., 1986).

To utilize the high detection power of TXRF in the case of very low element concentrations in the presence of high matrix content, chemical separation procedures are mandatory.

Fig. 4.   Dependence of the detection limits on the example of nickel in aqueous solution with increasing NaCl-concentration

Overall the requirements for sample preparation can be summarized as follows:

- preparation of thin film residues from solution and/for fine suspensions
- solvents which can easily be evaporated
- the necessity of internal standardization
- similar behaviour of the internal standard and the elements under investigation
- the amount of the sample spot should not exceed 10 µg
- separation of one element or an element group in excess in order to exploit the high detection sensitivity.

PROSPECTS FOR SAMPLE PREPARATION

For the diverse applications mentioned in the introduction, appropriate sample preparation techniques have already been developed and adapted to TXRF.

Figure 5 gives a survey of these preparation techniques for liquid and solid samples and demonstrates the differences in sophistication. Direct measurements and freeze drying are only applicable where no or only small amounts of matrix elements occur, for instance for rainwater, certain ultrapure fluids, or the measurement of radioactive solutions. In the presence of one or more elements in excess, however, the detection power suffers because of scattering and peak overlapping. For the quantitative separation of alkaline and alkaline earth elements and the enrichment of the trace elements, the reverse-phase technique with sodium dibenzyldithiocarbamate as complexing agent has proved to be suitable. For natural water with larger amounts of organic matrix, a digestion is needed prior to the separation and concentration step.

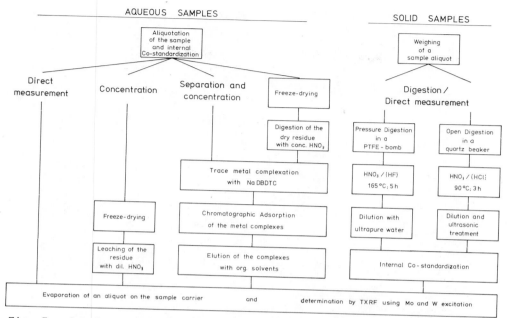

Fig. 5. Sample preparation techniques for aqueous and solid samples hitherto adapted to TXRF

For solid samples digestion procedures are necessary in most cases. This can be a pressure digestion with HNO_3/HF in a PTFE bomb, or an open digestion with HNO_3/HCl in a quartz vessel as used for AAS or ICP-OES. For TXRF measurements the digestion does not have to be complete.

In the following two examples of typical applications are explained, the first demonstrating the more or less instrumental use of TXRF in the course of which the sample support serves as sampling target for airborne particulates (Schneider, 1989), the second demonstrating the capability of TXRF in ultra trace analysis of estuarine water by combination with chemical pretreatment.

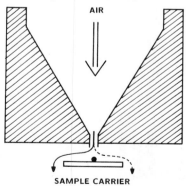

Fig. 6. Operation mode of one stage of the cascade impactor (according to Schneider, 1989)

Instrumental use of TXRF

For the use of carrier plates as targets for a size-separating cascade sampler of airborne particulates the sample carriers are mounted in a stage cascade impactor. Figure 6 shows the operation mode on one stage of the cascade impactor. A stream of particle-laden air is directed at the sample support for TXRF. Particles of sufficient inertia will impact upon the surface and smaller particles will follow the air streamlines to the next stage. The airborne particle spots are subsequently supplied with an internal standard, pipetting a drop of standard solution (e.g. Y) upon the spot and allowing to dry.

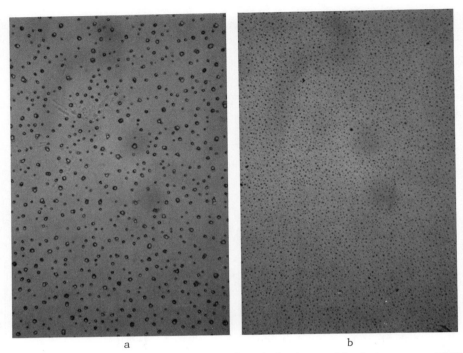

a   b

Fig. 7. Microscope pictures of airdust particulates on TXRF
        sample carriers with aerodynamic diameter of 2 - 4   m (a)
        and 0.5 - 1 m (b)

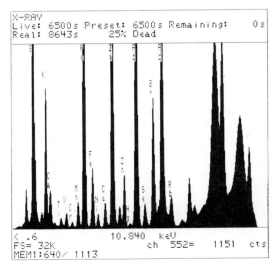

Fig. 8. TXRF-spectrum of an airdust sample collected directly on the
         sample carrier.

Figure 7 displays two microscope pictures from airdust particulates on TXRF sample carriers with aerodynamic diameters of 2 - 4 μm and 0.5 - 1 μm, respectively. An almost homogeneous distribution of the sample has been obtained. A TXRF spectrum of such a sample is shown in Figure 8.

This demonstrates the good peak to background ratio for this kind of sample. Elements like S, K, Ca, Ti, V, Cr, Mn, Fe, Ni, Cu, Zn, Hg, Pb, Se, Br and Rb are detected with concentrations for instance of V 2.7 ng/m³ and Pb 30.2 ng/m³. Compared to typical aerosol investigations using filters and/or digestions of particle loaded filters, the rate of air flow can be reduced considerably. This procedure enables the detection of e.g. lead concentrations of 1ng/m³ in 30 minutes collection time. There is certainly a great potential for applications using this more or less instrumental procedure, for example in the fields of operating conditions control, the analysis of smoke etc.

Combination with chemical preparation steps

The second example deals with the determination of trace elements in estuarine water. To take full advantage of the high detection sensiti-

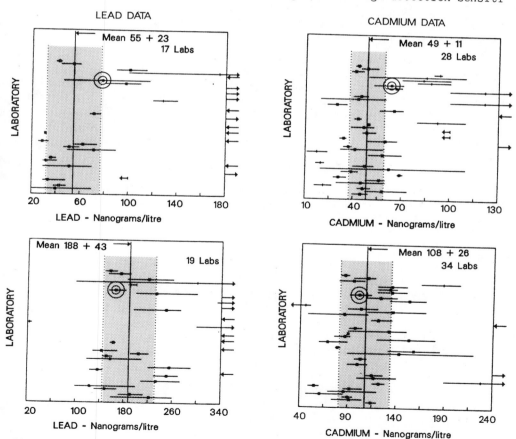

Fig. 9.   Lead and cadmium data from the ICES sixth round intercalibration for trace metals in estuarine water (JMG 6/TMSW)

vity of TXRF also for such matrix loaden samples the reverse phase technique in conjunction with sodium dibenzyldithiocarbamate as complexing agent has been used for the separation and concentration of the heavy metals. This procedure has already been described in the literature in detail (Prange et al., 1985).

To ensure the reliability of the results obtained with this procedure in estuarine field studies, interlaboratory tests have been performed. Some results from an ICES-intercalibration for trace metals in estuarine water are shown, organized by Dr. Shier Berman at the National Research Council in Canada.

About 45 of the 78 participating laboratories have provided results for the intercomparison of two samples from the river Schelde (Belgium) for the determination of the elements Cu, Zn, Cd, Ni and Pb. A third sample in a glass bottle was for the determination of Hg. Most of the laboratories were very experienced in this kind of analysis.

We were the lab 38 A and have reached 8 points out of the 11 possible. Overall, we were in position 5. Hg was not determined because of contamination problems. The results for Cu, Zn and Ni were all accepted, but one each only was accepted for Cd and Pb. Figure 9 shows these results in more detail.

From the 45 participants only 17 or 19 labs, respectively, were used for the evaluation of the Pb results, the others are out of the considered range. Our results are marked by a circle. From the upper graphs our two rejected values are recognizable, being just out of the standard deviation range. These values are very near the detection limits of about 50 ng/l for Pb and Cd for the procedure employed. No problems arose with the second sample where the concentrations are slightly higher.

CONCLUSION

The applications presented are exemplary for a series of appropriate sample preparation procedures which have been adapted to TXRF and have stood the test in manifold applications. Utilizing the special features of TXRF for trace element analysis (the high detection power, the multielement capability, the inherent calibration curve with internal standardization and the minute sample masses required) TXRF not only competes with other analytical techniques, but also offers new possibilities in tackling intricate analytical problems. Some potential applications are finally summarized which also exploit the special features of TXRF.

These are
- quantitative microanalyses of samples in the 10 - 100 µg region
  - special biological specimen
  - forensic applications
  - microinclusions
  - minute amounts of radioactive solutions
- multielement capability down to ppt for 100 ml samples in matrices which can be evaporated or otherwise separated
  - ultrapurity reagents like acids or organic solvents
  - envrionmental samples like rainwater, seawater etc.

REFERENCES

Aiginger H. and Wobrauschek P., 1985, Total Reflectance X-Ray Spectrome-
    try, Adv. X-Ray Anal. 28:1-10.
Bohlen, A.v., Eller, R. , Klockenkämper R., and Tölg, G., 1987, Micro-
    analysis of Solid Samples by Total-Reflection X-Ray Fluorescence
    Spectrometry, Anal. Chem. 59:2551-2555.
Bohlen, A.v., Klockenkämper, R., Otto, H., Tölg G., and Wiecken, B.,
    1987, Qualitative survey analysis of thin layers of tissue samples -
    Heavy metal traces in human long tissue, Int. Arch. Accup. Environ.
    Health. 59:403-411.
Gerwinski W. and Goetz, D., 1987, Multielement analysis of standard
    reference materials with Total Reflection X-ray Fluorescence (TXRF),
    Fresenius Z. Anal. Chem. 327:690-693.
Hentschke, U., Junge, W., and Rath, R., 1985, Chemical and Optical Pro-
    perties of Columbites, N. Jb. Miner. Abh. 152:113-121.
ICES cooperative research report No. 152.
Junge, W., Knoth J., and Rath, R., 1983, Chemische und optische Untersu-
    chungen von komplexen Titan-Niob-Tantalaten (Betafiten),
    H. Jb. Miner. Abh. 147:169-183.
Ketelsen P. and Knöchel, A., 1985, Multielementanalyse von größenklas-
    sierten Luftstaubproben, Staub Reinhaltung der Luft 45:175-178.
Knoth J. and Schwenke H., 1978, An X-Ray Fluorescence Spectrometer with
    Totally Reflecting Sample Support for Trace Analysis at the ppb
    Level, Fresenius Z. Anal. Chem. 291:200-204.
Knoth J. and Schwenke H., 1980, A New Totally Reflecting X-Ray Fluo-
    rescence Spectrometer with Detection Limits below 10^{-11} g, Fresenius
    Z. Anal. Chem. 301:7-9.
Leland, D.J,Dilbrey, D.B., Leyden, D.E., Wobrauschek, P., Aiginger H.,
    and Puxbaum, H., 1987, Analysis of Aerosols Using Total Reflection
    X-Ray Spectrometry, Anal. Chem. 59:1911-1914.
Michaelis W., Knoth, J., Prange A., and Schwenke H., 1985, Trace Analyti-
    cal Capabilities of Total-Reflection X-Ray Fluorescence Analysis,
    Adv. X-Ray Anal. 28:75-83.
Michaelis, W., 1986, Naß- und Trockendeposition von Schwermetallen,
    Technische Mitteilungen 79:266-271.
Prange, A., 1987, Totalreflexions-Röntgenfluoreszenzanalyse,
    GIT Fachz. Lab. 31:513-526.
Prange, A., Knöchel A., and W. Michaelis, 1985, Multielement Determina-
    tion of Dissolved Heavy Metal Traces in Sea Water by Total-Reflec-
    tion X-Ray Fluorescence Spectrometry, Anal. Chim. Acta 172:79-100.
Prange, A. and Kremling, K., 1985, Distribution of Dissolved Molybdenum,
    Uranium and Vanadium in Baltic Sea Waters, Mar. Chem. 16:259-274.
Prange, A., Knoth, J., Stößel, R.-P., Böddeker H., and Kramer, K., 1987,
    Determination of Trace Elements in Water Cycle by Total-Reflection
    X-Ray Fluorescence Spectrometry, Anal. Chim. Acta 195:275-287.
Rastegar, B., Jundt, F., Grallmann, A., Rastegar F., and Leroy, M.J.F.,
    1986, Sample Homogeneity in Energy-Dispersive XRF Trace Metal Analy-
    sis, X-Ray Spectrometry 15:83.
Schneider, B., 1989, The determination of atmospsheric trace metals con-
    centrations by collection of aerosol particles on TXRF-sample hol-
    ders. submitted to Spectrochim. Acta B.
Stößel, R.-P. and Prange, A., 1985, Determination of Trace Elements in
    Rainwater by Total Reflection X-Ray Fluorescence, Anal. Chem.
    57:2880-2885.

Wobrauschek T. and Aiginger H., 1975, Total-Reflection X-Ray Fluorescence Spectrometric Determination of Elements in Nanogram Amounts, Anal. Chem. 47:852-855.

Yoneda Y. and Horiuchi, T., 1971, Optical Flats for Use in X-Ray Spectrochemical Microanalysis, Rev. Sci. Instr. 42:1069-1070.

SAMPLE PREPARATION OPTIMIZATION FOR

EDXRF ANALYSIS OF PORTLAND CEMENT

Wayne Watson
Tracor Xray, Mountain View CA

Jim Parker
Manville Technical Center, Denver CO

Anthony R. Harding
Tracor Xray, Mountain View CA

ABSTRACT

Various sample preparation methods for Energy Dispersive X-ray Fluorescence (EDXRF) analysis of Portland cement were compared in order to evaluate improvement in analytical accuracy and precision. Sample preparation requirements for EDXRF are slightly different than for Wavelength Dispersive X-ray Fluorescence (WDXRF), and the methods commonly used in WDXRF are not optimized for EDXRF. Primarily, the work focuses on techniques for producing a fused sample with the lowest practical concentration of lithium borate flux. Determination of minimum detection limits were made from samples with varying proportions of flux in order to evaluate analytical optimization. Ease and reproducibility of preparation of the sample was also considered.

INTRODUCTION

X-ray Fluorescence analysis of Portland cement is a fast and accurate analytical method that provides excellent precision. Typically, sample preparation involves either grinding and pelletizing, or fusion with a low melting sodium or lithium borate glass[1]. Grinding has the advantages of being relatively quick and uncomplicated, but the disadvantage that it does not necessarily eliminate particle size effects and mineralogical effects. Fusion on the other hand is known to eliminate particle size and mineralogical effects, but requires more work and is somewhat more expensive.

In order to demonstrate the need for proper sample preparation, a set of NBS cement standards was pressed into pellets and analyzed. No grinding or fusion was used on the material as received from NBS. The calibration curves shown in Figures 1 and 2 show data for Al and Si plotted after matrix correction. The size and random nature of the residual errors indicate

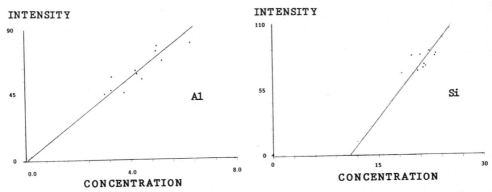

Figures 1 and 2. Matrix-corrected calibration curves for aluminum and silicon in pressed pellets of cement prepared without grinding.

problems due to either particle size effects or mineralogical effects. The goal of this work was to compare and optimize techniques for sample preparation of Portland cement for Energy Dispersive X-ray Fluorescence (EDXRF).

GRINDING

        Grinding was done using a Spex Shatterbox grinder with tungsten carbide holders. We studied the change in measured intensity vs. grinding time as well as improvement in the speed of grinding when cornstarch was added as a grinding aid. Results for both sulfur and silicon suggested that cornstarch did not reduce grinding time and that 3 minutes or more is required to produce a sample that does not continue to change with further grinding. Analysis of the particle size distribution shows that about 25% of the particles are larger than 10 microns. Figure 3 shows a typical distribution.

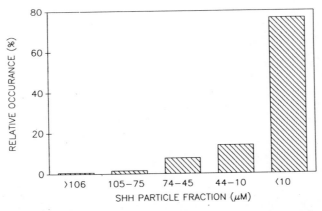

Figure 3. Particle size distribution of sample
ground for 6 minutes.

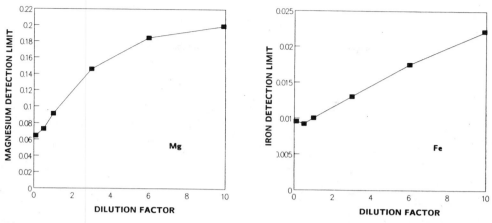

Figures 4 and 5. MDL vs. dilution for magnesium and iron with EDXRF of fused samples.

Information presented by Alyssa Malen in a XRF Sample Preparation Workshop at this conference indicates that there is a better approach. SEM analysis has shown that grinding without a grinding aid, or grinding using the wrong grinding aid, can actually cause formation of particles larger than the 10 micron size sought. Also, although the intensities do not change with further grinding, the particle size isn't necessarily small enough. Ethylene glycol is a much better grinding agent for cement, providing sufficient particle size reduction with 2 minutes in the grinder.

FUSION

The purpose of the fusion step is to create a homogenous solid solution with a smooth, flat surface. Dissolving the material eliminates the possibility of particle size and mineralogical effects as well as reducing matrix effects in the XRF analysis. Fusion methods used for Wavelength Dispersive X-ray Fluorescence (WDXRF) involve dilution of the sample in a sodium or lithium borate glass in order to diminish or eliminate matrix effects, as well as to eliminate particle size and mineralogical effects. By diluting the cement by 10:1 or more, relative changes in concentration of the major element is reduced to the point where linear calibration can be used for the analysis.

With EDXRF however, one doesn't have the luxury of analyzing a dilute sample of Portland cement, as some elements of interest would then be near or below the detection limit of the instrument. Figures 4 and 5 show how the detection limit increases as a sample is diluted. Furthermore, new fundamental parameters software allows matrix correction to be done quickly, easily and with excellent accuracy; eliminating the need for use of dilution to eliminate matrix effects.

Fusion Procedure

Our goal was to develop a technique for producing a fused sample with the lowest practical concentration of lithium borate flux. The first fluxes

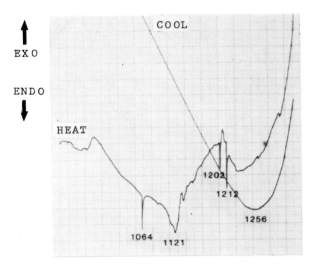

Figure 6. Differential Thermal
Analysis of the
fusion product.

tried were acidic LiF-containing mixtures with lithium tetraborate. They
did not work. Next we evaluated various mixtures ranging from lithium
metaborate to lithium tetraborate. Of these, the eutectic mixture[2], 4 parts
lithium metaborate and 1 part lithium tetraborate, allowed for formation of
a sample using 1/2 part flux to one part sample. Higher Li content helps
dissolve the silica and alumina in cement as well as reducing the viscosity,
which helps mixing. Higher borate content helps in formation of a stable
glass. Crystal formation was observed in some cases and several tests were
made to determine the composition of the crystals in order to minimize their
occurrence.

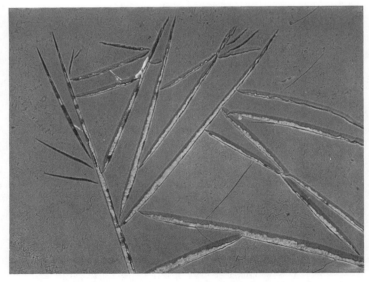

Figure 7. SEM micrograph of crystal formation.

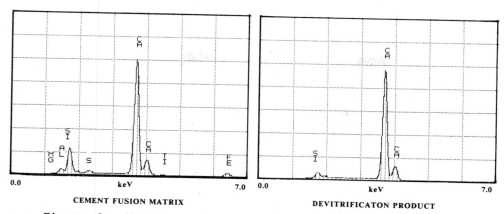

CEMENT FUSION MATRIX   DEVITRIFICATON PRODUCT

Figures 8 and 9. X-ray micro analysis of the fusion product and
devitrification product.

The three fluxes which formed stable glasses were lithium metaborate,
lithium tetraborate, and a eutectic mixture of the two. Evaluation of
crystallization of the sample indicated the need for rapid cooling in order
to maintain a solid solution for the final product.

Procedure used for fusion of the samples:

1)   Samples were ground for two minutes with Freon to prevent
caking, then dried overnight.

2)   Fusion was done with a Herzog HAG-12 with rocking, using
platinum crucibles at a temperature of 1200°C, in the presence
of 40 mg of HBr (non-wetting agent) in water.

3)   After cooling, the samples were ground and polished.

Selection of 1200°C was based on tests done using a gradient furnace,
which suggested that crystal formation stopped at 1175°C. Differential

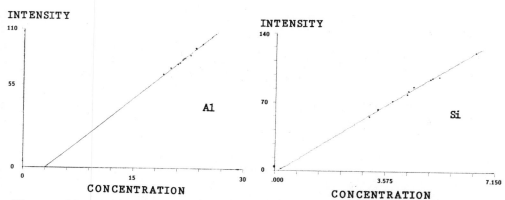

Figures 10 and 11. Matrix corrected calibration curves for Al and Si in
fused standards.

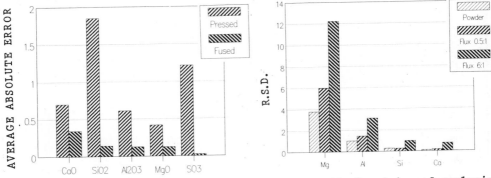

Figure 12. Comparing pressed powders
and fusion for accuracy
of unknowns.

Figure 13. Precision of analysis
for Mg, Al, Si, and Ca.

thermal analysis of a fused sample showed melting of the sample in 3 sharp
endotherms in the heating cycle. The highest, at 1212°C, indicating that
our selection at 1200° may not have been high enough. See figure 6.

The SEM micrograph, Figure 7, shows the presence of the crystals in
the fused disk. X-ray micro analysis of the fusion product, Figure 8,
shows the elements expected for a cement sample. Analysis of the
devitrification product, Figure 9, shows primarily calcium. The crystal
being formed is probably calcium borate. The presence of silicon in this
spectrum is probably originating from the dissolved cement in the
surrounding lithium borate-cement glass.

In spite of the crystallization problem, we were able to make 0.5:1
fusions of NBS cement standards with very good results, as shown in Figures
10 and 11. Evaluation of accuracy of the analysis was made by analyzing 7
NBS standards as unknowns. Improvement in accuracy over our unground
pressed powders is shown in Figure 12. Precision was evaluated by analyzing
one sample 10 times and finding the Relative Standard Deviation (RSD) in the
measurement. The value of a more concentrated fused sample can be seen in
Figure 13 in the results for Mg, Al, Si and Ca.

REFERENCES

1.   Clyde W. Moore, Spectrochemical Analysis of Portland Cement by Fusion
     with lithium tetraborate using an X-ray Spectrometer, in: "Methods for
     Emission Spectrochemical Analysis," ASTM Commitee E-2, Philadelphia

2.   H. Bennett and G. J. Oliver, Development of Fluxes for the Analysis of
     Ceramic Materials by X-ray Fluorescence Spectrometry, _Analyst_
     101:803-807 (1976)

THE VIABILITY OF XRF DETERMINATION OF GOLD

IN MINERAL RECONNAISSANCE

T. K. Smith & M. N. Ingham

British Geological Survey
London WC1X 8NG
United Kingdom

INTRODUCTION

Commercial interest in gold is persistent and resurgent, and there is a consequent need for reappraisal of its methods of analysis. Because its modes of occurrence often manifest themselves in irregular dispersion and low concentration, special care must be taken in sampling and analysis. A sufficient amount of the initial sample must be taken to ensure adequate representation, and preconcentration is often necessary to elevate the metal content to a confidence level high enough above the detection limit for the analytical technique.

The strategy and tactics of mineral reconnaissance and geochemical prospecting frequently involve initial sampling of materials originating from a large catchment area, such as stream waters and sediments and, where heavy minerals are sought, concentrates panned from the latter. Placer deposits may be discovered at this stage. Vegetation can also be a suitable medium for exploration of some types of terrain. Narrowing of the search will involve sampling of soils and later, rocks, when vein occurrences, for example, may be encountered. The size and shape of gold particles can be critical in stream sediment exploration, since small or flaky material may not be retained in the sample but washed out, perhaps adhering to organic material. Panning may therefore in some cases reduce the concentration of the element such that its discovery is made less likely.

In the latter stages of a survey, where rock-hosted gold is sought, accurate determination of the representative concentration in that medium becomes more important since extraction costs will be higher. A very large sample is required (perhaps at least one or two kilograms) and if the analysis is to take place on only a small subsample it is essential that this should give a true indication of the bulk concentration. For some techniques the detection limit may require preconcentration and this may be combined with the procedure of almost quantitative extraction from the large sample by, for example, partitioning into a metal in the molten state followed by volatilisation of the metal and then a gravimetric determination. Alternatively a solution of appropriate concentration may

Table 1. Analytical details

Instrumental parameters

kV: 80   Channel mask: large
mA: 30   Collimator: fine
Anode: Mo   Crystal: LiF(220) or LiF(200)
Window: side   Order: 1
Spinner: on   Detectors: scintillation + flow
Path: vacuum   Window: 25-70%

Line overlaps [LiF(220)]

Element	Factor
W	0.0586
Zn	-0.0005

Detection limits (three sigma confidence level)

Total counting time (excl. overlaps)(sec)	Detection limit (ppmw or grams/tonne)	
	LiF(200)	LiF(220)
400	0.8	1.2
600	0.7	1.0
800	0.6	0.9

be prepared for a technique such as AAS, ICPOES or ICPMS. Some of these techniques are slow and tedious and in the case of dry ashing of vegetation there may be loss by volatilisation when the metal is present as cyanogenic complexes. Neutron activation is sensitive but requires access to a reactor. In some exploration pathfinder elements may be sought (such as As, Sb, W and Cr) since they may have a wider dispersion or higher concentration but at some stage gold itself must be measured.

DETERMINATION BY XRF

An assessment of the potential of this technique was made in the interests of speed, and economy of staff time and materials. The method employed was to grind a large sample to at least minus 200 mesh and to press 12 grams of this powder with a methyl/n-butyl methacrylate copolymer binder in a 40mm die at a load of 25-30 tonnes. With some types of mill, smearing of metallic gold over the surface of the matrix may occur giving falsely high readings (and greater sensitivity). This could be partially corrected for by making standards in the same manner, as was the case in this study.

A conventional sequential Bragg spectrometer (Philips PW1404/10) was employed with the conditions shown in Table 1. The gold $L\alpha_1$ line was used (9.71 keV) with a background at 9.35 keV. Correction may be made as required for the line overlaps (Table 1).

The gold pathfinders arsenic, antimony and chromium cause little interference. The LiF(220) crystal is preferred because its higher dispersion reduces line overlaps but where tungsten (and to a lesser extent zinc) levels are low the benefit of the higher reflectivity of the LiF(200) crystal may be exploited. Correction may be applied for the matrix (depending on its type) by, for example, internal ratio to the background (Smith, 1984) or by use of an alpha coefficient correction for iron. The calibration graph (for various matrices) is shown in Fig. 1, whilst the pulse distribution for gold is shown in Fig. 2 and some potential line overlaps in Fig.3.

The lower limits of detection obtained in a silica matrix are shown in Table 1. Where there is contention between use of the two cuts of LiF crystal, automatic program selection based on a rapid preliminary analysis is possible with some spectrometers. If drift is a problem with long counting times, it may be minimised by splitting each of the peak and background measurements into several consecutive shorter periods. Adjustment of the concentration intercept may be necessary for different matrix types.

Some workers have used other analytical lines. Balaes (1984), for example, preferred gold $L\beta_1$ in spite of its lower intensity. Although the gold $K\alpha_1$ line (68.79 keV; K absorption edge 80.71 keV) is above the effective energy range of most X-ray spectrometers, Sharland (1986) reported a specially developed (energy dispersive) instrument with an excitation potential high enough for that purpose. This has an advantage in that the greater penetration of the more energetic X-rays means that the sample need not be so homogeneous and the analysis is more representative. In practice it is not weighed or pressed into a disk but poured into a plastic tube, and correction made for density variations by a scatter function. A further feature is that interfering elements may be determined simultaneously.

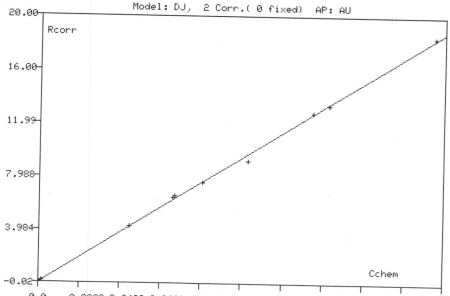

Fig. 1. Calibration graph. (X-axis = %Au; Y-axis = intensity ratio)

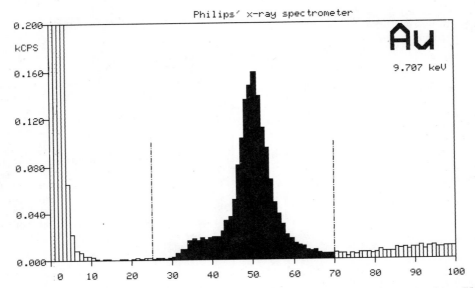

Fig. 2. Pulse distribution.

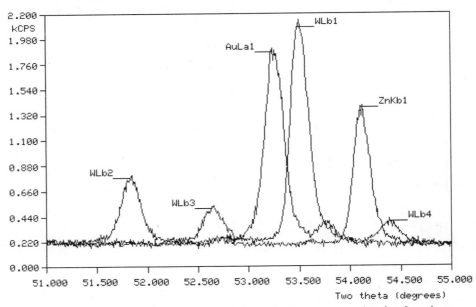

Fig. 3. Potential line overlaps   (values were obtained using
0.1% elemental concentrations in Grade I SiO₂)

CONCLUSION

Where applications can tolerate detection limits high enough to avoid the difficulties of preconcentration,   X-ray fluorescence analysis offers a rapid, cost-effective, non-destructive method requiring little sample preparation.   In geochemical prospecting and mineral reconnaissance the sensitivity of the technique is sufficient for many, but not all, exploration purposes and the equipment already exists in some laboratories. Use in exploitation of deposits is also feasible.

ACKNOWLEDGEMENTS

The authors would like to thank A S Robertson for assistance with the experimental work.   This paper is published with the permission of the Director, British Geological Survey (NERC).

REFERENCES

Balaes, A. M. E., 1984, The determination of gold in activated
    charcoal by use of a loose-powder technique and X-ray
    fluorescence spectrometry, Report No. M136, Council for
    Mineral Technology, Randburg.
Sharland, C. J., 1986, Aztec: a high precision automated instrument
    for gold assaying, in: "Proceedings of the International
    Conference on Gold: Vol 2: Extractive Metallurgy of Gold,"
    C. E. Fivaz and R. P. King, eds., South African Institute
    of Mining and Metallurgy, Johannesburg.
Smith, T. K., 1984, A versatile XRF analytical system for
    geochemical exploration and other applications, Adv. X-Ray
    Anal., 27:481

AN IMPROVED FUSION TECHNIQUE FOR MAJOR-ELEMENT ROCK ANALYSIS BY XRF

Rex A. Couture

Dept. of Earth and Planetary Sciences
Washington University
St. Louis MO 63130

ABSTRACT

A new apparatus and technique are described for borate flux fusion of rocks for x-ray fluorescence analysis. The method yields homogeneous, strain-free glass discs with flat, smooth surfaces that do not require polishing. The technique is adapted from several previous methods but has advantages over each in terms of sample uniformity, quality of the discs, or capital cost. The ignited rock powder is fused with flux over a burner mounted on a stock laboratory mixer, and is cast into a solid flat, polished Pt-Au mold. The very effective mixing action ensures homogeneity. An oxidizing atmosphere, which is necessary to prevent loss of iron to the crucibles, is maintained by injecting air during fusion.

There is no significant loss of alkali metals during fusion, and negligible loss of flux. Duplicate samples of several rock types show excellent reproducibility, approaching counting statistical errors, for 10 major elements.

INTRODUCTION

Routine whole-rock major-element analysis by x-ray fluorescence requires a sample dilution and fusion method that is rapid, accurate, economical, and useable without modification for a wide variety of samples. Flat, polished glass discs are required, and it is further preferred that the discs be useable as cast, without further surface preparation. The flux must not contain analyte elements (Na-Mn), and it is strongly desired not to use heavy elements that may compromise trace-element analysis. Considerable work has been done on fusion methods (cf. Tertian and Claisse, 1982; Pella, 1978; LeHouillier and Turmel, 1974; Baker, 1982; Thomas and Haukka, 1978; Taggart et al., 1987; Norrish and Chappell, 1967). Nevertheless, many previously existing methods lack reliability and ease of casting glass discs, yield discs of dubious flatness and polish, or have a high capital cost. However, it appeared to be straightforward to combine the best features of several methods. What follows is a description of newly designed fusion equipment and a demonstration of its performance.

233

With the exception of the platinum ware, all the equipment was readily
available or was inexpensively constructed.

FUSED DISC REQUIREMENTS

Fused discs should be flat and polished. For a target-sample distance
of 26 mm, an error of 0.013 mm changes the intensity by 0.1%. To maintain
approximately this tolerance, the platinum-gold mold of Taggart and
Wahlberg (1980) was adopted for casting. This mold consists of a solid,
polished plate and a ring. The mold can be dismantled for polishing, and
was shown to yield consistent results of the highest accuracy (Taggart and
Wahlberg, 1980). Unfortunately, since the method they use (Taggart et al.,
1987) requires fusion times of 40 minutes, a large amount of platinum is
required to achieve a reasonable productivity. In order to hasten the
process, it was decided to use a gas burner, which may heat the samples
faster, and which allows better access for agitation of the melt.

Unfortunately, without special precautions gas burners reduce ferric
iron, even in a covered crucible. The glass shows the green color of
ferrous iron, and the crucibles show evidence of removal of iron and/or
manganese from the melt. During the initial part of this investigation it
was found that if an empty crucible is heated in a furnace at 925°C after
use with a burner, a pronounced reddish-black oxide stain develops below
the melt line, showing that iron or manganese in the melt reacts with
platinum crucibles. Pella (1978) has shown that platinum crucibles can
remove a significant proportion of iron from a melt unless oxidizing
conditions are maintained. Possible preventative measures are as follows.

Fusion in a furnace will guarantee an oxidizing atmosphere but will
not necessarily oxidize the melt completely. Ignition of the samples
before fusion or addition of a mild oxidizing agent to the flux can be used
to oxidize samples and to protect the crucibles from sulfides and certain
other compounds (cf. Tertian and Claisse, 1982; Norrish and Chappell, 1967;
Pella, 1978). (Warning: powerful oxidizing agents such as Na_2O_2 strongly
attack platinum crucibles!) For this work, the samples were pre-ignited
(cf. Taggart et al., 1987), and air was injected through a hole in the
crucible lid to exclude burner gases.

AGITATION OF THE MELT

More-or-less continuous agitation of the melt promotes rapid
dissolution of the sample and ensures homogeneity. However, it was found
that agitation methods vary considerably in effectiveness. The
effectiveness of agitation is easy to evaluate by dropping a few milligrams
of Cr_2O_3 powder into the melt. The powder is easily visible until it is
dispersed.

Rocking the crucible about the horizontal axis was not found to be
very effective in our crucibles, even if the entire melt was tipped past
the mid-point. The undispersed powder simply rocked back-and-forth with
the melt, for up to several minutes.

Tipping the crucible and rolling the melt in a circle around the
bottom was found to be much more effective. For this purpose a low-form
crucible, which has a broad bottom that allows tipping of the melt past the
mid-point, is ideal. In several trials the powder was dispersed and

dissolved within a minute or two, and the discs appeared to be completely homogeneous. This method of mixing was chosen for the equipment described below.

EQUIPMENT

The flux fusion apparatus was constructed from a commercial biological laboratory mixer and an air-injected natural-gas Meker burner (Fisher blast burner). The burner is mounted on the mixer, and a basket made from two nichrome wire triangles holds the crucible in place and is mounted on an iron ring clamped to the burner. A nickel lid retains heat and excludes reducing gases from the flame. (Nickel and nichrome do not contaminate platinum appreciably if they are oxidized before use.)

Air is injected into the crucible through a nickel tube 3 mm in diameter (1/8 inch) which passes through a slot in the crucible lid. The air is provided by a small aquarium pump. The oxidizing atmosphere is demonstrated by observing the underside of the nickel lid while the flame is burning. A very low flow rate is sufficient to maintain an oxidizing atmosphere without cooling the melt or blowing the sample away.

Fig. 1 shows the burner, crucible, and melt. During fusion, the axis of the crucible precesses at about 20 rpm without rotating. The axis is inclined 30 degrees, so a 4-g charge is tipped slightly past the center of the crucible. The result is equivalent to rotation about an inclined axis, and the mixing action is nearly ideal, as described above.

The mold was described above. The mold and crucibles are constructed of 95% platinum, 5% gold.

PROCEDURE

The fusion technique is based on that of Taggart et al. (1987). The sample is ignited at $925^{\circ}C$ for 50 min. in an alumina crucible, and ground in a sapphire mortar and pestle (Diamonite), to ensure that the sample is pulverized. A 0.4-g portion of the ignited sample is mixed with 9 parts of $Li_2B_4O_7$, and 0.1 mL of a 44% LiBr solution is added. (The $Li_2B_4O_7$ is dried at $400^{\circ}C$ before use.) The sample is then fused for 10 minutes, poured into a red-hot mold, and allowed to cool. After about 5 min. the mold can be cooled by forced air to facilitate handling and reuse of the mold.

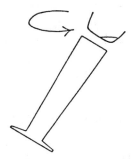

Fig. 1. The burner, crucible and sample, showing rotational mixing action.

The grinding step is crucial to ensure that the sample was completely ground initially, and to break apart any sintered lumps. In the early stages of this work it was found that when samples had not been well ground, failures followed. For effective grinding it is essential that the pestle not have a smaller radius than the mortar, in order to assure contact over a wide area. Unfortunately, most mortar and pestles fail this test.

The flame is the hottest that can be obtained with the natural gas-air burner. It was not possible with available equipment to measure the <u>true</u> temperature of the flame or melt, but an unsheathed type S thermocouple showed a temperature of 1630°C in the flame and 1570°C 3 cm above.

RESULTS

The method, and especially the mixer, are easy to use and are reliable. The effective mixing action was described above; typical reproducibility is shown in Table 1. Table 1 demonstrates good accuracy, based on several U.S. NBS standards and several other high-quality standards. More importantly, there is good agreement between counting statistical errors and reproducibility. As a further test of reproducibility, two or three samples each were made from an andesite, a granite, a silica, and two feldspar samples. Space does not permit showing the data, but the reproducibility was similar to that shown in Table 1.

There is no evidence of removal of iron from the melt by the crucible, as demonstrated by the color of iron-containing glasses, by the lack of staining of the crucibles (as described above), and by the linearity of the iron calibration curve shown in Fig. 2.

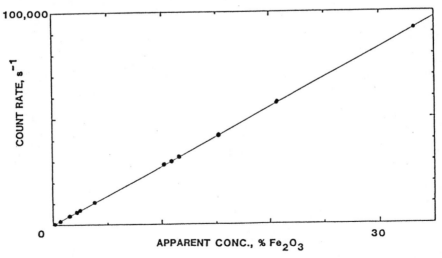

Fig. 2. Calibration curve for iron. Samples were diluted 10x in $Li_2B_4O_7$. Rh target, 44kv. Matrix corrections were done by program CILT (Rousseau, 1986.)

Table 1.   Reproducibility and Accuracy for Basalt NBS688

	Composition, Wt. %			
	Certified	XRF	SD[a]	CST[b]
SiO_2	48.4 $\pm$.1	48.6	.26	.13
TiO_2	1.17 $\pm$.01	1.187	.002	.02
Al_2O_3	17.36 $\pm$.09	17.36	.07	.08
Fe_2O_{3T}	10.35 $\pm$.04	10.295	.026	.02
MnO	0.167 $\pm$.002	0.174	.003	.005
MgO	(8.4)	8.51	.08	.08
CaO	(12.17)	12.19	.12	.04
Na_2O	2.15 $\pm$.03	2.07	.05	.12
K_2O	0.187 $\pm$.008	0.176	.014	.008
P_2O_5	0.134 $\pm$.003	0.119	.008	.017

[a]SD = Sample standard deviation, 3 discs.
[b]CST = Standard counting statistical error.

Volatility of melt components has been investigated and does not appear to be significant, except for LiBr. No loss of alkali metals has been observed in 10 min. to 30 min. of fusion. Gravimetric studies show loss of $Li_2B_4O_7$ on the order of 0.04% in 10 min. LiBr, which is added as a non-sticking agent, is more volatile. This is a minor but significant problem, as Br has a spectral interference with Al and also causes a matrix absorption effect. Consequently, variable Br content of the discs sometimes reduces the accuracy and precision. Use of less LiBr, and overlap and absorption corrections for Br are suggested.

The success rate, in terms of forming strain-free glass discs, is virtually 100% for lithium tetraborate and a wide variety of rock types. I have successfully fused granite, basalt, peridotite, bauxite, feldspars, carbonates, iron and titanium ores, and quartz. I have never observed devitrification, provided that the procedure, as described above, was properly followed. Devitrification was caused in a few instances by large, undissolved lumps of rock or aggregates of rock powder, or by a mold that was not hot enough. The main difficulty in fusion is that quartz, alkali feldspars, and some highly silicic rocks require about 15 minutes for complete fusion, compared to 5 minutes for most rocks.

Slow cooling of the glass for silicate rocks and many oxides leads to strain-free glass, but forced cooling of the mold or pouring into an insufficiently hot mold leads occasionally to devitrification or breakage of the glass. This observation appears to be contrary to that of Tertian and Claisse (1982), who reported that quenching may be required to prevent devitrification.

CONCLUSIONS

The method described is reliable and easy to use, and is relatively fast, even with only one mold and one burner. The success rate is close to 100% for making strain-free glass discs from a wide variety of rock types.

The essential elements of the method are ignition followed by thorough grinding of the samples before fusion, effective, continuous mixing of the melt with a simple mixing apparatus, maintenance of an oxidizing atmosphere by injecting air into the crucible, and use of a flat, polished platinum-gold mold.

Commercial equipment may be somewhat faster to use and less labor-intensive, but is more expensive and may give a less well prepared glass surface. The use of a furnace rather than a flame may offer more reproducible and more flexible temperature control, but appears to be slower to heat the samples and is less flexible in terms of choice of agitation methods.

ACKNOWLEDGEMENTS

I thank Joseph Taggart, Jr. for much valuable help and Richard Rousseau for donating valuable software. I thank F. Claisse, R. F. Dymek, T. Hasenaka, J. M. Rhodes, J. Sparks, S. Sylvester and L. Walter for many helpful discussions, and J. Luhr, B. Owens, J. Smith, and M. Smith for giving the method the acid test. Financial support for this project was provided by the Department of Earth and Planetary Sciences, Washington University, and the National Science Foundation (grant DPP-8619457 to R. F. Dymek).

REFERENCES

Baker, J. W., 1982, Volatilization of sulfur in fusion techniques for preparation of discs for x-ray fluorescence analysis, Adv. X-Ray Anal., 25:91.

LeHouillier, R. and Turmel, S., 1974, Bead homogeneity in the fusion technique for x-ray spectrochemical analysis, Anal. Chem., 46:734.

Norrish, K. and Chappell, B. W., 1967, X-ray fluorescence spectrography, in "Physical Methods of Determinative Mineralogy", J. Zussman, ed., Academic Press, London.

Pella, P. A., 1978, Effect of gas burner conditions on lithium tetraborate fusion preparations for x-ray fluorescence analysis, Anal. Chem., 50:1380.

Rousseau, R. M. and Bouchard, M., 1986, Fundamental algorithm between concentration and intensity in XRF analysis. 3 - experimental verification, X-Ray Spectrometry, 15:207.

Taggart, J. E., Lindsay, J. R., Scott, B. A., Vivit, D. A., Bartel, A. J., and Stewart, K. C., 1987, Analysis of geologic materials by wavelength dispersive x-ray fluorescence spectrometry, U. S. Geol. Survey Bull., 1770.

Taggart, E., Jr. and Wahlberg, J. S., 1980, New mold design for casting fused samples, Adv. X-Ray Anal., 23:257.

Tertian, R. and Claisse, F., 1982, "Principles of Quantitative X-Ray Fluorescence Analysis", Heyden, London.

Thomas, I. L., and Haukka, M. T., 1978, XRF determination of trace and major elements using a single-fused disc, Chem. Geol., 21:39.

MODERN ALLOY ANALYSIS AND IDENTIFICATION WITH A PORTABLE

X-RAY ANALYZER

Stanislaw Piorek

Columbia Scientific Industries
P.O. Box 203190
Austin, Texas 78720

ABSTRACT

The combination of an improved resolution, gas proportional detector with advanced microprocessor technology provides a new and unique solution to the problem of alloy analysis.

A field portable, microprocessor controlled, fully user-programmable x-ray analyzer is described for fast, reliable, on-site, positive alloy identification and assay. The radio-isotope based analyzer employs a modified Lucas-Tooth and Price model of intensity corrections for quantitative, multielement analysis. Applications are reviewed and include examples to show the superior performance of the instrument in such difficult cases as sulfur in carbon steels, and titanium and nickel in stainless steels.

Discussion of the unique alloy identification scheme and its underlying principles is followed by examples of applications illustrating the capabilities of the instrument in distingui-shing, in 5 seconds, alloys of closely similar compositions such as stainless steels 303, 304 and 321, and 410 and 416.

INTRODUCTION

Metal industry, supplying its products to all other areas of industry, has been for years using various methods of positive materials identification. Sophisticated technologies as well as specialization of products make it nowadays equaly important to know composition of the critical part of a pacemaker or of an ordinary foundation bolt.

Number of various techniques have been used over the years for alloy sorting and identification. The traditional ones include color recognition, spark testing, magnetism, differences in apparent density and chemical spot test. More sophisticated methods are based on thermoelectricity (the

Seebeck effect) and an optical emission spectroscopy [1,2].
All these methods have a common denominator of being operator
dependent. An experienced operator has to make the final identification decision himself, interpreting correctly the results
of measurements. The same is true for the conventional, full
scale chemical analysis of an alloy, which must be followed by
the search through composition tables to find the matching alloy.

The contemporary, portable, microprocessor-based x-ray
analyzer offers a real breakthrough in this respect by relieving
the operator from any decision making. All that is necessary is
to expose the sample to the probe and read the final
identification from the liquid crystal display.

The successful expansion of x-ray fluorescence analysis
from laboratory to plant environment was made possible by
progress in technology of the portable instrumentation. Three
factors were particularly responsible for this:

- the availability of powerful microprocessors;
- the use of small, sealed radioisotope sources to excite
  the characteristic x-rays of the sample, and
- the use of rechargeable batteries to make the instrument idependent of the AC power.

In particular, the developments in microprocessor
and liquid crystal technology made it possible for portable,
battery operated x-ray analyzers to perform complex analyses of
x-ray spectra followed by sophisticated data processing;
until recently the domain of large off-line computers.

ALLOY IDENTIFICATION - DEFINING THE PROBLEM

Generally identification can be defined as a process of
ascertaining those characteristics of a given material by
which it is definitively recognizable or known.

Alloys are best characterized by their chemical composition. While the presence of major alloying elements determine
principal properties of an alloy, such as resistance to corrosion or high temperature, minor elements as well as treatment
procedures will affect mainly mechanical characteristics of an
alloy. For example alloying iron with excess of 12% chromium
makes steel resistant to corrosion whereas minute amount of
carbon (tenths of percent) will affect tensile strenght of
the steel.

Examination of the specifications of thousands of alloys
reveals that overall there are about 50 elements involved in the
alloying process. However, any given alloy contains only about 10
to 20 elements and of those only about ten are responsible for
its main characteristics. Thus the number of elements to be
analysed for in alloy identification is significantly reduced
putting the problem of identification into much better
perspective.

The key elements of alloy composition are usually present
at concentrations of at least several percent, as is evidenced
by examples in Figs. 1 and 2. These elements can be relatively

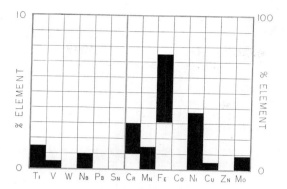

Fig. 1.   Concentration ranges; Stainless Steels.

easily analysed by most techniques including x-ray fluorescence
spectrometry.

       We can conclude that in order to positively identify any
alloy one has to monitor some 10 elements in an unknown sample.
The practical implementation of this idea is discussed below.

MODERN, PORTABLE X-RAY ANALYZER

       A contemporary, portable x-ray analyzer configured for
alloy analysis and sorting consists of a hand-held probe and
an electronic unit. The probe contains source of primary
radiation, detector and necessary electronics. An electronic unit
houses power supplies, I/O circuits and all electronics and
software needed to accept probe signal, process it and display
the final result. A rechargeable battery makes the analyzer truly
field-portable.

       Field conditions require that the lightweight probe has
a slim, easy to handle shape, enabling measurements in corners
(such as welds) and tight areas, and in any position.
Measurement time cannot be too long as often, in order to reach
the object with the probe, the operator has to assume an awkward,

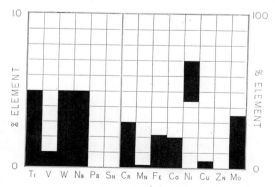

Fig. 2.   Concentration ranges; Nickel Based Alloys.

uncomfortable position. This means that a reliable measurement
has to be feasible in about 10 to 15 seconds maximum. Both
probe and electronic unit have to be rugged, shock resistant
and waterproof.

All these requirements automatically exclude designs based
on Si/Li detectors and/or x-ray tubes. Small, sealed radioisotope
sources are preferred means of providing primary radiation, while
the "old" but reliable gas-filled proportional counters remain
the detectors of choice. Progress in recent years brought us
proportional detectors with significantly improved resolution (10
to 13% for the Fe Kα line) enabling reliable analyses of
neighbouring elements in the periodic table [3].

The only other viable alternative to the proportional
counter is mercuric iodide detector which yet has to prove its
long term performance.

IDENTIFICATION ALGORITHMS

As it was pointed out earlier, to identify an alloy one has
to monitor some ten different elements in a sample that are
critical for its properties. In terms of an x-ray analysis this
means measurement of characteristic x-ray intensities of selected
elements, each intensity reflecting the concentration of the
corresponding element in a sample.

The intensities are usually determined as integrals in
nonoverlapping windows preset on digitized, multichannel x-ray
spectrum of the sample. The gross intensities, N_{kj}, obtained for
sample k, are converted to the net intensities, I_{ki}, by solving
the system of linear equations (1), in which the coefficients G_{ij}
are spectral overlap stripping factors. This calls for
manipulation of an n-by-n matrix equation which modern, 16-bit
microprocessors can handle in a fraction of a second:

$$I_{ki} = \sum_{j=1}^{n} (G_{ij} \cdot N_{kj}) \tag{1}$$

where n can be any number from 2 to 10 and i varies from
1 to n.

Simultaneously, the variance of each intensity $\sigma^2(I_{ki})$ due
to counting statistics is calculated:

$$\sigma^2(I_{ki}) = \frac{1}{T} \sum_{j=1}^{n} [(G_{ij})^2 \cdot N_{kj}] \tag{2}$$

where T is measurement time.

The net intensities can be utilized either in quantitative
analysis or in the identification algorithm.

Two approaches can be distinguished in practical realization of an alloy identification. One, the classical, is based on
assay analysis of the sample followed by search-match through
composition tables. The other is built around the modern method
of pattern recognition.

Early accounts of an assay-based approach have been reported for copper alloys [4], while the first use of direct comparison of x-ray intensities for material identification was described in [5].

An assay-based identification is a procedure that first of all calls for accurate analysis of an alloy sample. The accuracy and precision of the analysis have to be tight enough so that similar alloys can be distinguished from each other by differences in measured concentrations of elements. On the grounds of x-ray fluorescence analysis this is best accomplished by employing the method of fundamental parameters. Modern technology allows for the complex calculations to be programmed and executed on portable instruments; however, the limitations of the fundamental parameters method still remain. It is therefore, still necessary to know what elements are present in the measured unknown sample in order to obtain accurate results. To meet the requirements of precision the measurement time has to be usually of an order of 30 seconds at least. After all elements in an unknown sample are finally assayed a search through alloy composition data tables follows, supported with an appropriate match criterion. Only then can the instrument software decide on the sample identity.

While the method outlined above seems to be obvious and straightforward it is quite cumbersome and limited if judged from standpoint of speed, practicality, convenience and flexibility. Nevertheless, this method has been successfully implemented in commercialy available instruments [6].

Identification through pattern recognition does not have deficiencies of an assay-based method.

The set of net x-ray intensities, I_{ki}, of sample k, obtained as a result of measurement can be interpreted as an object vector I_k with the components I_{k1}, I_{k2}, ... I_{kn}. Alternatively, such a set of coordinates can also be considered as a point "k" in an n-dimensional pattern space.

The expectation is that points representing the samples of the same alloy group will cluster in one limited region of the space, separated from the points that correspond to another alloy group. Extending this reasoning further, within a specific cluster of points, we can expect the data points representing measurements of the same alloy to be very close to each other, whereas those representing different alloys to be noticeably distant from each other, although each being within its own alloy group.

An Euclidean distance coefficient, d_{kl}, can be used as a measure of similarity of the samples or, in terms of the pattern space, a measure of closeness of the data points k and l:

$$d_{kl} = \left\{ \sum_{i=1}^{n} (I_{ki} - I_{li})^2 \right\}^{1/2} \qquad (3)$$

where: I_{ki}, I_{li} - are the i-th coordinates of the points k and l, respectively.

In fact, the minimum-distance coefficient is one of the earliest methods suggested for solving pattern classification problems [7]. A fundamental question of how small the d_{kl} has to be for the two samples k and l to be called identical, has still to be answered. However, the answer is very often application dependent.

The X-MET 880 [a]) portable x-ray analyzer fully utilizes the pattern-recognition approach outlined here for alloy iden- tification. After initial calibration of the instrument, during which elemental windows are set up for integrating x-ray intensities of operator-selected elements, a library of known alloys is created in the instrument memory. Each sample of known alloy is measured for at least 100 seconds. After the measurement, the analyzer determines net intensities and their standard deviations according to formulae (1) and (2). These data are stored in memory along with the alloy name or code, the so-called library reference. Thus each known alloy measured is represented by its "intensity vector", I, of as many compo- nents as there are elements being measured. The maximum number of references the analyzer can store is 400.

When an unknown sample is measured - for practical reasons for not longer than 5 seconds - its spectrum is processed in the very same way as that of any reference sample; that is, net intensities are calculated along with their standard devia- tions due to counting statistics. In the next step a statistic t_r is created for each of the possible pairs, p, "unknown x - reference r", where p is the number of references in the given library:

$$t_r = \sum_{i=1}^{n} \frac{(I_{ix} - I_{ir})^2}{\sigma^2(I_{ix}) + \sigma^2(I_{ir})} \qquad (4)$$

where: r - index denoting r-th reference,
         x - index denoting unknown sample,
         t_r - statistic t for the x-r pair.

As can be seen, t_r is a squared Euclidean distance coefficient, d_{xr}, between the unknown sample and the r-th reference, weighted with the variances of measured intensities. The d_{xr} is thus transformed into its dimensionless equivalent, t_r, which is suitable for statistical testing.

The form of equation (4) suggests that the t_r statistic is a chi-square distributed variable. In fact, it is only a very good approximation to the chi-square distribution, as the net intensities in equation (4) are not strictly indepen- dent variables.

Tables of the chi-square distribution determine the probability of obtaining in a single measurement a t_r value not exceeding a predetermined value, which depends on the number of degrees of freedom, here the element windows. Thus, for a given probability (confidence) level one can select the corresponding

---

[a]) Manuf. by Outokumpu Electronics Oy, Box 27, Espoo 20, Finland

t_r value and use it as the criterion threshold for the match.
If the actually calculated t_r value for a given pair, "unknown x
- reference r", happens to be smaller than the criterion
threshold, then the unknown is considered as matching the
particular reference r.

In order to make the algorithm insensitive to sample size
and shape the intensities used in equation (4) are normalized
appropriately each time a new t_r value is calculated.

Performance of the identification algorithm. The pattern-
recognition-based identification algorithm has been tested on
countless occasions for various alloys. Table 1 summarises the
results of the tests.

For each alloy group a library of reference alloys was
created by measuring each reference for 200 seconds using a
surface probe fitted with a 5 mCi Cd-109 radioactive source.
Next, each reference sample was remeasured as an unknown for at
least 10 times for 5 seconds each, and the percentage of correct
identifications was recorded.

As can be seen the highest degree of positive identification was obtained with the group of nickel-based alloys, consisting of the most commonly used hastelloys and inconels.
Copper alloys were successfully identified with more than 90%
success. The alloys which were mixed up can be readily separated by including tin in the model for which, however, an
Am-241 source would have to be used. This can be accomplished

Table 1.   Summary of the Results of Identification Procedure

ALLOY GROUP	MEASURED ELEMENTS	IDENTIFICATION RESULTS (% feasible)
Nickel/Cobalt Alloys	Ti, Cr, Fe, Co, Ni, Cu, Nb, Mo, W, BS*	100
Copper Alloys	Mn, Fe, Ni, Cu, Zn, Pb, BS	90 - 100
Stainless and High Temper. Steels	Ti, Cr, Mn, Fe, Co, Ni, Cu, Nb, Mo, BS	90 - 100
Cr/Mo Steels	Cr, Fe, Ni, Mo	95 - 100
Carbon Steels	Cr, Fe, Ni, Mo	62
Titanium Alloys	Ti, V, Cr, Mn, Zr, Mo, BS	95 - 100
Aluminum Alloys	Ti, Cr, Mn, Fe, Cu, Zn, BS	90 - 100

*) Backscattered radiation

using a dual source probe equipped with two sources, 5 mCi Cd-109 and 30 mCi Am-241.

Another group consisted of stainless steels and high temperature alloys. The overall performance of at least 90% is good if we note that the mixups occured between such alloys as SS 409 and SS 410 which differ in their main elements by less than 1%.

The most difficult case is represented by carbon steels. The well known difficulty of separation of carbon steels stems from the fact that the concentrations of the alloying elements are very low in the presence of almost 100% iron, and that the difference in concentrations of the same elements between two grades is very small, approaching the detection limit of the method.

The power of the pattern recognition algorithm is perhaps best illustrated by its ability to separate alloys into distinguishable groups, such as Ni-based, Fe-based, etc. A so-called screening model can be created by building a library of references, each measured only for 30 seconds. Shorter measurement time widens the acceptance band around each reference to the degree that only several alloys of the same group can create in the pattern space a region covering all the alloys belonging to that group. This kind of sorting into alloy groups would be quite difficult to implement in an assay-based identification.

Another example of flexibility of the pattern-recognition approach makes also for an excellent proof of its superiority over the assay-based one. It is well known fact that the stainless steels 303 and 304 differ only by 0.3% sulfur, otherwise being almost identical. In an assay based identification both alloys would have to be analyzed for all their components including, of course, sulfur. However, the error of sulfur determination would likely be of an order of the sulfur content itself, rendering the identification questionable. Using the pattern recognition approach we are not bound by the constraints of the fundamental parameters method to analyse all elements. Instead, we can program the analyzer equipped with a light element probe and 20 mCi Fe-55 source to look for only sulfur, chromium, and backscattered radiation intensities, thus emphasizing the true difference between the alloys and achieving positive identification in just 5 seconds per sample. A similar example is the pair of 410 and 416 stainless steels which also differ by only 0.3% sulfur. It is worth of mentioning that sulfur and other light elements are measured in an air path.

Extreme flexibility of programming of the identification algorithm, as provided by the X-MET 880, accounts for some seemingly impossible results such as separation of pure titanium from titanium alloyed with 3.5% tin using a Cd-109 source and Xe/CO_2-filled proportional detector.

Implementation of a dual source probe improved the performance of sorting in such difficult cases as separation of SS 321 from SS 303 or 304. Combination of a 5 mCi Cd-109 source with a 40 mCi Fe-55 source is ideal for all steels applications with the Fe-55 part handling flawlessly the 0.45% difference in titanium between those alloys.

QUANTITATIVE ANALYSIS OF ALLOYS

Calibration

The x-ray intensities measured on the sample (Eqn. 1) can also be used for quantitative alloy analysis, although this application seems to be secondary to the fast, reliable pattern recognition algorithm.

A modified Lucas-Tooth and Price [8] intensity correction model was implemented in the analyzer discussed here, to complement its identification algorithm. The intensity correction model allows one to account for matrix effects by direct use of measured intensities of the elements without the need for complex calculations required for the fundamental parameters method. It also allows for calibration of the instrument in any concentration range as long as adequate standards are available. Calibration for a particular element does not require analysis of the sample for the other elements.

Equation (5) is an exact expression of the intensity correction concept while equation (6) illustrates its practical realisation, showing also all combinations of intensities allowed as independent variables in modelling for the optimum calibration equation:

$$C_i = B_i + I_i \cdot \{k_0 + \sum_{\substack{j=1 \\ j \neq i}}^{j=n} k_{ij} \cdot I_j \} \qquad (5)$$

where B_i is a background correction and k_0 and k_{ij} are correction coefficients;

$$C_i = r_i + \sum_{l=1}^{6} r_{il} \cdot f_l \qquad (6)$$

where intensity factor, f_l, can be any of the expressions below:

$$f_l = I_i \quad \text{or} \quad I_i/I_{BS} \quad \text{or} \quad 1/I_B$$

$$f_l = I_i \cdot I_j$$

$$f_l = (I_i \cdot I_j)/I_{BS}^2 \quad \text{or} \quad I_k/I_{BS}^2 \quad \text{or} \quad 1/I_{BS}^2$$

for any i different from j.

As can be seen from the Eqn (6) the calibration equation for any element i may have up to six different independent variables l selected from up to ten element intensities allowed in any calibration model. In turn any model set up for an assay analysis can be calibrated for quantitative readout of up to six elements.

The coefficients in Eqn (6) are determined from measurements taken on known standards. The form of the equation implies that at least (n + 1) standards with known concentrations C_i are needed for this task. However, both the measured intensities and chemical concentrations are never absolutely accurate.

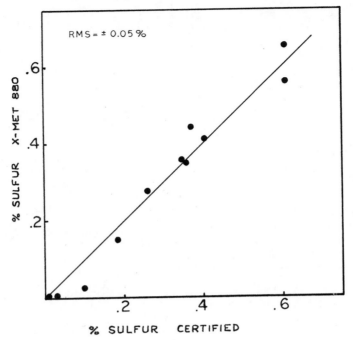

Fig. 3.   Calibration for sulfur in low alloy steels.

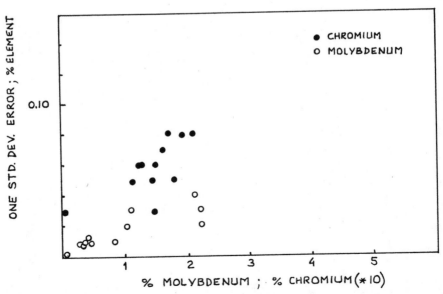

Fig. 4.   Errors reported for certified concentrations
          of Cr and Mo in Alloy Calibration Standards.

Therefore in practice a much larger number of specimens is used for calibration, and a method of least-squares is employed to calculate coefficients for equation (6).

After the calibration measurements are completed and con- centrations of analytes in standards are entered into analyzer memory, the operator can optimize the calibration equation for any desired element by iterative procedure of selecting the set of variables, executing a least-squares fit, analysing the coefficients and their statistical significance, then selecting another set of variables, and so on. During this procedure some intensity factors, f_i, are usually found superficial and are eventually dropped. Therefore the optimized calibration equation for a given element does not necessarily have $(n + 1)$ coefficients. It has been found that for all practical reasons n can be limited to 6.

Performance of an assay model

Despite its limitations and deficiencies, the intensity correction model proved to be a very valuable and reasonably accurate tool, complementing the identification algorithm.

As could be expected, precision of measurement for most of the elements ranging from titanium to molybdenum varies between 0.5 to about 3 percent relative. This is valid for typical measurement conditions of 100 sec. counting time with a 5 mCi Cd-109 source. Typical measurement accuracy varies from 0.4% for copper in brasses to 0.03% for molybdenum in steels.

Fig. 3 illustrates an interesting case of calibration for sulfur in low-alloy steels, using a probe equipped with a 20 mCi Fe-55 source and a Neon-filled detector. As can be seen, the calibration is quite good, especially if we note that the measurements were taken in the air path.

When discussing the errors of analysis it is helpful to have in mind the error introduced by uncertainty associated with certified concentrations of elements in the calibration standards. Fig. 4 shows standard deviations of each certified value plotted against certified values themselves for chromium and molybdenum in stainless steels [8]. It is obvious that certified data for chromium are associated with an error varying from about 0.05% Cr to 0.08% Cr. This error determines the ultimate accuracy of chromium analysis with the portable x-ray analyzer. In case of molybdenum the x-ray method easily achieves the accuracy level of the referee analyses.

In view of these facts one can better appreciate the benefits of the intensity-correction model and, perhaps, raise the question of whether pursuing more accurate correction concepts, such as those for fundamental parameters, is indeed necessary in an alloy sorting application.

CONLCUSION

It has been shown that the pattern recognition method can be successfully adopted for alloy sorting and identification. It allows for fast, reliable identification of alloys with virtually

no prior knowledge of alloy composition. Due to its extreme
flexibility and ease of programming, the pattern-recognition-
based identification algorithm can be fully user customized for
any sorting application. An added benefit is the ability of the
analyzer for quantitative analysis of alloys; this complements
its identification scheme to create a unique, state of the art
instrument for alloy analysis and sorting.

REFERENCES

1. Brown R.D., Riley W.D., Zieba C.A. - "Rapid Identification
    of Stainless Steel and Superalloy Scrap", U.S. Bureau of
    Mines Report, RI 8858, 1984.

2. Spiegel F.X., Horowitz E. - "Instruments for the Sorting and
    Identification of Scrap Metal", The John Hopkins University,
    Center for Material Research, Oct. 15, 1981. Published by
    John Hopkins University.

3. Jarvinen M-L., Sipila H. - "Effect of Pressure and Admixture
    of Neon Penning Mixtures on Proportional Counter Resolution"
    Nucl. Instr. Meth. 193 (1982), pp. 53-56.

4. Marr III, H.E. - "Rapid Identification of Copper-Base Alloys
    by Energy Dispersion X-Ray Analysis", U.S. Bureau of Mines
    Report, RI 7878, 1974.

5. Russ J.C. - "Elemental X-Ray Analysis of Materials Methods",
    EDAX Int'l, 1972.

6. Berry P.F. - "Developments in Design and Application of
    Field-Portable XRF Instruments for On-Site Alloy Identifi-
    cation and Analysis", presented at the 40th ASNT National
    Fall Conference, Atlanta, Oct. 12-15, 1981.

7. Bow Sing-Tze - "Pattern Recognition; Application to Large
    Data-Set Problems", Marcel Dekker, Inc., New York, 1984.

8. Lucas-Tooth H.J., Price B.J. - "A Mathematical Method for the
    Investigation of Interelement Effects in X-ray Fluorescent
    Analyses", Metallurgia, Vol.54, No.363, pp. 149-152, 1962.

9. Brammer Standard Reference Materials, Master Alloy Set, Data
    Sheets, Metals Analysis, Houston, Tx, 1986.

LOW LEVEL IODINE DETECTION BY TXRF SPECTROMETRY

F. Hegedüs and P. Winkler

Paul Scherrer Institut (PSI)
CH-5303 Würenlingen, Switzerland

Introduction

A special measurement technique has been developed to measure very low level iodine concentrations. The gas and water samples to be analysed are taken from the POSEIDON facility at PSI where retention of iodine in water pools in conjunction with light water reactor safety analysis is under investigation. The amount of iodine was measured by means of a Total Reflectance X-Ray Spectrometer (TXRF).

The aim of the POSEIDON experiment and the principle of the iodine detection was presented at the 1986 Denver conference (1).

Description of the POSEIDON Facility

The main component of the POSEIDON facility is a cylindrical tank (diameter:1.0 m, height:6 m) filled with demineralized water. The height of the water column is variable between 1 and 5 meters. The N2 gas-iodine vapor mixture is generated below the tank and then introduced into a glass bell (volume:10 litre) which is in the bottom of the water tank. The iodine concentration in the gas is about 3 mg/l. The size and the speed of the gas bubbles is variable. The aim of the experiment is to measure the attenuation factor, i.e. the ratio of the iodine concentration at the top of the column to that of the bottom.

The iodine concentration was measured at two locations: at the bottom of the water tank in the glass container and at the surface of the water where the outcoming gas was collected in a gasometer.

Preparation of the Samples

The gas samples were taken in preevacuated glass bottles (0.1 l and 1.0 l) and then bubbled through a liquid trap containing 4 ml 0.01N NaOH solution using nitrogen carrier gas. It was found that the trap efficiency for iodine absorption was higher than 95%.

Description of the TXRF Spectrometer

The principle of the spectrometer (Fig.1) is similar to that of the Atominstitut of Vienna (2). The elements of the spectrometer, the line focus X-ray tube, the two collimator slits, the mirror, the

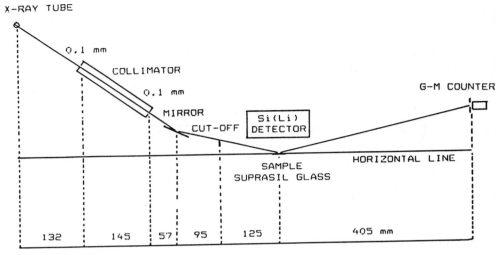

FIG.1. TXRF Spectrometer.

high energy cut-off, the sample holder and the vacuum chamber, were mounted on an optical bench. The X-ray spectrum, induced in the sample, was measured by means of Si(Li) detector and analysed with a PC coupled MCA using the PCRFA program of the Atominstitut of Vienna. As sample substrate a suprasil glass plate was used. The thin samples for TXRF were prepared by pipetting 5μl of solution on the substrate.

Measurements and Results

The spectrometer was calibrated with monoelemental samples. A constant amount of cobalt (20 ng) as internal standard was added to each sample and the intensity of the measured X-lines was

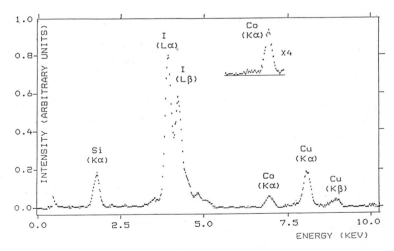

FIG.2. X-Ray Fluorescence Spectrum
Sample:2986 ng I + 20 ng Co
Cu Tube:30 kV,20 mA.

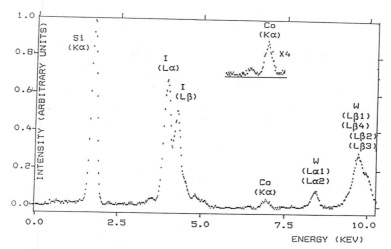

FIG.3. X-Ray Fluorescence Spectrum
Sample:3570 ng I + 20 ng Co
W Tube:30 kV,20 mA.

compared to that of the Co Ka-line. By that means an accurate weigth determination was possible.

Two X-ray tubes were used. In case of the Cu tube the optimum for iodine L-line detection was with 6.45' total reflection angle and with HV=20 kV. Fig. 2 shows a typical X-ray spectrum of an iodine sample. Using a W tube with the Cu tube setting of the spectrometer, similar detection performances of the iodine L-lines were obtained (Fig. 3). The minimum detection limit of iodine was 0.8 μg/litre in nitrogen gas and 20 μg/litre in water samples. In presence of Ca, Ti and Cs contamination, the detection limit will be higher.

To detect the iodine K-lines, the spectrometer was operated without mirror unit and with W tube. By setting 2.15' total reflection angle and HV=50 kV, the iodine K-lines were well measurable. The minimum iodine detection limit was however ten times higher than with L-lines but free of Ca, Ti or Cs interference.

Conclusion

The TXRF spectrometer was successfully used in the POSEIDON experiments. A large number of gas and water sample were analysed for iodine. The minimum detection limits for gas and water samples were respectivetly 0.8 and 20 μg /litre.

References

1.   F. Hegedüs, P. Winkler, P. Wobrauschek and Ch. Streli: Low Level Iodine Detection by TXRF in a Reactor Safety Simulation Experiment. Advances in X-Ray Analysis, Vol. 30 (85-88)

2.   H. Aiginger and P. Wobrauschek: Total Reflectance X-Ray Spectromery. Advances in X-Ray Analysis, Vol. 28 (1-10).

THE APPLICATION OF P-32 AND Sn-113 RADIONUCLIDES FOR THE DETERMINATION
OF NOBLE METALS

J. J. LaBrecque and P. A. Rosales

Instituto Venezolano de Investigaciones Cientificas, IVIC
Apartado 21827, Caracas 1020A, Venezuela

INTRODUCTION

Recently, a new radioisotope X-ray fluorescence technique was reported
(1) in which a small quantity (100μCi) of a selected radioisotope was directly
mixed with a small amount of the sample, as a source-sample. Many different
types of excitation radiation from various radioisotopes have been previously
studied: Na-22, S-35, Fe-55, Co-57, Ni-63, Zn-65, Cd-109, I-125, Cs-137,
Pm-147 and Am-241 (2).

Direct Beta excitation for producing characteristic X-rays for X-Ray
Fluorescence Analysis was shown in 1981 (3) by employing Pm-147 and Ni-63.
The β_{max} for Ni-63 is only 67 KeV while the β_{max} of PM-147 is 244 KeV; this
seems to be the reason why the Pm-147 radioisotope was able to produce higher
energy secondary X-rays more efficiently than Ni-63. Thus, it was one of the
objectives of this work to investigate if P-32 (β_{max}=1709 KeV) can excite
higher secondary X-rays more efficiently as well as to be employed for the
analysis of Au and Pt via their K -X-rays, even though its half-life is only
14.31 days.

Sn-113 was also studied in this work to see if it could be applied for
simultaneous measurement of Pd, Ag, Pt, and Au. Since the Cd-109 conventional
radioisotope cannot in most cases be used for the determination of Ag.

In this work we shall describe the applications of P-32 and Sn-113 for
the analysis of Noble metals. The radiation from P-32 is purely beta
particles which can excite the high energy K X-ray lines of Pd, Ag, Pt and
Au. The excitation radiation from Sn-113 is the In X-rays via electron
capture process. A comparison of the P-32 and Sn-113 radioisotopes will be
presented with the emphasis placed on the advantage of using the characteristic K-X-rays rather than the lower energy L-X-rays.

EXPERIMENTAL

About 20-40 mg of fine powder (< 1 μm) of various noble metal alloys
from Alfa Inorganic (Danvers, Massachusetts) were sealed between two pieces of
"Scotch Magic Tape" in a 2x6 cm piece of IBM card with a 1 cm aperture as the
sample-source holder. Before closing this sample-source 50μl of a 2 mCi/ml
solution of P-32 or Sn-113 were added and allowed to air dry overnight.

    The sample-sources were placed directly on the face of the detector
window which was protected by a thin polypropylene film from contamination
when the dead time was less than 50%. When the dead time was greater, the
sources were separated from the detector window by a distance so that the dead
time was just less than 50%. In all cases, this was less than 25 mm.

    The characteristic X-rays were collected with a high resolution (Si(Li)
detector with a thin window of 0.013 mm of Beryllium or with a planar pure Ge
semiconductor with a resolution of 230 eV at 6.4 KeV and 630 eV at 122 KeV
from a Co-57 calibration source. The Be window was 0.150 mm. The data
acquisition and analysis was performed by an Apple IIe microprocessor with a
Nucleus ADC/interface card. The peak areas were calculated by a simple
computer process, that is, summing the counts in the region the centroid of
the peak of interest plus and minus 0.6 FWHM of the peak. This routine is
part of the software package for XRF supplied by Dapple Systems. (Sunnyvale,
California).

RESULTS AND DISCUSSION

    Comparisons of not only the application of P-32 and Sn-113 for
excitation of the characteristic X-rays but also of the employment of a Si(Li)
detector and a hyper-pure germanium (HpGe) semiconductor to collect the data
were performed with a Pt-Pd-Au alloy (40:20:40). The resulting X-ray
fluorescence spectra are shown in Figs. 1-4. The first spectrum of the
Pt-Pd-Au alloy with P-32 as the excitation source was measured with the HpGe
detector for 600 seconds of fluorescent time. It can be seen that the P-32
excitation not only can excite sufficiently all the high energy K-X-rays
(65-78 KeV) of Pt and Au, but also the Pd K-X-rays (21-24 KeV). It can also
be clearly seen that the peak intensities of the K-X-rays are more suitable
than those of the L X-rays for analytical analysis. Actually, the Au L-X-rays
don't seem to appear in the spectrum. Possibly they are hidden under the
background radiation from the sample-source.

    The second spectrum (Fig. 2) of the same sample-source but with a high
resolution Si(Li) semiconductor only detects peaks of less than 30 KeV because
of the poor detector effeciency after 30 KeV. Thus, an analytical determination for Pt and Au need to employ the L-X-rays, but again the Au L-X-rays were
not observed. While in this spectrum the background radiation was small in
the region where the Au L-X-rays should have been observed.

    The use of the higher energy K-X-rays rather than the lower energy
L-X-rays also have the following inherent advantages: 1) the variation in
peak intensity due to surface roughness or particle size is minimized because
the emitting high energy K-X-rays are coming from a depth of millimeters
rather than microns. 2) Most of the spectral overlays between the multiple L
or M X-rays of high Z-materials with the K-X-rays of low Z-elements in a
sample are eliminated.

    In the third spectrum (Fig. 3), when Sn-113 was applied for excitation
of the characteristic X-rays only the data in the energy range of about 1-30
KeV was collected since the excitation is accomplished using the In K-X-rays
(24-27 KeV). In this spectrum, it can be clearly seen that the Pt and Au
L-X-rays can be employed for their respective determination as well as the Pd
K-X-rays. Previously, Pt and Pd have been determined in catalytic materials
(4) by conventional radioisotope induced X-ray emission but it was necessary
to measure the sample twice, once with a Cd-109 source for the Pt and Au
L-X-rays and an Am-241 source for the Pd K-X-rays. But it is shown here that
with the Sn-113 source they both can be determined simultaneously.

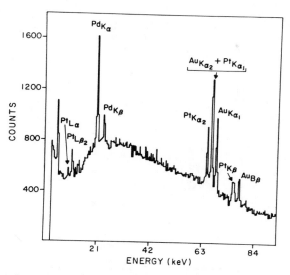

Fig. 1.   X-ray Spectrum of a Pt-Pd-Au alloy with P-32 excitation and
detection by a HpGe semiconductor for 600 s of counting time

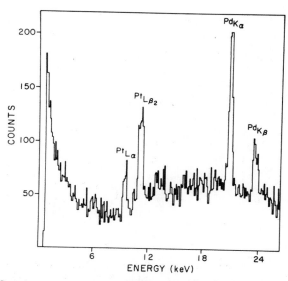

Fig. 2.   X-ray Spectrum of the same Pt-Pd-Au alloy with P-32 sample-source
but detection by a Si(Li) detector for 250 s of counting time

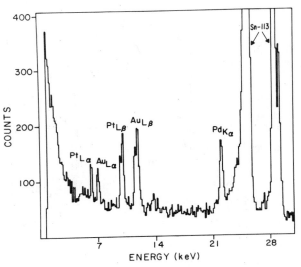

Fig. 3. X-ray Spectrum of the same Pt-Pd-Au alloy but with Sn-113 and detection by a Si(Li) detector for 250 s of counting time

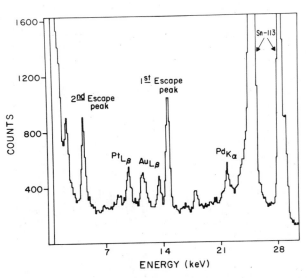

Fig. 4. X-ray Spectrum of the same Pt-Pd-Au alloy with Sn-113 sample source but detection by a HpGe detector for 600 s of counting time

Table 1. Determination of Pt, Pd and Au in a Pt-Pd-Au (40:20:40) alloy
employing MoO_3 as an internal standard for 1000 s counting time

Excitation source	Element	X-ray peak	Concentration (%)
P-32	Pt	K_{α_2}	39.1
	Pd	K_α	19.6
	Au	K_{α_1}	39.3
Sn-113	Pt	$L\beta$	38.4
	Pd	K_α	19.3
	Au	$L\beta$	38.7
Cd-109	Pt	$L\beta$	39.3
	Au	$L\beta$	39.2
Am-241	Pd	K_α	19.4

In the final spectrum (Fig. 4) of this series, the same sample-source as
in Fig. 3 was measured with the HpGe detector. In this spectrum not only the
escape peaks of Ge from the In K-X-rays appear but also the Pt and Au L_α-X-rays
didn't seem to be present. The peak areas are somewhat bigger but the
background is much more than when the Si(Li) detector was used.

Finally, Table 1 shows the measured values of the Pt-Pd-Au noble metal
alloy employing P-32 and Sn-113 as well as the conventional Cd-109 and Am-241
excitation sources with MoO_3 as an internal standard. Other noble metal
alloys, such as Ag-Pt, Ag-Pd, etc., were also investigated but will be
reported elsewhere.

CONCLUSIONS

It can be concluded that for P-32 excitation of noble metal materials
that the hyper-pure germanium detector should be employed since gold L-X-rays
were not observed in the X-ray spectra, as well as the other inherent
advantage of using the higher energy X-rays. When the Sn-113, sample-sources
are applied the Si(Li) detector is preferred for measuring the sample because
of presence of the escape peaks in the X-ray spectrum of the HpGe, as well as
for other inherent reasons.

REFERENCES

1. J.J. LaBrecque and W.C. Parker, Adv. X-ray Anal., Vol. 26 (1983) 337-340.
2. J.J. LaBrecque, P.A. Rosales and W.C. Parker, Journal of Radioanalytical
    Chemistry, Vol. 78, N° 1 (1983) 87-89.
3. J.J. LaBrecque, W.C. Parker, P.A. Rosales, Radiochem. Radioanal. Letters,
    49/4 (1981) 262-270.
4. W.C. Parker and J.J. LaBrecque, Adv. in X-ray Anal., Vol. 25 (1982) 151-155.

CHARACTERIZATION OF PERMALLOY THIN FILMS VIA VARIABLE SAMPLE EXIT ANGLE ULTRASOFT X-RAY FLUORESCENCE SPECTROMETRY

George Andermann, Francis Fujiwara
Department of Chemistry, University of Hawaii
Honolulu, Hawaii 96822

T.C. Huang
IBM-Almadden Research Center
San Jose, CA 95120

J.K. Howard, N. Staud
IBM-General Products Division
San Jose, CA 95193

INTRODUCTION

Recently variable sample exit-angle x-ray fluorescence spectrometry (VEA-XRF) has been shown to be a useful analytical tool for monitoring the oxidation of the surfaces of bulk Cu, Ni[1] as well as that of Fe[2]. In these studies advantage was taken of the well known phenomenon[3] that for each transition metal oxide (MO) the $L\beta/L\alpha$ intensity ratio value is higher than for the transition metal (M), itself. Within the limits of the photon-escape depth d_e, which for these photons are generally below 5000 Å, varying the sample exit-angle θ offers an opportunity for seeing whether or not the oxidation of the surfaces of bulk M belongs to one of the following two classes: (I) uniform oxidation throughout the entire observable sample-depth, (II) preferential oxidation of the top surface layer, i.e. depth dependent oxidation. Under the conditions of oxidation Fe appeared to oxidize more nearly as a Class I oxidizer and Cu and Ni were shown to belong to Class II. Figure 1a depicts ideal Class I behavior and Figure 1b shows typical Class II behavior.

In extending the variable angle $L\beta/L\alpha$ intensity-ratio method to the evaluation of the oxidation of thin films of permalloy (Ni-Fe) the following topics were pursued:

(i) Since Fe behaves approximately as a class I and Ni is a class II oxidizer, it appeared to be pertinent to ascertain about the nature of oxidation in a Ni-Fe alloy (permalloy) recognizing the well known fact that Ni is used in permalloy to inhibit Fe oxidation.

(ii) Since VEA-XRF via transition-metal L-radiation studies are capable of monitoring oxidation in the thickness region of about 50 to 5000 Å,

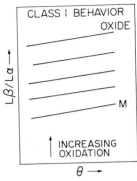
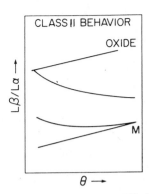

Fig. 1a.   Ideal Class I behavior and   1b.   Ideal Class II behavior of
           Lβ/Lα intensity ratio.

it appeared to be of interest to evaluate whether or not the differences
in morphological structure in a sputtered permalloy film as compared to
bulk permalloy would result in different oxidation mechanisms.

(iii)   Since the Lβ/Lα intensity ratio is related to the 3d valence
orbital occupancy, it was logical to determine whether or not the Lβ/Lα
technique could yield useful information about any possible valence-
electron structural changes in creating an intermetallic alloy out of Fe
and Ni.

(iv) Since at very low values of θ both the 200 Å and 400 Å thin film –
substrate interface represents an observable region, it appeared to be
relevant to evaluate the possibility of observing the effects of any
optical phenomena due to the presence of anomolous dispersion regions
for Fe and Ni L radiation.

(v)   Since   the   standard,   isolated – atom   model–based fluorescence–
production   equation   had   never   been   applied   to   the   theoretical
calculation of θ-dependent Lβ/Lα intensity–ratio monitoring of thin
films, it appeared logical to develop such theoretical calculations.
This challenge was especially interesting because the pertinent mass
absorption–coefficient values at Lβ are not available in the literature.

EXPERIMENT

     The sample studied included four permalloy (80% Ni – 20% Fe) thin
films (P-f) which were rf sputtered on a Si wafer.   Two of the P-f
samples had a thickness of 200 Å, hereafter referred to as P-f(200), and
two of them had a thickness of 400 Å, hereafter referred to as P-f(400).
One of the P-f(200) and P-f(400) samples was oxidized at 200° C in air
for 2 hours, thus yielding samples P-f*(200) and P-f*(400).   In addition
a thin film of pure Ni was prepared by electron-beam sputtering on a Si
substrate, hereby designated as sample Ni-f(500).   In order to gain
insight to challenges (i), (iii), and (v), bulk samples of Ni (99.6%),
NiO (reagent grade), Fe (99.5%), Fe₂O₃ (reagent grade), and a special
HyMu-80 permalloy (80.3% Ni, 14.5% Fe, and 4.3% Mo) was utilized to
yield a sample of ambient bulk H, and an oxidized bulk H*(heated at 400°
C for 3 days in air.)

The instrument utilized was the previously described[4], unique U. of Hawaii 5 M grating spectrometer equipped with a two-position push-pull variable-θ sample holder. The detection was based on the use of a pressure-tunable thin-windowed flow proportional counter. The excitation of the samples was obtained via a Henke-style x-ray tube employing a Cu anode operated at 8 KV and 180 mA. A bulk Ni standard was employed for intensity calibration. The Ni and Fe-L spectra were measured at an optical resolution of 17 and 12 eV, respectively. Pulse-height discrimination was utilized to eliminate second-order Si-Kα interference with the Ni-Lβ line.

RESULTS AND DISCUSSION

Fe-L Spectrum and Data Collection

Figure 2 depicts the Fe-L spectrum of sample P-f*(400) taken at an exit angle of 18° and at 12 eV instrumental resolution which corresponds roughly to the spin-orbit splitting of $2p_{3/2}$ (Lα) and $2p_{1/2}$ (Lβ) transitions. Also observed is the Ni-Lℓ transition just above the Lβ line. In order to obtain Lβ/Lα intensity-ratio values as a function of θ at a precision level of about 3 to 5%; photons were collected at Lβ to reach a level of at least 500 to 1000 and at Lα until about 2 to 3000 counts were collected. This meant that at θ value of 0.5°, counting times of at least 30 to 60 minutes were necessary.

Bulk Permalloy Study

Figure 3 shows the exit-angle θ-dependence of Fe-Lβ/Lα intensity for bulk samples Fe, H, H* and Fe_2O_3. Also shown is the theoretical curve for H. Figure 4 depicts the experimental Ni-Lβ/Lα θ-dependent plots for Ni, NiO, H and H*. There are several interesting observables in Figures 3 and 4. At θ = 18°, where the observation of Lβ/Lα intensity ratio is for sample depths of about 5000 Å under the experimental conditions used, the Lβ/Lα intensity-ratio values are identical for Fe and Ni, namely 0.158. However, as shown in Figure 3,

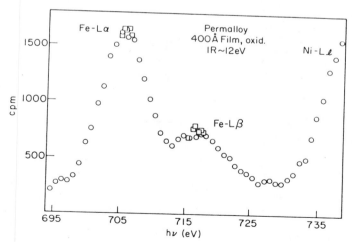

Fig. 2. Fe-L spectrum of sample P-f*(400)

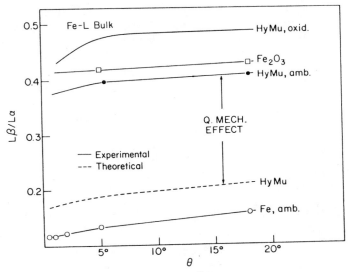

Fig. 3. Exit-angle dependence of Fe-L$_\beta$/L$_\alpha$ intensity ratio for Fe, HyMu (H), HyMu oxidized (H*) and Fe$_2$O$_3$ bulk samples.

the experimental 18° value of Fe Lβ/Lα for H is 0.41 and the theoretically calculated value is only 0.21. Note that this theoretical value merely takes into account standard isolated atom model Beer's-Law absorption calculations relating to the change in composition. The additional increase of 0.20 must, therefore, be due to either a drastic change in self absorption, and/or due to drastic changes in the 3d partial density of states, i.e., valence - electron structural modifications in forming the intermetallic alloy of Ni and Fe.

There are two other interesting observables in Figures 3 and 4, namely, (1) the θ-dependence of Fe-Lβ/Lα in H* and H are nearly parallel

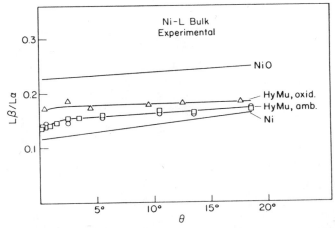

Fig. 4. Exit-angle dependence of Ni-L$_\beta$/L$_\alpha$ intensity ratio.

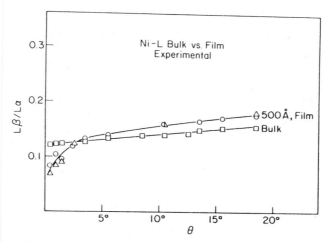

Fig. 5. Bulk vs. film θ-dependence of Ni-L$_\beta$/L$_\alpha$ ratio for Ni.

indicating Class I oxidation bulk behavior, but (2) the Ni Lβ/Lα θ
dependent response for H* indicates significant deviation from Class I
behavior. Whereas for pure bulk Ni, as shown in previous studies[1],
the Lβ/Lα intensity ratio approaches NiO values at low θ values, here
there is merely some increase in the Ni Lβ/Lα intensity ratio values as
θ is decreased. This indicates that, while the Ni oxidation in H is
greater near the surface than well below the surface, the oxidation of
Ni in bulk permalloy is somewhere between Class I and Class II behavior.
Given the data available, we do not know whether or not Fe oxidation is
complete, but the parallelness of Fe-Lβ/Lα curves for H and H* indicates
ideal Class I behavior.

Ni Thin Film Study

In order to see whether or not there may be any effects due to the
presence of a thin film on a Si substrate, the θ-dependence of Ni-f(500)
was investigated. As shown in Figure 5, at high θ values the Ni-Lβ/Lα
intensity is significantly higher for the 500 Å film than for the bulk,
with the film response decreasing monotonically as θ is decreased. The
film response reaches the bulk response at θ≈ 2.5° and sharply decreases
below 2.5° to a value at θ=0.5°, where the film value is about 40% lower
than the bulk value.

The high-θ film response may be explained for a self-supported thin
film on the basis of using finite limits of integration in the standard
fluorescence equation. The low-θ behavior of a Ni thin film, therefore,
must be due to having some sort of a Ni-Si interface effect. The
problem of the Lβ/Lα intensity response for a transition-metal Si-
substrate interface involves unusual interfacial-reflection effects due
to the presence of an anomalous dispersion region for x-ray optical
indices (constants) of transition-metal L lines. Accordingly, if we
assume that at and below 2.5° the Ni-Lα photon is totally reflected at
the interface but the Ni-Lβ photon is not reflected but penetrates into
the Si substrate, then the Lβ/Lα intensity ratio would sharply decrease
below 2.5°.

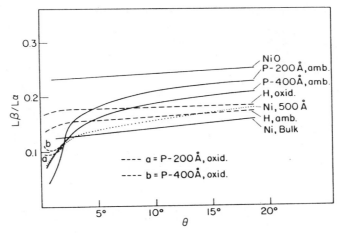

Fig. 6. Experimental Ni-L$_\beta$/L$_\alpha$ dependence on θ.

Permalloy Thin Film Study

The behavior of the oxidized and ambient permalloy thin films on Si substrates is illustrated in Figures 6 and 7.   Figure 6 shows the Ni-Lβ/Lα dependence on θ, and Figure 7 depicts the Fe-Lβ/Lα response. Accordingly, as shown in Figure 6, since the P-f values agree with the P-f* values for $\theta > 2°$, the oxidation of Ni is quite negligible. Below 2° the Ni-Lβ/Lα values increase significantly for the P-f* samples as compared with the P-f samples indicating typical Class II behavior for the P-f* samples.   Figure 6 also demonstrates that the interfacial optical effect below 2.5° increases with decreasing film thickness, as it should since for a 200 Å thin film a greater portion of internally

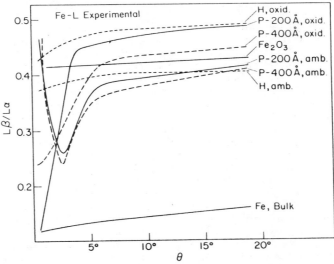

Fig. 7. Experimental Fe-L$_\beta$/L$_\alpha$ dependence on θ.

reflected Ni-Lα photons may be detected than for films of 400 and 500 Å thickness. The high-θ response as a function of film thickness is qualitatively correct. In fact, the Lβ/Lα method at high θ values can be used for film thickness determination. Whereas the estimation of the oxidation of Ni in the P-f samples presents a relatively straight forward story, the estimation of oxidation of Fe presents a far more complicated venture. First of all, the low-angle substrate-film interface effects are truly unusual. Second, there appears to be significant air oxidation for both P-f*(200) and P-f*(400) even under the mild conditions of 200° C for 2 hours. Third, a visual inspection of the P-f Lβ/Lα response at high θ cannot be used to estimate the degree of oxidation under ambient conditions, i.e. a rigorous quantitative approach needs to be used. A quantitative approach using Class I oxidation model towards high-θ response leads to the following degrees of oxidation P-f(200) – 4 to 3%; P-f(400) –2%; P-f*(200) – 9 1 to 99%; Pf*(400) – 36 to 32%. For each of the above films the first number represents H* calibration standard with only Fe oxidized, the second H* with both Ni and Fe oxidized. Thus, the different degrees of oxidation represent the use of different bulk calibration standards. As shown in Figure 7, the high θ performance of P-f(200) indicates nearly flawless Class I behavior, whereas there appears to be a descernible upward deviation for P-f*(400) with respect to P-f(400) as θ goes from 18° to 8° indicating mixed Class I and II behavior.

The unusual behavior of Fe-Lβ/Lα for the region $2.5 < θ \leq 5$ has several interesting aspects. All of the 4 films, namely, P-f(200), P-f*(200), P-f*(400), P-f(400) have an initial sharp down-turn in intensity with P-f(200), Pf(400) and P-f*(400) being around 5° and that of P-f*(200) being around 3.5°. Whereas the response for both P-f* films continue to decrease sharply all the way down to 0.5°, both P-f films show a minimum at 2.5°, and below 2.5° the unoxidized films' response dramatically increases for both films, to approach and perhaps even exceed the H* bulk value at θ = 0.5°. Thus, with the ambient 200 Å film the θ = 0.5° Lβ/Lα intensity response resembles the oxidized bulk value and with the oxidized film this 0.5° response equals that of pure bulk Fe. Interestingly enough the θ = 0.5° value for P-f*(400) reaches an intermediate value of 0.24, which translates to about 36% oxidation. The upshot of these findings is then as follows: The analytical sensitivity of monitoring oxidation of a permalloy film is far greater at very low angles of θ than at θ = 18°. While this increased sensitivity is a desirable phenomenon, it can become useful only if at least a qualitative explanation is available for this "upside-down" state of affairs at low values of θ.

In order to understand qualitatively the low-θ Lβ/Lα intensity responses for Fe the following model may be useful. As with the previously discussed Ni case, if one now assumes that the Fe-Lα photon is totally reflected at the Fe-Si substrate at θ=5°, but that the Fe-Lβ photon is not, then for P-f(200), P-f(400) and for P-f*(400) the Lβ/Lα intensity-ratio nose-dives below 5°. Note that for P-f*(200) this occurs below θ = 3.5° indicating that the 200 Å film is totally oxidized and the Fe-Lα photon interfacial reflection occurs at a lower value of θ for an Fe_2O_3-Si interface than for a Fe-Si interface. It also means that the oxidation of the 400 Å film is not a Class I type, i.e., there is considerable unoxidized Fe at the film-Si interface. This finding validates the finding from the previously discussed high-θ behavior.

The behavior of P-f(200) and P-f(400) in showing a minimum at θ = 2.5, and the upward trend below 2.5° can be explained by realizing that at θ = 2.5° not only the Fe-Lα but also the Fe-Lβ photons can be totally reflected. Moreover, for θ = 2.5° the Fe-Lα photon is reflectable from the film-air interface.

While the above discussion may be acceptable qualitatively, a rigorous quantitative evaluation awaits the availability of reliable high-resolution optical constants for Ni, NiO and Fe_2O_3 in the pure form as well as in bulk permalloy.

CONCLUSION

This study has demonstrated the following interesting capabilities for VAE-XRF Lβ/Lα spectrometry:

(a) Whereas Fe oxidation in bulk permalloy under drastic conditions of oxidation follows Class I behavior, Ni oxidation is somewhere between Class I and II.

(b) Under mild conditions of oxidation in permalloy thin films Ni remains essentially unoxidized with top-layer oxidation being of Class II type. For a 200 Å film Fe oxidation is of Class I type and for 400 Å the oxidation is more nearly a Class II type clearly showing that in bulk permalloy Fe oxidizes very differently from thin-film permalloy.

(c) For Ni and permalloy films on a Si substrate, the nature of the anomolous-dispersion region's optical constants associated with the Lβ and Lα lines provides a new non-destructive analytical tool for observing the effect of selective interfacial reflection at low angles. Conceivably, this technique could be turned into a useful analytical tool for non-destructive characterization of buried interfaces.

REFERENCES

1. G. Andermann, Appl. Surf. Sci. 31, 1-41 (1988).
2. T. Scimeca, Ph.D. Dissertation, Univ. of Hawaii (1988).
3. D. W. Fischer, J. Appl. Phys. 36, 2048 (1965).
4. G. Andermann, L. Berghnut, M. Karras, G. Grieschaber and J. Smith, Rev. Sci. Instr. 51, 814 (1980); G. Andermann, F. Burhard, R. Kim, F. Fujiwara, and M. Karras, Spec. Letters, 16, 851 (1983).
5. I. P. Bertin, Principles of X-Ray Spectrometric Analysis, 2nd Ed. (Pleneum Press) New York (1975).
6. L. Kaihola and J. Bremer, J. Phys. C: Solid State Phys., 14, L 43 (1981).

X-RAY DIFFRACTION ANALYSIS OF HIGH Tc SUPERCONDUCTING THIN FILMS

T. C. Huang, A. Segmüller*, W. Lee, V. Lee, D. Bullock, and R. Karimi

IBM Research Division, Almaden Research Center
650 Harry Road, San Jose, CA 95120-6099

* IBM Research Division, Thomas J. Watson Research Center
P.O. Box 218, Yorktown Heights, NY 10598

ABSTRACT

X-ray diffraction techniques have been used for the structure characterization of Y-Ba-Cu-O and Tl-Ca-Ba-Cu-O thin films. A powder diffraction analysis of Y-Ba-Cu-O films showed that the films deposited at 650°C on Si are polycrystalline and have an orthorhombic structure similar to that of the $YBa_2Cu_3O_7$ bulk superconductors. In addition to the conventional powder diffraction technique, both the rocking curve and the grazing incidence diffraction methods were used to characterize a $YBa_2Cu_3O_7$ film on (110) $SrTiO_3$ substrate. Results showed that the film was epitaxially grown and aligned with its substrate in a true epitaxy. Phase identification and line broadening analyses of Tl-Ca-Ba-Cu-O films showed that the films are comprised of one or more superconducting phases and probably contain stacking faults.

INTRODUCTION

The recent discovery of high temperature superconductivity in metallic oxides[1-4] has dramatically altered the prospects for applications of superconductivity. Superconductors prepared in thin film form are vital not only for fundamental studies but also for device applications in fast digital electronic circuits, picosecond pulse transmission lines, high frequency amplifiers and detectors, etc. Metallic oxide thin films can be produced by a variety of techniques, but the fabrication of high-quality superconducting films remains a challenge because of the problems of high temperature processing, controlling structure and stoichiometry, etc. In order to understand the preparation and properties of the films, it is important to characterize the structures of these materials.

X-ray diffraction (XRD) technique has been used extensively for the structure characterization of amorphous, polycrystalline, and epitaxial thin films.[5,6] Herein, results on the X-ray diffraction analysis of the structures of superconducting Y-Ba-Cu-O and Tl-Ca-Ba-Cu-O thin films on substrates are reported, and the deposition - structure - property relationships of the superconducting films will also be discussed.

EXPERIMENTAL

Thin Y-Ba-Cu-O and Tl-Ca-Ba-Cu-O films used in this study were deposited by sputtering from superconducting targets in Ar/O_2 atmospheres.[7-9] The temperature dependence of the film resistivity was measured using a low-frequency AC four-point probe technique[10], and the superconducting transition temperature (Tc) is defined as the temperature where the resistivity reaches zero. The structures of thin films were characterized by the XRD techniques. Powder diffraction data were used for phase identification, preferred orientation, and microstructure analysis of polycrystalline thin films.[5] The rocking curve measurement was used to determine the dispersion and the quality of oriented or epitaxial films.[5,11] The grazing incidence diffraction (GID) method was used to study the film/substrate epitaxial relationship.[6]

RESULTS AND DISCUSSION

Y-Ba-Cu-O Films on Si Substrates

Following the discovery of high Tc $YBa_2Cu_3O_7$ bulk superconductors[2], successful preparation of superconducting $YBa_2Cu_3O_7$ films on Al_2O_3, MgO, yttrium stabilized ZrO_2 (YSZ), and $SrTiO_3$ substrates were reported.[12] The growth of high Tc $YBa_2Cu_3O_7$ films directly on Si was, however, not successful because of the diffusion of Si caused by a high temperature post annealing. Recently, a new thin film deposition technique has been developed in our laboratory which makes the growth of $YBa_2Cu_3O_7$ films without a post deposition annealing possible.[7] Thin Y-Ba-Cu-O films were deposited on Si substrates from a superconducting $YBa_2Cu_3O_7$ target synthesized by solid state reactions at high temperatures from a mixture of Y_2O_3, $BaCaO_3$, and CuO. XRD results were used to study the entire thin film process from the preparations of starting powder mixtures and sputtering targets, to the growth of Y-Ba-Cu-O thin films. XRD patterns of a starting oxide mixture of 10% depletion in the barium stoichiometry and the sputtering target are shown in Figs 1a and 1b, respectively. A comparison between these two XRD patterns showed that the original oxide mixture underwent phase transformations and formed new compounds by solid-state reactions. A search of the Powder Diffraction File (PDF)[13] revealed that the sputtering target composes predominately of the $YBa_2Cu_3O_7$ phase (PDF #38-1433) with minor amounts of Y_2BaCuO_5 (PDF #38-1434) and CuO (PDF #5-661). The observed phases matched those of the $\frac{1}{2}(Y_2O_3)$-BaO-CuO phase-equilibrium diagram[14], and this indicated that the oxide mixture was fully reacted and the sputtering target has been properly prepared. A resistivity measurement showed that the target was superconducting at Tc=91K. It should be noted that $YBa_2Cu_3O_7$ is the only superconducting phase in the ternary Y_2O_3-BaO-CuO system.

Films deposited at low substrate temperatures (Ts<600°C) were nonsuperconducting down to 4.2K. Films prepared at Ts=600°C were superconducting, and values of Tc reached a maximum at Ts=650°C and then decreased at Ts≥700°C.[7] A typical XRD pattern of a nonsuperconducting Y-Ba-Cu-O film deposited at Ts=400°C is plotted in Fig. 1c. It shows a broad and diffuse diffraction pattern indicating a noncrystalline film. XRD patterns of the superconducting films have relatively sharp and strong diffraction peaks indicating polycrystalline $YBa_2Cu_3O_{7-x}$ materials (see, e.g., the diffraction pattern of a 68K film plotted in Fig. 1d). An analysis of the XRD data revealed that films deposited at 650°C with Tc=60-76K were orthorhombic, while those deposited at 600°C with lower values of Tc were mostly tetragonal. Least-squares refinements of the $YBa_2Cu_3O_7$ diffraction peaks of the high Tc films showed that the orthorhombic phase has lattice constants (a=3.85Å, b=3.89Å, and c=11.66Å) similar to those of the bulk superconductors.

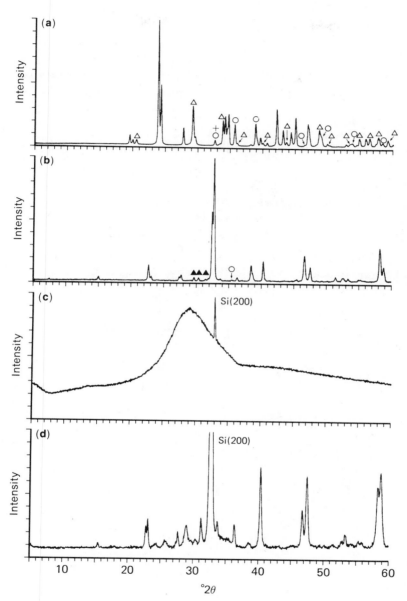

Fig. 1. Cu Kα XRD patterns: (a) starting oxide mixture of $BaCO_3$, Y_2O_3 (open triangle) and CuO (open circle); (b) superconducting $YBa_2Cu_3O_7$ target with minor amounts of Y_2BaCuO_5 (solid triangle) and CuO (open circle); (c) nonsuperconducting Y-Ba-Cu-O film deposited at Ts=400°C; and (d) superconducting $YBa_2Cu_3O_7$ film deposited at Ts=650°C.

Epitaxial YBa$_2$Cu$_3$O$_7$ Films on SrTiO$_3$ Substrates

Recently, there has been a large interest in the growth of superconducting epitaxial films because of its ability of carrying high critical current densities. It was reported that superconducting YBa$_2$Cu$_3$O$_7$ films deposited on SrTiO$_3$ (100) substrates have high critical currents in excess of 10^5Å/cm^2 at 77K.[15] XRD and transmission electron microscopy analyses showed that the YBa$_2$Cu$_3$O$_7$ films were epitaxially grown and contained twins and precipitates.[16] Herein, the results of XRD analysis of a new YBa$_2$Cu$_3$O$_7$ film grown on SrTiO$_3$ (110) substrate are presented.

A conventional powder diffraction pattern of the film is plotted in Fig. 2a. The pattern is dominated by two pairs of (110) and (220) peaks from both the YBa$_2$Cu$_3$O$_7$ film and its SrTiO$_3$ substrate.[8] This indicates that the film exhibits a high degree of [110] texture or epitaxial growth. An analysis of the powder diffraction data showed that the YBa$_2$Cu$_3$O$_7$ film had a preferred orientation factor (I_{110}/I_{001}) over 70 times higher than that of a randomly oriented powder specimen indicating a 98% [110] texture. In addition to the diffraction peaks from the bulk of the YBa$_2$Cu$_3$O$_7$ film, relatively weak peaks associated with the presence of Y$_2$BaCuO$_5$ and CuO have also been detected.

The [110] axis dispersion or the quality of the epitaxial YBa$_2$Cu$_3$O$_7$ film was measured by the rocking curve method. As shown in Fig. 3, the (110) rocking curve

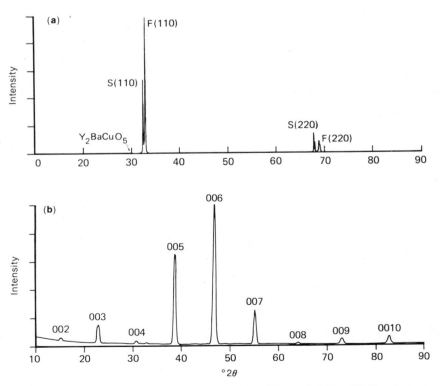

Fig. 2. Cu Kα XRD patterns of a YBa$_2$Cu$_3$O$_7$ film (F) on SrTiO$_3$ (110) substrate (S): (a) conventional θ-2θ scan; (b) GID θ-2θ scan.

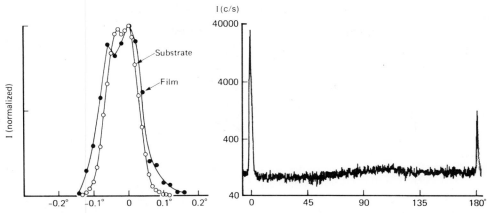

Fig. 3. Rocking curves of the (110) reflections of the $YBa_2Cu_3O_7$ film and the $SrTiO_3$ substrate.

Fig. 4. GID ω scan data of the (006) reflection of the $YBa_2Cu_3O_7$ film.

obtained from the film was almost identical in shape to that of the $SrTiO_3$ substrate. The double peaks of the substrate were reproduced in the film suggesting an epitaxial growth. The width of the rocking curve of the film is slightly broader than that of the single crystal substrate. The net orientational spread of the film is 0.03°, an order of magnitude smaller than those of the molecular beam epitaxial and laser evaporated $YBa_2Cu_3O_7$ films on $SrTiO_3$ (100) substrates.[11,17]

The GID technique[6] was used to study the film/substrate epitaxial relationship. The GID pattern shown in Fig. 2b was obtained from a θ-2θ (or radial) scan parallel to the [001] direction of the $SrTiO_3$ substrate. In GID, the diffraction vector is parallel and the diffraction planes are perpendicular to the specimen surface. The presence of nine multiple-order $YBa_2Cu_3O_7$ (00l) reflections in the GID pattern indicates that the film was epitaxially grown with its c-axis aligned parallel to that of the substrate. GID data obtained from a second θ-2θ scan parallel to the $SrTiO_3$ [1$\bar{1}$0] direction confirms that the film aligned with the single crystal substrate in a true epitaxy. The d-spacings measured parallel to the interface are: $d_{001} = 11.647$Å and $d_{1\bar{1}0} = 2.733$Å. A comparison with the d-spacings ($d_{001} = 11.6804$Å and $d_{110} = 2.7462$Å) of a strain-free $YBa_2Cu_3O_7$ powder[13] shows that the film was under compression and the parallel compressive strain were ~0.3% and 0.5% measured along the [001] and the [1$\bar{1}$0] directions, respectively. An ω scan of the $YBa_2Cu_3O_7$ (006) reflection was also obtained, and the pattern is plotted in Fig. 4. In an ω scan, the specimen is rotated around the normal of the film surface, and the detector is fixed at the Bragg angle 2θ of a selected reflection to record the diffracted X-rays. The presence of two (006) peaks in Fig. 4 with a 180° separation shows that the epitaxial $YBa_2Cu_3O_7$ film has a single domain.

Tl-Ca-Ba-Cu-O Thin Films on YSZ and SrTiO₃ Substrates

Since the recent discovery of a new class of Tl-Ca-Ba-Cu-O superconductors with Tc values up to 125K[4,18], there has been enormous efforts around the world to develop superconducting Tl-Ca-Ba-Cu-O thin films.[9,19-21] This new class of superconductors[22] has a chemical formula $Tl_mCa_{n-1}Ba_2Cu_nO_{m+2(n+1)}$, where m=1 or 2 and n=1, 2 or 3. The crystal structures consist of periodic arrays of 1 or 2 Tl-O

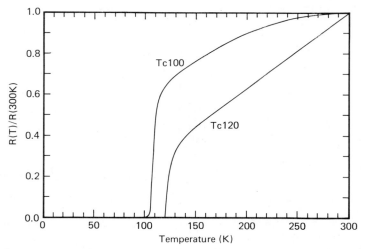

Fig. 5. Resistance, normalized to the resistance at 300K, versus temperature for the 100K and 120K films.

layers separated by 1, 2 or 3 CuO_2 layers. Depending on the structure, the supercon-ducting transition temperature varies from 0 to 125K. In general, the larger the numbers of Tl-O and/or CuO_2 layers, the higher the value of Tc.

Two superconducting Tl-Ca-Ba-Cu-O thin films with "triple digit" transition temperatures are first used to illustrate the effect of structure on superconducting properties. The resistance versus temperature curves for the 100K and 120K film are

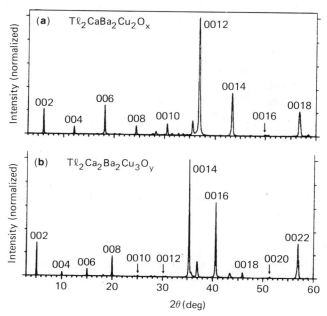

Fig. 6. Cu Kα XRD patterns of (a) the 100K film, and (b) the 120K film.

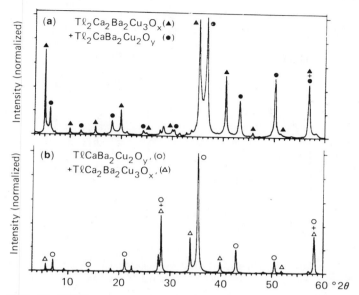

Fig. 7.Cu Kα XRD patterns of (a) the 101K film, and (b) the 93 film.

plotted in Fig. 5. To avoid a possible error due to the uncertainty in the sample geometry, the resistance data were normalized to the resistance at 300K. The resistance of the 100K film decreases with decreasing temperature indicating metallic behavior in a normal state, and then sharply drops at the on-set of superconductivity near 110K and finally reaches zero resistance at 100K. The 120K film has a resistance curve similar to that of the 100K film, but with lower values of normalized resistance and higher superconducted on-set and zero resistance temperatures at ~130K and 120K, respectively.

The X-ray diffraction patterns of the 100K and 120K films deposited on YSZ substrates are plotted in Figs. 6a and 6b, respectively. A comparison of the XRD data with those previously obtained from bulk superconducting Tl-Ca-Ba-Cu-O materials[22] allows an unambiguous determination of the structures of the films. As shown in Fig 6a, the first peak of the 100K film appears at $2\theta = 6.05°$ with a d-spacing close to 14.61Å indicating a (002) peak of the $Tl_2Ca_1Ba_2Cu_2O_8$ (2122) phase. The remainder of the pattern is dominated by a set of regularly spaced (00l) reflections indicating a highly textured film with its c-axis oriented perpendicular to the film surface. The value of c, determined by a least-squares refinement of nine observed (00l) peaks, is 29.229(4)Å. The XRD pattern shown in Fig. 6b is also dominated by a set of regularly spaced (00l) peaks. Compared to those of Fig. 6a, the diffraction peaks of Fig. 6b, however, appear at significantly lower 2θs indicating a different crystal structure with a larger c lattice dimension. A least-squares refinement of eleven observed (00l) peaks determined the value of c to be 35.662(4)Å. Phase identification results showed that the 120K film is mostly single $Tl_2Ca_2Ba_2Cu_3O_{10}$ (2223) phase with a small amount (<10%) of the 2122 phase. The analysis indicates that the films comprised almost solely of the 2122 and the 2223 phases superconduct at 100K and 120K respectively, in agreement with the properties of the bulk superconductors.[22]

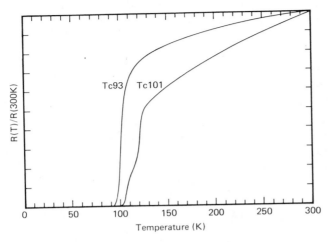

Fig. 8. Resistance, normalized to the resistance at
300K, vs temperature for
the 93K and 101 K films.

Because of a similarity in the crystal structures of Tl-Ca-Ba-Cu-O compounds,
c-axis preferentially oriented multiple-phase thin films have also been obtained. Two
Tl-Ca-Ba-Cu-O films superconducting at Tc=93K and 101K were analyzed by XRD.
Analysis of the diffraction pattern of the 101K film plotted in Fig. 7a reveals that
the film is a two-phase mixture of the 2223 and 2122 compounds. The presence of
approximate equal amounts of the two superconducting phases in the 101K film is
believed to have led to the observed double transitions with on-set temperatures at
125K and 110K (see Fig. 8). The diffraction of the 93K film plotted in Fig. 7b
corresponds to a different mixture of a major $Tl_1Ca_1Ba_2Cu_2O_7$ (1122) and a minor
$Tl_1Ca_2Ba_2Cu_3O_9$ (1223) phases. It should also be noted that both the 101K and the
93K films were prepared from the same sputtering target and annealing parameters,
but in different batches. This example demonstrates the strong dependency of the
superconducting properties of the Tl-Ca-Ba-Cu-O films on the actual annealing
conditions and the structures of the films.

The microstructure of thin films may also have an effect on the superconducting
properties. Two superconducting Tl-Ca-Ba-Cu-O films with different values of
Tc=116K and 102K were analyzed by XRD. Phase identification results showed that

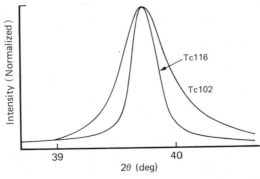

Fig. 9. X-ray diffraction profiles of the $Tl_1Ca_2Ba_2Cu_3O_9$ (007) peaks of the 116K
film deposited on YSZ and the 102K film on $SrTiO_3$.

both films are comprised mainly of the 1223 phase and small amounts of the 1122 phase. An examination of the diffraction pattern of the 102K film deposited on SrTiO$_3$ revealed that its diffraction peaks were asymmetrically broadened (see, e.g., the (007) peaks in Fig. 9).[21] The profile broadening was probably due to stacking faults (or intergrowths[23]) similar to those reported in metals.[24]

CONCLUSION

Conventional powder diffraction results were used to analyze the process of synthesis of superconducting YBa$_2$Cu$_3$O$_7$ target and the deposition of high Tc thin films. An analysis of the XRD pattern of the high Tc films deposited at 650°C on Si showed a polycrystalline YBa$_2$Cu$_3$O$_7$ film with an orthorhombic structure (a=3.85Å, b=3.89Å and c=11.66Å) similar to those of the bulk superconductors. The conventional powder diffraction, the rocking curve, and the GID techniques were used to determine the texture, the epitaixial quality and the film/substrate relationship of the YBa$_2$Cu$_3$O$_7$ film on SrTiO$_3$ (110) substrate. Results showed that the film was (110) epitaxially grown with its [001] and [1$\bar{1}$0] axes parallel to the substrate [001] and [1$\bar{1}$0], respectively. Phase identification analysis of Tl-Ca-Ba-Cu-O thin films showed a strong correlation between the structure and superconducting properties of the films. XRD data indicate that the 100K and the 120K films comprised almost solely of the 2122 and the 2223 phases, respectively. XRD results also revealed that the 101K films with a double transition is a mixture of the 2122 and 2223 phases, while the 93K film is comprised of the 1223 and 1122 phases. XRD patterns of the 116K and the 102K films showed that both films are comprised mainly of the 1223 phase, and the diffraction profiles of the 102K film were asymmetrically broadened, probably due to stacking faults.

REFERENCES

1 J. G. Bednorz and K. A. Müller, Z. Phys. B64, 189 (1986).

2 M. K. Wu et al., Phys. Rev. Lett. 58, 908 (1987).

3 H. Maeda, Y. Tanaka, M. Fukutami, and T. Asano, Jpn. J. Appl. Phys. 27, L209 (1988).

4 Z. Z. Sheng and A. M. Hermann, Nature 332, 138 (1988).

5 T. C. Huang and W. Parrish, Adv. X-Ray Anal. 22, 43 (1979).

6 A. Segmüller, Adv. X-Ray Anal. 29, 353 (1986).

7 W. Y. Lee, J. Salem, V. Lee, T. C. Huang, R. Savoy, V. Deline, and J. Duran, Appl. Phys. Lett. 52, 2263 (1988).

8 D. C. Bullock, G. Lim, J. Salem, W. Lee, R. J. Savoy, C. Hwang, V. Y. Lee, IBM Research Report, RJ-5918 (1987).

9 W. Y. Lee, V. Y. Lee, J. Salem, T. C. Huang, R. Savoy, D. C. Bullock, and S. S. P. Parkin, Appl. Phys. Lett. 53, 329 (1988).

10 W. Y. Lee, J. Salem, V. Lee, T. Rettner, G. Lim, and R. Savoy, and V. Deline, AIP Conf. Proc. 165, 95 (1988).

11 J. Kwo, T. C. Hsieh, R. M. Fleming, M. Hong, S. H. Liou, B. A. Davision, and L. C. Feldman, Phys. Rev. B36, 14039 (1987).

12 See, e.g., R. B. Laibowitz, R. H. Koch, P. Chaudhari, and R. J. Gambino, Phys. Rev. B35, 8821 (1987); M. Naito et al., J. Mater. Res. 2, 713 (1987); and references therein.

[13] Powder Diffraction File of the Joint Committee on Powder Diffraction Data (International Centre for Diffraction Data, Swarthmore, PA, 1988).

[14] W. Wong-Ng, R. S. Roth, F. Beech, and K. L. Davis, Adv. X-Ray Anal. 31, 359 (1988).

[15] P. Chaudhari, R. H. Koch, R. B. Laibowitz, T. R. McGuire, and R. J. Gambino, Phys. Rev. Lett. 58, 2684 (1987).

[16] P. Chaudhari, F. K. LeGoues, and A. Segmüller, Science 238, 342 (1987).

[17] A. M. De Santolo, M. L. Mandich, S. Sunshine, B. A. Davison, R. M. Fleming, P. Marsh, and T. Y. Kometani, Appl. Phys. Lett. 52, 1995 (1988).

[18] S. S. P. Parkin, V. Y. Lee, E. M. Engler, A. I. Nazzal, T. C. Huang, G. Gorman, R. Savoy, and R. Beyers, Phys. Rev. Lett. 60, 2539 (1988).

[19] D. S. Ginley, J. K. Kwak, R. P. Hellmer, R. J. Baughman, E. L. Venturini, and B. Morosin, Appl. Phys. Lett. 53, 406 (1988).

[20] M. Nakao, R. Yuasa, M. Nemoto, H. Kuwahara, H. Mukaida, and A. Mizukami, Jpn. J. Appl. Phys. 27, L849 (1988).

[21] T. C. Huang, W. Y. Lee, V. Y. Lee, and R. Karimi, Jpn. J. Appl. Phys. 27, L1498 (1988).

[22] S. S. P. Parkin, V. Y. Lee, A. I. Nazzal, T. C. Huang, G. Gorman, and R. Beyers, Phys. Rev. B 38, 6531 (1988).

[23] R. Beyers, S. S. P. Parkin, V. Y. Lee, A. I. Nazzal, R. Savoy, G. Gorman, T. C. Huang, and S. La Placa, Appl. Phys. Lett. 53, 432 (1988).

[24] B. E. Warren, X-Ray Diffraction (Addison-Wesley, New York, 1969).

THICKNESS MEASUREMENT OF EPITAXICAL

THIN FILMS BY X-RAY DIFFRACTION METHOD

J. Chaudhuri*, S. Shah*, and
J.P. Harbison**
*The Wichita State University
Mechanical Engineering Department
Wichita, KS   67208

**Bellcore
Red Bank, NJ

ABSTRACT

A method was described for determining the thickness of epitaxical thin films common to electronic materials. The equations were developed based on the kinematical theory of X-ray diffraction and effects of both primary and secondary extinctions were considered. As an example of the applications of this method, thickness measurement of AlGaAs thin films on GaAs was demonstrated. These films were grown by molecular beam epitaxy. The integrated reflected intensities from the film and the substrate were obtained by the X-ray double crystal diffractometer. An excellent agreement was obtained between the results from X-ray measurements and RHEED oscillation data.

INTRODUCTION

A substantial amount of current research activity in electronic materials is focused on thin films. It is desirable to know precisely the thickness to assist in the investigation of the growth and kinetics of these films. One way to measure the thickness is by using the Pendëllosung fringes (1). However, since the fringes disappear when the thickness is larger than the effective depth, this technique is often limited.

In this paper, a simple nondestructive method was demonstrated to determine the thickness of thin films from the ratio of the integrated reflected intensity from the substrate to the film. Thus, one does not need to measure the direct beam intensity. The kinematical expression of integrated intensity was corrected for primary and secondary extinction effects (2).

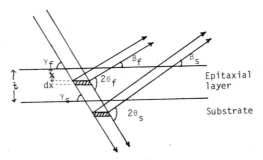

Fig. 1. Diffraction from an epitaxial thin film
and the underlying substrate.

THICKNESS CALCULATION

Figure 1 depicts diffraction from an epitaxial layer and the underlying substrate. The kinematic diffracted intensity from the epitaxial layer is given (3) by equation (1):

$$I_f = \left(\frac{e^2}{mc^2}\right)^2 \frac{I_o/F_f/^2\lambda^3}{\sin\gamma_f V_f^2} \frac{1 + \cos^2 2\theta_f}{2\sin 2\theta_f} \int_{x=0}^{x=t} e^{-\mu_f x \left(\frac{1}{\sin\gamma_f} + \frac{1}{\sin\beta_f}\right)} dx \qquad (1)$$

where

$\dfrac{e^2}{mc^2}$ = classical electron radius

I_o = incident x-ray intensity

F_f = structure factor for the film

λ = x-ray wavelength

V_f = unit cell volume of the film

θ_f = Bragg angle for the film

μ_f = mass absorption coefficient of the film

γ_f = angle between incident beam and the film

β_f = angle between diffracted beam and the film

t = thickness of the film

After integration, the above equation is expressed as:

$$I_f = \left(\frac{e^2}{mc^2}\right)^2 \frac{I_o/F_f/^2\lambda^3}{\sin\gamma_f V_f^2} \frac{1 + \cos^2 2\theta_f}{2\sin 2\theta_f} \frac{1 - e^{-\mu_f \alpha_f t}}{\mu_f \alpha_f} \qquad (2)$$

where

$$\alpha_f = \frac{1}{\sin \gamma_f} + \frac{1}{\sin \beta_f}$$

Similarly the diffracted intensity from the substrate is given by:

$$I_s = (\frac{e^2}{mc^2})^2 \frac{I_0/F_s/^2\lambda^3}{\sin\gamma_s V_s^2} \frac{1 + \cos^2 2\theta_s}{2 \sin 2\theta_s} e^{-\mu_f t(\frac{1}{\sin\gamma_s} + \frac{1}{\sin\beta_s})}$$

$$\int_{x=0}^{x=\alpha} e^{-\mu_s x(\frac{1}{\sin\gamma_s} + \frac{1}{\sin\beta_s})} dx \qquad (3)$$

where

F_s = structure factor for the substrate

V_s = unit cell volume of the substrate

θ_s = Bragg angle for the substrate

μ_s = mass absorption coefficient of the substrate

γ_s = angle between incident beam and the substrate

β_s = angle between diffracted beam and the substrate

$e^{-\mu_f t \alpha_s}$ = the amount by which the intensity is attenuated in the epitaxical layer

After integration, equation (3) is expressed as

$$I_s = (\frac{e^2}{mc^2})^2 \frac{I_0/F_s/^2\lambda^3}{\sin\gamma_s V_s^2} \frac{1 + \cos^2 2\theta_s}{2\sin 2\theta_s} \frac{e^{-\mu_f \alpha_s t}}{\mu_s \alpha_s} \qquad (4)$$

where

$$\alpha_s = \frac{1}{\sin \gamma_s} + \frac{1}{\sin \beta_s}$$

Applying the primary extinction correction of a mosaic crystal model, the integrated reflected intensity, I_p, can be written as shown by Zachariasen (2):

$$I^p = If(A) \qquad (5)$$

where I is the kinematic diffracted intensity, and the Bragg case primary extinction correction is:

$$f(A) = \frac{\tanh A + \cos 2\theta \, \tanh/A\cos 2\theta/}{A(1 + \cos^2 2\theta)} \qquad (6)$$

where

$$A = \frac{e^2}{mc^2} \frac{/F/\lambda t}{V \sin\theta} \left(\frac{1 + \cos 2\theta}{2}\right) \tag{7}$$

The secondary extinction correction of a mosaic crystal model is applied by replacing the mass absorption coefficient μ by $\mu + gQ$ in the expression for the integrated reflected intensity (2). Here,

$$Q = \left(\frac{e^2}{mc^2}\right)^2 /F/^2 \frac{\lambda^3}{V^2} \left(\frac{1 + \cos^2 2\theta}{2\sin 2\theta}\right) \tag{8}$$

$$g = \frac{1}{2\eta\sqrt{\pi}} \tag{9}$$

where η is the standard deviation of the block tilts in the mosaic crystal model.

After applying the appropriate primary and secondary extinction corrections to both film and substrate, the ratio of integrated reflected intensity is given by:

$$\frac{I_f^{ps}}{I_s^{ps}} = \frac{(\mu_s + g_s Q_s)\alpha_s}{(\mu_f + g_f Q_f)\alpha_f} \frac{f(A_f)/F_f}{f(A_s)/F_s} \frac{/^2 V_s^2}{/^2 V_s^2} \frac{1+\cos^2 2\theta_f}{2\sin 2\theta_f} \frac{2\sin 2\theta_s}{1+\cos^2 2\theta_s} \frac{1-e^{-(\mu_f+g_f Q_f)\alpha_f t}}{e^{-(\mu_f+g_f Q_f)\alpha_s t}} \tag{10}$$

If the integrated reflected intensity values are known, the thickness of the thin film may thus be determined by iteration.

APPLICATION

As an illustration of the application of the method, the thickness of AlGaAs thin films on GaAs substrates were determined. These films were grown by molecular beam epitaxy. The integtated reflected intensities from the films and the substrates were measured utilizing a Blake Industries double crystal diffractometer in the (+,-) parallel setting (2). A perfect crystal of germanium (001) was used as the first crystal. The rocking curves of (004) reflection were measured by using $CuK_{\alpha 1}$ and $MoK_{\alpha 1}$ radiations. Since the ratio of X-ray intensities from the film and the substrate for different positions in the samples varied about 5%, average values of 6 measurements were considered. The resulting ratios of measured intensities from the film and the substrate, I_f^e/I_s^e, are listed in Table I.

Table I. Intensity Ratios and Thickness Values for AlGaAs Thin Films

Sample	I_f^e/I_s^e $CuK_{\alpha 1}$	I_f^e/I_s^e $MoK_{\alpha 1}$	t in μm $CuK_{\alpha 1}$	t in μm $MoK_{\alpha 1}$	t in μm RHEED	% error $CuK_{\alpha 1}$	% error $MoK_{\alpha 1}$
1	0.76	0.89	1.55	1.55	1.6	3.1	3.1
2	1.00	1.09	2.02	1.94	2.0	1.0	3.0
3	1.56	1.61	3.33	3.57	3.5	4.9	2.0

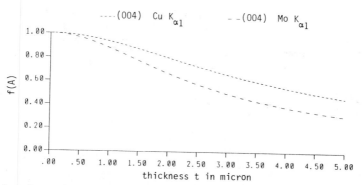

Fig. 2.   Dependence of Bragg case Primary Extinction Coefficent
of GaAs on the block thickness in the mosaic model.

Figure 2 shows the dependence of the Bragg case primary extinction coefficient f(A) of GaAs, on the block thickness in the mosaic crystal model. In the present study f(A) corresponding to 2 µm block thickness of GaAs was considered. The secondary extinction correction was applied considering the standard deviation in block tilts, η, as equal to the full width at half maximum of the X-ray rocking curves. The η values are approximately 11.0 and 13.8 arcsec for the substrate and the film, respectively. The thicknesses as calculated are listed in Table I.

An attempt was made to compare the results obtained from X-ray measurement with RHEED oscillation (4,5,6) data as shown in Table I. There was good agreement between the results from these two different methods, with the percentage difference being less than 5%. Without the extinction corrections the values were found to be off by more than 30%.

CONCLUSION

An X-ray method was described which is capable of measuring the thicknesses of epitaxical films with high precision. This method is non-destructive, straight-forward and rapidly performed. The technique utilizes the ratio of the integrated diffracted intensity from the film to the substrate. Both the primary and secondary extinction corrections were applied. X-ray results were in good agreement with RHEED oscillation data. It was anticipated that the results could be further improved using a weaker reflection or a shorter wavelength since the extinction corrections tend to become negligible in either case. This technique can be extended to measure the thicknesses of films in multi-heterostructure systems.

ACKNOWLEDGEMENT

This work was supported in part by the National Science Foundation, NSF Grant #DMR-8605564.

REFERENCES

1.   W.J. Bartels, J. Vac. Sci Technol. B1 (2), Apr-June, 338 (1983).
2.   W.H. Zachariasen, Theory of X-Ray Diffraction in Crystals (John Wiley and Sons, Inc., New York, 1945), pp. 147-175.

3.   B.D. Cullity, Elements of X-Ray Diffraction (Addison-Wesley Publ.
      Co., Inc., Reading, Massachusetts, 1978), pp. 133-135.
4.   J.J. Harris, B.A. Joyce, and P.J. Dobson, Surf. Sci. 103 (1), L90
      (1981).
5.   J.H. Neave, B.A. Joyce, P.J. Dobson, and N. Norton, Appl. Phys.
      A31, 1 (1983).
6.   J.P. Harbison, D.E. Aspnes, A.A. Studna, L.T. Florez, and M.K.
      Kelly, Apple Phys. Lett. 52 (24), 2046 (1988).

TEXTURE ANALYSIS OF THIN FILMS AND SURFACE LAYERS

BY LOW INCIDENCE ANGLE X-RAY DIFFRACTION

J. J. Heizmann, A. Vadon, D. Schlatter, J. Bessières

Laboratoire de Métallurgie Physique et Chimique
Centre de Métallurgie Structurale
57 045 - Metz University - France

INTRODUCTION

It is necessary to know the orientation of thin surface layers for the electronic industry as well as for different studies on interphases (epitaxy, topotaxy, phase transformation, reactivity of solids).

It is difficult to obtain information with a conventional Schulz goniometer (Bragg-Brentano geometry) because of the insufficient amount of diffracting material.

Generally, with a reflection Schulz goniometer [1], we use the Bragg-Brentano geometry which is interesting because it works with a parafocussing geometry and we do not need the intensity correction (except the defocussing phenomenon) when the sample rotates about its azimuthal axis $\vec{\phi}$ and its tilting axis ϕ. There is an exact balance between the increase of intensity, coming from the increase of irradiated area, and the decrease of intensity, owing to the absorption of the X-ray by the sample. This balance does not occur with thin layers of diffracting materials.

In the case of thin layers, the diffracted intensity is very weak and we can use the low incidence-angle diffraction technique [2, 3,] to increase the diffracting volume, and consequently the diffracted intensity. In this case, it is necessary to correct all the information coming from the detector. These corrections are:
- the absorption correction which depends on the thickness of the film, the irradiated area and the "position" of the sample,
- the new location of the information on the pole figure, which depends on the sample "position" (defined by the tilt angle ψ and azimuth angle ϕ of the goniometer),
- the enlargement of the diffracted beam arising from defocussing.

THEORY: correction laws

Absorption: Bragg-Brentano conditions

For an anisotropic specimen, in Bragg-Brentano conditions, when the sample is perpendicular to the incidence plane, the intensity detected depends on the thickness of the film,

$$I_{B,t} = I_\infty \cdot [\, 1 \cdot \exp(\frac{-2 \cdot \mu \cdot t}{\sin \theta})\,] \qquad (1)$$

where I_∞ is the intensity which the detector would measure for an infinite thickness of the sample, μ the linear absorption coefficient, t the thickness of the film and θ the Bragg angle. So we can see that for a thin film, the diffracted intensity can be very weak.

In texture measurement, when the sample rotates about its tilting axis $\vec{\psi}$ (the angle between the normal of the sample and the incidence plane (Fig. 1a), the intensity diffracted by the (hkl) lattice planes depends also on the ψ angle,

$$I_{B,t,\psi} = I_\infty \cdot [\, (\, 1 - \exp(\frac{-2 \cdot \mu \cdot t}{\sin \theta \cdot \cos \psi})\,)\,] \qquad (2)$$

We can see that for bulk sample (t = ∞), we do not need the intensity correction; the intensity is independent of the sample position ψ. On the other hand (if one except the defocussing phenomenon), for thin layers, the diffracted intensity increases as the tilt angle increases. So the intensity on the border of the pole figure will be too high compared to that at the center. For thin films, even if the goniometer works in Bragg-Brentano conditions, we must apply the intensity correction according to the formula (2).

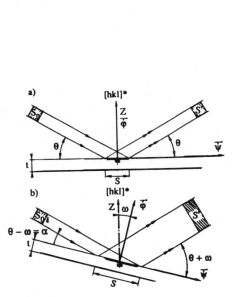

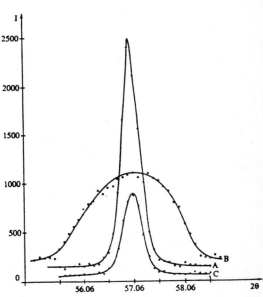

Figure 1 . Geometric arrangement, with a parallel X-ray beam, for $\psi = 0°$ ($\vec{\psi}, \vec{\phi}$ rotation axes of the texture goniometer) :

a) - Bragg Brentano geometry ($\omega = 0°, \alpha = \theta$)

b) - Low incidence angle ($\omega \neq 0°, \alpha < \theta$).

Figure 2 . Widening of the diffracted peaks (110) of a bulk isotropic iron sample, measured by 2θ step scanning for $\psi = 0°$:

A : Bragg Brentano arrangement
    ($\omega = 0°, \alpha = 28.53°$),
B : Low angle incidence
    ($\omega = -18.53°, \alpha = 10°$),
C : High angle incidence
    ($\omega = +18.53°, \alpha = 47.06°$).

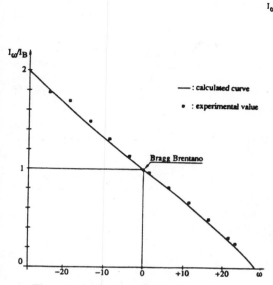

Figure 3 . I_ω/I_B (110) diffracted intensity ratio of a bulk isotropic iron sample versus the position ω of the sample.

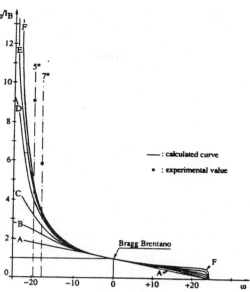

Figure 4 . I_ω/I_B (110) diffracted intensity ratio of a thin film of aluminium for different thicknesses ($\lambda Fe K\alpha$).

A : ∞ D : 2 μm
B : 10 μm   E : 1 μm
C : 5 μm   F : 0.1 μm

Absorption: Low incidence angle

In Fig. 1b, the geometric arrangement drawn for a parallel X-ray beam, and for $\psi = 0$, shows us the enlargement of the diffracted beam, the incidence angle α defined by the angle ω, the value of which is the gap between the Bragg-Brentano arrangement and the new one. We can see on this figure that the angular location of the (hkl) lattice planes depends on the angles (ω, ψ, ϕ).

In this case, the intensity detected is given by the relation,

$$I_{B, t, \psi} = I_\infty \cdot [\, 1 - \frac{tg\, \omega}{tg\, \theta}\,] \cdot [\, 1 - \exp(\frac{-\mu \cdot t}{\cos \psi} \cdot (\frac{1}{\sin(\theta + \omega)} + \frac{1}{\sin(\theta - \omega)}))\,] \qquad (3)$$

so whatever the thickness, the intensity must be corrected. For a bulk sample, for example, the last term is always 1 and the intensity varies between $2I_\infty$ and 0 when ω varies between $- \theta$ and $+ \theta$.

Verification of absorption correction

All the intensity results we present are given in relation to the Bragg-Brentano intensity, i.e., as the ratio I_ω/I_B.

1st sample

We have chosen a bulk iron sample coming from a compressed (8 ton/cm^2), 3 μm grain size iron powder. The intensity measurement on the radius of the

(110) Bragg-Brentano pole figure indicates the isotropic nature of the sample and a negligeable defocussing phenomenon (in our conditions of measurement the detector slits were side enough to get all information until ψ = 75° - 80°). The widening of the diffracted peak (110), measured by 2θ step scanning is presented in Fig. 2 for three values of the incidence angle. From these peaks, the integral values of the intensities are drawn with the ω variation as abscissa (Fig. 3). There is a very good agreement between the experimental values and the theoretical curve which confirms the 1st term of the formula (3).

2nd sample

The 2nd sample is an aluminium thin film, obtained by vacuum deposition on copper. Its thickness measured by interferometry is 0.6 µm. The theoretical intensity curves presented for (111) diffracted peaks versus ω indicate a great amplification of the intensity for low incidence angles. Two measurements of intensity (ψ = 0) were done for two low incidence angles (5° and 7°). These two measured intensities are located above the theoretical curve (Fig. 4). This can be explained by the texture of the film as we shall see later. We can notice that the lattice planes we detected had reflections that were at angles, respectively, of $\omega = \theta - \alpha = 17.47°$ and $\omega = 19.47°$ with the surface of the sample.

Location correction

The texture goniometer kept the same movement it would have in a classical pole figure measurement, i.e., a 0°<ϕ<90° rotation with a 5° step for each 0°<ψ<90° value, the step of the ψ scanning was 2.5°. With a low incidence angle, the goniometer works out of the Bragg-Brentano conditions. When the goniometer angles (Fig. 5) are ψ and ϕ, the detector will detect reflections from all the lattice planes (hkl), the normals [hkl]* of which are the bisectrix of the Bragg angle.

During the ϕ rotation, the lattice plane (hkl) reflections detected have their normals [hkl]* located on a cone of aperture ψ_2.

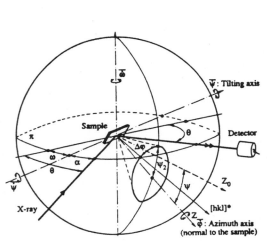

Figure 5 . For the (ψ, ϕ) angles of the goniometer,
the detector will detect the lattice planes
of which the normals [hkl]* are located
by the two angles ψ_2 and $\phi_2 = \phi + \Delta\phi$.

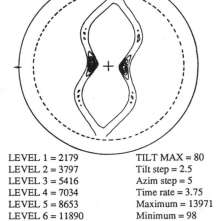

LEVEL 1 = 2179	TILT MAX = 80
LEVEL 2 = 3797	Tilt step = 2.5
LEVEL 3 = 5416	Azim step = 5
LEVEL 4 = 7034	Time rate = 3.75
LEVEL 5 = 8653	Maximum = 13971
LEVEL 6 = 11890	Minimum = 98

Figure 6 . (111) pole figure of a rolled nickel
sheet 5 µm thick (Bragg Brentano
arrangement, with intensity corrections).

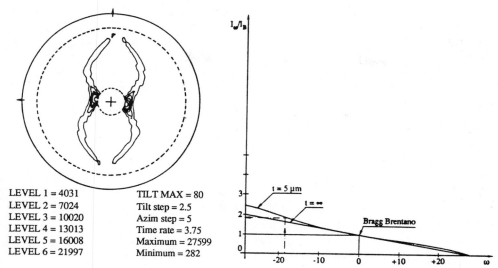

LEVEL 1 = 4031	TILT MAX = 80
LEVEL 2 = 7024	Tilt step = 2.5
LEVEL 3 = 10020	Azim step = 5
LEVEL 4 = 13013	Time rate = 3.75
LEVEL 5 = 16008	Maximum = 27599
LEVEL 6 = 21997	Minimum = 282

Figure 7 . (111) pole figure of the previous sheet of nickel (low incidence angle : $\alpha = 10°$, intensity and location corrections).

Figure 8 . Amplification coefficient of the pole figure for a 5 μm thick nickel sheet versus the angle ω.

Spherical trigonometry gives us the coordinates (ψ_2, ϕ_2) of the normal [hkl]* on the pole figure according to the (ψ, ϕ) angles of the goniometer [4]:

$$\cos \psi_2 = \cos \psi \cdot \cos \omega \quad \text{and} \tag{4}$$

$$\varphi_2 = \varphi + \Delta\varphi, \text{ with } \sin \Delta\varphi = \frac{\sin \omega}{\sin \psi_2} \tag{5}$$

Blind area

When $\psi = 0$, the normal of the sample Z becomes Z_0 (Fig. 5), which is not the bissectrix of the Bragg angle, so all the lattice planes (hkl), the normals [hkl]* of which are inside the cone of ω aperture cannot be detected. This brings about a blind area on the pole figure.

APPLICATION

Pole figure measurement

To confirm all we have stated above, we used two kinds of sample:
- a rolled nickel sheet, 5 μm thick, glued on a plate of glass,
- a thin film of aluminium (0.6 μm) deposited on a plate of copper under vacuum and a thin film of aluminium (0.06 μm) deposited on a plate of glass under vacuum.

Nickel sample

Two pole figures of the nickel are shown. The first one is made with $\omega = 0$, i.e., in Bragg-Brentano conditions (Fig. 6). The pole figure presented is corrected by the absorption coefficient which depends on the thickness and the tilt angle. We can observe the well-known texture of a sheet of nickel.

The second one is made with an incidence angle $\alpha = 10^\circ$ (Fig. 7). It is drawn with the location and absorption corrections. The levels were chosen 1.85 higher than the previous ones because the global increase of the intensities of the pole figure coming from the low incidence technique is 1.85. This pole figure is the same as the previous one except its intensity and the blind area we can observe at the center. These results indicate that the location corrections (Fig. 8), the intensity corrections and the softwares used were correct.

Aluminium sample

The Bragg-Brentano pole figure without correction presented in Fig. 9 shows a smooth texture of aluminium which can be described by a fiber texture, the [111] fiber axis of which makes 15° with the normal of the sample. Nevertheless, the intensities of the ring located at 70° from the maximum of the pole figure are too high. After intensity corrections, we obtain pole figure Fig. 10. The ratio of the intensities on the 70° ring to that of the maximum located near the center corresponds now to the ratio for a fiber texture.

We can now explain the high intensities measured previously (Fig. 4). The lattice planes observed were located near the maximum of the pole figure.

A low incidence angle measurement of the pole figure presented in Fig. 11 gives the same information (except in the blind area). The intensities are about twice as high as in the Bragg-Brentano pole figure even with a smaller counting time (3.75 seconds instead of 5.00 seconds).

We have tried to measure the texture of a 0.06 μm thick aluminium coating on glass. In Bragg-Brentano geometry, only the background was observed. With low incidence angle technique, after background subtraction, it is possible to observe a texture of aluminium (Fig. 12), but as the texture

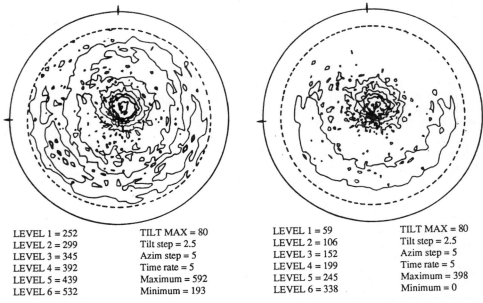

LEVEL 1 = 252	TILT MAX = 80
LEVEL 2 = 299	Tilt step = 2.5
LEVEL 3 = 345	Azim step = 5
LEVEL 4 = 392	Time rate = 5
LEVEL 5 = 439	Maximum = 592
LEVEL 6 = 532	Minimum = 193

LEVEL 1 = 59	TILT MAX = 80
LEVEL 2 = 106	Tilt step = 2.5
LEVEL 3 = 152	Azim step = 5
LEVEL 4 = 199	Time rate = 5
LEVEL 5 = 245	Maximum = 398
LEVEL 6 = 338	Minimum = 0

Figure 9 . (111) pole figure without correction of a 0.6 μm thick aluminium coating on copper (Bragg Brentano arrangement).

Figure 10 . The same as figure 9 with intensity correction.

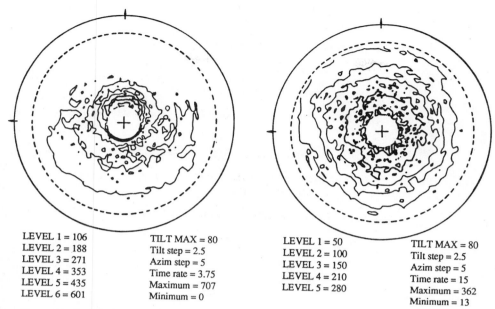

LEVEL 1 = 106	TILT MAX = 80
LEVEL 2 = 188	Tilt step = 2.5
LEVEL 3 = 271	Azim step = 5
LEVEL 4 = 353	Time rate = 3.75
LEVEL 5 = 435	Maximum = 707
LEVEL 6 = 601	Minimum = 0

Figure 11 . Low incidence angle (5°)
(111) pole figure of a 0.6 µm thick
aluminium with intensity and
location corrections.

LEVEL 1 = 50	TILT MAX = 80
LEVEL 2 = 100	Tilt step = 2.5
LEVEL 3 = 150	Azim step = 5
LEVEL 4 = 210	Time rate = 15
LEVEL 5 = 280	Maximum = 362
	Minimum = 13

Figure 12 . Low incidence angle (5°)
(111) pole figure of a 0.06 µm thick
aluminium with intensity and
location corrections.

of this thin film is extremely flat, the result obtained must be verified
later on. Nevertheless for strong texture film, it will be possible to get
the texture of films which are less than 1000 Å thick.

CONCLUSIONS

We have shown that it is possible to obtain the texture of a thin film
by low incidence X-ray texture goniometry. The intensity correction and the
correction of the location of the information have been checked and correspond
to the theory.

Nevertheless, we must take care:
- of the defocussing phenomenon because this technique is not a
parafocussing one,
- of the enlargement of the incident beam for the great tilting
angle.

The loss of information in the blind area could be known by the use of
the Vector Method and the M.P.D.S. (Minimum Pole Density Set) texture analysis
[5].

The use of a linear or curved position-sensitive detector will give an
answer to the defocussing phenomenon [6].

ACKNOWLEDGEMENTS

The authors would like to thank the L.P.M.M. of the Metz University for
the aluminium vacuum deposition and the Laboratoire de Cristallographie U.L.P.
Strasbourg for the thickness measurement.

BIBLIOGRAPHY

1.   L. G. Schulz, J. Appl. Phys., 20:1030-1036 (1949).

2.   A. Segmüller and M. Murakami, "Characterisation of Thin Film by X-Ray Diffraction" in "Thin Films from Free Atoms and Particles," J. Klabunde, ed., Academic Press, New York, 325-351 (1985).

3.   S. S. Hyiengar, M. W. Santana, H. Windischmann and P. Engler, in "Adv. in X-ray Anal., 30, 457-464 (1987).

4.   J. J. Heizmann and C. Laruelle, J. Appl. Cryst., 19, 467-472 (1986).

5.   A. Vadon, D. Ruer and R. Baro, 30th Annual Denver Conference on Application of X-ray Analysis, (1981).

6.   J. J. Heizmann, C. Laruelle and A. Vadon, Analysis, 16, n° 6, 334-340 (1988).

FAST THICKNESS MEASUREMENT OF THIN CRYSTALLINE LAYERS BY RELATIVE INTENSITIES IN XRPD METHOD

G. Kimmel, G. Shafirstein and M. Bamberger

Department of Materials Engineering
Technion, Haifa 32000, Israel

INTRODUCTION

In this work a continuous wave CO_2 laser was used for melting a layer of amorphous alumina obtained by anodizing 6061 aluminum plates. Melting the surface of the anodized plates led to the formation of a new uniform corundum layer at the expense of some of the amorphous coating, resulting in a double layer coating of crystalline on amorphous.

The characterization of the corundum layer is essential for process optimization study. Among the non-destructive methods X-ray diffraction is useful, because it provides us with selected data on the crystalline layer.

In the present paper we will concentrate on the ability to measure the texture and the thickness from the measurement of integrated intensities.

USE OF RELATIVE INTENSITIES FOR THICKNESS MEASUREMENT

It is possible to measure the thickness of polycrystalline thin samples by intensity measurements of a single diffraction line [1]. This method requires randomly oriented samples or epitaxial layers for absolute intensity measurement.

The reflection power which is received by the diffraction of a monochromatic X-ray beam from a thin flat sample in a Bragg-Brentano diffractometer, is proportional to the term:

$$I(hkl) = G.R(\theta).f.[1-\exp(-2\mu t/\sin\theta)]/\mu \qquad (1)$$

where G is the instrumental scaling factor, $R(\theta)$ is the total angular dependency factor, f depends on a number of factors, including the degree of crystal perfection, grain shape, packing density and sample size, μ is the linear absorption coefficient and t is the sample thickness normal to the reflecting planes.

For many thin samples it is impossible to measure the thickness in a single diffraction line either by rocking or by absolute intensity correlation, due to microabsorption and preferred orientation effects. Thus, a multiple line method [2,3] is applied.

For random samples, the best way to determine the μt value using relative intensities is through the refinement of μt from different orders of reflection using equation 1; but, in principle, we can use a pair of reflections in two Bragg angles, $θ_1$ and $θ_2$, comparing their intensity ratio: $i(t) = I_1/I_2$ to $i(∞)$ of bulk. From equation 1 we obtain:

$$y = \frac{i(t)}{i(∞)} = \frac{1-\exp(-2μt/\sinθ_1)}{1-\exp(-2μt/\sinθ_2)} \tag{2}$$

In the case of preferred orientation we can use the intensity distribution of the reflection lines which are generated from the same crystallographic plane. The refinement process is made for this case on selected reflecting lines (nh,nk,nl) which are along the [hkl] direction in the reciprocal space. As seen from figure 1, by measuring the intensity ratios, thickness values of solid corundum between 1 to 100 μm could be derived.

EXPERIMENTAL

Samples and characterization methods

The initial material consisted of anodized aluminum plates with a 60 μm external coating of amorphous alumina. The surface was melted by the laser beam, and a new crystalline layer (α-alumina) was formed during cooling. The crystalline layer parameters were: laser beam power (85-190 W); scanning speed (3-17 mm/s): shielded gas type (air or helium): shielded gas pressure (0.15-0.30 MPa). The α alumina thickness was determined by direct measurement on metallographic cross sections using SEM and by relative intensity of X-ray diffraction.

In our case we used several pairs of (hkl) and (2h,2k,2l) diffraction lines and derived the thickness t by the ratio of the integrated intensities using equation 2 (see fig.1).

The ratio i(∞) can be obtained through calculations, but with caution because some parameters still may affect the relative intensities. In particular, the instrumental aberrations and microabsorption [4-6]. Thus, in a manner

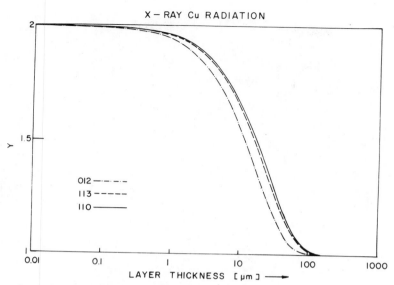

Fig.1. The ratio of relative intensities data obtained from
       pairs of corundum diffraction lines (hkl/2h 2k 2l)

similar to Brandt and van der Vliet [3], we obtained an i(∞)
value taken from a reference bulk samples.

The reference bulk sample was α alumina (corundum)
consisting of very fine and round crystallites, prepared by
the American National Bureau of Standards (NBS) as a
standard intensity material for powder X-ray diffraction
(SRM-674) [7].

When the thickness data was obtained the absorption
correction was used to reconstruct the bulk relative
intensity spectrum for each sample, as advised by Brandt and
van der Vliet [3], and then characterize the layer texture
by the inverse pole figure method [8].

The X-ray diffraction system

The data were collected by a commercial Philips PW-1820
automatic powder diffractometer including a long fine-focus
Cu X-ray tube powered by a PW-1730 generator and routinely
operated at 40 kV, 40 mA, fixed 1 deg divergent and anti-
scattering slits, a 0.2 mm receiving slit followed by a
curved graphite monochromator.

The counting electronics chain consisted of the standard
Philips PW-1710 system, which compensates for pulse height
variation, up to about 500,000 counts per second. The system
was operated by a PDP 11/53 micro computer.

The integrated intensities were measured by accumulation of

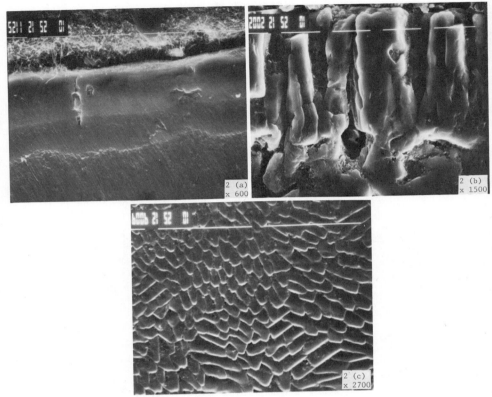

Fig.2:   SEM micrographs of the alumina layers. Low
         magnification (2a) and high magnification (2b)
         of cross section and surface view (2c).

the counts between two background points using straight-line
background subtraction.

RESULTS AND DISCUSSION

A SEM micrograph showing the general view over the cross
section after the laser treatment is given in figure 2a.
With higher magnification the crystalline external layer
shows a columnar growth (figure 2b).   The elongated
crystallites were observed also over the surface (figure
2c),

Figure 3 shows a diffraction pattern of the anodic (3a) and
the treated layer (3b), The anodic layer is transparent to
Cu Kα X-ray beam showing the diffraction pattern of the
aluminum. The absence of other diffraction lines shows that
the anodic coating is amorphous. After the laser treatment
the α alumina is the only crystalline phase added to the
aluminum as seen in figure 3b.   The X-ray diffraction

intensities are correlated with polycrystalline corundum with strong preferred orientation.

The reciprocal directions used to determine the thickness were (012), (113) and (110). It was found that the α alumina thickness is strongly dependent on the shielded gas type and pressure, and the laser intensity (power density). Under the optimal conditions, the corundum layer was $18\pm2\,\mu m$ thick and the length of the crystallites (figure 2c) was the shortest. When the laser treatment varied from these conditions, the thickness values were sharply decreased down to 9 μm and less.

There was a complete agreement between the thickness found by the X-ray diffraction and by the SEM as shown in Table 1.

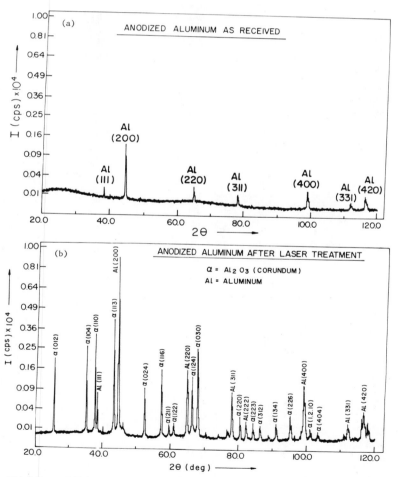

Fig.3. Diffraction pattern of an anodized aluminum
       surface before (3a) and after laser treatment (3b).

Table 1. Comparison between XRD and SEM thickness results
(μm)

XRD (averaged)	8.0	9.0	18.0
SEM (range)	9-10	7-9	16-19

Table 2. Comparison of integrated intensities between
random bulk reference sample and examples
of laser treated thin layers samples.
(*) corrected to bulk

				observed relative integrated intensity		
h k l	2θ	Ical	random	18 μm(*)	7 μm(*)
0 1 2	25.58	53.5	57.2	34.7	27.0
1 0 4	35.15	86.1	94.3	51.8	18.4
1 1 0	37.77	40.4	38.2	71.6	101.8
1 1 3	41.68	84.4	100.4	111.6	96.8
0 2 4	52.55	49.5	47.2	28.7	22.1
1 1 6	57.50	100.0	92.4	47.6	20.8
2 1 1	59.74	2.5	2.4	5.5	6.9
0 1 8	61.3	7.5	10.8	7.3	9.1
1 2 4	66.52	40.7	33.7	50.0	49.5
0 3 0	68.21	61.8	60,2	117.1	166.0
2 2 0	80.69	7.4	6.5	12.0	17.6
3 1 2	86.35	4.4	6.5	8.9	12.7
1 3 4	91.18	10.5	10.4	11.0	12.7
2 2 6	95.25	21.1	19.4	22.2	18.9

Table 3. P_{hkl} values calculated for samples treated under
various conditions. The thin sample intensities
were converted to bulk

h k l	2θ	t=18 μm Air 0.3 Mpa	t=8 μm Air 0.15 Mpa	t=9 μm Helium 0.3 Mpa	t=6 μm Helium 0.15 Mpa
0 1 2	25.58	1.8	2.1	2.0	2.3
1 0 4	35.15	1.8	5.1	5.3	5.6
1 1 0	37.77	0.5	0.4	0.4	0.3
1 1 3	41.60	1.0	1.0	1.0	1.3
0 2 4	52.55	1.6	2.1	2.0	2.4
1 1 6	57.50	2.0	4.5	4.5	6.1
2 1 1	59.74	0.4	0.4	0.3	0.3
0 1 8	61.30	1.5	1.2	1.4	1.5
1 2 4	66.52	0.7	0.7	0.7	0.7
0 3 0	68.21	0.5	0.4	0.4	0.3
2 2 0	80.69	0.5	0.4	0.4	0.3
3 1 2	86.35	0.7	0.5	0.6	0.6
1 3 4	91.18	0.9	0.8	0.8	0.8
2 2 6	95.25	1.0	1.0	0.9	1.2

Moreover, there were no significant differences among the results obtained from different crystallographic planes. However, the X-ray method is much faster. Moreover, whereas the SEM analysis was made only on local spots of selected samples, the X-ray method was applied for the entire surface of all the treated specimens providing us with a complete map of the corundum-layer thickness versus the laser parameters.

Using the found thickness values, we could obtain the relative intensity data corrected to bulk. The characteristic texture was observed as seen in Table 2.

Following Harris [8] we derived P_{hkl} (see appendix) values for each diffraction line as summarized on Table 3. It is clearly seen that the extent of deviation from isotropy as expressed by P_{hkl} values are the lowest for the thickest layer and higher for the thinner. This agrees with the SEM studies showing that the columnar grains increase when the external layer thickness decreases.

Using the inverse pole-figure we obtained a typical texture for the thin samples as shown in figure 4. We believe that the high P_{hkl} values around {116} and {104} planes shows a trend for a rapid crystal growth rate normal to these planes.

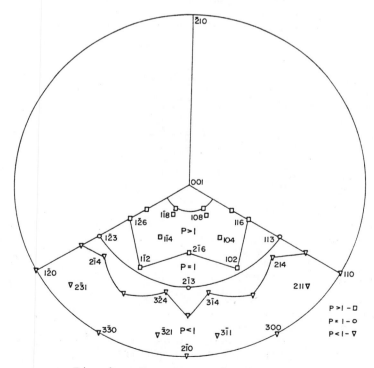

Fig.4.   Inverse Pole Figure

SUMMARY AND CONCLUSIONS

In this work, reflected intensities of X-ray diffraction were used to characterize crystalline surfaces after treatment with a CO_2 laser beam.

The thickness of the crystalline layers was determined by studying the integrated intensities of the corundum phase, taken by a Bragg-Brentano diffractometer. The analysis is based on relative intensities, as a function of the Bragg-angle and sample transparency. When the layer is thin, the angular distribution of the intensities is strongly dependent on the ratio between the actual sample thickness and the free path of the diffracted beam. This distribution is measurable by the relative integrated intensities of the reflected lines within the sample. Since the layers were with strong texture, the ratio of intensities was measured in a single direction of the reciprocal space.

It was found that fast and reliable results can be obtained by this method. Moreover, although changes in grain size caused large variability in the absolute counting rates, the relative intensities were not affected by this phenomenon, and this is one of the advantages of this method. The relative intensities measurements provide a fast tool to optimize the physical parameters of the laser beam surface treatment, namely, the laser power and the type of shield gas, which in turn affect the crystalline layer thickness.

Appendix

In the inverse pole-figure technique [8], the pole densities, P_i, are calculated from the relation:

$$P_i = \frac{I_i/I_{ri}}{(1/n) \; \Sigma \; I_i/I_{ri}}$$

I_i and I_{ri} are the intensities from the specimen and from a randomly oriented polycrystalline standard and n is the number of peaks taken into account.

Acknowledgement

The research was supported by the Technion V.P.R. Fund - Edelstein Research Fund.

REFERENCES

1. H. Hejdov and M. Cermak, "A new X-ray Diffraction Method for Thin Films Thickness Estimation", Phys. Stat. Sol. 72:K95 (1972).

2. L. S. Zevin, P. Rozenak and D. Eliezer, "Quantitative
    X-ray Phase Analysis of Surface Layers",
    J. Appl. Cryst. 17:18 (1984).

3. C.G. Brandt and G.H. van der Vliet, "Quantitative
    Analysis of Thin Samples by X-ray Diffraction",
    Adv. X-ray Anal. 29:203 (1985).

4. P. Suortti and L.D. Jennings, "Effects of Geometrical
    Aberrations on Intensities in Powder Diffractometry",
    J. Appl. Cryst. 4:37 (1971).

5. W.N. Schreiner and G. Kimmel, "Observed and Calculated
    XRPD Intensities for Single Substance Specimens".
    Adv. X-ray Anal. 30:351 (1987).

6. H. Herman and M. Ermrich, "Microabsorption of X-ray
    Intensity in Randomly Packed Powder Specimens",
    Acta Cryst. A43:4011 (1987).

7. C.R. Hubbard X-Ray Powder Diffraction Intensity Set, NBS
    Certificate, Standard Reference Material 674, Natl. Bur.
    Stand. (U.S) (1983).

8. G.B. Harris, "Quantitative Measurement of Preferred
    Orientation in Rolled Uranium Bars".
    Philos. Mag. 43:113 (1952).

X-RAY DIFFRACTION OF THIN OXIDE FILMS ON SOLDERED MODULE PINS

T. Paul Adl and H.F. Stehmeyer

General Technology Division
IBM Corporation
Manassas, Virginia

ABSTRACT

The presence of metal oxide films from wave solder baths on tinned module pins are partly responsible for non-wet problems in subsequent soldering steps. The cylindrical geometry of the pins lends itself to the characterization of thin oxide films by using the highly sensitive Debye-Scherrer camera method. As confirmed by Electron Microprobe Analysis (EMA), pins containing thin oxide films were used to obtain the diffraction patterns. A software program was developed that subtracts the diffraction angles of an oxide-free control pin from the pattern of the contaminated pin, and tabulates the residual d-spacing (interplanar distance) of the contaminant film.

INTRODUCTION

X-ray diffraction studies of thin films normally involves the use of the glancing angle method or Seeman-Bohlin geometry [1,2]. Both of these methods operate on the principle that the penetration of the x-ray beam into the sample is decreased at low Bragg angles. As depicted in Figure 1, the sample is set at a low fixed angle of incidence to the x-rays. The decrease in the beam's penetrating depth is tantamount to increasing the sample thickness by as much as an order of magnitude. The Seeman-Bohlin diffractometer in Figure 2 has added significant features such as a crystal monochromator and a focusing counter to overcome the line shape problems in the glancing angle method.

In the present work, a Debye-Scherrer camera [3] was used to obtain diffraction patterns from thin film deposits of palladium on glass rods. This was done to establish the detectability level for a thin film under standard Debye-Scherrer conditions for sample holder, geometry, and instru-

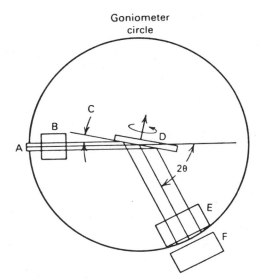

A = Line of X-Ray Source
B = Soller Slit
C = Glancing Angle of Incidence
D = Sample
E = Divergence Slit
F = Counter

Fig. 1.   Schematic Drawing of Glancing Angle Diffractometer

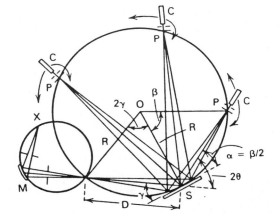

X = X-Ray Line Source
M = Monochromator Crystal
S = Sample
R = Radius of Focusing Circle
D = Focus to Sample Distance
P = Receiving Slit
C = Counter

Fig. 2. Schematic Drawing of the Seeman-Bohlin Geometry

mental parameters. Once the feasibility of the method for detecting thin films was established, the method was extended to characterize thin layers of contamination on tinned module pins.

EXPERIMENTATION

Palladium Films

     A Hummer VI[1] sputtering chamber was used to deposit approximately 600, 800, 1,000, and 2,000 angstroms of

<hr>

1   Trademark of Anatech Ltd.

Fig. 3.   X-Ray Diffraction Pattern of 2,000 Angstrom Pd film

palladium on 0.2 to 0.3 mm glass rods. The films were
deposited at 16 milliampere (ma) current and 100 millitorr
argon pressure. This deposition condition produced films
with well-defined diffraction patterns. The thickness measurements were recorded from the crystal detector gauge on
the sputtering chamber. The coated glass rod was placed in
a commercial Debye-Scherrer camera and exposed to varying
doses of x-rays in a Siemens D-500[2] x-ray system. Typical
patterns were taken at 25 kilovolt (kv) and 15 ma tube
parameters with 12 hours of exposure time.

Module Pins

Module pins from manufacturing showing a creamy white
film subsequent to a wave solder bath process were used in
the study. The pins were mounted on the camera and exposed
to x-rays typically for 14 hours at 20 to 25 kv and 15 to 19
ma tube parameters. Pins with similar histories, but clean
surfaces, were used as controls. Electron Microprobe Analysis (EMA) on the contaminated pins was performed by a Jeol
Superprobe 733[3] equipped with a backscatter electron
detector and wavelength dispersive x-ray spectrometer.

RESULTS

Palladium Films

Palladium film of 1,000-angstrom thickness on a glass
rod produced the four most intense lines of the diffraction
pattern reported in the powder diffraction files [4]. Figure
3 illustrates the observed pattern from a 2,000-angstrom
palladium deposit. Deposition conditions must be optimized
to produce well-defined diffraction lines; otherwise, small
particle size causes the lines to appear diffused due to
line broadening. The pattern from an 800-angstrom film contains only the strongest line of palladium. No diffraction
pattern could be obtained from a 600-angstrom deposit.

Module Pins

X-Ray Diffraction. A 24 mm module is shown in Figure
4. The module pins are 0.5 mm in diameter, have copper
cores, and are coated with eutectic solder in a wave solder
bath. Occasionally, a creamy white film (Figure 5) is
observed on the pins subsequent to this process. The con-

---

2   Trademark of Siemens Energy and Automation, Inc.

3   Trademark of Japan Electron Optics Laboratory (Jeol) Company.

Fig. 4. A 24 mm Module

taminant film could potentially cause non-wet problems in future soldering steps.

Figure 6a is the diffraction pattern from a contaminated pin taken over a 14-hour exposure time at 20 kv and 15 ma. Figure 6b is the diffraction pattern of a control pin obtained under identical conditions. Inspection of

Fig. 5. Optical Microscope
Image of the
Contaminant Film
on Module Pins

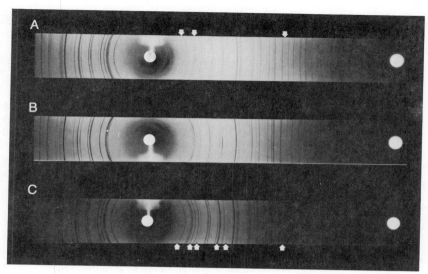

Fig. 6. X-Ray Diffraction Patterns. (A) Contaminated pin at
         20 kv/15 ma after 14 hours of exposure; (B) Control
         pin at 20 kv/15 ma after 14 hours of exposure;
         (C) Contaminated pin at 25 kv/18 ma after 14 hours
         of exposure.

these patterns reveals the presence of two lines with
d-spacings of 3.35 and 2.64 angstroms that are exclusive to
the contaminated pin. In Figure 6c, the pattern of the
contaminated pin under similar exposure time, but 25 kv and
18 ma tube parameters, shows three additional lines at 1.76
angstroms, 2.37 angstroms, and 1.03 angstroms d-spacings.
These lines are absent in the pattern of the control pin.
The five lines exclusively present in the thin film can be
assigned to tin oxide. Although changing the tube parameters enhanced the tin and lead lines, it produced no new
lines for these elements.

     A software program using Lotus 1-2-3[4] was prepared that
accepts as input the measured angles from the control and
contaminated pins. The program converts input data to
d-spacings and tabulates the d-spacings exclusive to the
contaminated sample. Table 1 shows the summary of our
results for diffraction angles up to 100 degrees.

     Electron Microprobe Analysis. Figure 7 shows a micro-
graph of the backscatter electron image of a contaminated
pin under EMA. Light areas in this micrograph are associated
with compositions of high average atomic number; dark areas
represent compositions of low average atomic number. The
black spots are due to elements with very low atomic
numbers. Carbon, presumably from carbonized flux material,
is the element responsible for producing the black spots in
this case. A higher magnification image of one area is

---

4   Trademark of the Lotus Development Corporation.

Table 1. Summary of Measured Line Positions and Input/Output
        Format of Software

	Input	Output
2-Theta Values For Contaminated Pin (in degrees)	2-Theta Values For Control Pin (in degrees)	d-Spacings For Residual Pattern* (in angstroms)
26.58	30.62	3.35
30.62	31.20	2.65
31.20	32.00	2.36
32.00	36.20	1.76
33.82	44.66	1.03
36.20	52.10	
38.00	62.20	
44.66	64.10	
51.70	72.40	
52.10	73.30	
62.20	77.00	
64.10	79.00	
72.40	85.00	
73.30	88.10	
77.00		
79.00		
85.00		
88.10		
97.00		

*Residual Pattern = Contaminated Pin's Pattern - Control Pin's Pattern

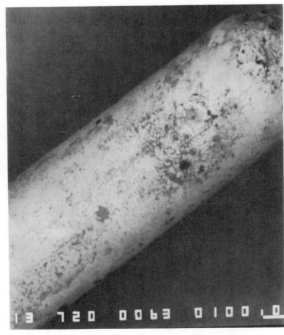

Fig. 7. Backscatter Electron Image of a Contaminated
        Pin at 72 X Magnification

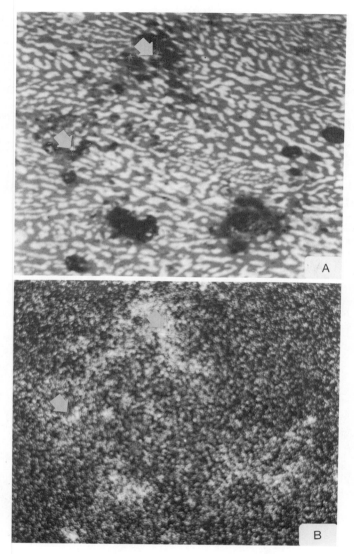

Fig. 8. (A) 860 X Magnification Image and (B) Oxygen Element
         Map of a Contaminated Pin

illustrated in Figure 8a. The accompanying wavelength
dispersive x-ray element map for oxygen of this area in
Figure 8b clearly indicates that the contaminant film is
rich in oxygen.

DISCUSSION

Palladium Films

     Under optimum conditions of deposition and x-ray
diffraction parameters, an 1,000-angstrom film of palladium

is detectable using the Debye-Scherrer method. Improper deposition conditions cause the diffraction lines to appear diffused due to line-broadening effects.

Module Pins

Five isolated and non-overlapping lines exclusive to the contaminated pins match the pattern for tin oxide. Presence of an oxygen-rich film is further confirmed by EMA.

Complications due to back reflection from a sample holder with high mass absorption does not interfere with the pattern interpretation. Although this effect manifests itself strongly by enhancing the intensities of the reflections with 2-theta values greater than 90 degrees, line positions can be used for correct assignment.

Despite the rich pattern of the matrix, sufficient non-overlapping lines can be isolated for identification. Pin dimensions and geometry make this a viable and efficient method for quality control of the pin surface.

ACKNOWLEDGEMENT

The authors wish to thank Mr. B. Leland of the Quality Assurance Support and Test Quality Engineering at IBM Manassas for his continued encouragement and support during this study.

REFERENCES

1. R. Feder and B. S. Berry, Seeman-Bohlin X-Ray Diffraction for Thin Films, J. Appl. Cryst. 3:372 (1970).
2. G. Wasserman and J. Wiewiorosky, Uber ein Geiger-Zahlrohr-Goniometer nach dem Seeman-Bohlin-Prinzip, Z. Metallk. 44:567 (1963).
3. B. D. Cullity, "Elements of X-Ray Diffraction," 2nd ed. Addison Wesley, Reading, Massachusetts (1978).
4. Joint Committee on Powder Diffraction Standards, Swarthmore, Pennsylvania (1974).

X-RAY DIFFRACTION STUDIES OF POLYCRYSTALLINE THIN FILMS USING GLANCING ANGLE DIFFRACTOMETRY

R.A. Larsen, T.F. McNulty, R.P. Goehner, K.R. Crystal

Siemens Analytical X-ray Instruments, Inc.
5225-1 Verona Road, Madison WI 53711

Abstract

The use of conventional $\theta/2\theta$ diffraction methods for the characterization of polycrystalline thin films is not in general a satisfactory technique due to the relatively deep penetration of x-ray photons in most materials. Glancing incidence diffraction (GID) can compensate for the penetration problems inherent in the $\theta/2\theta$ geometry. Parallel beam geometry has been developed in conjunction with GID to eliminate the focusing aberrations encountered when performing these types of measurements. During the past year we developed a parallel beam attachment which we have successfully configured to a number of systems.

It is the purpose of this paper to contrast a number of methods for the measurement of polycrystalline thin films. In addition to the above mentioned parallel beam modifications, we have also explored the use of Guinier type focusing systems with position sensitive detectors (PSD). The results of these experiments show, as would be expected, that practical considerations such as laboratory throughput, film thickness, and lineshape requirements dictate the method of choice.

Introduction

Characterization of polycrystalline thin films using x-ray diffraction techniques has a wide range of applications in materials science, the electronics industry, and chemical engineering.[1,2,3] These applications require that detailed information regarding thin film structure, composition, texture, and residual stress be determined. There are problems, however, when measuring thin films

311

using x-ray diffraction which stem from the relatively large penetration depth of hard x-ray photons in solids.

The expressions governing the penetration depth of x-rays in solids have been previously derived using a number of approaches.[4,5] The basis of the glancing angle experiment is to reduce the incident or "glancing" angle of the x-ray beam impinging on the sample surface. Figure 1 shows glancing angle scans of the Fe_3O_4 311 reflection recorded at a number of incident angles. Figure 1 illustrates that as the glancing angle is decreased we see less of the diffuse scattering from the amorphous substrate while the diffracted intensity of the 311 reflection increases.

In recent years a number of groups have exploited the glancing angle effect in the measurement of both polycrystalline and epitaxial thin films.[6,7] The experimentation in the epitaxial case is somewhat more complex due to the additional constraint that the glancing angle also satisfy the diffraction conditions of the lattice. In the case of polycrystalline films however the main problems stem from the focusing aberrations which occur due to the glancing angle geometry. During the past year we have developed a parallel beam attachment to analyze thin films and have successfully configured the attachment to both a curved incident beam monochromator system as well as a traditional $\theta/2\theta$ diffractometer. The next sections will describe a number of configurations we have used for looking

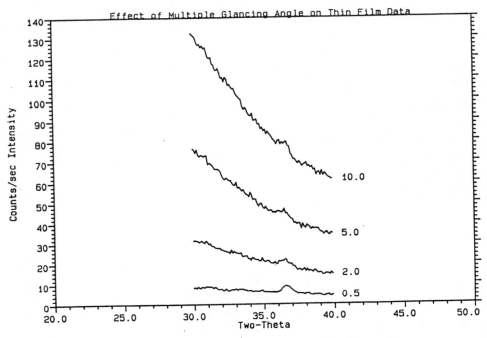

Figure 1. Multiple glancing angle scans of Fe_3O_4 on glass.

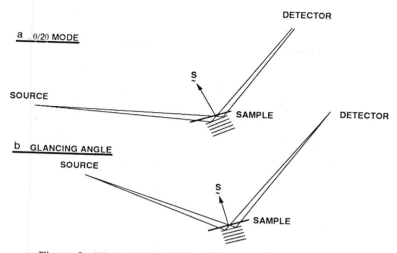

Figure 2. Illustration of defocusing in the glancing mode.

at thin films and present results which illustrate both the focusing effects and the parallel beam performance.

Experimental

A. Modification of a conventional $\theta/2\theta$ geometry to parallel-beam geometry

The focusing problems encountered with the traditional $\theta/2\theta$ diffractometer when performing experiments in the glancing angle mode may best be explained as an effective displacement error taking place within the sample. By keeping the incident angle fixed with respect to the sample surface we can see that each diffracted beam originates from a set of lattice planes specifically inclined to the sample surface such that the diffraction condition is maintained (Figure 2a). This contrasts the traditional $\theta/2\theta$ geometry in which the planes that contribute to diffraction are always parallel to the surface (Figure 2b). In the glancing angle geometry then, there is a specific focusing radius for each 2θ value. One way to maintain focusing over the entire 2θ range would be to translate the receiving slit and detector assembly parallel to the diffracted beam as the 2θ angle was varied. This method which resembles the Seemann - Bohlin geometry, has mechanical limitations when going to very small glancing angles.[8,9]

The simplest way to remove the focusing problems in the pattern is to do away with the focusing altogether. This is achieved by reconfiguring the system in a parallel beam geometry. This is done by placing a set of angular divergence soller slits in front of the detector. By doing this only the parallel rays diffracted from the sample will be accepted into the detector (Figure 3). While we would not expect to get either the intensity or the resolution we would obtain if we were able to maintain the focusing geometry, this method affords several positive

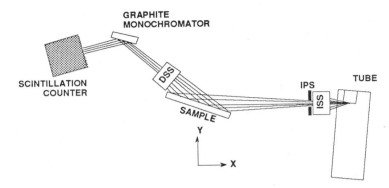

Figure 3. Illustration of parallel beam optical modification for a vertical goniometer with a flanged tube. DSS, ISS and IPS represent parallel beam soller slits, incident beam soller slits and incident beam primary divergence slits.

considerations. First, since we are accepting only the parallel rays, as dictated by the divergence of the slits, it is the slits themselves which control the resolution and lineshape. Since the slits form a very simple optical system, and divergence from a set of parallel plates is easily calculated, the observed lineshape is easily understood. This enables accurate peak parametrization for profile intensive measurements. Secondly, since the irradiated area is fixed by the glancing angle, and the slit acceptance is small with respect to the irradiated area, the profile shapes and intensities also remain relatively constant over the data range. Finally, since we are accepting only the parallel rays diffracted from the sample, the

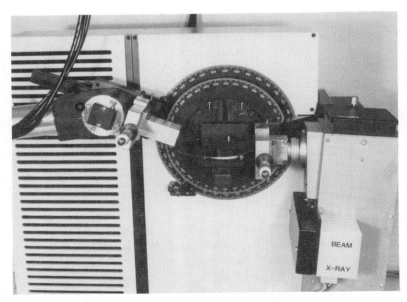

Figure 4. Photo of the Siemens I2/V diffractometer configured with parallel optics.

severity of sample displacement errors is significantly reduced. This last feature has important implications for experiments such as residual stress determination where very accurate positional information is required at varying θ angles.

Figure 4 shows a photo of the Siemens I2/V $\theta/2\theta$ diffractometer which has been configured with parallel optics. The soller slits that create the parallel acceptance path are mounted on the 2θ arm in front of the monochromator housing. Due to the fact that the geometry is no longer focusing, it is not necessary to have the detector assembly at any specific distance from the center of the sample. For this reason we have mounted the slits, monochromator, and detector on a dove tail coincident with the 2θ arm so the entire assembly may be slid up as close to the sample as its size will permit. This reduces air scatter and decreases counting time. The slits themselves are made of molybdenum foils roughly 0.025mm thick with a 0.25mm spacing. The slits have an acceptance of 10 X 15 mm and a calculated angular divergence of 0.3 degrees. Figure 5 shows data collected from a Fe_3O_4 film using the parallel-beam glancing angle arrangement. The data was collected at a 1.0 degree glancing angle, using a 30 second step time and a 0.1 degree stepwidth. The full pattern of the spinel structure is clearly visible and may be identified.

Figure 6 shows the 110 reflection recorded from a 50Å thick film of Cr on silicon. The region shown took 5 hours to collect and shows that even at this extremely low angle the signal from the film is still extremely weak.

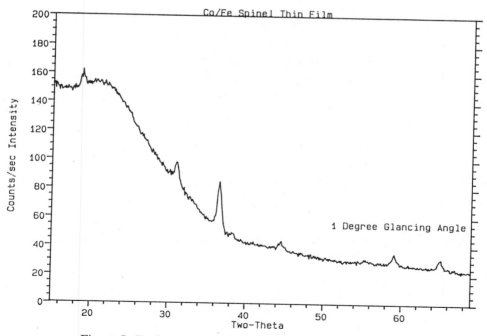

Figure 5. Fe_3O_4 raw data collected with parallel-beam optics.

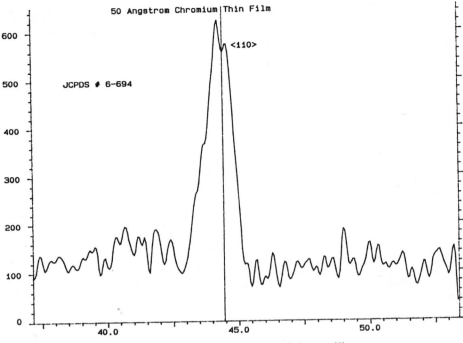

Figure 6. Glancing angle scan of Cr on silicon.

B. Use of an incident-beam monochromator focusing diffractometer with a curved position sensitive detector (PSD)

Another method that we have employed in the measurement of thin films is the use of a curved germanium incident-beam monochromator and 45 degree curved PSD. This system which was originally developed for transmission work also has some beneficial features when applied to the measurement of thin films.[10] Since the curved incident-beam monochromator concentrates more of the x-ray beam onto the sample, the beam produced has a higher intensity than the typical divergent beam in the $\theta/2\theta$ geometry. This is an advantage primarily if the sample is small. We have used this configuration when forced to measure small fragments of coated silicon wafers.

The advantage of the PSD is straight forward. A total counting time of 300 seconds for the full 45 degree acceptance of the detector corresponds to a 300 second per step counting time using traditional point counting methods.   As shown in the above example using the Fe_3O_4 film, the data collections involving thin films are usually quite long. In this respect the time saved when using the PSD for thin film work is even greater than when doing conventional powder work. Figure 7 shows the pattern obtained from a 3000Å thick Fe-Ni/Cu double-layer film . The data were collected in 600 seconds with a 1.0 degree glancing angle.

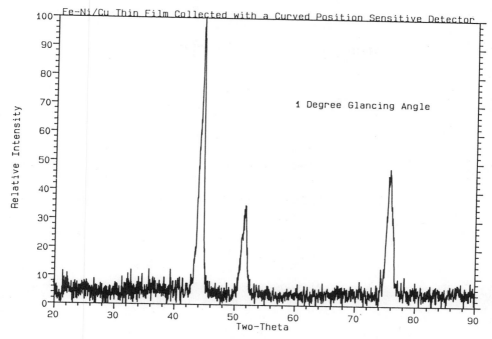

Figure 7. Fe-Ni/Cu film collected with curved PSD.

The problems associated with the use of incident beam monochromators, particularly curved ones, and PSD's for thin film experiments, again relate back to focusing considerations. Figure 8 shows the optical diagram for the above mentioned focusing system. The sample sits halfway between the monochromator and the detector, with the focal point being at the detector. Figure 9 shows the same diagram this time with the beam undergoing reflection at the sample. In this case it can be seen that the beam diverges from the sample. If we look back at Figure 7 we can see that this divergence of the beam is evident in the observed peakwidth (FWHM ~ 1.7 degrees).

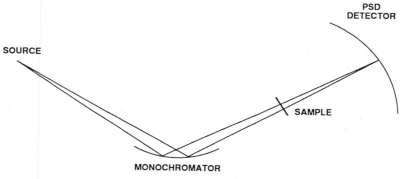

Figure 8. Optical diagram of focusing system.

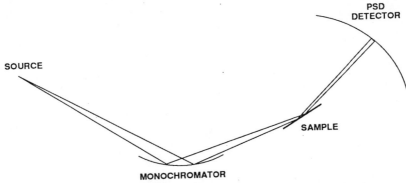

Figure 9. Illustration of the defocusing in reflection mode.

The problem when using the curved crystal monochromator for glancing angle work is that this defocusing in reflection mode is most severe at low angle. Figure 10 shows a comparison of the peak widths as a function of glancing angle for the first line of Figure 7. This comparison shows that as the glancing angle gets larger, i.e. closer to 90 degrees which is the case of transmission, the resolution gets better. The defocusing can be reduced by using a flat monochromating crystal instead of the curved one, but the diffracted beam crossfire produced by the polycrystalline film still causes peak broadening when using the "wide open" PSD.

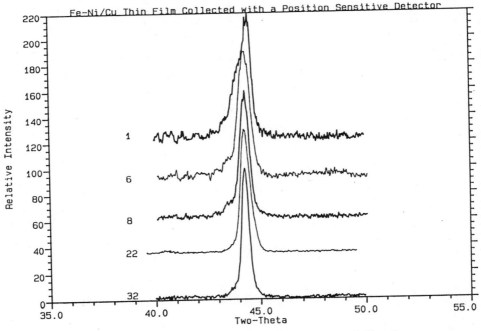

Figure 10. Multiple scans of Fe-Ni/Cu showing defocusing.

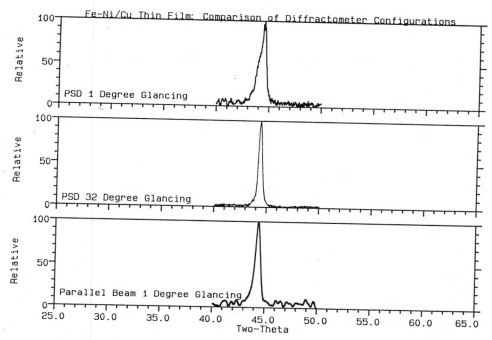

Figure 11. Comparison of PSD and parallel-beam profiles.

These defocusing effects thereby limit the usefulness of this geometry to applications in which the lineshape is not important or to films which are reasonably thick, such that the glancing angle does not have to be too low.

C. Use of parallel-beam optics with a curved incident-beam monochromator

In order to try and improve the resolution of the incident–beam monochromator system at low glancing angles we have constructed a mounting block which allows the parallel-beam slits to be attached to this system. The mount attaches to the 2θ arm via a dove tail previously used to mount the PSD. The slits attach to the front of the mounting block and a scintillation counter fits in the back. Figure 11 compares the resolution obtained in the parallel-beam mode to that when using the PSD. The scan is again of the first line of the Fe-Ni / Cu pattern shown in Figure 7. The parallel-beam scan was recorded at a 1.0 degree glancing angle, using a 0.1 degree stepwidth and 30 second step time. The figure shows that the resolution obtained with the parallel-beam optics is in between that of the two PSD scans. In this case, since the film is so thick, it would be possible to collect the data at a high glancing angle and use the PSD.

Perhaps the most rigorous experiment   that we have attempted to date involves the measurement of an oxidized silicon wafer. Figure 12 shows the pattern obtained from a small fragment of of a wafer intended to have a deposited

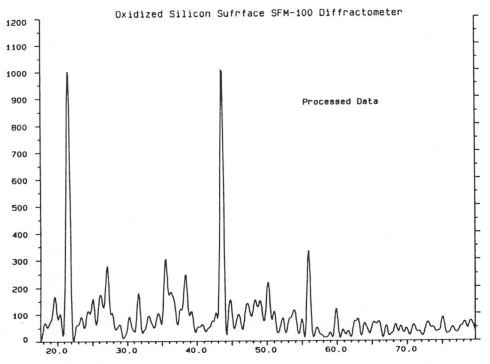

Figure 12. Scan of oxidized silicon surface using curved incident-beam monochromator system configured with difracted-beam parallel optics.

film. The observed pattern is a mixture of the two phases of cristobalite. The conclusion drawn from this experiment was that the deposition process was done incorrectly, yielding an oxidized silicon surface with no trace of the intended film. This scan was recorded using the focusing geometry in order to concentrate as much of the x-ray flux as possible on to the small fragment. The data collection time for this scan was 10 hours.

Discussion and Conclusions

During the past year our laboratory has had a tremendous number of thin film samples submitted for analysis. This has given us the opportunity to work with a wide variety of materials, and collect a great deal of data using a wide variety of experimental configurations. The data presented has contrasted a number of methods for the investigation of thin films via x-ray diffraction. The question naturally arises, which one is best? At this time we would like to present some ideas which summarize the above discussion and reflect our experiences over the past year.

The parallel-beam system seems to be of the most general use, the profiles being better at low angles, the primary area of interest for thin film work. The conversion of a conventional θ/2θ system to one employing parallel-beam optics is relatively straight forward. The reduction of a rigorous sample height requirement is attractive to those applications such as residual stress and depth profiling which require that data be collected at varying θ angles without a corresponding displacement error. The draw back with this configuration is the data collection times. The collection of good quality data typically involves per point counting times of 30 seconds or more when working with the majority of films we have investigated (500 Å or less). This may present a problem to laboratories with large throughput requirements.

The PSD based system on the other hand collects the data very fast, a characteristic which may be used effectively to increase per point counting times. It has limitations however at low glancing angles due to the focusing considerations mentioned previously. The PSD system may be used effectively when speed rather than absolute accuracy and resolution are a priority. Applications such as the study of phase transitions where collection speed is a must, also would point to a PSD based system. One important consideration to note is that PSD systems also involve a significant cost consideration, and require a more sophisticated interface to an existing diffractometer than a simple parallel-beam modification.

In general what we have done in our lab is to run the PSD during the day to investigate a large number of materials in routine fashion. Those materials which then require either better resolution or a very low glancing angle are run over night to get improved data quality.

References

1. B. A. Bellamy , 1982, The APEX goniometer as a glancing angle x-ray powder diffractometer for the study of thin films: United Kingdom Atomic Energy Unclassified Report AERE-R-10687 Materials Development Division, AERE HARWELL 13 p.
2. N. S. Choudhury, R. P. Goehner, N.Lewis and R. W. Green, Thin Solid Films, 122 1984 231-241
3. T. C. Huang, Advances in X-Ray Analysis, 31 1987 107-112
4. G. H. Vineyard, Physical Review B, 26 No. 8 4146 - 4159
5. S. S. Iyengar, W. M. Santana, H. Windischmann and P. Engler, Advances in X-Ray Analysis, 30 1986 457- 464
6. A. Segmuller, Thin Solid Films, 154 1987 33 - 42
7. D. W. Berreman, A. T. Macrander, Advances in X-Ray Analysis, 31 1987 161 - 165
8. C.N. J. Wagner, M. S. Boldrick and L. Keller, Advances in X-Ray Analysis, 31 1987 179 - 143
9. R. Feder and B. S. Berry, Journal of Applied Crystallography, 3 1970 372 - 379
10. E. R. Wolfel, Journal of Applied Crystallography, 16 1983 341 - 348

DENSITY MEASUREMENT OF THIN SPUTTERED CARBON FILMS

G. L. Gorman, M.-M. Chen*, G. Castillo, R. C. C. Perera**

IBM Research Division
Almaden Research Center
650 Harry Road
San Jose, Ca. 95120

*IBM General Products Division

**Center for X-ray Optics
Lawrence Berkeley Laboratory
University of California
Berkeley, Ca. 94720

ABSTRACT

The densities of sputtered thin carbon films have been determined using a novel X-ray technique. This nondestructive method involves the measurement of the transmitivity of a characteristic soft (low energy) X-ray line through the carbon film, and using the established equation $I_1 = I_0 e^{-\mu\rho t}$ where I_1/I_0 is the transmitivity, μ the photoabsorption cross section, t the independently measured thickness, the density ρ can be easily solved for. This paper demonstrates the feasibility of using this simple technique to measure densities of carbon films as thin as 300 Å, which is of tremendous practical interest as carbon films on this order of thickness are used extensively as abrasive and corrosive barriers (overcoats) for metallic recording media disks. The dependence of the density upon film thickness for a fixed processing condition is presented, as also its dependence (for a fixed thickness) upon different processing parameters (e.g., sputtering gas pressure and target power). The trends noted in this study indicate that the sputtering gas pressure plays the most important role, changing the film density from 2.4gm/cm^3 at 1 mTorr to 1.5gm/cm^3 at 30 mTorr for 1000 Å thick films.

INTRODUCTION

Sputtered carbon films are used extensively as protective overcoats of metallic magnetic recording disks. The role of the carbon is twofold: it has to protect the metallic recording film from abrasive wear, and it has to act as a barrier against chemical corrosion. Therefore there is considerable interest in pursuing a deeper understanding of the material, mechanical and chemical properties of sputtered thin carbon films.

Historically, material characterization of the carbon films has been extraordinarily difficult, mainly due to two reasons: one is the amorphous nature of the films which

323

makes structural characterization difficult, and secondly, because the presence of carbon is prevalent in almost all experimental environments, making quantitative analytical measurements is very difficult or even impossible. The necessity of obtaining quantifiable material characteristics of the films, however, is obvious when trying to correlate film properties with functional properties.

One characteristic which plays an important role in determining film effectiveness is the density ρ (measured in gm/cm³). The denser the films, the more likely they are to be free of microcracks and microvoids. Therefore, one would expect them to be more resistant to abrasion. The density is also directly related to the porosity of the film which in turn affect the ability of the film to act as a corrosion barrier. In the past the density of thicker films (typically on the order of a few microns in thickness) has been measured by the so-called sink and float method.[1] The question always arises whether these measured densities can be correlated to films with thicknesses of a few hundred angstroms.[1] More recently, however, density measurements have been made on films ranging in thicknesses from 400 Å to 2000 Å using a combination of profilometry with Rutherford backscattering,[2,3] proton recoil[3] and/or nuclear resonance shift[2] measurements.

An established and commonly used nondestructive technique to determine areal density (ρt measured in gm/cm²) for thin films is X-ray Fluorescence Analysis.[4,5] The method involves the direct excitation of a radiation line characteristic of the element under observation. Quantitative analysis involves a known bulk standard of the element and precise values for photoabsorption cross section values of the element of interest. This fundamental parameters technique has been proven effective for many material systems, including both bulk and thin film multi-element (and multi-layer in the case of thin films) complexes.[6,7]

Attempts to use this technique to measure light elements, and in particular carbon thin films, however, met with no success. This is due to the proximity of the K-absorption edges to the characteristic Kα emission lines. Specifically, for carbon, the K-absorption edge is at 284 eV and the Kα line is at at 277 eV.[8] Furthermore, the imprecise photoabsorption cross section values in the near vicinity of the high and low energy sides of the absorption edge (see Fig. 1), and the discontinuous

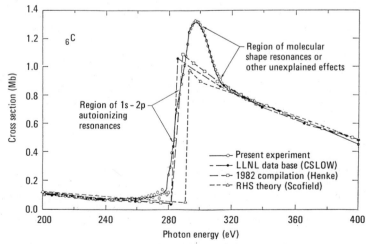

Figure 1.   Experimental and theoretical photoabsorption cross section for amorphous carbon in the region of the K-edge (N. K. Del Grande et al., Ref. 9).

nature of the absorption edge itself makes the accurate monitoring of the C Kα emission extremely difficult. If, however, radiation lines above the C Kα absorption edge is used, the photoabsorption values are sufficiently away from the edge to be well behaved, and thus well determined.

This paper presents density measurements of sputtered carbon films with thicknesses ranging from 300 Å to 2600 Å using absorption of a selected X-ray emission line from the substrate by the carbon film and not by monitoring the direct C Kα radiation. The experimental techniques and selection of a suitable X-ray emission line will be discussed and the dependence of density on various sputtering parameters will be presented.

EXPERIMENTAL

Using the well-known equation

$$I_1 = I_0 e^{-\mu\rho t}$$

where I_1 is the intensity of the monochromatic radiation after it passes through the carbon film, I_0 the incident intensity and μ the photoabsorption cross section (in cm²/gm) for the monochromatic radiation and ρt is the area concentration (in gm/cm²), the density ρ can be determined when the film thickness t is measured by an independent technique such as ellipsometry and profilometry.

The carbon films were sputtered from a D.C. planar magnetron cathode. Two identical cathodes with target size of 3.5 × 8.25 inches were installed in a diffusion pumped vacuum system. The base pressure of the system was in high 10^{-8} Torr region. Up to two layers of film can be deposited in one pump down cycle. This allows us to prepare recording thin film media with Co-alloy magnetic and carbon overcoat layers onto substrates of up to 5.25 inch diameter. The pyrolytic graphite target was obtained from Degussa Corp. with a density of 2.2 grams/cm³. In order to deposit films with excellent thickness uniformity over the 5.25 inch disk substrates, the samples were rotated at 20 to 100 RPM during deposition. A shaped mask opening was used between the targets and samples for uniformity optimization.

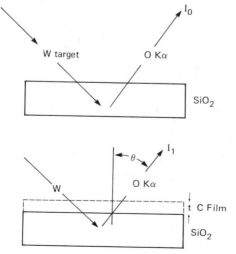

Figure 2.   Diagram of actual experiment to measure the absorption of O Kα by a carbon film on SiO₂ substrate.

For these experiments the carbon films were deposited onto Si, quartz and CoCr/Si substrates. The use of the SiO_2 substrate was particularly convenient since O Kα characteristic radiation can be utilized to measure the X-ray absorption by the carbon film. Since the SiO_2 substrate is on the order of 0.5 mm thick, it can be treated essentially as a bulk sample. In this method, I_0 is the O Kα intensity measured from a pure SiO_2 substrate whereas I_1 is the O Kα intensity measured through the carbon film of known thickness on an identical substrate.

The X-ray measurements were obtained using a Rigaku 3070 X-Ray Spectrometer system consisting of an end window tungsten tube target, a RX35 multilayer analyzing crystal (2d = 55.4 Å) and a flow proportional counter. Over 5×10^4 total counts were collected at peak intensity in all measurements and results presented in this work were repeated at least twice with the individual measurements reproducible to within statistical deviations. The precision of the intensities measured this way is better than 1%. The takeoff angle θ of the counter position is 40°, thus modifying the effective thickness by $1/\cos\theta$. The thicknesses of the films were measured by a Dektek II profilometer. The overall precision of the technique for a 1000 Å film is about 5%.

The density ρ of the carbon films were obtained using the relation

$$\rho = (\mu t / \cos \theta)^{-1} \ln (I_0/I_1)$$

where μ of carbon for O Kα radiation[8] is 1.24×10^4 cm²/gm.

RESULTS AND DISCUSSION

The properties and microstructure of the thin films depend greatly on the detailed deposition conditions.[10] The density of carbon films has been reported[1] ranging from 1.4 gm/cm³ to 2.2 gm/cm³ depending upon the film preparation method. These include the direct sputtering of a graphite target in an argon gas environment and reactive sputtering in a hydrocarbon gas environment. Using the density measurement technique described above, the densities of carbon films were studied as functions of film thickness, substrate rotation speed, D.C. power and argon gas pressure.

A comparison of calculated film densities from X-ray fluorescence analysis[4,5] using identical data but two different analysis programs, XRF and LAMA3, are presented in Table 1. The film thicknesses of these samples were measured using a profilometer. Note that ρ for diamond is 3.515 gm/cm³ and for graphite is 2.260 gm/cm³. The XRF program indicates that the carbon films are all denser than diamond, and that the density of the films decrease with increasing film thickness while the LAMA3 program suggests that all the carbon films are lighter than graphite, and that the density of the films increase with increasing thickness, a direct contradiction to the results using XRF. Inconsistency between analysis programs using identical experimental data is a result of different mathematical models used in calculating the "self-absorption" resulting from the near edge structure. Furthermore, all these calculations rely on available μ values even though near the absorption edges where both theory and experiments are sensitive to the particular chemical or solid state properties of the materials these μ values will represent only an average through any absorption fine structure. This in turn suggests that results from both the XRF and LAMA3 are unreliable. The LAMA3 program also failed to calculate the density for the 2600 Å thick film.

Also listed in Table 1 are calculated densities from the present technique indicating that the densities of the carbon films are nearer to that of graphite which agrees with independent Rutherford Backscattering measurements.[11]

Using the present technique, variation of film density with thickness for a fixed processing condition, and also as functions of other processing parameters were studied.

Table I

Comparison of Calculated Film Densities Versus Thickness of Carbon Films
by XRF and LAMA3 Programs Utilizing Identical Data and by Present Work

		density (gm/cm³)		
	t (Å)	XRF	LAMA3	Absorption
#23	320	4.31	1.47	2.04
#35	680	4.22	1.64	2.18
#36	1150	3.89	1.85	2.12
#37	1415	3.90	2.26	
#56	2600	3.75	*****	2.21

Figure 3 shows the dependence of density upon film thickness for films deposited in
Argon at 6 mTorr of pressure, rotating at 100rpm with 1500 Watts of target power. Film
thicknesses ranged from 318 Å to 2600 Å. The density increased from 2.08 gm/cm³ at
318 Å to 2.22 gm/cm³ at 2600 Å, a 7% increase, indicating that the effect of film
thickness upon density is slight.

Films of 2800 Å to 3000 Å were also deposited at 6 mTorr of gas pressure and 1500
Watts of target power, but with the substrate palate rotating at 20 rpm and at 100 rpm.
The measured densities were found to be 1.97 ± 0.02 gm/cm³ when the rotation speed
was 20 rpm, and 2.08 ± 0.01 gm/cm³ for the latter case of 100 rpm. This shows that the
densities were practically independent of the rotation speed.

The dependence of density upon the target power during sputtering has also been
studied. The observed trend is that at higher power, the films are denser which disagrees

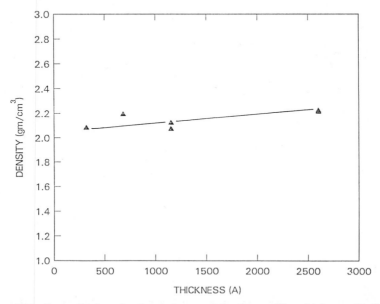

Figure 3.   This figure depicts the dependence of density on film thickness for fixed
deposition condition of 1500 Watt target power, 100 rpm substrate rotation,
and 6 mTorr of argon pressure.

with previously reported data.[1] Figure 4 illustrates the film density dependence upon target power for a fixed Ar pressure of 6 mTorr and a rotation speed of 100 rpm.

The most dramatic dependence of density, however, occurs with change in the Argon gas pressure during sputtering where an appreciable drop in density is observed with increasing pressure. Keeping the power fixed at 1500 Watts with the rotation speed at 100 rpm, the density drops from 2.29 gm/cm³ at 1 mTorr of Ar pressure to 1.48 gm/cm³ at 30 mTorr of pressure. Figure 5 illustrates this dependence.

Densities of carbon films deposited upon CoCr were also determined by measuring the transmitivity of the Co Lα line. Samples deposited at 1.5 kW target power and in Argon gas pressures of 30 mTorr (957 Å) and 6 mTorr (1000 Å) resulted in transmitivities of 0.8785 and 0.8555 and densities of 2.04 gm/cm³ and 2.53 gm/cm³ respectively. This can be compared with films deposited under similar conditions but on SiO_2 substrates which gave corresponding O Kα transmitivities of 0.7688 (30 mTorr pressure, 1100 Å) and 0.6799 (6 mTorr pressure, 1150 Å), and densities of 1.48 gm/cm³ and 2.07 gm/cm³ respectively.

This demonstrates the necessity of optimizing the experimental condition for increased sensitivity to the desired X-ray energy. As I/I_0 is an exponential function, the best region to work in is when $I/I_0 = e^{-\mu\rho t} = 0.30$ to 0.70 as ρt would be least sensitive to slight variations in the experimental precision of I/I_0. Outside this region, an error of 1% in I/I_0 can result up to 20% error in the calculated density, i.e. $\ln (0.93) = -0.073$ and $\ln (0.94) = -0.062$.

CONCLUSIONS

An accurate and precise method for determining the carbon film densities is essential. This paper describes such a technique using absorption of long wavelength

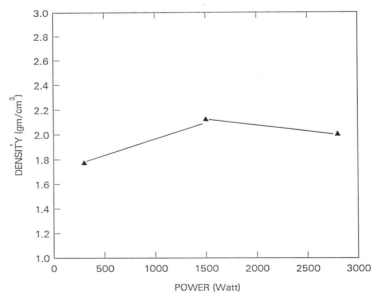

Figure 4.   This figure shows the dependence of films density on sputtering target power for 1000 Å thick films deposited in 6 mTorr of argon pressure and 100 rpm substrate rotation.

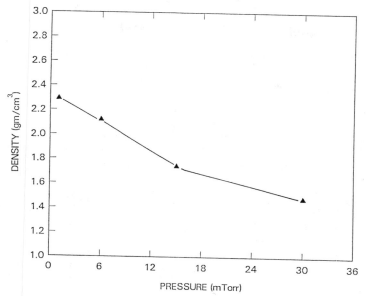

Figure 5. This figure shows the dramatic dependence of density on the argon gas pressure while keeping the target power fixed at 1500 Watts and the substrate rotation at 100 rpm.

X-rays through carbon films. From this characteristic transmitivity, the density is determined. Specifically, the transmitivity of O Kα through carbon films with thicknesses ranging from 300 Å to 2600 Å is demonstrated. The overall precision is about 5%.

This nondestructive technique can be further refined and expanded, by the selection of an X-Ray target which more efficiently excites the desired radiation as mentioned above, or the use of other soft X-ray lines with appropriate photoabsorption cross sections.[8] For example, V Lα ($\mu = 1.34 \times 10^4$), Cr Lα ($\mu = 9.9 \times 10^3$), Co Lα ($\mu = 4.45 \times 10^3$) (example given above) and Ni Lα ($\mu = 3.46 \times 10^3$) radiation lines may also be used in situations where there are magnetic layers in between other substrates and the carbon overcoats. The use of a larger emergence angle can also be designed between the sample and detector to increase the effective sample thickness (as $1/\cos\theta$) and thus increase the effective thickness of the films. In this way, the I/I_0 ratio can be optimized to yield the most precise values for density.

The carbon films in this paper showed slight dependences of density upon target power and film thickness, while being independent of substrate rotation speed. However, the films showed a drastic dependence of density upon the Argon sputtering gas pressure during film deposition, with the low pressure depositions producing the denser films. Furthermore, this technique is not restrictive to studying the carbon films, and can be applied to films of any element, and in particular for other light element films.

ACKNOWLEDGEMENTS

The authors would like to thank A. Wu, D. Palmer and R. Lovell for the initial design and modification of the sputtering system, and C. R. Brundle for his encouragement in this work. One of the authors (R.C.C.P.) acknowledges the support by the U. S. Department of Energy under Contract No. DE-AC03-76SF00098.

REFERENCES

1. H. C. Tsai and D. B. Bogy, J. Vac. Sci. Technol. A 5:3287 (1987).
2. A. Anttila, J. Koskinen, M. Bister, and J. Hirvonen, Thin Solid Films 136:129, (1986).
3. D. C. Ingram, J. A. Woollam, and G. Bu-Abbud, Thin Solid Films 137:225 (1986).
4. D. Laguitton and M. Mantler, Adv. X-ray Anal. 20:515 (1977).
5. T. C. Huang, X-Ray Spectrom. 10:28 (1981).
6. M. Mantler, Adv. X-Ray Anal. 27:433 (1984).
7. T. C. Huang and W. Parrish, Adv. X-ray Anal. 29:395 (1986).
8. B. L. Henke, P. Lee, T. J. Tanaka, R. L. Shimabukuro, and B. K. Fujikawa, Atomic Data and Nuclear Data Tables 27 1, Jan. (1982).
9. N. K. Del Grande, K. G. Tirsell, M. B. Schneider, R. F. Garrett, E. M. Kneedler, and S. T. Manson, J. de Physique 48:C9-951 (1987).
10. L. Maissel and R. Glang, "Handbook of Thin Film Technology," McGraw-Hill Inc., New York (1970).
11. A. Spool, private communication.

DETERMINATION OF ULTRA-THIN CARBON COATING THICKNESS BY X-RAY FLUORESCENCE TECHNIQUE

R. L. White
T. C. Huang*

IBM General Products Division
San Jose, CA 95193

*IBM Research Division
Almaden Research Center
San Jose, CA 95120

ABSTRACT

A technique for high-precision measurement of carbon thin-film thickness using X-ray fluorescence (XRF) is described. A quadratic calibration procedure is used for carbon thin films on silicon. Measurement of carbon-film thickness in a double-layer structure of carbon and CoCrX alloy is complicated by interference effects from the underlying layer. The dependence of the relative precision in measuring thickness (σ_T/T) on the counting time has been derived. It shows that a precision of 2% for a 25-nm carbon coating can be obtained using a W/C crystal and counting time of 4 minutes. Intensity and resolution advantages provided by the recently developed Ni/C and V/C multilayer synthetic crystals are also described.

INTRODUCTION

X-ray fluorescence is a technique widely used in industry for chemical analysis. Irradiating X-rays eject core electrons from atoms within the sample. Characteristic radiation is emitted when the atoms relax back to their ground state. Chemical composition can be deduced from a quantitative analysis of the relative intensities of the emitted X-rays.[1] Thin-film analysis of both composition and thickness has also been well developed using either bulk or thin-film standards.[2]

Detection of elements lighter than sodium ($Z = 11$) is made more difficult because their characteristic radiation is readily absorbed in air. In addition, synthetic crystals with large interplanar spacing are required to diffract such long wavelength radiation.[3] Other techniques such as total reflection have also been used to quantitatively measure radiation produced by these elements.[4]

331

EXPERIMENTAL

The Rigaku 3070 spectrometer used in this study operates under rough vacuum (5 Pa) to minimize the problem of air absorbance of soft carbon X-rays. Samples are loaded through a load-lock using small cylindrical cans. The maximum diameter of the sample that can be mounted in these cans is 2 inches. A rhodium end-window X-ray tube operating at a voltage of 45 kV and current of 40 mA was used to provide the irradiating X-rays for all the data provided below. A number of crystals can be used with the instrument. Choice of the analyzing crystal depends on which elements are being analyzed. For the carbon analysis a synthetic crystal was used, composed of alternating layers of tungsten and carbon with a 2d spacing of 16.0 nm. A proportional counter was used with a thin mylar window and a mixture of 90% Ar-10% CH_4 (P10) as the ionizing gas. Relevant equations are given in the Appendices.

RESULTS AND DISCUSSION

Figure 1 shows the XRF spectrum obtained using this crystal with a sample of 100 nm of sputtered carbon on a silicon wafer. An area of the sample 25 mm in diameter was irradiated to produce this trace. The figure demonstrates both the symmetry of the peak and the relatively good signal to noise that is obtained even at these relatively fast scan rates. Quantitative data are gathered by choosing the peak 2θ position and two background positions on either side of the peak. The net intensity is calculated by subtracting a background intensity, obtained by interpolation of the intensities at the two background positions, from the intensity at the peak. This net intensity is proportional to the areal density ($\mu gm/cm^2$). A density assumption is necessary to translate this quantity into thickness.

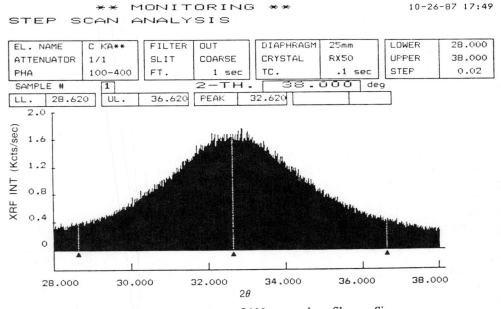

Figure 1.   Step scan of 100-nm carbon film on Si.

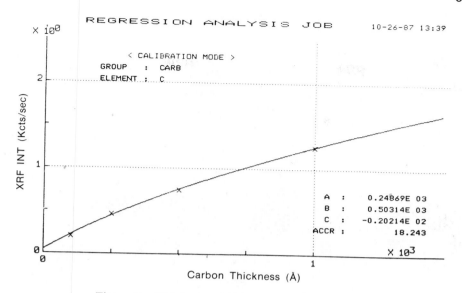

Figure 2.   Thickness calibration for carbon on Si.

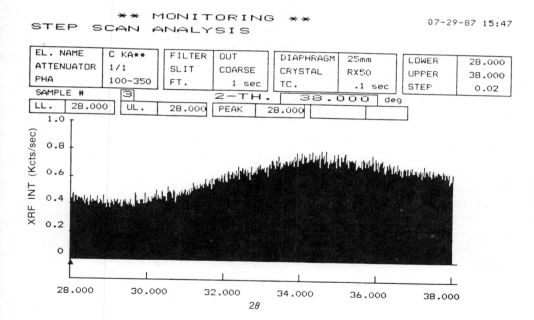

Figure 3.   Step scan of CoCrX on Si.

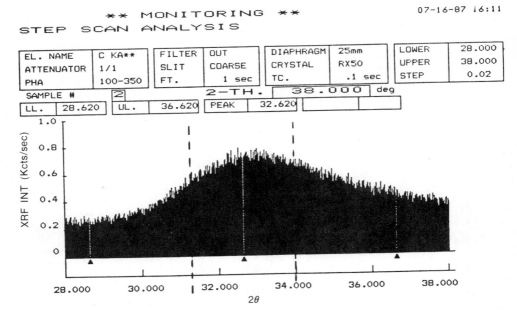

Figure 4.   Step scan of carbon /CoCrX on Si.

The calibration curve obtained from a set of sputtered carbon thin films of 10, 25, 50, and 100 nm thickness appears in Figure 2. The thickness of these films was assigned by the time sputtering at fixed power. The deposition rate had previously been determined using stylus measurements of thicker films. The peak and background positions used are those indicated in Figure 1.

Figure 3 illustrates step scan data taken for a CoCrX film on Si. The broad peak positioned at $2\theta \simeq 35°$ is the third-order Co $L_{\alpha\beta}$ peak. Figure 4 illustrates the carbon Kα peak superimposed over the broad interfering peak of the underlying CoCrX layer. The peak and background positions chosen for the quantitative analysis are indicated in the figure by the vertical dashed lines. These positions were chosen to best linearize the background contribution from the CoCrX layer. To the extent the background is linear, the background 2θ positions can be chosen as close to the peak position as desired and still maintain proportionality between the reduced and full peak heights. Derived using a Gaussian peak profile, Equation 1 describes this proportionality between the peak intensity (I_k) when the background positions are chosen at some multiple (k) of σ from the peak position, and the peak intensity (I_o), when the background positions are sufficiently removed from the peak such that the element contributes no intensity (see Appendix 1).

$$I_k = I_o \left(1 - \exp \frac{-k^2}{2} \right) \qquad [1]$$

This technique was used to generate the calibration data presented in Figure 5. Because of the reduction in signal necessitated by the narrow range of the background positions, relatively long counting times are used at the peak and background positions (240 seconds). As above, the standards used for this calibration were obtained by varying the

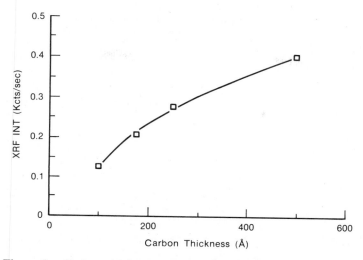

Figure 5.   Carbon thickness calibration for carbon/CoCrX on Si.

time of deposition. The slight positive x intercept of this calibration curve suggests possible error in the assigned carbon thickness (positive curvature from the interfering Co would be expected to produce a positive y intercept).   The background positions are chosen considering the trade-off between increased counting time and nonlinearity in the background.

Because the X-ray generation process is random in nature, the number of counts recorded at a given 2θ position for a fixed counting time is a normally distributed random variable.[1]   It then follows that the net intensity is a function of random variables, each normally distributed, and the variance of that function can be analytically derived (see Appendix 2). The relative precision (σ_T/T) varies inversely with the square root of the counting time and is plotted in Figure 6. This figure can be used to deter-

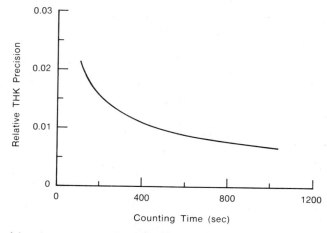

Figure 6.   Precision in measurement of carbon thickness resulting from increased XRF counting time.

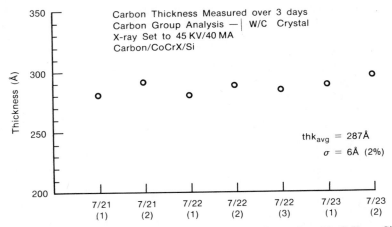

Figure 7.   Thickness measurement repeatability for carbon/CoCrX on Si.

mine the counting time necessary to achieve the desired precision.   Figure 7 represents
carbon-film thickness data taken over a period of three days for the same C/CoCrX
sample.   The error in these measurements includes effects such as instability of the incident X-rays and gain variations in the detector.   Nevertheless, that error ($\sigma = 2\%$) is
close to that predicted from the analysis described above (1.5%).

Figure 8 compares measurements of carbon thickness for a series of films sputtered for
different lengths of time at constant power.   The ellipsometry thickness is calculated by
assuming a constant real index of refraction for each of the films.   This must be supplied
by another independent measurement.   The absorbance measurement assumes that both
the real index and absorption of the films do not change with thickness and must be

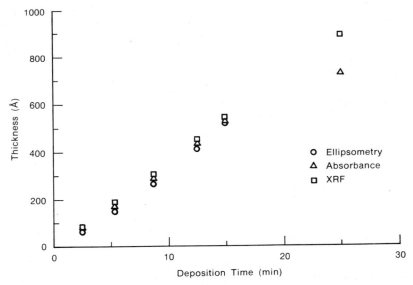

Figure 8.   Thickness correlation among XRF, ellipsometry, and absorbance techniques.

Table I. Comparision of Performance of Synthetic Multilayer Crystals

Multilayer	2d (Å)	I_{P-B} (cts/sec)	$\dfrac{I_{P-B}}{I_B}$ (cts/sec)	$\dfrac{\Delta\lambda}{\lambda}$ (%)
W/C	161.2	400	3.3	12.1
Ni/C	158.8	700	5.1	9.6
V/C	121.8	120	3.1	5.9
W/Si	80.2	45	3.0	5.9

- 300-Å carbon film on silicon
- Tungsten end-window tube
- 45 kV/60 mA
- Proportional counter (P10)

calibrated with independent measurements of thickness. Even with these assumptions, Figure 8 demonstrates that the correlation among these techniques is very good, with the observed errors probably due to misassignment of the optical constants described above. For the thickest films the absorption is too great for either the absorbance or ellipsometry technique.

The intensity, signal to noise, and/or resolution can be improved significantly using recently developed Ni/C and V/C crystal. Table 1 compares the performance of four different synthetic crystals used for soft X-ray analysis. The data indicate that the Ni/C crystal is superior in both intensity and resolution to the W/C crystal used in the present study. Further experimentation is required to determine the relative advantage of trading off the intensity of the V/C crystal against its improved resolution for this particular application.

SUMMARY

An effective procedure for measurement of carbon-coating thickness on silicon substrates and over CoCrX alloys has been described. Measurement of carbon thickness in the latter case is made more complicated by the interference of the underlying CoCrX layer (Co $L_{\alpha\beta}$). An exponential reduction in intensity results when the background 2θ positions are positioned within the C K_α XRF profile. This procedure was required because of the CoCrX interference. The relative precision of the measurement varies inversely with the square root of the counting time and has been analytically derived for the thin carbon coating measurement over a CoCrX underlayer. Excellent correlation is obtained between XRF measurement of carbon thickness and two different optical techniques. Measurements using the recently developed Ni/C and V/C synthetic crystals indicate that both crystals would be superior to the W/C crystal. The Ni/C provides improved intensity and resolution, while the V/C crystal has superior resolution.

REFERENCES

1. Cullity, B. D., Elements of X-Ray Diffraction, 2nd ed. (Reading, Mass.: Addison-Wesley, 1978).

2. Huang, T. C., Thin Solid Films, 157, 283 (1988); and references therein.

3. Nicolosi, J. A., Groven, J. P., and Merlo, D., Adv. X-Ray Anal., 30, 183 (1987).

4. Arai, T. , Adv. X-Ray Anal., 30, 213 (1987).

APPENDIX 1

Gaussian representation of diffracted intensity where μ = peak 2θ:

$$I_C = I_O \exp -\left(\frac{x-\mu}{\sigma\sqrt{2}}\right)^2 \qquad [1]$$

Linear background:

$$I_B = K_2 + K_3(x-\mu) \qquad [2]$$

Total intensity:

$$I_T = I_O \exp -\left(\frac{x-\mu}{\sigma\sqrt{2}}\right)^2 + K_2 + K_3(x-\mu) \qquad [3]$$

Net peak intensity after substracting background:

$$I_N = I_T(\mu) - \frac{I_T(\mu+\delta)}{2} - \frac{I(\mu-\delta)}{2} \qquad [4]$$

$$I_T(\mu) = I_O + K_2 \qquad [5]$$

Substituting $k\sigma$ for δ:

$$I_T(\mu+\delta) = I_O \exp -\left(\frac{\mu+k\sigma-\mu}{\sigma\sqrt{2}}\right)^2 + K_3(\mu+k\sigma) + K_2 - K_3\mu \qquad [6]$$

$$I_T(\mu-\delta) = I_O \exp -\left(\frac{\mu-k\sigma-\mu}{\sigma\sqrt{2}}\right)^2 + K_3(\mu-k\sigma) + K_2 - K_3\mu \qquad [7]$$

Substitute 5, 6, and 7 into 4:

$$I_N = I_O\left(1 - \exp\frac{-k^2}{2}\right) \qquad [8]$$

APPENDIX 2

T $\quad$ = Carbon thickness

N_μ $\quad$ = Counts recorded at peak after time t

$N_{(\mu+\delta)}, N_{(\mu-\delta)}$ = Counts recorded at respective background positions after time t

$$T = k \left[N_\mu - \frac{N_{\mu+\delta}}{2} - \frac{N_{\mu-\delta}}{2} \right]$$

N is a random variable that is approximately normally distributed with a standard deviation given by $\sqrt{N}$:

Var of T $= \sum$ Var of N:

$$\sigma_T^2 = k^2 \left[\sigma_{N_\mu}^2 + \frac{1}{4} \sigma_{N_{(\mu+\delta)}}^2 + \frac{1}{4} \sigma_{N_{(\mu-\delta)}}^2 \right]$$

$$\sigma_T = k \left[N_\mu + \frac{1}{4} N_{(\mu+\delta)} + \frac{1}{4} N_{(\mu-\delta)} \right]^{\frac{1}{2}}$$

N = Intensity (I) $\times$ time (t):

$$\sigma_T = k \left[I_\mu t + \frac{1}{4} I_{(\mu+\delta)} t + \frac{1}{4} I_{(\mu-\delta)} t \right]^{\frac{1}{2}}$$

Relative error:

$$\frac{\sigma_T}{T} = \frac{1}{\sqrt{t}} \left[\frac{I_\mu + \frac{1}{4} I_{(\mu-\delta)} + \frac{1}{4} I_{(\mu-\delta)}}{I_\mu - \frac{1}{2} I_{(\mu+\delta)} - \frac{1}{2} I_{(\mu-\delta)}} \right]$$

For 25-nm carbon over thin-film CoCrX:

$$I_{(\mu+\delta)} = .89 \, I_\mu$$

$$I_{(\mu-\delta)} = .78 \, I_\mu$$

$$\frac{\sigma_T}{T} = \frac{.218}{\sqrt{t}}$$

Separation of the Macro- and Micro-Stresses in Plastically Deformed 1080 Steel

R.A. Winholtz and J.B. Cohen

Department of Materials Science and Engineering
The Technological Institute
Northwestern University
Evanston, IL 60208

ABSTRACT

The stress tensors were measured with x-ray diffraction in both the ferrite and cementite phases of a 1080 steel specimen deformed in uniaxial tension. The macro- and micro-stresses were separated for all the components of the stress tensors. All the components except the hydrostatic components of the micro-stresses could be determined without an accurate value of the unstressed lattice parameter. Tensile stresses were found in the cementite and compressive stresses in the ferrite along the deformation direction. The tensile stresses in the cementite were quite large.

INTRODUCTION

Stresses have long been measured in the matrix phase of steels with diffraction. Steels, however, are generally two-phase materials containing carbides to enhance the properties of the material. Because the diffraction peaks from the carbide phase are so weak, the stresses in this phase and the stress system set up by its presence have not often been examined[1,2]. Recently, theories have been developed and used to investigate the stress system set up in a two-phase material[3,4]. The sensitivity of triaxial stress

341

measurements to errors in d_o, the unstressed lattice parameter, has been a drawback to their use[5]. In this work diffraction measurements of the stresses in both the cementite and ferrite phases of a 1080 steel will be described, as well as an analysis method that eliminates the sensitivity to d_o.

THEORY

Macro- and Micro-stresses in a Two-Phase Material

In an inhomogeneous material, the stresses from point to point will be different from those predicted by assuming the material to be homogeneous. Almost all engineering materials are inhomogeneous on a small scale and hence will have stresses different than those calculated from a simple theory. In a two-phase material the stresses will not distribute themselves uniformly between the two phases. This gives rise to macro- and micro-stresses within the material.

Macro-stresses vary slowly compared to a part's dimensions in at least one direction, and are by definition the same in both phases of a two-phase material. These stresses may originate, for instance, due to surface layers of a piece elongating more than the interior, for example during peening. Micro-stresses are the difference between the total stress at a point and the macro-stress value. Micro-stresses arise from microstructural properties that cause the stresses to deviate from the macro-stress value from point to point such as the yielding of one phase preferrentially to the other, or differences in the elastic response of two phases mutually constraining each other. The total stress at any point is the sum of these components[6]. For example:

$$^t\sigma_{ij}^\alpha = \ ^M\sigma_{ij} + \ ^\mu\sigma_{ij}^\alpha \ . \tag{1}$$

The superscripts t, M, and μ refer to the total, macro-, and micro-stresses respectively. The total stress in a phase is the quantity determined from a diffraction measurement on a crystalline material, but on occasion it may be important to separate this into its macro- and micro-stress components. This separation is now discussed.

Equilibrium Conditions

Stresses in a material must obey several equations of equilibrium. Both the macro-stress and micro-stress at any point must obey the differential equation of equilibrium[3]

$$\frac{\partial ^M\sigma_{11}}{\partial x_1} + \frac{\partial ^M\sigma_{12}}{\partial x_2} + \frac{\partial ^M\sigma_{13}}{\partial x_3} = 0 \ . \tag{2}$$

The macro- and micro-stresses $^M\sigma_{13}$, $^M\sigma_{23}$, and $^M\sigma_{33}$ must be zero at a free surface to satisfy the boundary conditions. If we can assume that the macro-stresses $^M\sigma_{11}$ and $^M\sigma_{12}$ do not vary in the surface, Equation 2 requires that the macro-stress $^M\sigma_{13}$ cannot vary with depth. Since it must be zero at the surface $^M\sigma_{13}$ is everywhere zero. The same hold true for $^M\sigma_{23}$ and $^M\sigma_{33}$. The micro-stresses $^\mu\sigma_{13}^\alpha$, $^\mu\sigma_{23}^\alpha$, and $^\mu\sigma_{33}^\alpha$ are not required to follow this last condition for equilibrium because the gradients in the surface are not required to be zero. From the requirement that the average stress over the whole body must be zero,

$$\int_D {}^t\sigma_{ij} \, dD = 0 \ , \tag{3}$$

it can be shown that the average micro-stresses in a two-phase material must obey the relation[3]

$$(1-f)\langle ^\mu\sigma_{ij}^\alpha\rangle + f\langle ^\mu\sigma_{ij}^\beta\rangle = 0 \ , \tag{4}$$

where f is the volume fraction of the β phase. Since a diffraction measurement samples a volume of material the stresses measured will be an average over that volume. Using Equation 1 for the stresses in each phase, and Equation 4, we have three equations in three unknowns, and hence we may separate the macro- and average micro-stresses.

Determination of Stresses with Diffraction

To measure stresses with diffraction we establish two coordinate systems as shown in Figure 1. The d-spacing is measured with diffraction in the L-coordinate system along the L_3 direction . The stresses in the S-coordinate system are then to be determined with these measurements. The strain along the L_3 axis in terms of the sample stresses is[7]

$$\langle \epsilon_{\phi\psi}^{\alpha} \rangle = \frac{d_{\phi\psi}^{\alpha} - d_{o}^{\alpha}}{d_{o}^{\alpha}} = \langle {}^{t}\sigma_{11}^{\alpha} \rangle\, S_{2}^{\alpha}/2 \, \cos^{2}\phi \, \sin^{2}\psi \; + \; \langle {}^{t}\sigma_{22}^{\alpha} \rangle S_{2}^{\alpha}/2 \, \sin^{2}\phi \, \sin^{2}\psi$$

$$+ \; \langle {}^{t}\sigma_{33}^{\alpha} \rangle S_{2}^{\alpha}/2 \, \cos^{2}\psi \; + \; (\langle {}^{t}\sigma_{11}^{\alpha} \rangle + \langle {}^{t}\sigma_{22}^{\alpha} \rangle + \langle {}^{t}\sigma_{33}^{\alpha} \rangle) S_{1}^{\alpha}$$

$$+ \; \langle {}^{t}\sigma_{12}^{\alpha} \rangle S_{2}^{\alpha}/2 \, \sin 2\phi \, \sin^{2}\psi \; + \; \langle {}^{t}\sigma_{13}^{\alpha} \rangle \, S_{2}^{\alpha}/2 \, \cos\phi \, \sin 2\psi$$

$$+ \; \langle {}^{t}\sigma_{23}^{\alpha} \rangle S_{2}^{\alpha}/2 \, \sin\phi \, \sin 2\psi \quad . \tag{5}$$

By measuring $d_{\phi\psi}^{\alpha}$ at a sufficient number of ϕ and ψ values and knowing d_{o}^{α}, the stresses in the sample coordinate system may be obtained by a least squares solution of this equation[8]. However, an accurate value of d_{o}^{α} must be available to determine the stresses accurately. This problem can be ameliorated by substituting

$$\langle {}^{t}\sigma_{ij}^{\alpha} \rangle = \begin{bmatrix} \langle {}^{t}\tau_{H}^{\alpha} \rangle & 0 & 0 \\ 0 & \langle {}^{t}\tau_{H}^{\alpha} \rangle & 0 \\ 0 & 0 & \langle {}^{t}\tau_{H}^{\alpha} \rangle \end{bmatrix} + \begin{bmatrix} \langle {}^{t}\tau_{11}^{\alpha} \rangle & \langle {}^{t}\tau_{12}^{\alpha} \rangle & \langle {}^{t}\tau_{13}^{\alpha} \rangle \\ \langle {}^{t}\tau_{12}^{\alpha} \rangle & \langle {}^{t}\tau_{22}^{\alpha} \rangle & \langle {}^{t}\tau_{23}^{\alpha} \rangle \\ \langle {}^{t}\tau_{11}^{\alpha} \rangle & \langle {}^{t}\tau_{12}^{\alpha} \rangle & \langle {}^{t}\tau_{13}^{\alpha} \rangle \end{bmatrix} , \tag{6}$$

where τ_{H} is the hydrostatic stress and τ_{ij} is the deviatoric stress tensor.

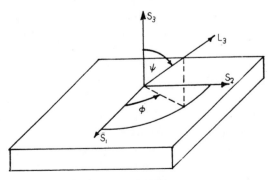

Fig. 1. Definitions of the coordinate systems used to determine stresses with diffraction.

Making use of the relation

$$({}^t\tau_{11}^\alpha + {}^t\tau_{22}^\alpha + {}^t\tau_{33}^\alpha) = 0 \quad , \tag{7}$$

which applies to both the stress at a point and averaged stresses, and solving for $\langle d_{\phi\psi}^\alpha \rangle$ gives the equation

$$\langle d_{\phi\psi}^\alpha \rangle = \langle {}^t\tau_H^\alpha \rangle d_o^\alpha [S_2^\alpha/2 + 3S_1^\alpha] - (\langle {}^t\tau_{11}^\alpha \rangle + \langle {}^t\tau_{22}^\alpha \rangle) d_o^\alpha S_2^\alpha/2 + d_o^\alpha$$

$$+ \langle {}^t\tau_{11}^\alpha \rangle d_o^\alpha S_2^\alpha/2 (1 + \cos^2\phi)\sin^2\psi$$

$$+ \langle {}^t\tau_{22}^\alpha \rangle d_o^\alpha S_2^\alpha/2 (1 + \sin^2\phi)\sin^2\psi$$

$$+ \langle {}^t\tau_{12}^\alpha \rangle d_o^\alpha S_2^\alpha/2 \sin2\phi \sin^2\psi$$

$$+ \langle {}^t\tau_{13}^\alpha \rangle d_o^\alpha S_2^\alpha/2 \cos\phi \sin2\psi$$

$$+ \langle {}^t\tau_{23}^\alpha \rangle d_o^\alpha S_2^\alpha/2 \sin\phi \sin2\psi \tag{8}$$

Written in this way we see that the d-spacing has five terms with different angular dependencies and three terms that are angularly independent. From a collection of $\langle d_{\phi\psi}^\alpha \rangle$ the six deviatoric components of the total stress tensor may be determined by least squares using Equation 8 and then using Equation 7. By using Equations 7 and 8 instead of Equation 5, we eliminate the need for an accurate d_o^α in the analysis, but still need information about the hydrostatic stresses.

Only the sum of the first three terms on the right of Equation 8 may be determined by least squares, and an accurate value of d_o^α appears necessary to determine the hydrostatic component of the stress. If we take values that might be typical for the ferrite phase in a steel:

$$S_2/2 = 5.77 \times 10^{-6} \text{ MPa}^{-1}$$

$$S_1 = -1.25 \times 10^{-6} \text{ MPa}^{-1}$$

$$d_o^\alpha = 1.1703 \text{ Å}$$

and assume an error in d_o^α of 0.0001 Å, we see that an error in the hydrostatic stress of 42 MPa would be needed to keep the sum of the first three terms on the right of Equation 8 a constant. (This corresponds to an error in 2θ of $0.03°$ at $156°$ 2θ in determining d_o^α.)

By using least-squares fitting and Equations 7 and 8 we may determine the total deviatoric stress tensor ${}^t\tau_{ij}^a$ in a phase from a collection of d vs. ϕ and ψ for that phase. Using Equations 1 and 4 we may separate the total stress tensors into the macro- and average micro-stress tensors for each component of the stress tensors. Then, if an accurate value of d_o is not available for each phase, the hydrostatic component of the macro-stress tensor may still be determined. We know from the equilibrium relations that ${}^M\sigma_{33}$ must be zero. We may thus write

$$ {}^M\sigma_{33} = {}^M\tau_H + {}^M\tau_{33} = 0 \ . \tag{9} $$

The hydrostatic macro-stress is then

$$ {}^M\tau_H = -{}^M\tau_{33} \ . \tag{10} $$

Thus, without accurate values of d_o, we may determine all the components of the macro-stress tensor and all the components of the micro-stress tensors in both phases except for the average hydrostatic micro-stress in each phase.

EXPERIMENTAL

Samples were cut from a plate of hot rolled 1080 steel. Heat treatment consisted of austenitizing in argon at 1073 K for 15 minutes followed by a slow cool (in argon) outside of the hot zone of the furnace. This heat treatment produced a pearlitic microstructure in the sample. The sample was then pulled in tension to a true plastic strain of 0.118. The final stress on the sample when it was released was 818 MPa.

Cementite has an orthorhombic crystal structure[9], which gives rise to many diffraction peaks, none of which are terribly strong. Not coincidently, the strongest diffraction peaks are often near to the ferrite peaks and overlap them considerably.

In order to obtain sufficient diffracted intensity from the carbide phase, a chromium rotating anode x-ray target was constructed. A copper target was chrome plated to a thickness of 0.001 inches in the area that the electrons strike it, so that chromium characteristic radiation would be produced. Chromium is widely used for stress analysis on steel because it gives a ferrite peak at a high angle. For the cementite, the long wavelength chromium radiation has the benefit of keeping the many peaks from bunching up and overlapping each other and keeps the distance in reciprocal space low so that the peak broadening is minimized. The x-ray generator was run at 47.5 kV and 200 mA for the measurements. Diffraction measurements were carried out on a Picker diffractometer with a quarter-circle goniometer and a graphite diffracted beam monochromator. The monochromator was needed to reduce the background counts so that a good peak to background ratio was obtained for the carbide peak. Psi goniometry was used to obtain the necessary ψ-tilts[10].

The 211 ferrite peak at about 156° 2θ was used for the stress measurements in the ferrite phase, while the 250 cementite peak at about 148° 2θ was used for the stress measurements in the carbide phase. Figure 2 shows the relative intensities of the ferrite and carbide peaks at ϕ and ψ equal to zero. The ferrite peaks were measured by counting 20 seconds at each point while the cementite peaks were counted for 300 seconds a point. The total measurement time for all the peaks measured was about 120 hours.

The sample was mounted on the diffractometer such that the tensile deformation axis corresponded to the direction $\phi = 0$. The peaks were point counted and fitted with pseudo-Voigt functions[11] including both the K_{α_1} and K_{α_2} components. The peak positions were then obtained from the functional parameters. For the ferrite peaks, a linear background was assumed, while for the cementite peaks the background was assumed to be an exponential function plus a constant, because the 250 carbide peak sits on the tail of the 211 ferrite peak. Figure 3 shows the functional fit to the cementite diffraction peak at $\psi=0$. Peak positions were determined in both phases over a ψ range of -45 to +45 degrees for ϕ equal to 0, 60, and 120 degrees.

The volume fraction of the carbide phase was determined from the ratio of the integrated intensity of the 121, 210, 112, 211, and the 250 carbide peaks to the intensity of the 100, 110, and the 211 ferrite peaks[12]. It was found to be 11 ± 2 percent. Averaging over a number of peaks helps to eliminate any errors due to preferred orientation. This measurement is consistent with the volume percent expected for a 1080 steel.

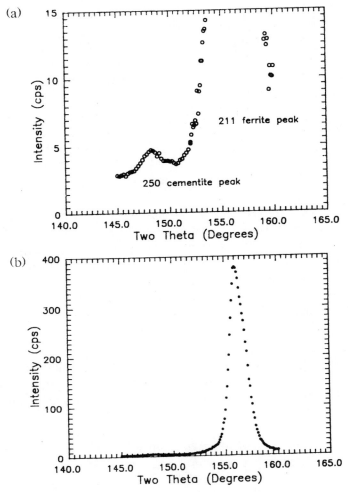

Fig. 2. Diffraction peaks used for the stress analysis at $\psi = 0$, empasizing the ferrite peak (a) and the cementite peak (b).

RESULTS AND DISCUSSION

Figure 4 shows a plot of d vs. $\sin^2\psi$ for both phases at $\phi = 0$. The stress tensors were determined from the above analysis to be

$$
\mu_T{}^c \;=\; \begin{bmatrix} 445 & 59 & 7 \\ 59 & -304 & 20 \\ 7 & 20 & -142 \end{bmatrix}
$$

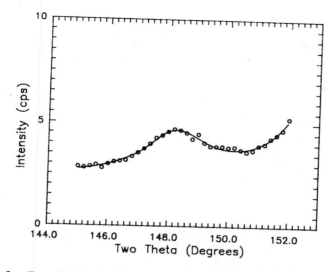

Fig. 3. Psuedo-Voigt functional fit to cementite peak for $\psi = 0$.

$$
\mu_T{}^\alpha = \begin{bmatrix} -55 & -7 & -1 \\ -7 & 38 & -2 \\ -1 & -2 & 17 \end{bmatrix}
$$

$$
{}^M\sigma = \begin{bmatrix} -39 & 5 & 2 \\ 5 & -28 & 0 \\ 2 & 0 & 0 \end{bmatrix}
$$

A value of 5.77×10^{-6} MPa^{-1} was used for $S_2/2$ for the ferrite phase[13]. This same value was used for the cementite phase as no better value is as yet available.

The deviatoric micro-stress $\mu_T{}^c_{11}$ along the deformation axis is tensile and very big in the carbide phase and the corresponding stress is compressive in the ferrite phase. The large tensile stress in the cementite is important to note as it could lead to crack initiation. The deviatoric micro-stresses perpendicular to the deformation direction $\mu_T{}^c_{22}$ and $\mu_T{}^c_{33}$ are smaller

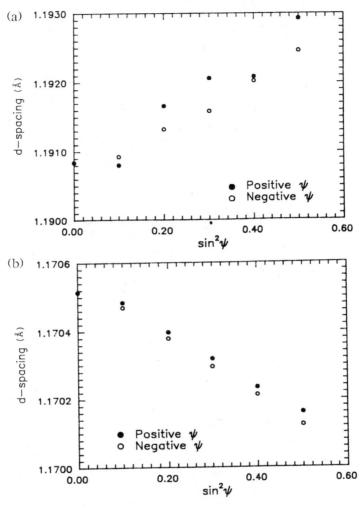

Fig. 4. d vs. sin$^2\psi$ plots for the cementite (a) and ferrite (b) phases.

and opposite to $^\mu\tau_{11}^c$. This is consistent with the deformation of a harder carbide phase in a softer ferrite matrix. The value of $^\mu\tau_{33}^c$ is smaller and consistent with the fact that this stress must be zero at the surface and the measured value is an average over the penetration depth of the x-rays. The macro-stress $^M\sigma_{33}$ is zero because it was set to that value to do the analysis. The non-zero value of $^M\sigma_{13}$ is probably indicative of the errors in the measurements.

The hydrostatic stress $^M\tau_H$ is shown as a function of the value of d_o^c used in Figure 5. For comparison the hydrostatic micro-stress, $^\mu\tau_H^c$, is also

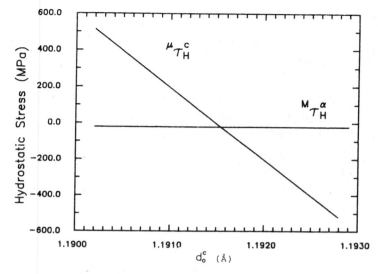

Fig. 5. The hydrostatic stresses $^{\mu}\tau_H^c$ and $^M\tau_H^\alpha$ as a function of the unstressed d-spacing d_o^c used.

shown. The range of d_o^c used was the range of d-spacings found in the data. We see that the hydrostatic macro-stress is indeed insensitive to the value of d_o^c, while the hydrostatic micro-stress in the cementite phase varies considerably. Using the lattice parameters in Reference 9 to determine d_o^c gives a hydrostatic macro-stress of -22 MPa and a hydrostatic microstress of -44 MPa in the cementite phase. This would require the hydrostatic micro-stress in the ferrite phase to be 5 MPa to satisfy Equation 4.

SUMMARY

1) A chromium rotating anode target was constructed, which was used to generate sufficient incident intensity to allow the measurement of stresses in the cementite phase of a steel using the 250 reflection.

2) The sensitivity of a measured triaxial stress tensor to the value of the unstressed lattice parameter was revealed to be an inability to determine the hydrostatic component of the stress tensor.

3) By making diffraction measurements in both phases of a two phase material the deviatoric components of the macro- and micro-stresses can be determined, as well as the hydrostatic component of the macro-stress

without accurate values of the unstressed lattice parameters. This leaves only the hydrostatic component of the micro-stresses undeterminable without the unstressed lattice parameters.

4) The macro- and average micro-stresses in a pearlitic 1080 steel sample plastically deformed in tension were determined using the above procedures. The results are consistent with a the stresses expected when a soft matrix with a hard second phase is deformed. The ferrite is in compression along the deformation axis and the cementite in tension. The large value of the latter stress could lead to crack initiation.

ACKNOWLEDGEMENTS

        This research was supported by the Office of Naval Research under contract No. N00014-80-C-116.

REFERENCES

1. D.V. Wilson and Y.A. Konnan, Work hardening in a steel containing a coarse dispersion of cementite particles, Acta Met. 12:617 (1964).

2. T. Hanabusa, J. Fukura, and H. Fujiwara, X-ray stress measurement on the cementite phase in steels, Bull. JSME 12:931 (1969).

3. I.C. Noyan, Equilibrium conditions for the average stresses measured by x-rays, Met. Trans. A 14A:1907 (1983).

4. I.C. Noyan and J.B. Cohen, An x-ray diffraction study of the residual stress-strain distributions in shot-peened two-phase brass, Mat. Sci. Engr. 75:179 (1985).

5. I.C. Noyan, Determination of the usnstressed lattice parameter "a_o" for (triaxial) residual stress determination by x-rays, Adv. X-Ray Anal. 28:281 (1985).

6. J.B. Cohen, The measurement of stresses in composites, Powder Diffraction 1:15 (1986).

7. H. Dölle, The influence of multiaxial stress states, stress gradients and elastic anisotropy of the evaluation of (residual) stresses by x-rays, J. Appl. Cryst. 12:489 (1979).

8. R.A. Winholtz and J.B. Cohen, A generalised least-squares determination of triaxial stress states by x-ray diffraction and the associated errors, Aust. J. Physics, in press.

9. H. Lipson and N.J. Petch, The crystal structure of cementite, Fe_3C, J. Iron Steel Inst., 142:95 (1940).

10. I.C. Noyan and J.B. Cohen, "Residual Stress: Measurement by Diffraction and Interpretation," Springer-Verlag, New York (1987).

11. G.K. Wertheim, M.A. Butler, K.W. West, and D.N.E. Buchanan, Determination of the Gaussian and Lorentzian content of experimental line shapes, Rev. Sci. Instrum., 45:1369 (1974).

12. L.H. Schwartz and J.B. Cohen, "Diffraction from Materials," Springer-Verlag, New York (1987).

13. H. Dolle and J.B. Cohen, Residual stresses in ground steels, Met. Trans. A, 11A:159 (1979).

EFFECT OF PLASTIC DEFORMATION ON OSCILLATIONS IN "d" vs. $\sin^2\psi$ PLOTS

A FEM ANALYSIS

I. C. Noyan and L. T. Nguyen

I.B.M. T. J. Watson Research Center
Yorktown Heights, NY 10598

ABSTRACT

Oscillations in "d" vs. $\sin^2\psi$ plots are due to the inhomogeneous partitioning of strains within the diffracting volume. In polycrystalline specimens, such inhomogeneity can be caused by the elastic incompatibility of neighboring grains or by the inhomogeneous partitioning of plastic deformation within the diffracting volume. There is, however, little work on the degree of inhomogeneity required to cause a given oscillation, and the relative contribution from the elastic and plastic deformation components to a given oscillation.

In a previous paper, we modeled the effect of elastic incompatibility on the "d" vs. $\sin^2\psi$ from Cu 311 planes. In the present study, the model is expanded to accommodate the effect of local plastic yielding within grains in the diffracting volume. The effects of such yielding on the oscillatory behavior in "d" vs. $\sin^2\psi$ are investigated.

THEORY

X-ray diffraction techniques are widely used in the non-destructive determination of residual stress/strain in engineering materials. Currently there are several x-ray methods, utilizing different assumptions as to the stress state existing in the diffracting region, that can be used for residual stress analysis[1-6]. All of these techniques utilize an internal atomic spacing, d_{hkl}, in the material as a strain gage. Once the plane spacing is measured with diffraction, it is converted into strain:

355

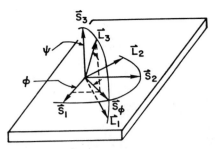

FIGURE 1. *Definition of the angles ϕ and ψ and orientation of the laboratory system L_i with respect to the sample system S_i and the measurement direction S_ϕ.*

$$< \epsilon'_{33} > = \frac{d_{\phi\psi} - d_0}{d_0} \tag{1}$$

where d_0 is the unstressed plane spacing and "$< >$" implies an x-ray average over the diffracting crystallites. The strain ϵ'_{33} is along the L_3 direction and makes the angles ψ and ϕ with the surface as shown in *Figure 1*.

In this figure, the $\tilde{S}$ coordinate system defines the specimen surface and the $\tilde{L}$ coordinate system defines the laboratory axes in which diffraction is occuring. The strains in the $\tilde{L}$ and $\tilde{S}$ axes are related through the second rank tensor transformation:

$$< \epsilon'_{33} > = a_{3k}a_{3l}\epsilon_{kl} \tag{2}$$

where the primed and unprimed strain components are referred to the $\tilde{L}$ and $\tilde{S}$ axes respectively and a_{3k}, a_{3l} are the direction cosines between these sets of axes. *Equation (2)* defines a set of linear equations in the unknown strains ϵ_{kl} and can be solved for these strains if a sufficient number of $< \epsilon'_{33} >$ are measured at various (independent) ϕ, ψ combinations. Once the strains ϵ_{kl} are determined, the residual stress tensor can be calculated from Hooke's law:

$$\sigma_{ij} = C_{ijkl}\epsilon_{kl} \tag{3}$$

where C_{ijkl} are the stiffness coefficients of the material in question. This procedure implicitly assumes that a homogeneous stress/strain distribution exists within the irradiated area from which the lattice parameter data is obtained, since the components obtained from the above treatment do not exhibit dependency on position within the irradiated volume. For such a "homogeneous" distribution, *Eqs (2)* and *(3)* predict a linear dependency of $d_{\phi\psi}$ on $\sin^2\psi$ if the shear strains ϵ_{13}, ϵ_{23}, or the corresponding shear stresses, are zero (*Figure 2 (a)*). In such a case, the slope of the d_ψ vs. $\sin^2\psi$ line is proportional to the product of the elastic constants of the material and the residual stress in the ϕ direction. Thus, classical methods are generally based on the determination of the slope of this line[1]. If either or both of these shear strains are non-zero, a "psi-split" variation is predicted with opposite curvature for the branches measured at positive and negative ψ (*Figure 2 (b)*). On the other hand, oscillatory variation of $d_{\phi\psi}$ with $\sin^2\psi$ indicates the presence of inhomogeneous stress/strain gradients within the irradiated volume, where the strain at a point A (x,y,z) is expressed as:

$$\varepsilon_{ij}^t = \varepsilon_{ij}^0 + \varepsilon_{ij}^{in} + \varepsilon_{ij}^r \tag{4}$$

ε_{ij}^0 is the elastic strain that would be observed if the material was homogeneous ε_{ij}^{in} is the interaction strain term due to inhomogeneous distribution of elastic properties and ε_{ij}^r is a residual strain term that is caused by inhomogeneous partitioning of plastic deformation.

From *Eqs (2)*, *(3)*, and *(4)*, the relationship between the stress within the diffraction volume and the strain measured by the x-rays can be obtained.[7] For example, for $\phi = 0$:

$$<\varepsilon_{33}>_\psi = \frac{d_\psi - d_0}{d_0} = \{\sigma_{11}^0\{\frac{1+v}{E} + K_1(\psi) + K_3(\psi)\} + \varepsilon_{11\psi}^r - \varepsilon_{33\psi}^r\} \sin^2\psi +$$

$$\sigma_{11}^0\{-\frac{v}{E} + K_3(\psi)\} + \varepsilon_{33\psi}^r \tag{5}$$

In *Eq (5)*, the stress/strain terms are the x-ray averages of the local variations over the diffracting volume. Since the diffracting volume itself is a function of ψ, these terms are expressed here as variables in ψ. It can be seen from *Eq (5)* that $d_{\phi\psi}$ is linear in $\sin^2\psi$ if $(\varepsilon_{ij}^r)_\psi$ and $K_i(\psi)$ are independent of ψ. If these terms are both zero, the material is equivalent to an isotropic homogeneous substance, and the x-ray stress

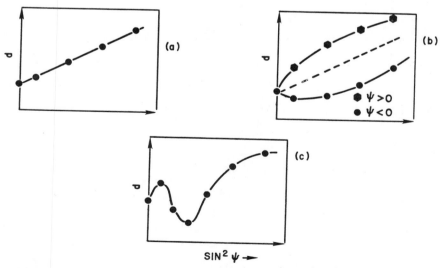

SIN² ψ →

FIGURE 2. *Types of "d" vs. $\sin^2\psi$ plots encountered. (c) cannot be explained by conventional analysis.*

analysis using Eq (5) will yield σ_{11}^0. If these terms are finite, but independent of ψ, the $d_{\phi\psi}$ is still linear in $\sin^2\psi$ but the analysis will yield a stress value with an average term that is proportional to the inhomogeneity effects added to σ_0^{11}. If, on the other hand, $K_i(\psi)$ or $(\varepsilon_{ij}^r)_\psi$ are non-linear in ψ, oscillations will occur in the $d_{\phi\psi}$ vs. $\sin^2\psi$ curve. Various methods have been proposed to analyze such data. Some of these methods yield a single number for the stress within the material, ignoring the inhomogeneity of stress/strain[4]. Others try to determine an average stress within each diffracting set of crystallites[5]. These latter methods, on the other hand, ignore the possibility of strain/stress varying drastically within the grains diffracting at a given ψ.

In the present sudy, the distribution of stress within the diffracting grains are obtained from a finite element mesh loaded to elastic and plastic loads. These stresses are then used to calculate the "d" vs. $\sin^2\psi$ data from which the residual stress value is obtained by the classical x-ray analysis. From such data, the contribution of elastic incompatibility and plastic inhomogeneity to oscillations can be deduced. Furthermore, stress variations within each diffracting sub-set are determined and compared to the obtained values.

FINITE ELEMENT MODELING

Material Definition

A hypothetical Cu specimen with 100 grains was assumed for this study. The 311 reflection was chosen for the stress analysis calculations. The grains in the mesh that would diffract at various ψ -tilts satisfy the equation:

$$\cos\{90 - \psi\} = \frac{ha + bk + c\ell}{\sqrt{h^2 + k^2 + \ell^2}\sqrt{a^2 + b^2 + c^2}} \qquad (6)$$

where a,b,c are the indices of the crystallographic direction along $\vec{S}_1$ in the diffracting grain and h,k,l are the indices of the x-ray reflection (311 in this case) which define the normal $\vec{D}$ to the diffracting planes, namely, $\vec{D} = h\vec{x}_1 + k\vec{x}_2 + l\vec{x}_3$.

TABLE I. Crystallographic orientation along $\vec{S}_1$ for the grains diffracting at various ψ -tilts. Type A grains have orientations closer to the 111 direction, while type B grains are closer to the 100 direction.

$\sin^2\psi$	TYPE A	TYPE B
0.0	$(3,\bar{4},\bar{5})$	$(\bar{1},0,3)$
0.1	$(27,\overline{20},\overline{20})$	$(\bar{2},9,38)$
0.2	$(17,\overline{21},19)$	$(7,\bar{8},18)$
0.3	$(25,\overline{23},17)$	$(18,\bar{4},\overline{11})$
0.4	$(7,6,\bar{5})$	$(10,7,29)$
0.5	$(14,31,22)$	$(10,7,22)$
0.7	$(23,21,30)$	$(32,3,37)$

TABLE II. Distribution of Young's moduli and Schmid factors for the grains defined in *Table I*. Type A grains, being closer to a close packed direction, are elastically stiffer. Furthermore, they also have lower Schmid factors, which means they will have smaller resolved shear stress on their primary slip planes compared to type B grains.

$\sin^2\psi$	TYPE A		TYPE B	
	E_{S1} (GPa)	Sch. F.	E_{S1}(GPa)	Sch. F.
0.0	167.0	0.39	80.9	0.49
0.1	175.9	0.34	74.3	0.48
0.2	186.0	0.33	112.2	0.44
0.3	176.8	0.37	116.1	0.49
0.4	179.1	0.36	90.2	0.48
0.5	145.8	0.44	106.7	0.47
0.7	174.5	0.36	128.7	0.45

The lattice directions along $\tilde{S}_2$ and $\tilde{S}_3$ are then determined from the cross products:

$$\tilde{S}_2 = \vec{D} \times \tilde{S}_1 \quad ; \quad \tilde{S}_3 = \tilde{S}_1 \times \tilde{S}_2 \tag{7}$$

From these equations, the specimen coordinate system in any grain diffracting at a particular ψ-tilt can be defined in terms of lattice directions. The numerical solution of *Eq (6)* for $\tilde{S}_1$ is shown in *Table I* for the diffracting grains in the finite element mesh.

There are two grains per ψ-tilt, one close to having a 111 orientation along the $\tilde{S}_1$ direction (A type grains), and the other closer to having a 100 orientation along this axis (B type grains). One can calculate the elastic constants in each grain, referred to the S coordinate system, from the fourth rank tensor transformation:

$$S_{ijkl} = a_{im}a_{jn}a_{ko}a_{lp}\tilde{S}_{mnop} \tag{8}$$

where $\tilde{S}_{mnop}$ are the compliances in the crystal axes (100, 010, 001) and the a_{im} are the direction cosines between the crystal axes and the lattice directions in the diffracting grain along $\tilde{S}_i$. One can also define an effective Young's modulus along each of the specimen axes for cubic materials[6]:

$$\frac{1}{E_{\tilde{S}_i}} = s_{11} - 2\{(s_{11} - s_{12}) - \frac{1}{2}(s_{44})\}[(a^2_{i1}a^2_{i2}) + (a^2_{i2}a^2_{i3}) + (a^2_{i3}a^2_{i1})] \tag{9}$$

where s_{ij} are the stiffnesses in the contracted notation and, as before, a_{ij} are the direction cosines. The variation of the effective Young's modulus along $\tilde{S}_1$ in the diffracting grains is shown in *Table II*.

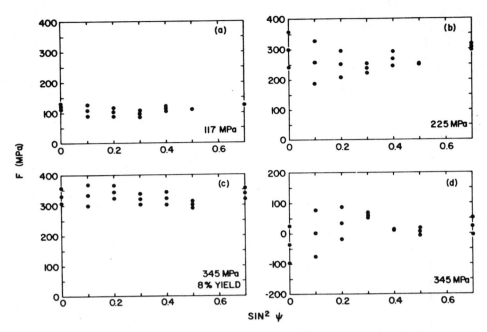

FIGURE 3. *Variation of load within the grains diffracting at various ψ-tilts for various load conditions: (a) 117 MPa; (b) 255 MPa; (c) 345 MPa. At this load, 8% plastic flow has taken place; (d) Material unloaded from 345 MPa to 0 MPa.*

Such a variation will cause elastic incompatibility strains to keep displacements constant across grain boundaries and can, thus, give rise to oscillations. Furthermore, since plasticity is also taken into account in the current model, the Schmid factor, $m = \cos \chi \cos \lambda$, which depends on the orientation of the slip plane (defined by the angle χ) and the slip direction (defined by the angle λ to the load axis), must be computed for these grains so that yield will start when the total stress at a point exceeds the critical resolved shear stress. These calculations are also summarized in *Table II*. From this table, it is seen that grains diffracting at the same ψ-tilts can have very different elastic moduli in the direction of the load. By comparison, the Schmid factors are grouped closer together, reflecting the large number of primary slip systems ($\{111\}, <110>$) in the FCC lattice. However, since elastic interaction stresses, which depend on the local inhomogeneity in elastic moduli, can also cause local plastic flow[7-9], plastic inhomogeneity is dependent on both factors.

FEM Analysis

The grains listed in *Table I* with the properties shown in *Table II* were placed at random within the mesh of 100 grains (via a random number generator program) and loads applied to the edge of the mesh along the $\bar{S}_1$ direction. The non-linear finite element code ANSYS (Swanson Analysis Systems, Houston, PA), version 4.3B, was used to determine the total stress in each grain at the surface. The mesh is composed of 4-node, 2-D isoparametric plane strain elements (STIF42). These elements have plastic, creep, swelling, strain-hardening and rotation capabilities. For analysis purposes, it was assumed that the mesh material had a yield point of 255 MPa and a strain hardening exponent of 0.3. The mesh was then loaded to 345 MPa (corresponding, approximately, to a total strain of 8%), and then relaxed to zero load, with stress data in each grain obtained at 117, 255, and 345 MPa, and after relaxation. These stresses were transformed into strains through Hooke's law using the single crystal compliances. Once the strains in the $\bar{S}$ coordinate system were obtained, Eq (2) was used to calculate the strains in the laboratory system $\bar{L}$. The lattice spacing along $\bar{L}_3$ was then obtained from Eq (1).

RESULTS AND DISCUSSION

Variation of Load within the Diffracting Grains

The variation of the stress along the $\bar{S}_1$ direction for the diffracting grains (*Table I*) is shown in *Figures 3 (a)* and *(b)* for the 117, 255 MPa elastic loading steps and in *Figures 3 (c)* and *(d)* for the 345 MPa load with plastic flow and after relaxation.

One can readily observe the following from these figures:

1. In the elastic regime, the interaction stresses increase with increasing applied load and can be quite large (*Figures 3 (a)* and *(b)*).
2. Even within grains of a subset diffracting at a given ψ -tilt, large variations in stress can exist (*Figure 3 (b)*).
3. Once uniaxial plastic flow starts, the variations in stress from grain to grain becomes less, even though the net load carried by the specimen is large (*Figure 3 (c)*).
4. The interaction stresses, however, are not small. This can be seen by relaxing the material to zero load (*Figure 3 (d)*). It is seen that a substantial stress field is "locked" in some of the grains. This field is due to the inhomogeneous partitioning of yield and has both tensile and compressive values depending on the grain. Because of this, at high applied loads, elastic and plastic interaction stresses may partially cancel. The net deviation can still be as high as 100 MPa for the cases investigated.

It must be noted here that, the residual stress field within the material is bi-axial even

TABLE III. Comparison of applied loads and stresses determined from x-ray analysis of *Figures 4 (a)-(d)*.

Applied Load (MPa)	σ_{x-ray} (MPa)	Error (MPa)
117	104	13
255	243	12
345	297	48
0 (relaxed)	7	7

TABLE IV. The total stress state in selected grains at various applied loads.

Appl. Load (MPa)	255	345 ·	0
$\sin^2\psi$	$(\sigma_{11}\sigma_{22}\sigma_{12})$	$(\sigma_{11}\sigma_{22}\sigma_{12})$	$(\sigma_{11}\sigma_{22}\sigma_{12})$
0.0	(243,13,7)	(307,27,1)	(125,5,9)
0.1	(328,-6,7)	(368,5,9)	(-75,104,0)
0.3	(223,-9,0)	(303,18,-6)	(70,61,-13)
0.4	(293,-16,11)	(343,-35,3)	(9,-33,-11)

though a uniaxial load is applied along the boundary. The complete stress field in some grains are shown in *Table III* for various load steps. It can be seen that the bi-axial stress field is quite pronounced in some grains after relaxation to zero load.

X-ray Stress Analysis

The d_ψ vs. $\sin^2\psi$ plots for the (complete) stress states depicted (partially) in *Figures 3 (a)-(d)* are shown in *Figures 4 (a)-(d)*. In these figures, the d_ψ from individual grains at each ψ-tilt are shown to illustrate the effect of inhomogeneous stress distributions on the oscillations. Of course, in a real experiment only the average value of the two grains would be seen since they will both diffract at the same time. Thus,

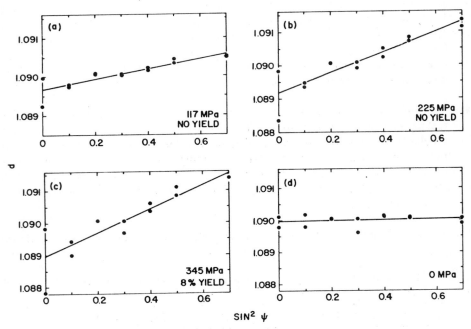

FIGURE 4. *Variation of lattice parameters with tilt for various load conditions: (a) 117 MPa. No plastic flow is permitted; (b) 255 MPa. No plastic flow; (c) 345 MPa. 8% plastic flow; (d) Specimen relaxed from 345 MPa to 0 MPa. The oscillations are caused by inhomogeneous partitioning of plastic flow.*

the least-squares line is fitted to the average of the two values over the entire ψ range. The results of the stress analysis are shown in *Table IV*.

It can be seen that, given the range of stress variation within the material (*Figure 3 (a)-(d)*), the results are quite reasonable. It seems that, for this reflection at least, using the entire ψ range and the classical analysis, one gets reasonable values for the applied stress. The following observations can also be obtained from *Figures 3 (a)-(d)* and *4 (a)-(d)*.

1. Deviation of a given d_ψ point from linearity can not be directly related to the deviation of the average stress in that diffracting subset from the applied stress, similar to the conclusions from earlier work[7].
2. One cannot neglect either plastic or elastic interaction effects when deciding on a method to use for stress analysis.
3. The variation of stress within grains of a given subset can be quite large and can not be determined from x-ray data alone. Those methods that purport to determine average stress within each grain set are, thus, to be treated with some caution.

SUMMARY

A well characterized net of grains were subjected to simulated elastic and plastic deformation using finite element methods. The deformation data from the finite element calculations were then used to calculate x-ray lattice parameter information at various ψ -tilts. This x-ray data was then analyzed in the light of the deformation fields causing it. It was seen that:

a. Interaction stresses caused by elastic and plastic inhomogeneity and incompatibility effects can be comparable in magnitude even for simple uniaxial testing involving plastic deformation.
b. Such inhomogeneity effects cause oscillations in d_ψ vs. $\sin^2\psi$ plots.
c. Even in the presence of oscillations, diffraction data from some peaks, such as 211, 311 may be analyzed by classical methods and seem to yield reasonable stress values when compared to the applied stress.
d. Such treatment, however, ignores the inhomogeneous distribution of strain within the material. On the other hand, since, even within grains diffracting at a particular ψ-tilt, just as large variations exist as between different subsets, no significant data is lost.
e. Finite element programs that can more accurately model polycrystalline materials, along the lines described by Deve et al.[8,9] are needed to analyze large deformation fields that cause changes in texture through changes in local orientation.

REFERENCES

1) Society of Automotive Engineers, "Residual Stress Measurement by X-Ray Diffraction," *SAE 784a*, 2nd ed. (1971).

2) J. B. Cohen, H. Dolle, and M. R. James, "Determining Stresses from X-Ray Powder Patterns," *NBS Special Publication 567*, 453 (1980).

3) I. C. Noyan, "Equilibrium Conditions for the Average Stresses Measured by X-Rays," *Met. Trans*, **14A**, 1907 (1983).

4) H. Dolle and V. Hauk, "System of Possible Lattice Strain Distributions on Mechanically Loaded Metallic Materials (in German)," *Z. Metall.*, **68**, 725 (1977).

5) V. Hauk, "Evaluation of Macro and Micro-Residual Stresses on Textured Materials by X-Ray, Neutron Diffraction, and Deflection Measurements," *Adv. in X-Ray Anal.*, **29**, 17 (1986).

6) I. C. Noyan, "Determination of the Elastic Constants of Inhomogeneous Materials with X-Ray Diffraction," *Mat. Sci. and Eng.*, **75**, 95 (1985).

7) I. C. Noyan and L. T. Nguyen, "Oscillations in Interplanar Spacing vs. $\sin^2\psi$. A FEM Analysis," *Adv. in X-Ray Anal.*, **31**, 191 (1988).

8) S. V. Harren, H. E. Deve, and R. J. Asaro, "Shear Band Formation in Plane Strain Compression," *Acta Metall.*, in press (1988).

9) H. Deve, S. Harren, C. McCullough, and R. J. Asaro, "Micro and Macroscopic Aspects of Shear Band Formation in Internally Nitrided Single Crystals of Fe-Ti-Mn Alloys," *Acta Metall.*, **36**, 341 (1988).

X-RAY DIFFRACTOMETRIC DETERMINATION OF LATTICE MISFIT
BETWEEN γ AND γ' PHASES IN Ni-BASE SUPERALLOYS

Conventional X-Ray Source vs. Synchrotron Radiation

Katsumi Ohno, Hirosi Harada, Toshihiro Yamagata,
Michio Yamazaki and Kazumasa Ohsumi*

National Research Institute for Metals
2-3-12 Nakameguro Meguro-ku, Tokyo, 153, Japan
*National Laboratory for High Energy Physics
1-1 Oho Tsukuba-shi Ibaragi-ken 305, Japan

ABSTRACT

The lattice misfits between γ and γ' phases in Ni-base
superalloys(single crystal) were accurately determined for filings of
specimens by using both a conventional X-ray tube focusing
diffractometer(CXRFD) and a synchrotron-radiation parallel beam X-ray
diffractometer (SRPXRD).   All reflection peaks measured with the CXRFD
were in a cluster of overlapping peaks because of the very small
differences in the lattice parameters of both phases and the instrumental
broadening due to X-ray optics including the spectral distribution of X-
ray source such as CuK$_\alpha$ doublet. The deconvolution method was applied to
remove the instrumental broadening from the peaks measured with the CXRFD.
The window functions for the deconvolution method were calculated from
CuK$_\alpha$ doublet reflection of Si standard by a nonlinear least-square method.
The instrumental broadening of SRPXRD was much smaller than that of
CXRFD since the monochromatic X-rays produced single peak profiles and
constant profile shape over a wide 2θ range.   A profile fitting with a
pseudo-Voigt function was used to determine 2θ angles to 0.0005 deg. for
the synchrotron powder data. The peak angle and shape reflected from γ'
phases in γ-matrix and those from electrochemically extracted γ'-phase
were significantly different.

Introduction

A computer aided alloy design system has been established in the
National Research Institute for Metals for γ' precipitation hardening
nickel-base superalloys[1-3].   A series of Ni-base superalloys, which
have γ' precipitate, an ordered FCC phase based on Ni$_3$Al in γ-matrix
having a disordered FCC structure, has been developed for gas turbine
blades by using the alloy design system.
The durability of single crystal blades is usually two or three times
higher than that of the columnar grained materials[4].   Creep strength is

Table I Chemical composition and micro-structure factors of alloys

Sample	Co	Cr	Mo	W	Al	Ti	Nb	Ta	Ni	V_{γ}'	$\delta(\%)$
NASAIR-100	--	10.32	0.62	3.41	12.71	1.49	--	1.00	Bal.	80	-0.36
TMS-1LA	8.23	6.86	0.03	5.67	10.29	--	0.01	1.68	Bal.	63	0.08
Alloy-B	4.99	9.63	--	3.04	12.00	2.70	--	0.91	Bal.	55	0.16
TMS-12	--	8.10	--	4.47	11.73	--	--	2.88	Bal.	76	0.17
TMS-19	--	8.17	--	2.67	10.96	--	--	4.47	Bal.	74	0.38

V_{γ}' : Volume(%) of γ', $\delta(\%)$: Misfit calculated

of prime importance for blading materials. The creep behavior of $\gamma-\gamma'$ alloys strengthened by the coherent γ' phase depends primarily on the following factors: (1) the volume fraction of γ' phases, (2) the size of γ' precipitate, (3) the composition of the γ and γ' phases, and (4) the coherency strains due to the lattice misfits.

The first two factors can be easily analyzed by electron microscopy combined with image processing[5]. The third factor is to be determined by electron probe X-ray micro-analysis(EPMA)[5]. However, the measured lattice misfits in the alloys by powder X-ray diffractometry were not so reliable, because the misfit values are usually so small that most of the main diffraction lines overlap in the patterns obtained with CXRFD.   In addition, the superlattice lines were usually very weak. Many attempts have been made to measure the lattice misfits between both phases on the filings of specimens with CXRFD[4,6,7]. No satisfactory results have been obtained, because the principal reflection peaks from γ and γ' phase could not be resolved into individual peaks.

The lattice misfits between γ and γ' phases determined by applying the profile fitting method to the diffraction pattern measured by using a synchrotron-radiation parallel-beam diffractometry were compared with those misfits determined by applying the deconvolution method to diffraction patterns measured by CXRFD.   The diffraction data occurring as folded peaks are separated into individual peak profiles, by applying the profile fitting[9] or the deconvolution method[8,9,10]. Here, the lattice misfit(δ) is defined as $2(a\gamma' - a\gamma)/(a\gamma' + a\gamma)$, where, $a\gamma'$ and $a\gamma$ are the lattice parameters of γ and γ' phases, respectively.

Specimen preparation

Five single crystal alloys having various lattice misfit values, from negative to positive, were provided from experimental and commercial alloys by means of the alloy design program for the system mentioned above[1-3]. The chemical compositions of the alloys are shown in Table I, together with the γ' volume fractions and the lattice misfit values calculated by the program.

A single crystal rod, 6mm in diameter, was prepared for each alloy by a directional solidification method. Each rod was heated in a temperature range from 1300 to 1350 deg. C. for 32h. to dissolve γ' precipitates and to prepare a highly homogenized γ structure. This heat treatment was followed by cooling in air (AC).   Those rods were then aged at 900 deg. C. for 120h. to complete precipitation of γ'. The heating conditions are shown in Table II.

Powder of each alloy was obtained by filing and was used for the lattice misfit measurement after annealing in a vacuum quartz capsule at 900 deg. C. for 1 h. followed by quenching in water (WQ).

Tabel II   Heating conditions of the specimens

Specimen	Homogenizing	Aging
NASAIR-100	1308°C x 32h, AC	
TMS-1LA	1348°C x 32h, AC	900°C x 120h
Alloy-B	1308°C x 32h, AC	WQ
TMS-12	1348°C x 32h, AC	
TMS-19	1336°C x 32h, AC	

AC : air cooled,   WQ : water quenched.

γ' phases were electrochemically extracted from the specimens with an aqueous solution containing 1% of ammonium sulphate and citric acid and were collected on micro-filter (Millipore Filter VCWP-4700).

Instruments

The synchrotron-radiation X-ray diffractometer having parallel beam optics[11] and the conventional X-ray tube focusing diffractometer(JDX-8020) were used.

The synchrotron-radiation parallel-beam X-ray diffractometer, which was installed on the beam line 3B at the Photon Factory in the National Laboratory for High Energy Physics, was equipped with a vertical-scanning silicon (111) channel-monochromator, a monochromatated beam monitor, a rotating specimen device with 24 rev/min rotation around an axis normal to the specimen surface, and step scanning mechanism.   Scintillation counters with single-channel analyzers were used as detectors for the monitor and powder specimen.   The path of X-rays excluding slits was made vacuous.

The conventional diffractometer was equipped with a diffracted beam focusing graphite monochromator and a step scanning mechanism.

A conventional sealed-off Cu-tube was operated at 40kV and 50mA.

Both instruments were controlled by a 16-bit personal computer; the collected data were stored in MS-DOS formatted files for processing.

DECONVOLUTION and 2Θ –MEASUREMENT WITH CXRFD

Measured peak profile shape $P(\Theta)$ results from the convolution $(W*G)*S$ + background.   Where W is the spectral distribution of X-ray source such as CuKα doublet received at detector, G is the convolution of all aberrations arising from the diffractometer geometry and process, S is the pure intrinsic diffraction of the specimen and * indicates the convolution operator.   The deconvolution method consists of two major steps: 1) Evaluation of the W*G function by measuring a number of profiles of standard specimens using an identical set of instrumental parameters as will be used in the later analysis.   This step determines the instrumental window function of the diffractometer.

2) Determination of the S by deconvolution with the window function $(W*G)$.   The specimens used to measure the window function must be carefully prepared and free of any line broadening (due to strain, small crystallite size, etc.).

The peak shape reflected from only the specimen itself can be calculated by deconvolution of $P(\Theta)$ with $(W*G)$ after background subtraction, if the $(W*G)$ is known.   Fourier transform theorem states that the convolution of three function, W, G and S, is equivalent to finding the product of their

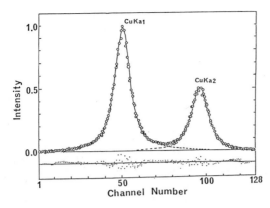

Fig. 1 (533) reflection of
Si-stand, CuKα doublet,
for determination of W*G,
experimental data points, O;
fitted pseudo-Voigt function,
dashed lines; sum of these
curves, solid lines.

Fourier transform, $W(\omega)$, $G(\omega)$ and $S(\omega)$. Therefore, the pure intrinsic
diffraction of the specimen is calculated by the equation,
$S(\omega) = P(\omega)/[W(\omega)G(\omega)]$, in Fourier domain.

It is necessary to first determine experimentally the convolution W*G by
accurate measurement of line profiles in the 2Θ range of interest. This
provides the window function as a function of the reflection angle and
this is the "Instrument Window Function" occurring in all powder patterns
recorded with these experimental conditions.

A typical final fitting to determine W*G for silicon standard (533)
reflection of $CuK\alpha_1$ and $K\alpha_2$ is shown in Fig. 1. The solid line is the
sum of two calculated profiles(dotted line), the experimental points are
presented by O's and 5 times amplified residuals are shown in the lower
part of Fig. 3 by •'s; the step width(2Θ) = 0.008 deg., counting time = 3
sec/step, the $K\alpha_1$ peak intensity = 12804 counts. Both of the measured
peaks were accurately expressed by Lorentzian/Gaussian function having
their mixing ratio to be 4. However, the full width at half maximum (FWHM)
of $CuK\alpha_2$ peak was about 10% wider than that of $CuK\alpha_1$. Therefore (W*G)
was, hereafter, approximated as the Lorentzian/Gaussian function(Pseudo-
Voigt function) having the FWHM of weighted average of both peak heights.

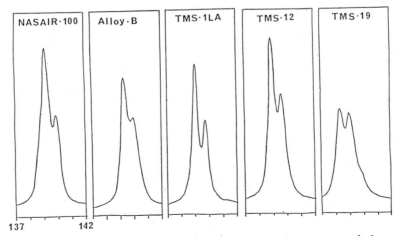

Fig. 2 Specimen powder (331) reflections recorded
with conventional plotter.

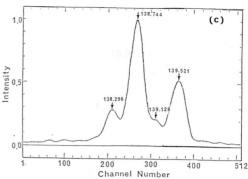

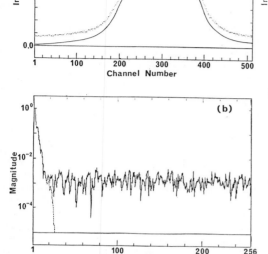

Fig. 3 Enlarged reflection peak
profile(331) of NASAIR-100,
(a) Experimental data in upper
part; curve after smoothing
experimental data by Fourier
analysis in lower part.
(b) Plot of experimental data
in frequency domain; smoothed
curve in Fig. 3(a) corresponds
to plot of inverse transform
after filtering off frequency
region higher than dotted line.
(c) Deconvoluted reflection
profile of NASAIR-100 with
instrumental window function
with indicated 2θ peak angles.

The X-ray intensity data of the specimens mentioned above were
collected on CXRFD in the 2θ range 137 to 142 deg. with CuK$_\alpha$ radiation.
Each specimen was packed into the cavity of a standard glass specimen
holder, then pressed, and the surface was flattened with a glass slide.
The step-scan technique was used at 0.008 deg. intervals in 2θ and with a
fixed time of 5 sec.   A divergence and receiving slit of 1 deg. were
used. The measured (331) reflected profiles are shown in Fig. 2.

All peak profiles shown in Fig. 2 were unresolved clusters of peaks.
It suggests that the true reflection peaks of the specimen were masked by
the instrumental broadening.   The enlarged (331) peak profile of NASAIR-
100 is shown in Fig.3(a).   The experimental points are represented •' in
the upper part; the solid line is the curve removed for statistical
fluctuation (errors) by the Hanning filter (cut off frequency = 19, 25) in
the frequency domain. The experimental data in the frequency domain is
presented in Fig. 3(b).

Fig. 3(b) shows the experimental data including statistical errors in
the high frequency region, and.then the statistical errors were removed by
a low pass filter named the Hanning filter (dotted line) in the frequency
domain.   The lower solid curve in Fig. 3(a) is represented by the smoothed
profile obtained by the inverse transformation to the time domain of the
filtered curve in the frequency domain.

The profile of NASAIR-100 deconvoluted with the instrumental window
function and smoothed with the Hanning filter is presented in Fig. 3(c).
In the peaks shown in Fig. 3(c), it is possible to identify two CuK$_\alpha$
doublets reflected from the γ and γ' phases.

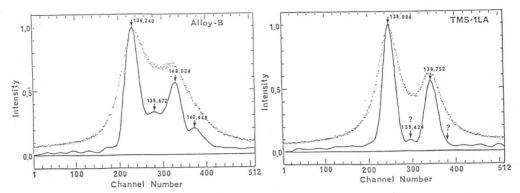

Fig. 4 Profiles obtained by deconvolution of experimental data with
        instrumental window function combine with hanning filter,
        together with indicate 2Θ peak angles,
           • : experimental data,   solid line : deconvoluted profile

A couple of peaks at 138.744 deg. and 129.527 deg. 2Θ angle were CuK$_{\alpha 1}$ and K$_{\alpha 2}$ lines reflected from the γ' phase, respectively. Another couple of peaks appeared at a lower angle than those peaks reflected from the γ phase. These were identified as CuK doublets reflected from the γ phase. The greatest advantage of this method is that no prior knowledge of the number of reflection peaks is needed.

The deconvoluted reflection profile of Alloy-B is also shown in Fig. 4(a). Contrary to NASAIR-100, the peaks reflected from the γ phase Alloy- B appeared at higher angles than that of the γ' phase. Therefore, the lattice misfits between the γ and γ' phases were calculated from the reflection peaks mentioned above.

The FFT algorithm[14) was used in the deconvolution process. Every peak position in the deconvoluted profiles was determined by the second order derivative of the deconvoluted profiles.

The other deconvoluted profiles were also easily identified; however it was difficult to identify the deconvoluted patterns of TMS-1LA shown in Fig. 4(b).

PROFILE FITTING AND 2Θ –MEASUREMENT WITH SRPXRD

The reflections from NBS 640b standard packed in the glass specimen holder were symmetrical and their shapes were presented by the pseudo-Voigt function as in the work by Parish, Hart and Huang[11).
The wave length of the monochromated X-ray source(1.5435A) was calculated from the reflections of the silicon standard mentioned above.

The step scan technique was used at 0.01 deg. intervals in 2Θ and fixed time of 10 sec to measure the reflection data of the specimens. The entrance and the receiving slits of 0.2mm were used. The reflection peaks of (111), (200), (220), (311) and (400) of the specimens were measured. The (311) reflection profile of NASAIR-100 measured with SRPXRD was compared with the same reflection profile measured with the CXRFD in Fig. 5.

The peaks reflected from γ and γ' phases overlapped though the angular resolution of SRPXRD was much higher than that of the CXRFD as shown in Fig. 5.   The (311) reflection peaks before and after deconvolution are shown in Fig. 6.   Fig.6 shows that the deconvolution method was not

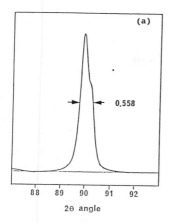 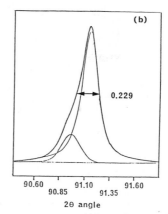

Fig. 5 Comparison of (311) reflection peak profiles measured by CXRFD
       and SRPXR (a) : CXRFD, (b) : SRPXRD

effective for separation of the overlapping peaks into individual peaks
because the instrumental broadening of the SRPXRD was much smaller than
that of the CXRFD.
      Therefore, the profile fitting method was used to resolve the
overlapping peaks into individual peaks.
      The resolved peaks of NASAIR-100 AND TMS-19 by the profile fitting
method are shown in Fig. 7.   Higher and lower peaks are reflected from γ'
and γ phases in both alloys.   The (311) peak profiles reflected from γ and
γ' phases in TMS-1LA measured with SRPXRD and then resolved into individual
peak profiles with the profile fitting method are shown in Fig. 8.
      Fig. 8 shows that the folded peaks measured with SRPXRD were resolved by
the profile fitting but the peaks(331) measured with CXRFD were impossible
to resolve into individual peaks (Fig. 4.).

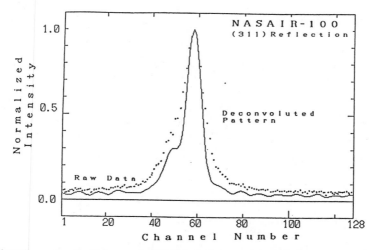

Fig. 6 (311) peak profiles of NASAIR-100 before and after deconvolution
       dotted line : experimentally obtained profile
       solid line : deconvoluted profile

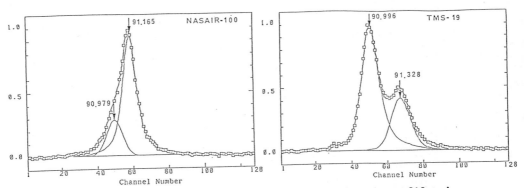

Fig. 7 Experimentally obtained and resolved peak profiles by
        profile fitting method.
        open rectangle : experimental data points,
        solid line : unfolded peaks

RESULTS AND DISCUSSION

The lattice parameters of both phases were not calculated but the
lattice misfits were calculated because the amount of each specimen used
for the measurement was under 0.3g and thus accurate alignment of the
specimens was very difficult.   In addition, all diffraction patterns were
measured without internal standards.
The plots of closed circles in Fig.9 show the relations between the
lattice misfits measured with SRPXRD and those calculated from the alloy
design system.   The plots of open circles also show the relation

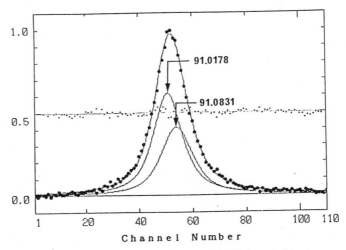

Fig. 8 (311) reflection peaks of TMS-1LA and their resolved
        peaks by profile fitting
        closed circle : experimental data point
        dot : 5 times amplified residuals
        thin solid line : unfolded peaks
        thick solid line : sum of unfolded peaks

between the calculated and the determined misfits by CXTFD with the deconvolution method. Fig. 9 shows that the values calculated by the alloy design program have some bias although the correlation coefficient between these measured and calculated misfit values was 0.993.

The agreement of the lattice misfits determined from those two different measurements was excellent with exception of misfits smaller than 0.06%. The accuracies of the misfits measured by CXRFD with the deconvolution method were a little poorer than those by SRPXRD with the profile fitting method, because the peak positions calculated by the second order derivative were affected by the nearest peaks.

The bias could be attributed to the elastic interaction between γ and γ' phases which usually decreased the magnitude of the lattice misfit[12]. On the contrary, the calculation model was based on the lattice parameter values assuming that γ and γ' phases were separated and unconstrained by each other. Thus, the calculation gives lattice misfit values larger in magnitude than in-situ measurement values.

The diffraction patterns of the γ' phase extracted from the γ-matrix were measured for NASAIR-100 and TMS-19 by SRPXRD to probe the magnitude of elastic coherency strains between both phases.

The (311) reflection peak profiles obtained from both the γ' phases electrochemically extracted from the sample rod and from the filings are shown in Fig. 10.

The peak position and shape reflected from the γ' phase extracted and from that in the γ-matrix were significantly different. The full width at half maximum (FWHM) of the peaks reflected from γ' extracted phases were less than 50% of those in γ-matrix. These peak position and shape changes suggest the magnitude of the elastic coherency strains between both phases.

The calculation of in-situ misfit values will be made possible in the next step by taking the interaction factor into account by using Poisson's ratio and shear modulus of the γ and γ' phases.

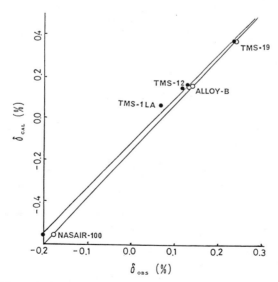

Fig. 9 Relations between measured and calculated lattice misfits experimental point determined by X-ray diffractometry open circle : CXRFD, closed circle : SRPXRD.

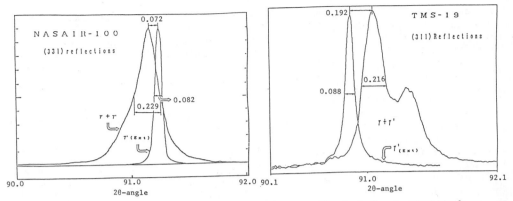

Fig. 10 Comparison of peak profiles reflected from extracted
γ phases and from specimens of filings
γ'(Ext) : extracted γ' phase,
γ + γ' : powder specimen

CONCLUSION

The lattice misfits between γ and γ' phases of the single   crystal
superalloys were accurately determined at room temperature from the filings
of the specimens by using the conventional X-ray tube focusing
diffractometry with deconvolution and the synchrotron-radiation parallel
beam diffractometry with profile fitting.   Both methods were powerful
tools for understanding the relations between the lattice misfit of γ-γ'
phases and the volume fractions, the compositions of γ' phase, the
magnitude of the elastic coherency strains between both phases, the
morphology of the γ' precipitate, etc. The elastic coherency strains
between both phases in the alloys were qualitatively observed.
The synchrotrn-radiation parallel-beam method had the following
advantages:
1) Small instrumental broadening(High angular resolution).
2) High precision 2θ determination by the simplified profile fitting.
3) Low detection limit of the lattice misfits between both phases.
The detection limit of the lattice misfits between γ and γ' pases in
Ni-base superalloys with this method was much smaller than that of the
conventional x-ray tube focusing diffractometry.

Acknowledgments

This work was performed as a part of "The advanced alloys with
controlled crystalline structures" in a project from "The basic technology
for future industries" sponsored by the Agency of Industrial Science and
Technology of MITI.
Authors would like to express their thanks to Professor. M. Ando of the
National Laboratory for High Energy Physics and to Professor. K. Uno of the
Nihon University   for their helpful cooperation.

References

1) H. Harada, M. Yamazaki: Tesu-to-Hagane, 65, (1979), 1059.
2) T. Yamagata, H. Harada, S. Nakazawa, M. Yamazaki and Y. G. Nakagawa:
Proceed. 5th Inter. Sympo. on Superalloys, (1984), 157.

3) H.Harada and H.Yamagata: Proceed. Fall Meet. JISI,(1986), 513.

4) P. Caron and T. Kahn : Mat. Sci. Eng., 61, (1983), 173.

5) K.Ohno and M.Yamazaki : Adv. X-ray Anal., 30, (1897), 67.

6) M.V.Nathal, R. A. Mcckay and R.G.Galick: Met. Sci. Eng., 75,(1985), 195.

7) R.A.Mackay and L.J.Ebert: Scropta Met., 17, (1983), 1217.

8) S. Ergun : J. Appl. Cryst., 1, (1968), 19.

9) W. Parrish, T.C. Huang and G.L.Ayers,
   Trns. Am. Cry tallogr. Assoc, 12, 55, (1976).

10) H.Berger : X-ray Spectrom., 15, (1986), 241.

11) W. Parrish, M. Hart, T.C. Huang and M. Bellotto, J. Appl. Cryst.
   19, 92, (1986).

12) D.A. Grose and G.S. Ansell, Metal. Trans. 12A, 1631,(1981).

13) K. Ohno, H. Harada, T. Yamagata and M. Yamazaki, Trans. ISIJ, 28,
   219,(1987).

14) J.W. Cooley and J.W.Tukey, Math. Comput. 19, 297,(1965).

STANDARD DEVIATIONS IN X-RAY STRESS AND ELASTIC CONSTANTS
DUE TO COUNTING STATISTICS

Masanori Kurita

Nagaoka University of Technology

Nagaoka, 940-21 Japan

ABSTRACT

X-ray diffraction can be used to nondestructively measure residual stress of polycrystalline materials. In x-ray stress measurement, it is important to determine a stress constant experimentally in order to measure the stress accurately. However, every value measured by x-ray diffraction has statistical errors arising from counting statistics. The equations for calculating the standard deviations of the stress constant and elastic constants measured by x-rays are derived analytically in order to ascertain the reproducibility of the measured values. These standard deviations represent the size of the variability caused by counting statistics, and can be calculated from a single set of measurements by using these equations. These equations can apply to any method for x-ray stress mesurement. The variances of the x-ray stress and elastic constants are expressed in terms of the linear combinations of the variances of the peak position. The confidence limits of these constants of a quenched and tempered steel specimen were determined by the Gaussian curve method. The 95% confidence limits of the stress constant were -314 ± 25 MPa/deg.

INTRODUCTION

X-ray diffraction can be used to nondestructively measure residual stress in a small area of a specimen surface. The stress measurement by x-ray diffraction is based on the continuum mechanics for macroscopically isotropic polycrystalline materials. In this method, the stress value is calculated selectively from strains of a particular diffraction plane in the grains which is favorably oriented for the diffraction. In general, however, the elastic constants of a single crystal depend on the plane of the lattice, that is, a single crystal is elastically anisotropic. Various analyses taking into account the influence of elastic anisotropy of constituent crystals in randomly oriented polycrystals have also been carried out[1-3]. In any case, it is desirable to determine a stress constant experimentally in order to measure the stress accurately. Additionally, every value measured by x-ray diffraction has variability arising from x-ray

377

counting statistics. Therefore, it is important to ascertain the reproducibility of the measured values[4].

The equations for calculating the standard deviations of the x-ray stress and elastic constants are derived which represent the size of the variation caused by counting statistics. The standard deviations can be calculated from a single set of measurements by using these equations. Although these equations can apply to any method for x-ray stress measurement, the Gaussian curve method was used in the present study because it can determine the stress precisely and rapidly[5-7].

PEAK POSITION OF DIFFRACTION LINE BY GAUSSIAN CURVE METHOD

To obtain net diffracted x-ray counts, observed diffracted counts above the background should be corrected for the LPA (Lorentz-polarization and absorption) factor. For practical purposes, however, omission of the background subtraction is desirable in order to reduce the measurement time. Fortunately, it has been proved that the background subtraction does not affect the stress value and can be omitted in x-ray stress measurement by the Gaussian curve method[5]. Therefore, observed counts y are corrected only for the LPA factor, that is,

$$z = ly \qquad (1)$$

where z is corrected counts and l is the reciprocal LPA factor given by

$$l_i = \frac{1 - \cos x_i}{(3 + \cos 2x_i)[1 - \tan\psi \cot(x_i/2)]}$$

where x_i is the diffraction angle (2θ), and ψ is an angle between the specimen and diffraction plane normals.

The peak position by the Gaussian curve method is defined as the main axis of the Gaussian function fitted to n data points (x_i, z_i) around the peak of the diffraction line by using a least squares analysis. Since the detailed derivation of the equation for calculating the peak position p is given in previous papers[5,6], only the final result is given here as

$$p = \frac{x_1 + x_n}{2} - d \frac{\Sigma t_i w_i}{\Sigma T_i w_i} \qquad (2)$$

where Σ denotes the summation from $i=1$ to n, and

$$d = 0.4c(n^2 - 4)$$
c = angular interval in degrees
$$t_i = i - (n + 1)/2 \quad (i = 1, 2, \ldots, n)$$
$$T_i = 12t_i^2 - n^2 + 1$$
$$w_i = \ln z_i \quad \text{(ln denotes a natural logarithm.)}$$

The following well known relationship holds between x-ray counts y and the variance of y[8,9];

$$\sigma_y^2 = y \qquad (3)$$

Applying Eq.(41)(Appendix) to Eq.(2) and using Eq.(3), we obtain the standard deviation, σ_p, of peak position p as[5]

$$\sigma_p = d\sqrt{\Sigma G_i^2/y_i}\Big/(\Sigma T_i \ln z_i)^2 \tag{4}$$

where

$$G_i = t_i \Sigma T_i \ln z_i - T_i \Sigma t_i \ln z_i$$

The standard deviation, σ_p, in Eq.(4) represents the size of the variation in the peak position caused by counting statistics.

X-RAY STRESS AND ELASTIC CONSTANTS

According to the theory of elasticity for isotropic materials, application of a uniaxial stress τ in the x direction produces a strain ε in the direction making an angle ψ to the specimen normal as shown in Fig.1;

$$\varepsilon = \frac{s_2}{2}(\tau + \sigma_x')\sin^2\psi + s_1(\tau + \sigma_x' + \sigma_y') \tag{5}$$

where

$$\left.\begin{array}{l} s_1 = -\nu/E \\ s_2 = 2(\nu + 1)/E \end{array}\right\} \tag{6}$$

E = Young's modulus
ν = Poisson's ratio
σ_x' and σ_y' = initial residual stresses in the x and y directions

The values s_1 and s_2 are elastic constants called compliances.

Various analyses taking into account the influence of elastic anisotropy of constituent crystals in randomly oriented polycrystals have given the same form of the equation, giving the different values of s_1 and s_2 in Eq.(5)[1-3]. In this sense, the compliances s_1 and $s_2/2$ in Eq.(5) are some-

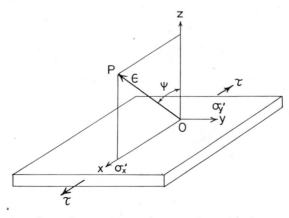

Fig.1   Directions of strain ε and applied stress τ.

times written as $(s_1)_{\text{x-ray}}$ and $(s_2/2)_{\text{x-ray}}$, and are called x-ray elastic constants; they should be determined experimentally in order to measure the stress accurately by x-rays. The compliances s_1 and s_2 depend on the measured lattice planes according to the Reuss model assuming a constant stress in all grains. On the contrary, the analysis based on the Voigt model assuming a constant strain in all grains gives the compliances independent of the lattice planes[1-3]. However, the values calculated from both models almost coincide for the (211) plane of alpha-iron. Various analyses and experimental observations have shown that the measured values are very close to the mean of the values calculated from the Reuss and Voigt models.

By using Bragg's law, the strain ε in Eq.(5) is transformed to the peak position p as

$$p = -\beta s_2(\tau + \sigma_x')\sin^2\psi - 2\beta s_1(\tau + \sigma_x' + \sigma_y') + 2\theta_0 \tag{7}$$

where θ_0 is a Bragg angle of a stress-free specimen, and β is a constant given by

$$\beta = (180\tan\theta_0)/\pi \tag{8}$$

Equation (7) shows that the peak position p varies linearly with $\sin^2\psi$ as shown in Fig.2. This is called a $\sin^2\psi$ diagram.

From Eq.(7), we get the slope M of the $\sin^2\psi$ diagram as

$$M = \frac{\partial p}{\partial \sin^2\psi} = -\beta s_2(\tau + \sigma_x') = (\tau + \sigma_x')/K \tag{9}$$

where K is a stress constant given by

$$K = -\frac{1}{\beta s_2} \tag{10}$$

Substitution of Eqs.(6) and (8) into Eq.(10) gives

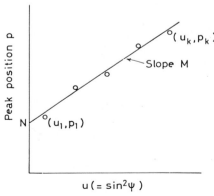

Fig.2 $\sin^2\psi$ diagram.

$$K = -\frac{\pi}{180} \frac{E \cot\theta_0}{2(\nu + 1)}$$

(11)

This is the stress constant based on the theory of elasticity for isotropic materials.

Equation (9) shows that the slope M of the $\sin^2\psi$ diagram varies linearly with the applied stress τ as shown in Fig.3, and that the slope B of the straight line in Fig.3 is given by

$$B = \frac{\partial M}{\partial \tau} = -\beta s_2 = \frac{1}{K}$$

(12)

Therefore, the compliance s_2 and stress constant K can be obtained by x-ray method from the slope B in Eq.(12).

From Eq.(7), the intercept N of the $\sin^2\psi$ diagram is given by

$$N = (p)_{\sin^2\psi=0} = -2\beta s_1(\tau + \sigma_x' + \sigma_y') + 2\theta_0$$

(13)

Equation (13) shows that the intercept N has a linear relationship with the applied stress τ as shown in Fig.4, and that the slope C of the straight line in Fig.4 is given by

$$C = \frac{\partial N}{\partial \tau} = -2\beta s_1$$

(14)

Thus, the compliance s_1 can be determined from the slope C in Fig.4.

STANDARD DEVIATIONS OF X-RAY STRESS AND ELASTIC CONSTANTS

Let

$$u_i = \sin^2\psi_i$$

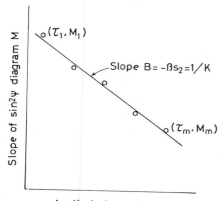

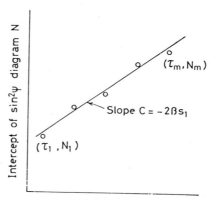

Fig.3 Linear relationship between slope M of $\sin^2\psi$ diagram and applied stress τ.

Fig.4 Linear relationship between intercept N of $\sin^2\psi$ diagram and applied stress τ.

The slope M of the straight line fitted to k data points (u_i, p_i) on the $\sin^2\psi$ diagram, as shown in Fig.2, by using the least squares method is given by

$$M = \frac{\Sigma'(ku_i - \Sigma'u_i)p_i}{k\Sigma'u_i^2 - (\Sigma'u_i)^2} = \Sigma'a_i p_i \tag{15}$$

where Σ' denotes the summation from $i=1$ to k, and

$$a_i = \frac{ku_i - \Sigma'u_i}{k\Sigma'u_i^2 - (\Sigma'u_i)^2} \tag{16}$$

From Eqs.(9) and (15), the stress S $(= \tau + \sigma_x')$ in the x direction in Fig.1 is given by

$$S = KM = K\Sigma'a_i p_i \tag{17}$$

The intercept N of the straight line in Fig.2 determined by the least squares analysis is given by

$$N = \bar{p} - M\bar{u} \tag{18}$$

where
$$\bar{p} = \Sigma'p_i/k, \quad \bar{u} = \Sigma'u_i/k \tag{19}$$

Substituting Eqs.(15) and (19) into Eq.(18), we get

$$N = [\Sigma'p_i - (\Sigma'a_i p_i)\Sigma'u_i]/k$$
$$= \Sigma'(1 - a_i\Sigma'u_j)p_i/k = \Sigma'c_i p_i \tag{20}$$

where
$$c_i = (1 - a_i\Sigma'u_j)/k \tag{21}$$

The slope B of the straight line fitted to m data points of (τ_i, M_i) by the least squares method, as shown in Fig.3, is given by

$$B = \Sigma''b_i M_i \tag{22}$$

where Σ'' denotes the summation from $i=1$ to m, and

$$b_i = \frac{m\tau_i - \Sigma''\tau_i}{m\Sigma''\tau_i^2 - (\Sigma''\tau_i)^2} \tag{23}$$

Substituting Eq.(15) into Eq.(22), we obtain

$$B = \Sigma''\Sigma'b_i a_j p_{ij} \tag{24}$$

where Σ'' and Σ' denote the summations from $i=1$ to m and $j=1$ to k, respectively, and p_{ij} denotes the jth peak position on the $\sin^2\psi$ diagram for an applied stress τ_i. Substitution of Eq.(24) into Eq.(12) gives

$$s_2 = -\Sigma''\Sigma'b_i a_j p_{ij}/\beta \tag{25}$$

$$K = \frac{1}{\Sigma''\Sigma' b_i a_j p_{ij}} \tag{26}$$

The slope C of the straight line fitted to m data points of (τ_i, N_i) by the least squares method, as shown in Fig.4, is given by

$$C = \Sigma'' b_i N_i \tag{27}$$

Substitution of Eq.(20) into Eq.(27) gives

$$C = \Sigma''\Sigma' b_i c_j p_{ij} \tag{28}$$

Substituting Eq.(28) into Eq.(14), we get

$$s_1 = -\frac{\Sigma''\Sigma' b_i c_j p_{ij}}{2\beta} \tag{29}$$

From Eq.(6), we get Young's modulus E and Poisson's ratio ν as

$$\nu = -2s_1/(2s_1 + s_2)$$
$$E = 2/(2s_1 + s_2) \tag{30}$$

Substituting Eqs.(25) and (29) into Eq.(30), we get

$$E = -\frac{2\beta}{\Sigma''\Sigma' B_{ij} p_{ij}} \tag{31}$$

$$\nu = -\frac{\Sigma''\Sigma' A_{ij} p_{ij}}{\Sigma''\Sigma' B_{ij} p_{ij}} \tag{32}$$

where

$$A_{ij} = b_i c_j$$
$$B_{ij} = b_i(a_j + c_j)$$

As can be seen from Eqs.(15),(17),(20),(25),(26),(29),(31), and (32), the slope M and intercept N of the $\sin^2\psi$ diagram, stress S, compliances s_1 and s_2, stress constant K, Young's modulus E and Poisson's ratio ν can be expressed as functions of k or km statistically independent random variables of peak positions. Therefore, applying Eq.(41) or (42) to these equations, we obtain the variances of these values as

$$\sigma_M^2 = \Sigma' a_i^2 \sigma_{p_i}^2 \tag{33}$$

$$\sigma_N^2 = \Sigma' c_i^2 \sigma_{p_i}^2 \tag{34}$$

$$\sigma_S^2 = K^2 \Sigma' a_i^2 \sigma_{p_i}^2 \tag{35}$$

$$\sigma_{s_1}^2 = \frac{\Sigma''\Sigma' b_i^2 c_j^2 \sigma_{p_{ij}}^2}{4\beta^2} \tag{36}$$

$$\sigma_{s_2}^2 = \Sigma''\Sigma' b_i^2 a_j^2 \sigma_{p_{ij}}^2 \Big/ \beta^2 \tag{37}$$

$$\sigma_K^2 = K^4 \Sigma''\Sigma' b_i^2 a_j^2 \sigma_{p_{ij}}^2 \tag{38}$$

$$\sigma_E^2 = \frac{4\beta^2}{(\Sigma''\Sigma' B_{ij}p_{ij})^4} \Sigma''\Sigma' B_{ij}^2 \sigma_{p_{ij}}^2 \tag{39}$$

$$\sigma_\nu^2 = \frac{\Sigma''\Sigma' [A_{ij}\Sigma''\Sigma' B_{ij}p_{ij} - B_{ij}\Sigma''\Sigma' A_{ij}p_{ij}]^2 \sigma_{p_{ij}}^2}{(\Sigma''\Sigma' B_{ij}p_{ij})^4} \tag{40}$$

where σ_p^2 is the variance of the peak position, and it is given by Eq.(4) in case of the Gaussian curve method. The standard deviations, i.e., square roots of variances, of these values can be calculated from a single set of measurements by using Eqs.(33) to (40). The variance of the peak position is inversely proportional to the x-ray counts, which is a product of x-ray intensity(cps) and preset time(s)[5]. The same conclusion holds for the variances given by Eqs.(33) to (40), because the variances given by these equations are linear combinations of the variances of the peak position. Equations (33) to (40) can apply to any method for x-ray stress measurement if we can obtain the variances σ_p^2 of the peak position.

The peak position p has a normal distribution because it is a function of normally distributed x-ray counts[9]. Therefore, every value that is expressed as a function of peak positions has a normal distribution. Thus, the 95% confidence limits of values measured by x-rays, such as x-ray stress and elastic constants, are given by

(measured value) ± 1.96σ

where σ is the standard deviation of measured value given by Eqs.(33) to (40).

TEST PROCEDURES

The confidence limits of x-ray stress and elastic constants for a steel specimen were determined by using Eqs.(33) to (40) in order to evaluate the reproducibility of the measured values. A specimen of a structural steel JIS type S45C (carbon content is 0.45%) was ground to the shape shown in Fig.5. It was quenched at 860°C and tempered at 400°C. The surface layer of the specimen was removed electrolytically by about 100 μm. The specimen was set in a loading device and a uniaxial stress was applied.

The stress in the specimen was measured by the Gaussian curve method with an automated x-ray stress analyzer developed in our laboratory[10]. At

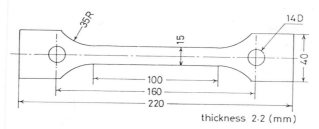

Fig.5 Specimen.

Table 1 Conditions of x-ray stress measurement.

Characteristic x-rays	Chromium Kα
Diffraction plane	(211) plane of ferritic iron
Filter	Vanadium foil
Divergence angle of slit	1°
Tube voltage and current	30 kV and 9 mA
Irradiated area	5 x 11 mm^2
Preset time	1 s
Angular interval	0.2°

the same time, the stress was measured with a strain gauge attached on the surface of the specimen. Table 1 shows the conditions for x-ray stress measurement. The number of data points for calculating a peak position was eight to ten.

TEST RESULT

Figure 6 shows the $\sin^2\psi$ diagram for the various applied stresses. The maximum and minimum 95% confidence intervals of the peak position, $\pm 1.96\sigma_p$ where σ_p is given by Eq.(4), are shown in Fig.6. Figure 6 shows that the peak position varies linearly with the $\sin^2\psi$ as Eq.(7) indicates.

Figure 7 shows a linear relationship between the applied stress τ and the slope M of the $\sin^2\psi$ diagram shown in Fig.6. In Fig.7, the 95% confidence intervals of M calculated from Eq.(33) are also ahown. Figure 8 shows a linear relationship between the applied stress τ and the intercept N of the $\sin^2\psi$ diagram shown in Fig.6. The 95% confidence intervals of N, $\pm 1.96\sigma_N$ calculated from Eq.(34), are also shown in Fig.8.

Table 2 shows the 95% confidence limits of x-ray stress and elastic constants which were calculated from Eqs.(36) to (40). The values of s_1 and $s_2/2$ calculated from the Voigt model are -12.3 and 56.2, and the values from the Reuss model are -13.0 and 58.3 $(10^4 GPa)^{-1}$, respectively[1]. These calculated values almost agree with the measured values shown in Table 2. Young's modulus E and Poisson's ratio ν in Table 2 are close to the mechanically determined values for steels. The stress constant K calculated from Eq.(11), which is based on the theory of elasticity for elastically isotropic materials, by using 206 GPa and 0.28 for E and ν is -293.4 MPa/deg. This value is also close to the stress constant measured by x-rays shown in Table 2.

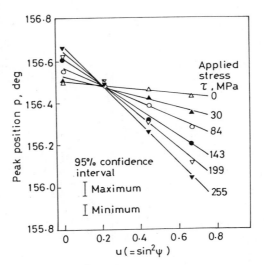

Fig.6   Sin²ψ diagram for various
applied stresses τ.

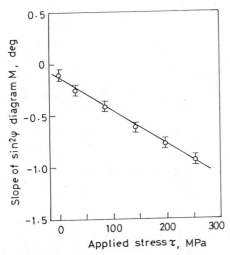

Fig.7   Change in slope M and its
95% confidence interval
with applied stress τ.

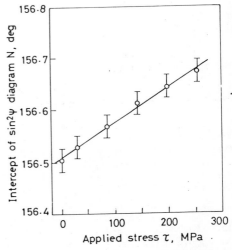

Fig.8   Change in intercept N and
its 95% confidence interval
with applied stress τ.

Table 2 95% confidence limits of x-ray elastic constants.

Compliances $(10^4\text{GPa})^{-1}$	s_1	-12.3 ± 1.9
	$s_2/2$	58.5 ± 4.7
Young's modulus E (GPa)		216 ± 16
Poisson's ratio ν		0.27 ± 0.035
Stress constant K (MPa/deg)		-314 ± 25

CONCLUSIONS

The equations for calculating the standard deviations of the x-ray stress and elastic constants are derived which represent the size of the variation due to counting statistics. The standard deviations of the slope M and intercept N of the $\sin^2\psi$ diagram, stress S, compliances s_1 and s_2, stress constant K, Young's modulus E and Poisson's ratio ν are given by Eqs.(33) to (40).

The confidence limits of x-ray stress and elastic constants of a quenched and tempered steel specimen were determined in order to evaluate the reproducibility of these values measured by x-rays. The Young's modulus E, Poisson's ratio ν, and the stress constant K which were determined by x-rays almost agreed with the mechanically determined values for steels.

APPENDIX[11]

If x_1, x_2, ..., x_n are statistically independent random variables having the variances σ_1^2, σ_2^2, ..., σ_n^2, respectively, the variance of the function

$$y = f(x_1, x_2, \ldots, x_n)$$

is given by

$$\sigma_y^2 = \Sigma(\frac{\partial f}{\partial x_i})^2 \sigma_i^2 \tag{41}$$

where Σ denotes the summation from $i=1$ to n.

Similarly, the variance of the function

$$y = \Sigma a_i x_i$$

is given by

$$\sigma_y^2 = \Sigma a_i^2 \sigma_i^2 \tag{42}$$

where a_1, a_2, ..., a_n are constants.

REFERENCES

1. Macherauch, E., X-Ray Stress Analysis, Experimental Mechanics, Vol.6 (1966), pp.140-153.
2. Macherauch, E. and Wolfstieg, U., Recent German Activities in the Field of X-Ray Stress Analysis, Materials Science and Engineering, Vol.30 (1977), pp.1-13.
3. Noyan, I. C. and Cohen, J. B., "Residual Stress Measurement by Diffraction and Interpretation," Springer-Verlag (1987), pp.62-74.
4. Kurita, M., Confidence Limits of Stress Values Measured by X-Ray Diffraction, Journal of Testing and Evaluation, Vol.11, No.2(1983), pp. 143-149.
5. Kurita, M., A Statistical Analysis of X-Ray Stress Measurement by the Gaussian Curve-Fitting Method, Journal of Testing and Evaluation, Vol.9, No.5(1981), pp.285-291.
6. Kurita, M., Residual Stress Measurement by X-Ray Diffraction with the Gaussian Curve Method and its Automation,"Role of Fracture Mechanics in Modern Technology, edited by Sih, G. C., Nishitani, H., and Ishihara, T.,"pp.863-874, Elsevier Science Publishers B.V.(1987).
7. Kurita, M., A New X-Ray Method for Measuring Residual Stress and Diffraction Line Broadness and its Automation, NDT International, Vol.20, No.5(1987), pp.277-284.
8. Klug, H. P. and Alexander, L. E.,"X-Ray Diffraction Procedures for Polycrystalline and Amorphous Materials,"John Wiley(1974), pp.360-364.
9. Kurita, M., Statistical Variation in Diffracted Intensity in Residual Stress Measurement by X-Rays, Transactions of the Japan Society of Mechanical Engineers, Vol.43, No.368(1977), pp.1358-1360 [in Japanese].
10. Kurita, M., Miyagawa, M., Sumiyoshi, M., and Sakiyama, K., JSME International Journal, Vol.30, No.260(1987), pp.248-254.
11. Bowker, A. H. and Lieberman, G. J., "Engineering Statistics," Prentice-Hall(1959), pp.62, 48, 49.

ELASTIC CONSTANTS OF ALLOYS MEASURED WITH NEUTRON DIFFRACTION

B. D. Butler[*], B. C. Murray, D. G. Reichel[**] and A. D. Krawitz

Dept. of Mechanical and Aerospace Engineering, University of Missouri, Columbia, MO 65211

[*]Present address: Dept. of Materials Science and Engineering, Northwestern University, Evanston, IL 60201

[**]Research Reactor Facility, University of Missouri, Columbia, MO 65211

ABSTRACT

Elastic constants as a function of crystallographic direction have been measured in polycrystalline alloy samples of 17-4PH stainless steel, Ni-Cr-Fe, and Ti-6%Al-4%V using a neutron diffraction technique. The results compare best with the constant stress model of Reuss. It is demonstrated that measurements of stress can be made sampling the bulk of the material using neutrons with an accuracy comparable to more conventional x-ray methods.

INTRODUCTION

Diffraction methods have been used for many years to determine the residual stress state in polycrystalline materials[1]. The basis of this method is well-established. The spacing of hkl planes, d_{hkl}, is measured as a function of orientation of the diffraction vector with respect to the surface normal of the sample and the variation in spacing of the planes is used to determine the strain tensor and, via the elastic constants, the stress tensor.

Elastic constants can be calculated using the theories of Reuss[2], Voigt[3], or Kroner[4] or they can be measured using diffraction by applying known loads to the material[1,5]. Reuss showed that if the stress is uniform across all grains the elastic constants equal the single crystal values. Voigt showed that if the strain is uniform in all grains, all crystallographic directions will have the same elastic constants, which can be determined from the compliance tensor of the single crystal. Kroner performed a more complete, self-consistent analysis in which both stress and strain are continuous across grain boundaries. The result is near the simple average of the Reuss and Voigt values, suggesting that the Reuss and Voigt models provide upper and lower limits between which actual values can be expected to lie.

Elastic constants measured by diffraction can differ substantially from isotropic or calculated values due to elastic anisotropy, plastic anisotropy and/or texture[6,7]. It is now appreciated that measured elastic constants are not related in a simple way to calculated values so that measurements on standard samples provide the most reliable values for use in converting measured strain values to stress. To date x-rays have been almost exclusively used to determine elastic constants in polycrystalline materials via diffraction methods. In the x-ray work, elastic constants have been determined using tensile and four-point bending devices and the d vs $\sin^2\psi$ method. In this method, the elastic constants $S_1 = -v/E$ and $S_2/2 = (1+v)/E$ are determined, where E is Young's modulus and v is Poisson's ratio. Preferred orientation can cause variation in the elastic constants with ψ (the angle between the sample normal and the diffraction vector) so that S_1 and S_2 are average values over the measured ψ range and the depth sampled, on the order of microns.

In this study the elastic constants E and v/E were determined for Ti-6%Al-4%V and Ni-Cr-Fe, and E for 17-4PH stainless steel using neutron diffraction. Because of the low absorption coefficient of neutrons in these materials, the entire 6.35 mm diameter of the specimens was sampled. The low absorption also enabled use of the entire ψ range, i.e., from 0 to 90°. Unlike the usual x-ray procedure, the d vs $\sin^2\psi$ technique was not used. Rather, the lattice spacing was measured at particular values of ψ while the load (stress) was increased on the sample through use of a tensile device. This method allows for the direct determination of E at $\psi = 90°$ and v/E at $\psi = 0°$ from plots of $\Delta2\theta$ vs σ (applied stress). The values obtained are compared to the Reuss and Voigt values calculated from the compliance tensors for pure Fe, Ni and Ti.

EXPERIMENTAL

All measurements were made using the 2XD powder diffractometer at the University of Missouri Research Reactor. It is equipped with a three-counter array of linear position sensitive detectors (PSD) that span a useable angular range of 25° at a distance of 1220 mm from the sample position. (Part of the measurements were made using only two of the PSDs due to electronic instability.) The data are rebinned into 0.1° increments from a basic channel width of 0.03° and the intensity of the detectors is added. The wavelength was about 0.129 nm. Further details of the PSD can be found in ref. 8.

A tensile device, designed specifically for this diffractometer, was used to apply load to the samples[9]. The device can apply 1300 MPa to a 6.35 mm diameter sample which has a gage length of 50 mm. Load is applied through a worm shaft and worm gear and is monitored through a load cell located in the load train. Each revolution of the worm gear elongates the sample 7.03×10^{-5} mm; the maximum achievable elongation is 10%. The applied strain was monitored with an electrical resistance strain gage located at the center of the gage length. The neutron beam was about 12.7 mm square located at the center of the gage length and intersecting the entire cross-section of the sample.

Bragg peak position was determined by fitting Gaussian or pseudo-Voigt functions to individual peaks. The typical statistical error in the peak position calculated by the fitting programs was about 0.003° 2θ. The estimated experimental error due to alignment and other factors is estimated at about 0.01° 2θ. Elastic constants were determined from the slopes of best-fit lines on plots of peak position vs stress.

17-4PH Stainless Steel

Measurements of peak position (°2θ) vs. applied stress (σ) were made from 0 to 690 MPa, in increments of 138 MPa, for the 211, 220 and 310 peaks. The diffraction vector was parallel to the tensile axis ($\psi = 90°$) so that the diffracting planes were perpendicular to the applied load. In this orientation Young's modulus, E_{hkl}, is determined directly from a plot σ vs. 2θ. The sensitivity of this measurement is about three times that for a measurement at $\psi = 0°$ since the direct normal strain is measured instead of the Poisson contraction. Load on the sample, sample strain, and the powder spectrum were recorded at each stress level. The peaks of interest span an angular range of about 25° 2θ. To avoid detector end effects, two arm positions were utilized; a counting time of about 30 min per arm position was used.

Ti-6 wt.% Al-4 wt.% V

This material is predominantly hexagonal (α), with a minor amount of the body-centered cubic (β) phase present. Measurements of σ vs. 2θ were made from 0 to 690 MPa in increments of 138 MPa for the 21·1, 11·4, 30·0, 21·2, 10·5, 21·3, and 30·2 peaks. At each stress level measurements were first made perpendicular to the tensile axis ($\psi = 0°$) and then the tensile device was moved to an orientation for which the diffraction vector was parallel to the tensile axis ($\psi = 90°$) and the measurement repeated. Two settings of the detector were required to cover the peaks of interest; counting time at each arm position was about 1.5 h.

Ni-Cr-Fe

Measurements of the Inconel X750 alloy were made from 0 to 690 MPa in increments of 69 MPa for the 311, 222 and 400 peaks. Measurements were made for the diffraction vector both parallel and perpendicular to the tensile axis. Only one setting of the detector arm was required.

RESULTS AND DISCUSSION

17-4PH Stainless Steel

The relative response of the three peaks is shown in Fig. 1 with the elastic constant $1/E_{hkl}$ plotted against the crystallographic orientation parameter $\Gamma = (h^2k^2 + k^2l^2 + h^2l^2)/(h^2 + k^2 + l^2)^2$. The values of the elastic constants are

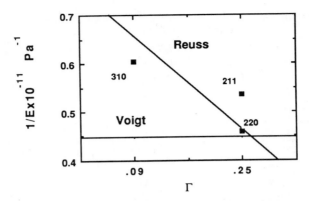

Fig. 1.   The elastic constant 1/E$_{hkl}$ vs the crystallographic orientation parameter
$\Gamma = (h^2k^2 + k^2l^2 + h^2l^2)/(h^2 + k^2 + l^2)^2$ for 17-4PH stainless steel.

presented in Table 1. The Reuss and Voigt limits for pure Fe are also shown.
The compliance constants S$_{11}$, S$_{12}$, and S$_{44}$ used to calculate the limits were
0.765, -0.282, and 0.878x10^{-11} Pa^{-1}, respectively. The 310 and 220 peaks fall
within the Reuss and Voigt limits and seem to follow the Reuss behavior closely.
The 211 falls above the Reuss limit, which may be due to the fact that it is at the
lowest Bragg angle and is thus the least sensitive of the three measurements.

Ti-6 wt. % Al-4 wt. % V

The relative response of the peaks for the $\psi = 90°$ and $\psi = 0°$ orientations
are presented in Figs. 2 and 3, respectively. Values for 1/E are shown in Table
I. Since hexagonal materials are isotropic in the basal plane, the data are

TABLE 1   Measured and calculated values of elastic constants for 17-4PH
stainless steel, Ti-6%Al-4%V, and Ni-Cr-Fe

Material	Peak	°2θ	1/E$_{exp}$ (x10^{-11} Pa^{-1})	1/E$_{Reuss}$ (x10^{-11} Pa^{-1})	1/E$_{Voigt}$ (x10^{-11} Pa^{-1})
17-4PH	211	66.5	0.535	0.461	0.448
	220	78.6	0.458	0.461	0.448
	310	90.2	0.604	0.656	0.448
Ti-Al-V	21·1	86.8	0.960	0.953	0.861
	11·4	89.8	0.874	0.825	0.861
	21·2	93.2	0.900	0.934	0.861
	10·5	94.6	0.744	0.741	0.861
	30·0	99.4	0.955	0.959	0.861
	21·3	104.3	0.910	0.908	0.861
	30·2	108.4	0.998	0.939	0.861
Ni-Cr-Fe	311	73.5	0.512	0.541	0.414
	222	77.4	0.395	0.330	0.414
	400	92.4	0.901	0.729	0.414

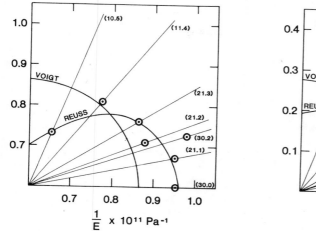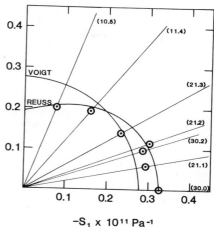

Fig. 2.   The elastic constant $1/E_{hkl}$ as a function of orientation with respect to the basal plane in the hexagonal phase of Ti-6 wt. % Al-4 wt. % V.

Fig. 3.   The elastic constant v/E_{hkl} as a function of orientation with respect to the basal plane in the hexagonal phase of Ti-6 wt. % Al-4 wt. % V.

presented in polar plots of the elastic constants v/E and $1/E$ vs. the angle to the basal plane. The stiffness constants C_{11}, C_{33}, C_{44}, C_{12}, and C_{13} used to calculate the Reuss and Voigt limits were 0.959, 0.699, 2.14, -0.462, and -0.190×10^{-11} Pa, respectively. A plot of σ vs. 2θ for both orientations is shown for the 21·1 peak in Fig. 4 to demonstrate the increased sensitivity of the $\psi = 90°$ orientation.

The data indicate substantial agreement with the Reuss model. This is especially apparent for the constant S_1 ($=v/E$). These data contrast with the rather more scattered data obtained using x-rays on the same material[10], which

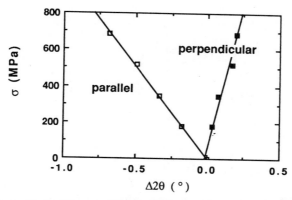

Fig. 4.   Peak shift vs applied stress for measurements parallel and perpendicular to the tensile axis for the hexagonal phase of Ti-6 wt. % Al-4 wt. % V.

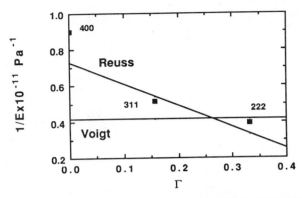

Fig. 5.   The elastic constant $1/E_{hkl}$ vs the crystallographic orientation parameter
$\Gamma = (h^2k^2 + k^2l^2 + h^2l^2)/(h^2 + k^2 + l^2)^2$ for Ni-Cr-Fe.

may indicate the difficulty of eliminating near surface effects due to sample preparation in the x-ray method and/or the advantage of bulk sampling of many grains afforded by neutrons.

Ni-Cr-Fe

Results for $1/E_{hkl}$ and ν vs. the crystallographic orientation parameter are shown in Figs. 5 and 6, respectively for the peaks measured. 1/E values are shown in Table I. The results for $1/E_{hkl}$ indicate a response that generally follows the behavior predicted by the Reuss model. The Poisson ratio data clearly does not follow the Voigt model and, in fact, two of the three points lie outside the bounds of the two models.

CONCLUSIONS

The utility of neutron diffraction in the measurement of diffraction elastic constants has been demonstrated. The results for the systems studied indicate

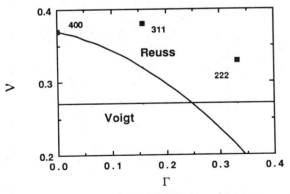

Fig. 6.   Poisson's ratio ν_{hkl} vs the crystallographic orientation parameter
$\Gamma = (h^2k^2 + k^2l^2 + h^2l^2)/(h^2 + k^2 + l^2)^2$ for Ni-Cr-Fe.

a rather clear tendency toward the behavior predicted by Reuss compared to that predicted by Voigt. A previous study that reported measurements with x-rays and neutrons on the same material[11] indicated that, within experimental error, the same results were obtained. Thus, the observed deviations from both Voigt and Reuss behavior appear to indicate that the elastic constants of the diffracting subset of planes in a polycrystal can deviate from the behavior predicted by models and should be obtained by direct diffraction experiments for use in conversion of strain data to residual stress.

ACKNOWLEDGEMENTS

The beam time for this study was provided by MURR and is gratefully acknowledged. The Ti-6 wt.% Al-4 wt.% V material was provided by Prof. G. Welsch, Case-Western University.

REFERENCES

1. I. C. Noyen and J. B. Cohen, "Residual Stress", Springer-Verlag, New York, 1987.

2. A. Reuss, Calculation of Flow Limits of Mixed Crystals on Basis of Plasticity of Single Crystals, Z. Agnew. Math. Mech., 9:49-58 (1929).

3. W. Voigt, "Lehrbuch der Kristallphysik", Teubner, Leipzig/Berlin, 1928.

4. E. Kroner, Berechnung der Elastischen Konstanten des Vielkristalls aus den Konstanten des Einkristalls, Z. Physik, 151:504-508 (1958).

5. A. D. Krawitz, P. J. Rudnik, B. D. Butler, and J. B. Cohen, Stress Measurements with a Position Sensitive Detector, Adv. X-Ray Anal., 29:163-171 (1986).

6. I. C. Noyen and L. T. Nguyen, Oscillations in Interplanar Spacing vs. $Sin^2\psi$: A FEM Analysis, Adv. in X-Ray Anal., 31:191-204, (1988).

7. I. C. Noyan, Determination of the Elastic Constants of Inhomogeneous Materials with X-Ray Diffraction, Mat. Sci. and Engrg., 75:95-103, (1985).

8. C. W. Tompson, D. F. R. Mildner, M. Mehregany, J. Sudol, R. Berliner and W. B. Yelon, A Position Sensitive Detector for Neutron Powder Diffraction, J. Appl. Cryst., 17:385-394 (1984).

9. B. D. Butler, In-Situ Stress Measurement by Neutron Diffraction, M.S. Thesis, University of Missouri-Columbia, 1985.

10. V. Bollenrath, W. Frohlich, V. Hauk, H. Sesemann, Rontgenographische Elastizitatskonstanten von Titan und TiAl6V4, Z. Metallkd., 62:790-794 (1972).

11. P. J. Rudnik, A. D. Krawitz, D. G. Reichel and J. B. Cohen, A Comparison of Diffraction Elastic Constants of Steel Measured with X-Rays and Neutrons, Adv. in X-Ray Anal., 31:245-204254, (1988).

STRESS MEASUREMENTS WITH A TWO-DIMENSIONAL REAL-TIME SYSTEM

G.M. Borgonovi
Science Applications International Corporation
San Diego, CA

and

C.P. Gazzara
U.S. Army Materials Technology Laboratory
Watertown, MA

INTRODUCTION

Conventional methods of determination of residual stress in polycrystalline samples use either diffractometers[1] or one-dimensional position-sensitive detectors[2,3]. The most commonly used technique, the so-called "$\sin^2 \psi$" method, requires several measurements at different angular positions of the sample. With diffractometers, two rotations are required, while with one-dimensional detectors, one rotation is required (except for the so-called single exposure technique, which requires two one-dimensional position-sensitive detectors). Rotation can be a potential source of errors if the sample is not aligned very carefully.

The conventional methods collect only a small part of the diffracted radiation from the sample. Thus, if a sample has large grains and therefore produces a spotty ring, use of a one-dimensional system may result in little diffracted intensity being collected, and therefore in large errors.

In both the one-dimensional and two-dimensional approach, a monochromatic beam of soft X-rays is diffracted at large angles. In the one-dimensional approach, the shift of the peak (or peaks) resulting from the intersection of the diffraction cone with one (or two) one-dimensional detector(s) is measured. In the two-dimensional approach, the shift of the entire ring (or a large portion of it) is measured.

There are several potential advantages of the two-dimensional method, namely:

- The method can simultaneously determine the two components of stress, σ_1 and σ_2, and their orientation in the sample plane.

397

- The method can be used successfully even with materials having relatively large grains.

- A system based on the method does not need any rotation.

- A variation of the distance of the sample from the detector will cause a recognizable variation in the diffraction ring. This variation can, therefore, be corrected for and discriminated from the variation due to stress.

For the method to work, it is necessary to be able to observe a large portion of the diffraction ring. In the system to be described, a crystal monochromator has been used, which has permitted the collection of up to 95 percent of the diffraction ring. The theory of the method has been described previously[4]; therefore, only a limited summary of it is presented in the next section, with emphasis on the procedure used for determination of the stress parameters.

THEORY AND COORDINATE SYSTEMS

The theory behind the two-dimensional stress analyzer consists of two parts. The first part consists of a relationship which connects the shift at each point of the diffraction ring to the stress components. The second part consists of procedures which permit one to derive the stress components from the observed shifts.

The relationship is best described in terms of a laboratory system of coordinates, defined by the detector and by the incident beam, and a sample system of coordinates. Figure 1 shows the relative orientation of the two systems. OXYZ is the laboratory system, with the OZ axis oriented in the opposite direction as the incident beam. The origin O is centered at the point of intersection of the incident beam with the sample surface. The sample system is defined by the three orthogonal vectors σ_1, σ_2, σ_3, with σ_3 normal to the surface of the sample, and the other two vectors parallel to the principal stresses in the surface of the sample. The stress component in the direction of σ_3 at the surface of the sample is equal to zero. Thus, the state of stress at the surface of the sample, at point O, is completely described by σ_1, σ_2 and the angle χ that the direction of σ_1 forms with a reference direction in the plane of the sample.

The incident X-ray beam is diffracted into a cone, which intersects the detector, forming a ring. A generic point on the ring corresponds to a diffracted beam identified by the polar coordinates α, β with respect to the laboratory system. Once the relative position of detector and incident beam is known, it is easy to derive the angular variables α, β from the coordinates of the spot on the detector.

The normals to the crystal planes responsible for the diffraction ring also describe a cone. A generic normal is also characterized by the azimuthal angle β. In the sample system of coordinates, however, the same normal is characterized by the azimuthal angle ϕ.

Because of the stress field, the diffracted beam at β undergoes a shift $\Delta\alpha$ given by:

$$\Delta \alpha = \frac{2}{E \tan \frac{\alpha_0}{2}} \left\{ (1+\nu) [\sigma_1 \cos^2 \phi + \sigma_2 \sin^2 \phi] \sin^2 \psi - \nu (\sigma_1 + \sigma_2) \right\} \qquad (1)$$

In Equation (1), E is the Young's modulus of the material, ν is Poisson's ratio, and α_0 is the value of α for no stress ($\alpha_0 = 180° - 2\theta_B$, where θ_B is the Bragg angle). The angle ψ between the normal to the crystal plane and the normal to the sample surface is given by:

$$\cos \psi = \sin \frac{\alpha_0}{2} \sin \epsilon (\cos \beta \cos \gamma + \sin \beta \sin \gamma) + \cos \frac{\alpha_0}{2} \cos \epsilon \qquad (2)$$

where ϵ and γ are the polar coordinates, in the laboratory system, of the normal to the sample surface at point O.

The angle ϕ can be calculated, as a function of β, from the relative orientation of the two systems of coordinates. The orientation of the sample system with respect to the laboratory system is completely defined by the three angles ϵ, γ, and χ.

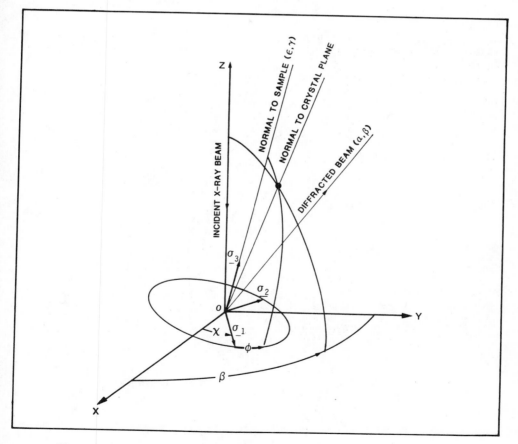

Fig. 1. Relative Orientation of Laboratory and Sample Systems.

Because of Equations (1), (2) and the geometric relationship between ϕ and β, the shift $\Delta\alpha$ is a function of β and of the parameters σ_1, σ_2, and X.

The analysis of the data has the purpose of determining the values of the parameters σ_1, σ_2, and X. In principle, this could be done by a least squares analysis over all the pixels in a region of interest surrounding the diffraction ring. Since this approach would be time-consuming, it is better to perform a sector analysis, as illustrated in Figure 2. The pixels in each sector are grouped so as to form a one-dimensional distribution (peak), which can be corrected if the response of the imager is not constant for all points in the image. After correction, the center of the peak is found with a parabolic fit to the central part of the distribution. In this way, the sector is replaced by a point, and the least squares analysis which provides the parameters is carried out on a relatively small number of points, rather than on a large number of pixels.

The parameters are obtained by satisfying the condition

$$\sum_{\beta}\left[\Delta\alpha\left(\beta,\sigma_1,\sigma_2,x\right)-\Delta\alpha_{exp}(\beta)\right]^2 w\left(\beta\right) = \text{minimum} \qquad (3)$$

where $\Delta\alpha_{exp}(\beta)$ is the experimental value of the shift found from the sector analysis and $W(\beta)$ is a weighting factor related to the intensity of the peak.

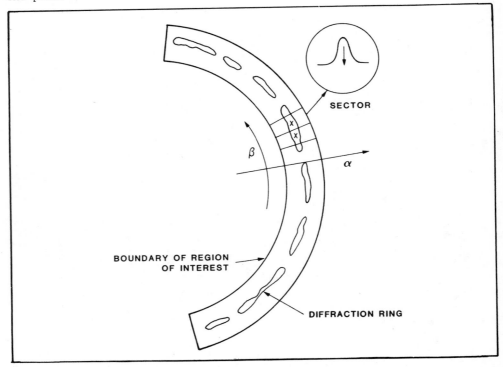

Fig. 2.   Sector Analysis: Average over β and Peak Determination Along α.

The function $\Delta\alpha(\beta)$ is linear in σ_1 and σ_2 but the dependence on the angle χ is highly nonlinear. The estimates of the parameters can be obtained by linearizing $\Delta\alpha$ as a function of χ and using an iterative process to simultaneously determine the three parameters. Another alternative is to try different values of χ, solving by least squares on σ_1 and σ_2, and then accept the triplet with minimum residual sum. Both approaches have been implemented in the two-dimensional stress analyzer software.

DESCRIPTION OF THE EXPERIMENTAL SYSTEM

The prototype two-dimensional stress analyzer consists of a data acquisition subsystem and a data storage and analysis subsystem. The data acquisition subsystem includes the detector, with its auxiliary equipment, the X-ray source, with its auxiliary equipment, and the mechanical components of the stress analyzer. The data storage and analysis subsystem includes the position analyzer, the computer with its interfaces, and the video monitor. The X-ray tube is a Kevex model K3052S, water-cooled, with 30 kV maximum high voltage, 6.7 mA maximum current, and chromium target. The tube was chosen because ot its small size, which minimized interference with the cone of diffracted radiation. Subsequently, a mosaic crystal monochromator was used, thus allowing collection of up to 95% of the diffraction ring. A small single crystal of iron, in the shape of a disc, with faces cut parallel to the 100 plane, has been used to reflect the K_α line of chromium.

The detector consists of an electro-optical system. The X-ray photons scattered from the sample impinge on a scintillator, which is Gadolinium Oxysulfide deposited on a disc of optical glass. The scintillator is covered by an opaque shield that prevents entrance of light. A fraction of the light photons emitted by the scintillator are imaged by a lens (Nikon 35 mm, f:1.4) on the photocathode of a Photon Image Analysis System (PIAS) built by Hamamatsu Corporation. The PIAS is a very sensitive, low noise detector that is capable of detecting and determining the coordinates of individual light photon events. The electrons emitted at the photocathode are multiplied through three microchannel plates, and collected on a position-sensitive detector. The total gain is 10^9, which allows detection of individual photons. The imaging tube is cooled thermoelectrically, and a water chiller is used to remove the heat from the thermoelectric cooler.

The individual photon events, after multiplication through the microchannel plates, produce charge pulses which are collected at the four corners of the position-sensitive detector plate. These pulses are analyzed by the position analyzer, and provide the coordinates x and y of the event. The coordinates are digitized and stored in a 16-bit 512 x 512 acquisition memory which is interfaced to a DEC 11-73 computer. Thus, the image of the diffraction ring builds gradually in the acquisition memory as a sum of individual events. Because the detector is cooled, the thermal noise at the photocathode is extremely low, about 5-6 counts per second over the entire image.

An 8-bit video memory card has also been inserted on the computer bus. The video memory is used to display the image of the acquired diffraction pattern on a color monitor, and also, because of its fast access time, to retrieve the intensity values at each pixel during the analysis.

Figure 3 shows a view of the acquisition subsystem of the two-dimensional stress analyzer.

DATA ANALYSIS

Software for the two-dimensional stress analyzer consists of a system of interactive computer codes which control acquisition, analysis, and processing of diffraction patterns for the purpose of stress calculation. A total of six programs is utilized as follows:

XRDSIM   Used to simulate diffraction patterns in presence of stress. This program was used during development, and is useful to become familiar with the system.

XRDACQ   Used to control acquisition of data and storage of images.

XRDSCT   Used to determine the centerline of a diffraction ring by a sector analysis. The number and width of the sectors can be controlled. This is mainly used for calibration.

XRDGEO   Used to determine, by a least squares method, the geometry parameters; that is, the parameters defining the relative position of the incident beam, detector, and sample.

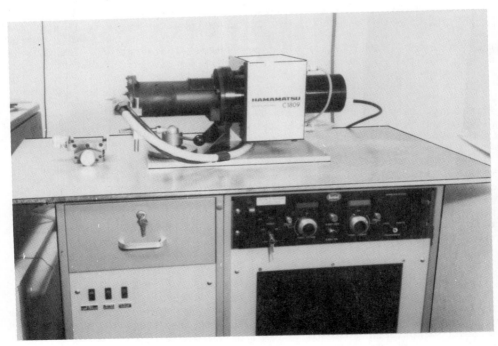

Fig. 3. View of Data Acquisition Subsystem.

XRDSTS   Used to determine the stress components σ_1 and σ_2 and the direction that σ_1 forms with a reference direction in the plane of the sample.

XRDPRC   Used to perform general image processing, visualization, and measurement functions on the diffraction images.

The codes run interactively on a DEC 11/73 minicomputer using the RT-11 operating system.

EXPERIMENTAL RESULTS

The two-dimensional stress analyzer has been tested on unidirectional stress fields created through both tensile and bending tests.

Figure 4 shows an image of the 211 diffraction ring from a sample of ferritic steel. A tensile specimen of ferritic steel, whose thinnest section was 0.0375 mm by 6.25 mm was gradually tensioned by attaching variable loads at one end. The normal to the surface of the sample formed

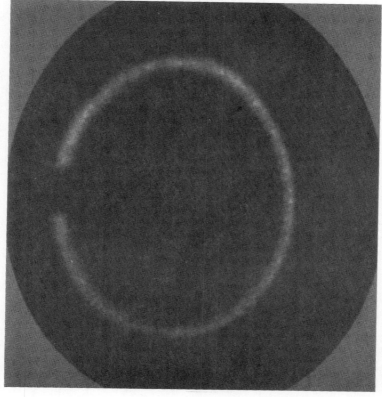

Fig. 4. Diffraction Ring Image Using the Crystal Monochromator.

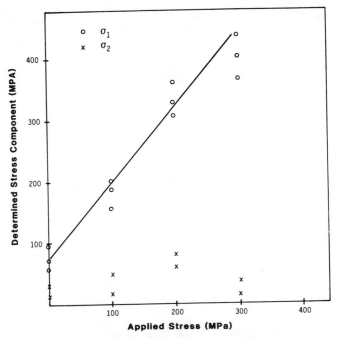

Fig. 5. Results of Tensile Test Stress Measurements in Ferritic Steel.

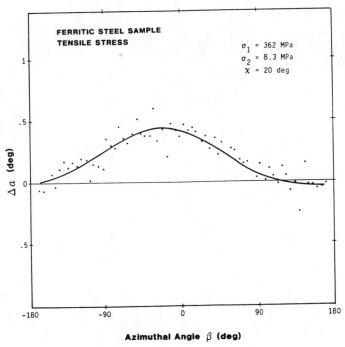

Fig. 6. Comparison of Theoretical and Measured Shifts Along the Ring.

a 40 degree angle to the direction of the incident X-ray beam.   The
measured stress components are shown in Figure 5.   The reproducibility of
the stress measurements is about 40-55 MPa for ferritic steel at a stress
level of 400 MPa.

        The results in Figure 5 show that the σ_1 component, which is
parallel to the direction of applied stress, increases with the applied
load, while the σ_2 component remains approximately constant.   Figure 6
shows an example of measured shifts for the ferritic steel sample, as a
function of azimuth, and the fit which resulted in the estimate of the
value and orientation of the stress components.

        Experiments were also performed by bending samples in form of thin
strips and determining the stress components at the point of maximum
bending.   Figure 7 shows the determined components of stress for different
deflections from analysis of the 222 diffraction ring of aluminum.   The
sample was a strip of 6061-T6 aluminum, 100 mm long, 22.8 mm wide, and 3.2
mm thick.   This sample had a coarser grain size than the ferritic steel
sample, and the diffraction ring had a spotty appearance.   The determined
stress components, however, agree with the unidirectional stress field
applied.   Reproducibility in aluminum is of the order of 15-20 MPa at a
stress level of 175 MPa.

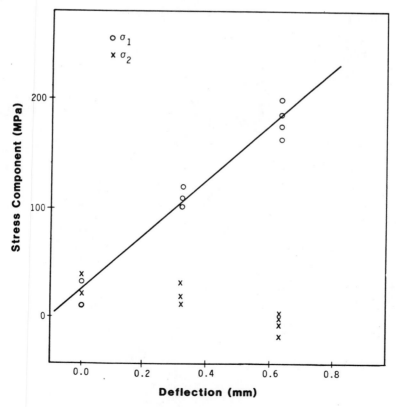

Fig. 7. Results of Bending Test Stress Measurements in Aluminum.

CONCLUSIONS

A prototype two-dimensional residual stress analyzer system has been designed, constructed, and tested. The system has been used both with and without a crystal monochromator.

Analysis of measurements carried out in samples of ferritic steel and aluminum has shown that the two-dimensional approach for determining residual stress is a viable one. The precision of the system in measuring the stress components is presently estimated to be about 15-20 MPa for aluminum and about 40-55 MPa for ferritic steel.

The possibility of simultaneous determination of both stress components in the plane of the sample has been demonstrated using samples of ferritic steel and both tensile and bending tests. Also, the possibility of determining stress from the analysis of the diffraction ring from materials having large grain size has been demonstrated using an aluminum sample.

The detector system used has proven to have sufficient spatial resolution and stability, and exceptionally low noise. It is, however, limited in sensitivity, due to the lens coupling of the scintillator to the photocathode and to an inherent count rate limitation.

ACKNOWLEDGEMENTS

This work has been supported by the U.S. Army Materials Technology Laboratory.

REFERENCES

1. P. S. Prevey, "A Comparison of X-Ray Diffraction Residual Stress Measurement Methods on Machined Surfaces", Advances in X-Ray Analysis, 19, 709 (1976).

2. M. Jones and J. B. Cohen, "PARS - A Portable X-Ray Analyzer for Residual Stresses", Journal of Testing and Evaluation, 6, 91 (1978).

3. C. O. Ruud and C. S. Barrett, "Use of Cr K-Beta X-Rays and a Position Sensitive Detector for Residual Stress Measurement in Stainless Steel Pipe", Advances in X-Ray Analysis, 22, 247 (1979).

4. G. M. Borgonovi, "Determination of Residual Stress from Two-Dimensional Diffraction Patterns", Nondestructive Methods for Material Property Determination, Plenum Publishing Corporation, 47 (1984).

APPLICATION OF A NEW SOLID STATE X-RAY CAMERA TO STRESS MEASUREMENT

M.A. Korhonen, V.K. Lindroos and L.S. Suominen[*]

Helsinki University of Technology,
Laboratory of Materials Science, 02150 Espoo, Finland

[*]Mexpert Instrument Technology Ltd.,
Otaniemi Science Park, 02150 Espoo, Finland

INTRODUCTION

X-ray camera methods of stress measurement are inherently flexible and easy to apply in different situations because of the low weight, portability and maneuverability of the equipment. However, the digital intensity recording with the resulting objectivity and better statistical accuracy has given the diffractometer methods a distinct advantage over the conventional camera methods[1].

In the present contribution a versatile and lightweight solid state X-ray camera system, MEXSTRESS[**], is applied to stress measurement. This new camera unifies the desirable features of the traditional camera and diffractometer techniques into one single equipment. Moreover, the equipment is well adapted to an improved exposure geometry to be dealt with. As an application example of the new camera system, calibration measurements on a quenched and tempered steel sample are described.

SOLID STATE X-RAY CAMERA

In a solid state camera the film has been replaced by photosensitive microchips, as linear or matrix charge-coupled devices (CCD's) and photodiode arrays. Significantly, our new camera is based on direct detection of X-rays, without applying any fluorescent coating on the surface of the chip. An experiment with a linear CCD-element has been described previously[2]. The present camera system applies linear photodiode arrays of 512 pixels/0.5", which are characterized by an exceptionally large pixel width. Quantum efficiencies of photodiode arrays for X-rays peak at $\geq 80\%$, at about 5 keV, and decrease very steeply on the high energy side, and more moderately on the low energy side[3,4]. Thus, these chips effectively filter the white X-ray spectrum, and no external filtering is necessary.

_____

[**]MEXSTRESS is a trademark of Mexpert Instrument Technology Ltd.

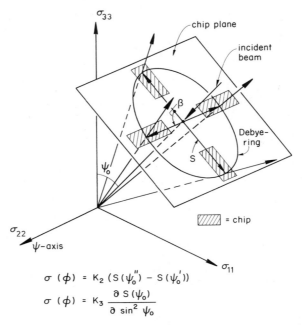

$$\sigma\,(\phi) \;=\; K_2\,(S(\psi_o'') - S(\psi_o'))$$

$$\sigma\,(\phi) \;=\; K_3\,\frac{\partial\,S(\psi_o)}{\partial\,\sin^2\psi_o}$$

Fig. 1.   Recording of diffracted rays in the one ($\beta=0$)
           and improved two ($\beta=\pi/2$) exposure methods

        Two photodiode arrays, positioned symmetrically around the collimator in
analogy to the conventional back-reflection flat film technique[5], Fig. 1, are
driven by a custom made controller/data acquisition board, interfaced through
RS 232 serial port to a PC-AT type microcomputer. This arrangement resulted
in a great flexibility in displaying and analysing the data. As depicted in
Fig. 1, the photodiode arrays can be positioned to record either the vertical
rays, $\beta = 0$ and π, or the axial rays, $\beta = +\pi/2$ and $-\pi/2$. A general layout of
the camera system is shown in Fig. 2: in order to facilitate measurement,
stepper motor-driven settings for the Ψ_o-inclination, see Fig. 1, and for the
specimen to photodiode-board distance are provided.

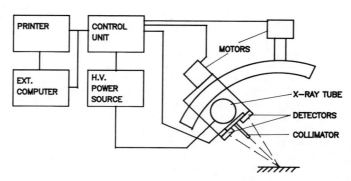

Fig. 2.   General layout of the MEXSTRESS X-ray
           camera system

The inclusion of refrigeration into the present solid state camera would have made it more bulky and vulnerable in field work. To reduce the effect of the resulting high black current, the background obtained in otherwise identical conditions but with the X-ray source off, is subtracted from the result of an actual measurement. Moreover, in order to increase the dynamic range, the total exposure is composed piecemeal of a large number of short exposures. This procedure has been found to lead to a net intensity to background ratio well sufficient for the present purposes.

IMPROVED CAMERA TECHNIQUE

As detailed elsewhere[6] the information on stresses and strains provided by a single Debye-ring, recorded at an inclination Ψ_o, Fig. 1, can advantageously be analysed as

$$\varepsilon(\Psi_o,\beta) = \frac{1+\nu}{E} (\sigma_{11}A^2+\sigma_{22}B^2+2\sigma_{12}AB) - \frac{\nu}{E} (\sigma_{11}+\sigma_{22}) \tag{1}$$

where σ_{ij} are the stress components of the assumed isotropic two-axial state of stress, ν and E the Poisson's ratio and the Young's modulus, respectively, and A and B the angular factors defined by

$$A = \sin\Psi_o \sin\Theta - \cos\Psi_o \cos\Theta \cos\beta \quad \text{and} \quad B = \cos\Theta \sin\beta. \tag{2}$$

Here Θ is the Bragg angle, Ψ_o is the angle between the primary beam and the surface normal, and β measures the rotation around the primary beam, Fig. 1. It can be readily shown from eq. (1) that the four strains at $\beta = 0$, $\pi/2$, $-\pi/2$ and π provided by a single Debye-ring, are sufficient, in principle, to yield the the whole two-axial state of stress in addition to the stress free lattice spacing. However, at $\Psi_o = 45^o$, for example, the inaccuracy in the transverse stress σ_{22} is found to be about six times to that in σ_{11}. Consequently, the determination of the entire state of stress from a single exposure appears useful only for cases where direct measurements in the transverse direction are not possible.

The strains at $\beta = 0$ and π yield the stress component σ_{11} according to the well-known one exposure method[1,5] as

$$\varepsilon(\Psi_o,\pi)-\varepsilon(\Psi_o,0) = \frac{1+\nu}{E} \sigma_{11} \sin2\Psi_o \sin2\Theta. \tag{3}$$

The one exposure method, however, is known to suffer from poor focusing, Fig. 3a, in addition to the need for absorption correction. These difficulties will be surmounted by evaluating only the horizontal diameters of Debye-rings at $\beta = +\pi/2,-\pi/2$. The resulting average strains $\varepsilon(\Psi_o,\pi/2)$ yield now the stress component σ_{11} according to

$$\frac{\partial\ \varepsilon(\Psi_o,\pi/2)}{\partial\ \sin^2\Psi_o} = \frac{(1+\nu)}{E} \sigma_{11} \sin^2\Theta. \tag{4}$$

This improved exposing geometry offers ideal focusing, Fig. 3b, in analogy to the Schultz technique used in texture studies[1] or the side inclination method of diffractometer studies of stress[7]. Moreover, no absorption corrections are needed, and further, many additional errors are reduced when applying this improved camera technique either in two or multiexposure modes[6,8].

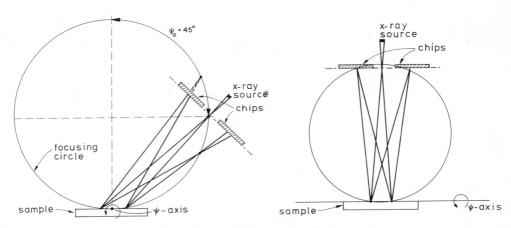

Fig. 3.   Focusing in the one (a) and improved
          two (b) exposure methods

EXPERIMENTAL

    The new camera system was tested by using a miniature flat tensile test
bar machined of quenched and tempered Cr-steel. The test bar was elongated in
a miniature frame, and the mechanical stress readings were taken by strain
gages glued to it. In the measurement Cr-Kα radiation at 30 kV and 7 mA from
a point source was used, and the specimen to chip board distance D was 40 mm.
The same mechanical stress levels of the test bar were recorded by using both
the one exposure method, eq. (3) and the improved two exposure method, eq.
(4). No absorption corrections were applied. The connection between the
strains appearing in eqs. (3) and (4) and the recorded line shifts ΔS is
given by Cullity[5], page 439, as

$$\Delta S = 2D \ \sec^2 2\Theta \ \tan\Theta \ \varepsilon.$$ (5)

    Exposure times of 40 s per a recorded X-ray line, Fig. 4, were found to
yield a fair statistical accuracy of the line position as determined from
repeated measurements, which most usually varied in the linear scale within 1

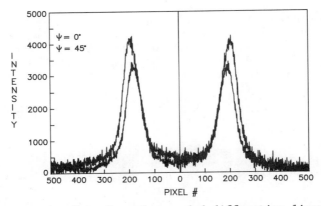

Fig. 4.   Examples of recorded diffraction lines

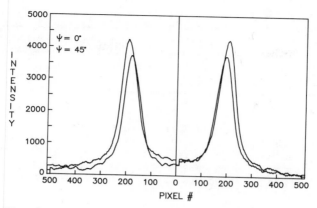

Fig. 5.   Examples of recorded diffraction lines
          after smoothing

pixel. However, very reasonable line shapes could be obtained already in 2
seconds of exposure. Accordingly, when using the one exposure technique,
tentative stress readings can be arrived in a couple of seconds by the new
solid state camera system.

The peak shifts were evaluated by two methods. Firstly, the conventional
parabola fit[1] was applied through the pixels above 85% of the maximum intensity. The fit was first made without smoothing the data, as shown in Fig. 4,
and then after an application of a running cubic polynomial fit through 25
pixels, as shown in Fig. 5. Secondly, the recently proposed cross correlation
method[9] was applied to the unsmoothed data, resulting in sharply defined peak
shifts, Fig. 6. Mechanical values of ν and E were used in converting the
measured strains to stresses.

RESULTS AND DISCUSSION

The results of the calibration measurements in the form of the observed
standard deviation of the X-ray stress are given in Table 1. Moreover, Fig. 7

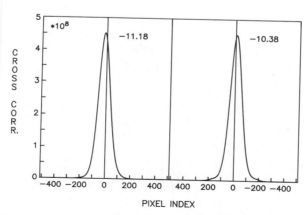

Fig. 6.   Cross correlation of the data in Fig. 4.

Table 1.   Estimated standard deviations of the X-ray stress in MPa

	parabola fit	parabola fit+smoothing	cross correlation
1 exp.	60.8	12.6	8.6
2 exp.	34.6	10.5	10.0

gives an example of the measured X-ray stresses against the imposed mechanical ones in case the parabola fit is applied both to the original and to the smoothed data, when using the improved two exposure technique; same kind of results were obtained also in case of the one exposure technique. As seen, without smoothing the X-ray stress readings fluctuate to a large degree. The most probable cause is the rather ill-defined peak intensity, used to select the points to be included in the parabola fit: intensity > 85% peak intensity. Because the lines are slightly asymmetrical, the cut-off intensity level, evidently, affects the apparent peak positions.

Table 1 shows that the best accuracies, about 10 MPa, were obtained in cases of (i) smoothing and parabola fit and (ii) direct cross correlation of data. Further, in both cases the one and the two exposure techniques resulted in about the same accuracies. This may seem somewhat surprising, because the improved two exposure method can be expected to yield a better accuracy owing to the better focusing and absorption conditions and slightly better stress factor. Evidently, in the present case, the X-ray lines still are so narrow and well defined that no essential differences arise. The small edge of the two exposure method may be lost because of the variation in the distance D between different exposures. However, the present accuracy of the distance setting, about .02 mm, warrants that for broad and ill-defined diffraction lines the improved two exposure method will retain its accuracy to a much larger degree than the conventional camera methods[6,8]. Table 1 also highlights the fact that the cross correlation method is "self smoothing", i.e. it tends to suppress the random fluctuations in the raw data. In case of grossly asymmetrical diffraction lines, or lines otherwise very different from a parabola shape, the cross correlation is expected to be superior to the parabola fit, even if applied to smoothed data.

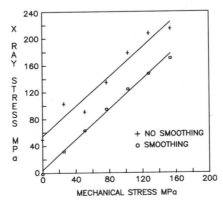

Fig. 7.   Examples of calibration measurements
          when using the parabola fit. Points (+)
          are displaced by + 50 MPa for clarity.

CONCLUSIONS

Exposure times of a few seconds yield a tentative stress reading when applying the new solid state camera to the stress study of quenched and tempered Cr-steel. An accuracy of 10 MPa in stress can be achieved in half a minute. The cross correlation method gives as good results as the combination of smoothing and parabola fit, while the direct parabola fit is inferior to these two. In this study no essential differences were found between the accuracies obtainable by the one and two exposure methods of stress measurement. However, in case of broader diffraction lines, the improved two exposure method is expected to yield better results, because of the ideal focusing and no need for the absorption correction.

REFERENCES

1. B.D. Cullity, Elements of X-ray Diffraction, 2nd ed., Addison-Wesley, 1978
2. K. Simomaa, V. Kelhä, T. Tuomi, M.A. Korhonen, and L. Suominen, Proc. Conf. Image Detection and Quality, July 16-18, 1986, Paris, 203
3. L.N. Koppel, Adv. X-Ray Anal. 19:587 (1976)
4. J. Launspach, J.L. Bourgade, C. Cavailler, J. de Mascureau, A. Mens and R. Saunef, SPIE 689:244 (1986)
5 B.D. Cullity, Elements of X-ray Diffraction, Addison-Wesley, first ed. 1956
6. M.A. Korhonen and V.K. Lindroos, Scand. J. Metallurgy 2:101 (1973)
7. E. Macherauch and U. Wolfstieg, Adv. X-Ray Anal. 20:369 (1977)
8. M.A. Korhonen, D.Sc. Thesis, Helsinki Univ. of Technology, 1980
9. K. Tönschoff, E. Brinksmeier, and H.H. Nölke, Z. Metallkde 72:349 (1981)

ADVANTAGES OF THE VECTOR METHOD TO STUDY
THE TEXTURE OF WELL TEXTURED THIN LAYERS

Albert Vadon and Jean-Julien Heizmann

Laboratoire de Métallurgie Physique et Chimique
Université de Metz, Ile du Saulcy
57045 Metz Cedex 1, France

INTRODUCTION

Depositing a thin layer of noble matter on a less noble material is a standard means of improving the qualities of the material. The characterization of the deposit is essential to understand the physical phenomena involved and to improve the final product. Two physical parameters of the layer deposited are of great consequence on the behavior of the material as a whole :

- the crystallographic and morphological textures related or not related to those of the substratum
- the internal stresses.

In collaboration with Centre d'Etudes Nucléaires de Grenoble (DMG-Mr Morlevat) we used the Vector Method [1-2-3-4] to study the feasability of a quantitative analysis of the crystallographic textures of molybdenum deposits on stainless steel by PVD. In this paper — short by necessity when compared to the importance of the problem — we shall try to show how the Vector Method allows to obtain very quickly the few strong texture components of the deposit and to follow their evolution according to the PVD parameters (carrier gas and polarization voltage) and to the thickness of the deposit.

We shall then approach the problem raised by the presence of internal stresses which cause distorsion up to 2% in the crystal net and consequently the conditions of a good analysis of the texture.

THE SAMPLES STUDIED

We studied 4 samples of Mo deposited on ferritic stainless steel, named R40, R59, R61, R62. For all of them the vector gas was argon (2% in atomic weight) and the thickness of the layer deposited was 10 μm.

The polarization voltages were respectively :

voltage	75	200	225	275	300
name of sample	R59	R63	R61	R62	R40

The Mo crystal is of the m3m class and the ASTM card shows that the diffracted intensity is maximum on the {110} planes. The M.P.D.S.[5] in the m3m class being 64.76° for those planes, one incomplete pole figure is sufficient for the analysis.

CONDITIONS OF THE MEASUREMENT OF THE POLE FIGURES

INEL goniometer with the TEXTURINEL data collection program
Planes aimed at : {110}
BRAGG angle 2 theta = 40.50°
Anticathode:copper
Monochromator:graphite
Circular collimator of diameter 0.8 mm
Azimuth step = 5°
Tilt step = 2.5°
Maximum tilt = 75°
Height of the detector slit = 7.5 mm

Thus, on the pole figure measured in reflexion, there are 30 rings of 72 boxes or 2160 intensitiy values. The measurement is made by scanning with a constant motion speed, the number of photons being counted in a given time chosen for each measurement box.

ORIENTATION DISTRIBUTION FUNCTION (ODF)

The O.D.F. is not represented in the usual form of contour levels but in the form of a spectrum. A spectrum is a graph showing the intensities of all the texture components according to their numbers.
The orientation space is discretized in :

28 fiber axis boxes of equal areas on the cubic standard triangle
72 rotation slices about the fiber axis, from -2π to 2π

representing all in all a right prism built on the m3m cubic standard triangle and partitionned in 2016 boxes of equal volume. Each texture component is given the weight that must be given to the corresponding Elementary Pole Figure (EPF) such as :

$$X = \sum_n (Y_n \cdot D_n)$$

with X = pole figure
            n = rank or number of the texture component
            Y_n = weight of the texture component of rank n
            D_n = elementary pole figure.

HOW TO MAKE THE MACROSCOPIC SYMMETRIES OF THE SAMPLE OBVIOUS

The spectrum is drawn so as to show the macroscopic symmetries of the sample (symmetry with respect to rolling direction RD and symmetry with respect to transverse direction TD). Figure 1 explains the representation.

The X and Y axes divide the sheet in 4 quadrants.
Each graduation along the X axis represents the number of the fiber axis box to which the orientation belongs (28 graduations on each side of the origin).
The texture components are distributed by quadrants inside the fiber axis boxes according to the 3rd parameter ζ of the orientation.

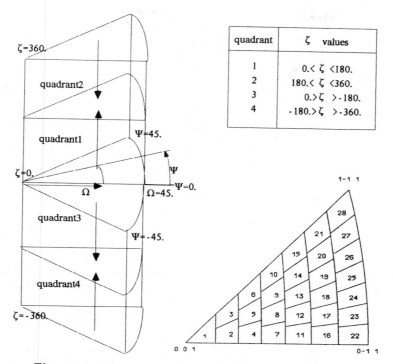

quadrant	ζ values
1	$0. < \zeta < 180.$
2	$180. < \zeta < 360.$
3	$0. > \zeta > -180.$
4	$-180. > \zeta > -360.$

Figure 1. This figure shows the volume of the orientations.
The upper prism at the left of this figure corresponds to unit triangle T1 and the lower one to T2 of the accompanying paper by Tidu, Vadon and Heizmann. The symmetry of the sample cuts the volume in 4 parts corresponding to the 4 quadrants of the texture spectrum. The triangle on the right shows the 28 boxes of the partitionning in equal areas. The arrows in the prisms belonging to the same fiber axis show the correspondence of the orientations in the two symmetries.

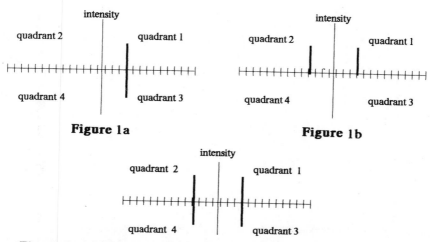

Figure 1a   Figure 1b

Figure 1c . 4 components of the texture spectrum of an orthotropic sample.

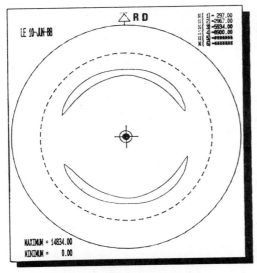 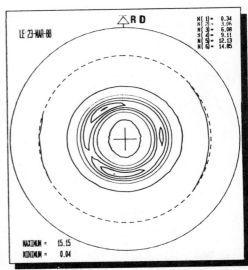

Pole figure (110) at 75 Volts   Pole figure (110) at 300 Volts

Figure 2

The intensities of the components are laid off as ordinates, in the axis direction for quadrants 1 and 2 and in the opposite direction for quadrants 3 and 4.

In this way :

- 2 components symmetrical with respect to **RD** are represented by 2 lines of equal lengths, opposite with respect to **RD** (figure 1a).

- 2 components symmetrical with respect to **TD** are represented by 2 lines which are symmetrical with respect to the origin and of equal lengths (figure 1b).

- The components of an orthotropic sample are symmetrical with respect both to **RD** and **TD** (figure 1c)

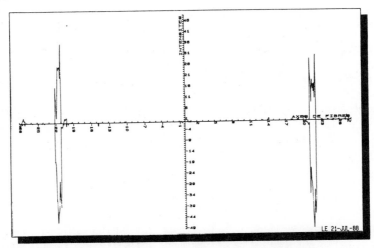

Figure 3 .Texture spectrum of R59 at 75 Volts.

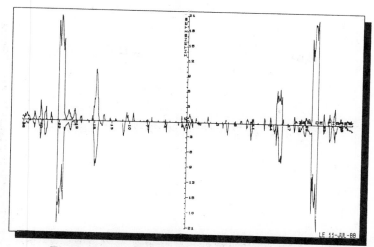

Figure 4 .Texture spectrum of R63 at 200 Volts.

RESULTS

Figure 2 shows the pole figure obtained with the two extreme samples (75 and 300V). Figures 3,4,5,6,7 show the five spectra obtained after analysis.

The use of the spectra is made easier by using our program which calculates the part of volume of the 28 fiber axes delimited by the partitioning of the standard triangle. This partitionning is shown in figure 1.

When comparing the spectra, we can see the strong influence of the polarization voltage on the texture of the deposit formed. In low voltage, component (110) dominates strongly. As the voltage increases, component (110) gradually disappears and component (111) appears.

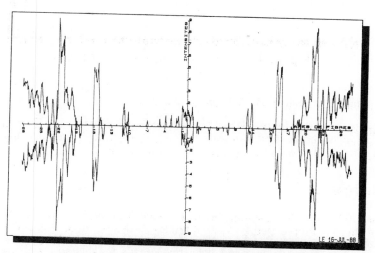

Figure 5 .Texture spectrum of R61 at 225 Volts.

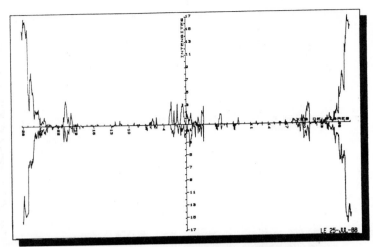

Figure 6 .Texture spectrum of R62 at 275 Volts.

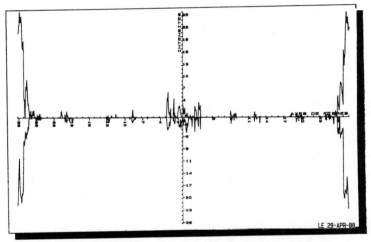

Figure 7 .Texture spectrum of R40 at 300 Volts.

The quantitative evolution is summarized in table 1.

The table shows that for this deposit whose orientation is practically (110) in low voltage, that component gradually decreases in favour of component (111) which dominates strongly at 300 volts. The two components are equal about 250 V.

CONCLUSION

Those results are only the starting point of a much larger and complete study of thin deposits (2 μm to 10 μm) carried at Centre d'Etudes Nucléaires de Grenoble (CENG).

We only meant to show that the Vector Method is an efficient tool to analyse the thin, usually strongly structured deposits and that in this case the quantitative results are easily interpreted.

Table 1 . Part of volume in % for components (110), (100), (111) .

sample	tension volts	part of the volume in %			
		(110) boxes 16+22+23	(100) box 1	(111) boxes 21+28+27	others
R59	75	93.26	0.26	5.49	negligible
R63	200	72.98	2.30	7.23	24+25+26=
R61	225	39.89	3.87	27.52	7.68 25+26=
R62	275	0.66	4.79	76.40	20.37 2+3 = 4.55
R40	300	0.	4.20	84.39	negligible

AKNOWLEDGMENTS

We would like to thank the texture team at CENG for providing the pole figures which made this work possible and for the talks we had on the subject.

BIBLIOGRAPHY

1—D. RUER. "Méthode Vectorielle d'Analyse de la Texture". Thesis. University of METZ. (1976).

2—D. RUER, R.BARO. "Méthode Vectorielle d'Analyse de la Texture des Matériaux Polycristallins de Réseau Cubique". J. Appl. Cryst.,10 : 458, (1977).

3—A. VADON. "Généralisation et Optimisation de la Méthode Vectorielle d' Analyse des Textures". Thesis. University of METZ. (1981).

4— H.SCHAEBEN, H.R. WENK, A. VADON
Vector Method. Chapter 6 in "Preferred Orientations in Deformed Metals and Rocks : an Introduction to Moderrn Texture Analysis". Academic Press.(1985).

5—D. RUER, A.VADON, R.BARO. "Refinements of the Vector Method" in Advances in X-Ray Analysis, 23 : 349, (1979).

TAKING INTO ACCOUNT THE TEXTURE EFFECT IN THE MEASUREMENT OF RESIDUAL STRESSES BY USING THE VECTOR METHOD OF TEXTURE ANALYSIS

Albert Tidu, Albert Vadon and Jean-Julien Heizmann

Laboratoire de Métallurgie Physique et Chimique
Université de Metz, Ile du Saulcy
57045 Metz Cedex 1, France

INTRODUCTION

Crystallographic texture induces an anisotropic mechanical behavior of the poly-crystal, so that the current analysis of the macro residual stress by X-ray diffraction cannot be used, because the lattice-strain distribution versus $\sin^2(\psi)$ presents a non-linear behavior. In order to take into account the influence of the texture, several authors have proposed theoretical explanations using the orientation distribution function. Many of them use the ODF to calculate the X-ray elastic constants /1,2,3/, another one uses Bunge's texture representation to obtain analytical expressions of the strain for textured specimens /4,5/. In this paper, we propose a treatment using the description of the texture by means of the Vector Method /6,7/. This treatment is based on the following constraints :

- the Reuss model is used, assuming a constant stress average over the diffracting volume,
- the X-rays do not penetrate beyond the deformed surface layer,
- the texture is homogeneous over the penetration depth,
- the influence of the plastic anisotropy on the mechanical behavior of textured material is not taken into account,
- the residual stresses are measured on cubic materials.

The different frames and rotations are given in figure 1. The rotation about the (L_3) direction, given by angle φ'_2, is free. We can see that the direction of measurement **MD** is parallel to the (L_3) direction and to the <hkl> direction which is perpendicular to the (hkl) plane.

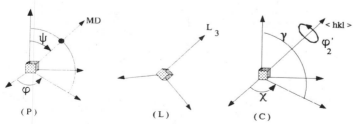

Figure 1 . Specimen, Measurement and Crystal frames.

STRAIN EXPRESSION

In a single crystal, the strain in the measurement direction (L_3) is given by the relation :

$$\varepsilon'_{zz} = S'_{zzij} \cdot \sigma'_{ij} \tag{1}$$

where S'_{zzij} is the compliance tensor and σ'_{ij} is the stress tensor in terms of the measurement frame (L).

S'_{zzij} can be obtained by using S_{mnop}, the single-crystal compliance with respect to the crystal frame (C) :

$$S'_{zzij} = c_{zm} \cdot c_{zn} \cdot c_{io} \cdot c_{io} \cdot S_{mnop} \tag{2}$$

$$= S_{1122} \cdot \delta_{ij} + 2 S_{1212} \cdot \delta_{3i} \cdot \delta_{3j} + S_o \cdot c_{zk}^2 \cdot c_{ik} \cdot c_{jk}$$

where c_{ij} are the elements of the inverted matrix which transform the measurement frame (L) into the crystal frame (C).(Note that another expression of S'_{zzij} can be obtained in terms of Miller's indices h,k,l for the generalization of the method). By X-ray diffraction, the strain measured in the (L_3) direction represents the average value of the strain of all the grains satisfying Bragg's law and having a <hkl> direction parallel to (L_3) for the imposed (φ, ψ) angles. From relation (1), we can write, assuming the Reuss approximation :

$$<\varepsilon'_{zz}> = \overline{S'_{zzij}} \cdot \sigma'_{ij} \quad \text{with} \quad \overline{S'_{zzij}} = \frac{\int_0^{2\pi} S'_{zzij} \cdot f(g) \cdot d\varphi'_2}{\int_0^{2\pi} f(g) \cdot d\varphi'_2} \tag{3,4}$$

where f(g) represents the value of the ODF for an orientation g. To evaluate the integral in (4), the dependency of S'_{zzij} with respect to φ'_2 and the (hkl) plane (or angles χ and δ), should be known. To illustrate the method proposed, we will describe the case of the (211) reflexion, keeping in mind that it can be extended to all the reflexions and generalized for an (hkl) reflexion.

CASE OF THE (211) REFLEXION

According to Brackman, it can be argued that it is sufficient to treat only one <hkl> direction parallel to the direction (L_3). From relation (2) we have :

$$\overline{S'_{zzxx}} = S_{1122} + \frac{1}{4} S_o - \frac{1}{12} S_o \cdot K_o(\varphi'_2, g) \qquad S'_{zzxy} = -\frac{1}{12} S_o \cdot I_o(\varphi'_2, g)$$

$$S'_{zzyy} = S_{1122} + \frac{1}{4} S_o + \frac{1}{12} S_o \cdot K_o(\varphi'_2, g) \qquad \overline{S'_{zzxz}} = \frac{\sqrt{2}}{6} S_o \cdot J_o(\varphi'_2, g) \tag{5}$$

$$\overline{S'_{zzzz}} = S'_{zzzz} = S_{1122} + \frac{1}{2} S_o + 2 S_{1212} \qquad \overline{S'_{zzyz}} = -\frac{\sqrt{2}}{6} S_o \cdot L_o(\varphi'_2, g)$$

where we defined $K_o(\varphi'_2, g)$, $I_o(\varphi'_2, g)$, $J_o(\varphi'_2, g)$ and $L_o(\varphi'_2, g)$ from the suitable integrals :

$$K_o(\varphi'_2, g) = \frac{\displaystyle\int_0^{2\pi} f(g) \cdot \cos 2\varphi'_2 \cdot d\varphi'_2}{\displaystyle\int_0^{2\pi} f(g) \cdot d\varphi'_2} \qquad\qquad L_o(\varphi'_2, g) = \frac{\displaystyle\int_0^{2\pi} f(g) \cdot \cos \varphi'_2 \cdot d\varphi'_2}{\displaystyle\int_0^{2\pi} f(g) \cdot d\varphi'_2}$$

$$(6)$$

$$I_o(\varphi'_2, g) = \frac{\displaystyle\int_0^{2\pi} f(g) \cdot \sin 2\varphi'_2 \cdot d\varphi'_2}{\displaystyle\int_0^{2\pi} f(g) \cdot d\varphi'_2} \qquad\qquad J_o(\varphi'_2, g) = \frac{\displaystyle\int_0^{2\pi} f(g) \cdot \sin \varphi'_2 \cdot d\varphi'_2}{\displaystyle\int_0^{2\pi} f(g) \cdot d\varphi'_2}$$

From relation (5), it can be seen that all the S'_{zzij}, except S'_{zzzz} are formed by a combination of two parts :

- an "isotropic" part, independent of the texture,
- an "anisotropic" part, depending on the texture.

All the same, the strain can be divided into two parts :

$$< \varepsilon'_{zz} > = < \varepsilon'_{zz} >_{iso} + < \varepsilon'_{zz} >_{anis}$$

From relation (5), and after re-arrangement, we obtain the following expression for the strain :

$$< \varepsilon'_{zz} > = S_1(hkl) \{ \ldots \} + \frac{1}{2} S_2(hkl) \{ \ldots \}$$

$$+ \frac{1}{12} S_o \cdot K_o(\varphi'_2, g) \{ \ldots \} + \frac{1}{6} S_o \cdot I_o(\varphi'_2, g) \{ \ldots \}$$

$$+ \frac{\sqrt{2}}{6} S_o \cdot J_o(\varphi'_2, g) \{ \ldots \} + \frac{\sqrt{2}}{3} S_o \cdot L_o(\varphi'_2, g) \{ \ldots \}$$

where the expressions in brackets are function of σ_{ij}, the stress tensor in the specimen frame (P), and of the angles (φ, ψ). $S_1(hkl)$ and $S_2(hkl)$ are the X-ray elastic constants for a given (hkl) reflexion. The problem is yet to evaluate the integrals given in (6): the Vector Method of quantitative texture analysis provides an analytical tool which allows the computing of these integrals.

THE VECTOR METHOD USED TO CALCULATE INTEGRALS K_o, L_o, I_o, J_o

To calculate an integral of the type:

$$I = \int_0^{2\pi} f(g) \cdot \cos \varphi'_2 \cdot d\varphi'_2$$

by expressing $f(g)$ by means of the components of texture vector **Y** of the Vector Method, we must find all the orientations of the crystallites having a plane (hkl) belonging to the family of diffracting planes and normal to the measurement direction (L_3). All the orientations can be deduced one

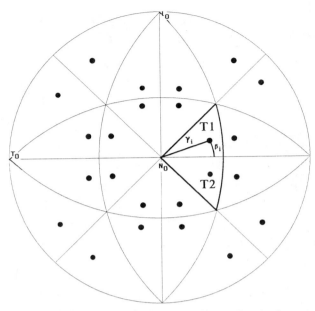

Figure 2. The 24 poles in reference position and axis (L_3) in the macroscopic
plane related to sample (OL_0, OT_0, ON_0).

from another by a rotation (L_3, φ'_2). The (hkl) plane aimed at remains invariable in the rotation.
The problem can be solved geometrically by using the stereographic projection.

Let us place on the pole hemisphere Σ^+_E :

- the location of (L_3) in the macroscopic frame (OL_0, OT_0, ON_0) by its polar coordinates
(ψ, φ)
- the reference orientation (0.,0.,0.) of the cubic crystal and the position of the m/2 poles of the
family of (hkl) diffracting planes.

Figure 2 shows that stereographic projection in the case of the (hkl) family.

A plane of that family, located by pole P_i^o will be in diffracting position about (L_3) if it is superposed
to (L_3) in the macroscopic frame.

The most direct orientation bringing pole P_i^o to position (L_3) is defined by the two rotations of the
Vector Method:

1. rotation ($OD, -\lambda_v$) which brings P_i^o onto the circle around axis ON_0 having angular radius
ψ. If P_i^o has for coordinates in the crystal frame (χ_i, β_i), the rotation is a rotation around axis OD,
which has polar coordinates $(\pi/2, \beta_i)$, a rotation of amount $|\psi-\chi_i|$. Then P_i^o is in P_i^1 on circle (Ψ).

2. rotation (ON_0, ζ) which brings P_i^1 to (L_3).
The two rotations are shown in figure 3. The fiber axis ON of the orientation is symmetrical of P_1^o
with respect to circle (δ) of axis ON_0 and of angular aperture $\psi/2$. The polar coordinates of ON are
$(\psi-\chi, \beta)$. To find the set of orientations bringing P_i^o on (L_3) we only have to make a rotation about
P_i^o of an angle φ'_2 varying from 0 to 2π, thus scanning all the possible orientations which bring P_i^o
on (L_3).

The rotation is shown in figure 4.

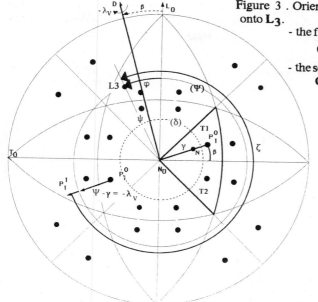

Figure 3. Orientation $(\beta, \gamma - \psi, 3\pi/2 + \varphi - \beta)$ brings P_i^0 onto L_3.

- the first rotation is the rotation around axis **OD** and of amount $-\lambda_v = \psi - \gamma$

- the second rotation is the rotation around the **ON$_0$** axis which brings P_i^1 onto L_3.

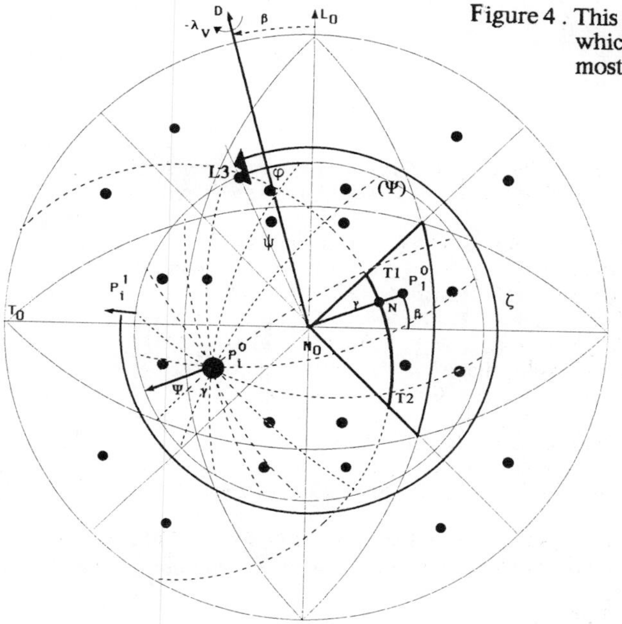

Figure 4. This figure shows the set of the orientations which bring P_i^0 onto L_3, deduced from the most direct rotations.

During that rotation, N, the fiber axis, scans a certain path in [T1+T2]. That path is the circle around axis P_i^o which has a angular radius Φ, brought back to the set [T1+T2] by the symmetry operations of the cubic m3m crystal. For each N, the third parameter ς being known, point I, figurative of an orientation volume, is known. Our program then allows to scan the locus of N, during the rotation φ'_2, calculating at the same time the length of the arc scanned in each orientation box and the limit $\varphi'_{2\,lim}$ at the borders of each box. So, the integral of type I given above, is replaced by the discrete sum :

$$I = \sum_i y_i \cdot (\sin \varphi'^{i}_{2\,lim} - \sin \varphi'^{i+1}_{2\,lim})$$

written in the matrix form :

$$I = [\,^{1}\ S_{hkl}\ ^{n}] \cdot [Y]^{1}_{n}$$

The matrix $[\ S_{hkl}\ ]$, or row vector, is composed of the terms

$$(\sin \varphi'^{i}_{2\,lim} - \sin \varphi'^{i+1}_{2\,lim})\quad \text{ordered according to i .}$$

For an isotropic vector that integral must be zero, so : $\Sigma_i s_i = 0$. Then we only have to calculate the matrices $[\ S_{hkl}\ ]$ for the usual measurement positions. The dependency of the integrals on the texture is then only a matrix multiplication.

At the moment, the program is being tested. The complete results of that new method for taking into account the texture in stress measurements will soon be published at ICRS 2 in Nancy (France) in November 1988.

BIBLIOGRAPHY

/1/ M.Barral, Mesure des contraintes résiduelles par diffraction X sur des matériaux présentant une texture cristallographique, Thesis University Pierre and Marie Curie, Paris.
/2/ M.Barral, J.L.Lebrun, J.M.Sprauel, and G.Maeder, X-Ray macrostress determination on textured material: use of the o.d.f. for calculating the X-ray compliances, Metallurgical Transactions A, **18A** :1229 (1987)
/3/ S.Nagashima, M.Shiratori and R.Nakagawa, Estimation of anisotropy of X-ray elastic modulus in steel sheets, Adv. X Ray Analysis, **29**:21(1986)
/4/ C.M.Brackman, Residual stresses in cubic materials with orthorhombic or monoclinic symmetry: influence of texture on ψ splitting and non-linear behaviour, J. Appl. Cryst., **16**:325 (1983)
/5/ C.M.Brackman, A general treatment of X-ray (residual) macro-stress determination in textured cubic materials: general expressions, cubic invariancy and application to X-ray strain pole figures, Cryst. Res. Technol., **20**:593 (1985)
/6/ D.Ruer, A.Vadon and R.Baro, Analysis of orientation distribution plots obtained with the vector Method for cubic polycrystals, Texture of Crystalline Solids, **3**:245 (1979)
/7/ A.Vadon, Thesis University of Metz (1981)

X-RAY DIFFRACTION STUDIES ON SHOCK MODIFIED Y $Ba_2Cu_3O_7$ SUPERCONDUCTORS

Lynn E. Lowry, Daniel D. Lawson, and Wayne M. Phillips

Mechanical And Chemical Systems Division
Applied Sciences And Microgravity Experiments Section
Jet Propulsion Laboratory
California Institute Of Technology
Pasadena, California 91109

INTRODUCTION

Y $Ba_2Cu_3O_7$, a high T_c superconductor powder, was shock compacted and explosively welded inside a copper matrix using the explosive fabrication methods described by Murr, Hare and Eror.[1] The shock compression fabrication technique provides the ability to process the superconductor powders into useable structures that will minimize environmental degradation and will not negatively affect the physical or mechanical properties. Additionally, the introduction of shock induced defects are known to increase solid-state reactivity in ceramic materials.[2] For this reason, shock compression fabrication of the superconductor/copper system offers the possibility of enhancing the superconducting properties of the $YBa_2Cu_3O_7$ powders.

X-ray diffraction line broadening analysis was used to examine the changes in the structure of the explosively compacted powders and quantify the crystallite size and residual lattice strain. These results were compared with those of $YBa_2Cu_3O_7$ powders that had undergone isostatic pressure.

EXPERIMENTAL

The superconductor powder was prepared using the solid-state reaction method of repeated mixing and grinding of Y_2O_3, $BaCO_3$ and CuO. The powder was heated in flowing oxygen for 9 hours at $900^\circ C$, reground and reheated,

cooled to 400°C at 100°C per hour and annealed in flowing oxygen for 3 hours.
A sample of the final powder was pressed into a pellet which was tested for
superconductivity using a four probe electrical measurement. The results
indicated a superconductivity temperature of 92°K. An x-ray diffraction
scan showed the sample to be a single phase with unit cell values of:
a_o = 3.819, b_o = 11.691 and c_o = 3.895. The observed interplanar spacings
and their relative intensities are listed in Table I.

   The explosive fabrication technique described by Murr, Hare and Eror[1]
was used to shock compact the superconductor powder and explosively weld
the surrounding copper to form a superconductor/metal composite. Figure 1
gives a schematic view of the basic welding system and the components. The
superconductor powder was poured into the area milled out of the copper base
plate and was compacted using a hydraulic press prior to the explosive
welding and shock compaction. The estimated shock compaction pressures
were from 22GPa to 35GPa.

   To prepare the isostatically pressed samples the superconductor powder
was poured into copper tubes of .375 inch diameter with .016 inch wall
thicknesses. The tubes were mechanically sealed and placed in latex bags.
The samples were pressed in an Autoclave Engineers Inc. isostatic press with
pressures ranging from 5GPa to 15GPa.

   The x-ray diffraction scans on the powders were obtained with the Siemens
D-500 automated diffractometer with CuKα radiation and a diffracted beam

Table 1.   Observed Interplanar Spacings
           for As-Prepared $YBa_2Cu_3O_7$ Powder

HKL	D$\overset{o}{A}$	Relative Intensity
020	5.8610	2
001/030	3.8973	6
100	3.8203	3
021	3.2382	5
120	3.1991	5
040	2.9241	5
031	2.7517	62
130/101	2.7277	100
121	2.4699	3
041-050	2.3382	25
131	2.2338	15
002-060	1.9475	42
200	1.9098	18

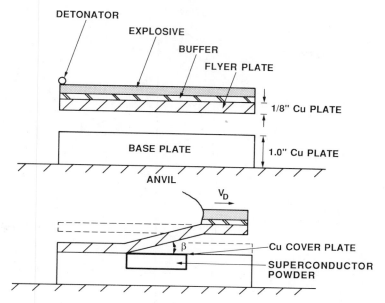

Figure 1.   Schematic of the Basic Welding System.

monochromator.   The scans were taken from $10°$ to $80°$ 2θ with $.01°$ steps and 60 second count times.   Figures 2 and 3 show x-ray diffraction line broadening in the principal crystal plane reflections for various shock compaction pressures.

X-RAY DIFFRACTION LINE BROADENING ANALYSIS

The x-ray diffraction peak profile analysis was done using the Warren-Averbach method[3] to obtain crystallite size and residual lattice strain in the superconductor powder.   First, the raw data was corrected for instrumental broadening using the Stokes method[4] and then the $K\alpha_2$ component was stripped using a Rachinger correction.[5]   Once the data was corrected the Fourier components of the diffraction line profiles were generated for multiple orders of reflections and the values for residual lattice strain and average crystallite sizes were determined.

RESULTS

Figure 4 shows a comparison of the [200] x-ray line width ratios for the isostatically pressed and the shock compacted powders.   The x-ray scans consistently showed greater line broadening in the isostatically pressed samples than in the shock compacted samples.

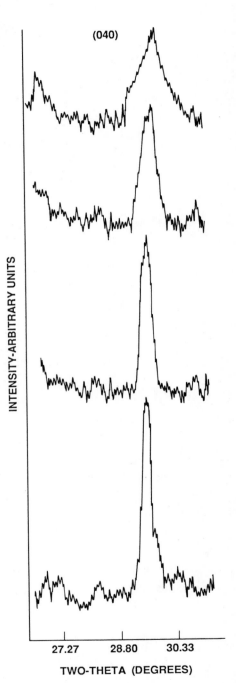

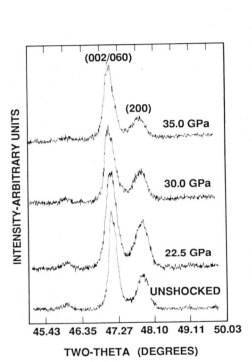

Figure 2.   X-Ray Diffraction Line
            Broadening in the (002/
            060) and (200) Reflec-
            tions.

Figure 3.   X-Ray Diffraction Line
            Broadening in the (040)
            Reflection.

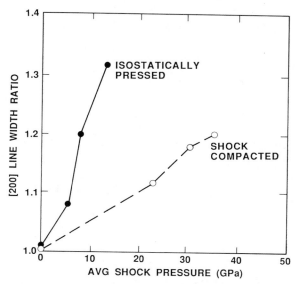

Figure 4. X-Ray Line Width Ratios.

The results of the Warren-Averbach analysis shown in Figures 5-8 indicate that higher residual strains and smaller crystallites are produced by lower pressures in the isostatically pressed samples. Preliminary transmission electron microscopy results support the conclusion that the microstructure of the superconductior powder is more severely affected by the isostatic pressure experiments.

Figures 5-8 also show the dependence of the residual strain and crystallite size on the crystal directions for each pressure. These differences are indicative of crystal and elastic anisotropies. For both sets of samples the [040] crystals exhibited more extreme changes in strain and crystallite size over the pressure ranges.

CONCLUDING REMARKS

The $YBa_2Cu_3O_7$ powder was shock compacted and the surrounding copper was explosively welded using the explosive fabrication method described. The superconductor/metal composite samples displayed a strong Meissner effect post fabrication.

The x-ray diffraction line broadening analysis on the modified materials showed the dependence of the resulting residual strain and crystallite

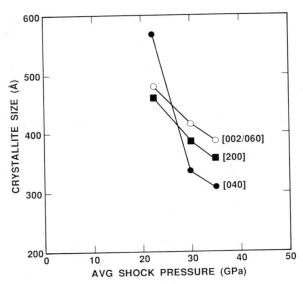

Figure 5.   Crystallite Size vs. Pressure
for Shock Compacted Samples.

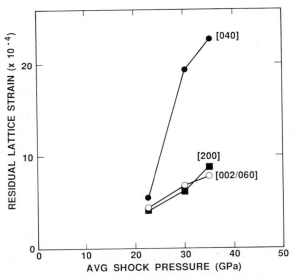

Figure 6.   Residual Lattice Strain vs.
Pressure for Shock Compacted
Samples.

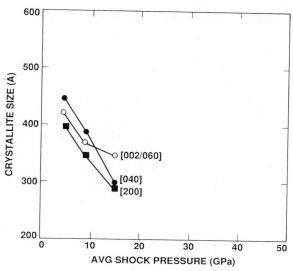

Figure 7.   Crystallite Size vs. Pressure
for Isostatically Pressed
Samples.

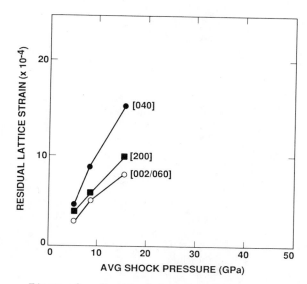

Figure 8.   Residual Lattice Strain vs.
Pressure for Isostatically
Pressed Samples.

size on the crystal plane for each pressure. This analysis also indicated that the shock modification had less of an effect on the microstructure of the powder than isostatic pressures that were four to six times smaller in magnitude.

REFERENCES

1. Murr, L.E., Hare A. W. and Error N. G., Nature 329:37-39 (1987).
2. Graham, R. A., Ann. Rev. Mat. Sci. 16:315-341 (1986).
3. Warren, B. E., "X-Ray Diffraction", Addison-Wesley, Reading, MA, (1969).
4. Cullity, B. D., "Elements Of X-Ray Diffraction", Addison-Wesley, Reading, MA. (1976).
5. Warren, B. E. and Averbach, B. L., J. Appl. Phys. 21:497-515 (1952).

THE CHARACTERIZATION OF A SOLID SORBENT WITH CRYSTALLITE

SIZE AND STRAIN DATA FROM X-RAY DIFFRACTION LINE BROADENING

Frank E. Briden and David F. Natschke*

U.S. Environmental Protection Agency
Air and Energy Engineering Research Laboratory
Research Triangle Park, NC 27711

*Acurex Corporation
4915 Prospectus Drive
Durham, NC 27713

INTRODUCTION

EPA's Air and Energy Engineering Research Laboratory is currently investigating the injection of dry calcium hydroxide [$Ca(OH)_2$] into coal fired electric power plant burners for the control of sulfur dioxide (SO_2) emissions. The overall chemistry for the process is:

HYDROXIDE PRODUCTION   BURNER REACTIONS

$CaCO_3 \rightarrow CaO + CO_2$ $Ca(OH)_2 \rightarrow CaO + H_2O$

$CaO + H_2O \rightarrow Ca(OH)_2$ $CaO + SO_2 \rightarrow CaSO_3$

                                      $2CaSO_3 + O_2 \rightarrow 2CaSO_4$

It has been found that the reactivity of the $Ca(OH)_2$ can vary considerably, depending on the limestone source. Furthermore, the time, temperature, and pressure conditions under which the $Ca(OH)_2$ is produced can have an effect on reactivity. Finally, it has been found that the addition of surfactants, during the production of the $Ca(OH)_2$, can enhance the reactivity. With all of these variables, the choice of materials and conditions for maximum efficiency is a formidable challenge. The use of full-scale furnaces, for the evaluation, is infeasible because of the expense. Bench-scale furnaces can produce relative ratings for sorbents, but are still very resource intensive for the level of testing needed. Laboratory characterization of sorbent material with techniques such as microscopic mineralogic examination, scanning electron microscopy, Brunauer, Emmett, and Teller surface measurements, thermogravimetric analysis, and differential thermal analysis has been explored, but none have proven to give conclusive results.

This situation has led to the investigation of x-ray diffraction (XRD) line broadening for sorbent characterization. It has long been known that the shape of XRD peaks is influenced by instrument factors, crystallite size, and lattice deformations. It was theorized that these microstructure factors could be a major factor in the differences between the reactivity of $Ca(OH)_2$ sorbents. The Warren-Averbach method[1] accomplishes separation of size and strain effects by first doing a Fourier transformation of the XRD peaks from the subject sample and a standard material which has relatively little peak broadening, due to crystallite size and microstrain. The Fourier coefficients of the sample are divided by the coefficients of the standard to deconvolute the component shape contributed by the instrument factors. The contributions due to size and strain can then be separated because the coefficients for strain effects are dependent on reflection order while those for size are not.

EXPERIMENTAL

For this study, seven different $Ca(OH)_2$ sorbent materials were evaluated for percent conversion (100 x moles of calcium reacted with SO_2 divided by moles of calcium available) in an isothermal flow reactor operating at 1000 °C with 0.3 percent SO_2 and 1 second residence time. The XRD analysis was done on a Siemens D-500 diffractometer with a copper target source running at 50 kV and 40 mA. The entrance aperture was 1 deg., and the detector slit was 0.05 deg. A scintillation detector, equipped with a graphite monochrometer, was used. The subject $Ca(OH)_2$ phase was identified as JCPDS 4-733 which has a hexagonal structure. The standard material was zinc oxide, JCPDS 36-1451, which also has a hexagonal structure. The peak data are given in Table 1.

A total of 257 steps were counted for each peak. The data were processed with the CRYSIZ program which was written by Gerhard Zorn and is furnished as part of the Siemens DIF-500 operating system, version 1.1. Before running CRYSIZ, the raw data were smoothed by using the FIT program to fit it with a split Pearson seven function. The output from the CRYSIZ program consists of microstrain as a function of column length from 1 to a maximum of 100 nm, the average column length, the column length occurring with the maximum frequency, the width at half maximum of column length distribution, the relative frequency of column lengths as a function of column length, and the cumulative frequency of column lengths as a function of column lengths. The three functions of column length can be plotted on the system plotter.

Table 1. Experimental Conditions

PHASE	MILLER INDICES	PEAK DEGS.	SCAN RANGE-DEGS.	FWHM DEGS.	I/Io	COUNT SECS.	STEP DEGS.
$Ca(OH)_2$	101	34.08	30.20-37.88	0.3-0.6	100	5	0.03
	202	71.76	68.00-75.68	0.4-0.9	12	30	0.03
ZnO	101	36.26	33.54-38.66	0.116	100	5	0.02
	202	76.77	74.34-79.46	0.121	5	20	0.02

Table 2.   Experimental Data

SAMPLE nm	PERCENT CONVERSION	202 PEAK HALF WIDTH DEGS.	AVERAGE COLUMN LENGTH nm	STRAIN @ AVERAGE COLUMN LENGTH	STRAIN @ MAX FREQ COLUMN LENGTH	MAXIMUM DETERMINABLE COLUMN LENGTH
Reagent	14.6	0.448	20.1	1.33	2.03	64
Koping	21.7	0.67	15.0	1.63	2.32	48
Linwood	22.1	0.582	13.6	1.76	2.19	44
0% Lig.	23.5	0.693	13.9	1.83	2.54	44
4% Lig.	25.6	0.78	11.9	2.06	2.74	36
1.5% Lig.	27.8	0.835	11.9	2.29	2.92	32
Fredonia	31.4	0.875	11.1	1.95	3.54	44

RESULTS AND CONCLUSIONS

The experimental data in Table 2 were taken from the output of the CRYSIZ program.   Seven samples with reactivity data were available. The samples labeled Lig. had various amounts of Lignosite (a by-product of the paper industry) added in processing.   The relation of the microstructure data to sorbent reactivity was tested by plotting the percent conversion for the samples versus the XRD data (Figs. 1-5).

The reactivity was seen to be directly proportional to the RMS strain at average column length and RMS strain at the column length of maximum frequency of occurrence (Figs. 3 and 4). The strain is a measure of potential energy stored in the crystal lattice due to disorder.   This strain could provide the potential energy necessary to overcome the energy of reaction.   The reactivity was seen to be inversely proportional to the crystallite column dimensions (Figs. 2 and 5).   This would be reasonable because the smaller dimensions could indicate higher surface area and consequently more intimate gas

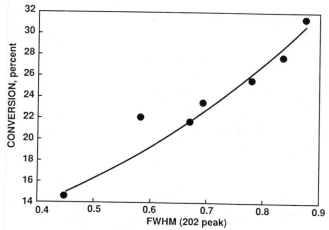

Fig. 1.   Percent conversion vs. diffraction peak half width.

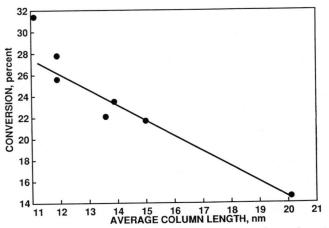

Fig. 2.   Percent conversion vs. average column length.

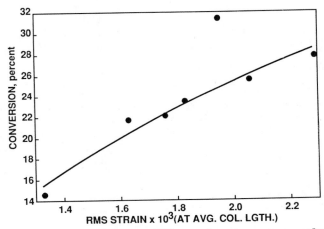

Fig. 3.   Percent conversion vs. RMS strain at average column length.

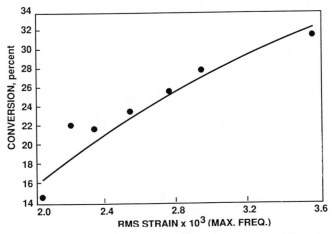

Fig. 4.   Percent conversion vs. RMS strain at column length of
maximum frequency of occurrence.

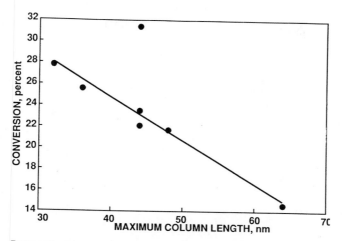

Fig. 5.   Percent conversion vs. maximum determinable column length.

contact.   The diffraction peak FWHM is directly proportional to reactivity because it reflects the contributions of increasing strain and decreasing crystallite dimensions (Fig. 1).   In the two plots, Figs. 3 and 5, one sample lies too far above the regression line.   The diffraction peaks for this sample apparently have been seen to be made up of contributions from a broad peak and a narrow peak, indicating the presence of two discrete phases.   It is expected that future work will reveal an acceptable method for weighting the contributions.   Future work will also be directed toward evaluating the relative contributions of size and strain as a function of column length.

REFERENCES

1.   B. E. Warren, "X-ray Diffraction," Addison-Wesley, Reading, Massachusetts, 1969.

X-RAY MEASUREMENT OF GRINDING RESIDUAL STRESS IN ALUMINA CERAMICS

Hiroyuki Yoshida, Yukio Nanayama, Yoshitaka Morimoto

Industrial Research Institute of Ishikawa
RO-1, Tomizu, Kanazawa 920, Japan

Yukio Hirose

Department of Material Science, Kanazawa University
1-1, Marunouchi, Kanazawa 920, Japan

and Keisuke Tanaka

Department of Engineering Science, Kyoto University
Yoshida Honmachi, Sakyo-Ku, Kyoto 606, Japan

INTRODUCTION

Machine parts made of ceramics are usually finished by grinding.
Residual stresses as well as defects introduced by grinding will influence
the fracture strength and the function of ceramics parts. Although several
investigations used the X-ray diffraction method to measure the grinding
residual stresses, their grinding conditions were rather limited.[1]

In the present study, sintered alumina ceramics of 99% purity were
ground with a resinoid diamond wheel (#140 grain size number) under various
grinding conditions. The effects of depth of cut and stock removal on the
residual stress was measured with the X-ray method. The X-ray diffraction
from the (1, 0, 10) plane by Cr-Kα radiation was used for stress
measurement. The profile of the residual stress distribution is discussed
on the basis of cutting mechanisms.

EXPERIMENTAL PROCEDURE

Material and Grinding Tests

The material used was an alumina ceramic with 99% purity made by slip
casting. Young's modulus was 358 GPa. The bending strength was 294 MPa.
The specimens measuring 10mm x 15mm (Fig. 1) were ground with a 140
grain-size resin-bonded diamond wheel under the conditions given in Table I.
A dynamometer was emloyed to measure the horizontal and vertical
components of force on the workpiece during traverse grinding.
An experimental setup is shown in Fig.2.

443

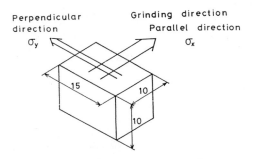

Perpendicular   Grinding direction
direction   Parallel direction
σ_y σ_x

Fig.1 Dimensions of specimen

Table I Grinding Condition

Traversing Table Type
Surface grinder (3.7 kW)

Wheel : SDC140P100B7
        (Resin bonded diamond)

Wheel diameter(D)	300 mm
Wheel speed(V)	26 m/s
Wheel depth of cut(d)	5,10,20, 30,40,50, 80, 100 μm
Table speed (v)	0.1 m/s
Water base coolant flow	20 l/min

Tool : GC220N87V
       (Vitrified bonded wheel)

Tool diameter (D')	122 mm
Speed ratio(Vs/Vt) Vs : Tool speed Vt : Wheel speed	0.3
Depth of cut	5 mm
Table speed	10 mm/s
Fluid dribble	

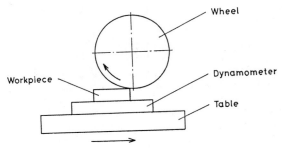

Fig.2 Grinding setup

Table II X-ray residual stress measurement condition

Characteristic X-ray	Cr-K$_\alpha$
Diffraction plane	(1.0.10)
Detector	PSPC (Effective length 100μm)
Radiation area	φ4mm
Peak position	Parabora Peak Top

X-Ray Observation

X-ray equipment used for the measurement of residual stress was a stress analyzer (Rigaku PSPC-MSF). A position sensitive detector was utilized to record X-ray diffraction. The diffraction profile of (1, 0, 10) plane was obtained by Cr-Kα X-rays. The residual stress on the ground surface was determined by the $\sin^2\psi$ method. The distribution of the residual stress beneath the ground surface was obtained by irradiating X-rays on the new surface revealed by successive chemical polishing. The conditions of X-ray observation are given in Table II. The X-ray stress

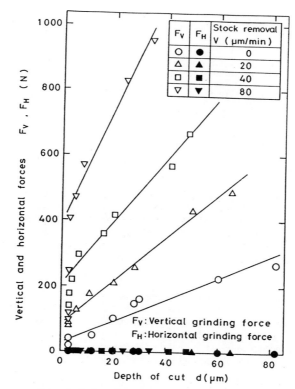

Fig.3 Relation between grinding
forces and depth of cut
after various stock removal

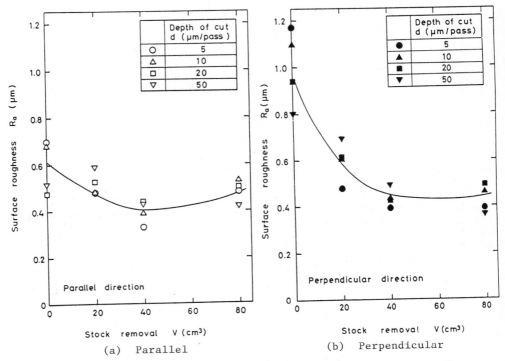

Fig.4 Relation between surface roughness and stock removal

constant H for the $\sin^2 \psi$ method was determined in our previous papers.[2,3] The obtained value was H = -1217 MPa/deg.

EXPERIMENTAL RESULTS AND DISCUSSION

Grinding Results

Fig. 3 shows vertical and horizontal forces, F_V and F_H, as a function of the depth of cut for various cases of stock removal. The vertical grinding force was much larger than the horizontal force. The vertical force increased linely with the depth of cut when the same amount of ceramic was removed after the dressing. The larger force was required to cut the same depth when the cutting edge became blunter due to the larger amount of stock removed.

The roughness of ground surfaces was very much dependent on the stock removal, while it was rather independent of the depth of cut. Fig. 4 shows the change in the roughness (averaged around the center line) in the grinding direction. The surface becomes smoother when the stock removal increases. The roughness in the perpendicular to the grinding direction is larger than that in the parallel direction only after dresssing.

Scanning electron micrographs of ground surfaces are presented in Fig. 5. The surface ground with a wheel just after dressing shows several scratching ridges parallel to the grinding direction. As the amount of removed ceramic increases, the scratching ridges are dispersed.

X-Ray Residual Stress Measurement

The relation between diffraction angle 2θ and $\sin^2\psi$ is well approximated by a straight line for all the cases examined as shown in Fig. 6. The residual stresses measured on ground surfaces were all in compression. All the data measured on the ground surfaces are plotted against the vertical grinding force in Fig. 6. They are scattered between 0 and -300 MPa, uncorrelated with the grinding force. The compressive residual stress in the direction parallel to the grinding direction was larger than that in the perpendicular direction. The compressive residual stress measured on the ground surface tends to decrease with increasing stock removal as shown in Fig. 7. The influence of the depth of cut on the residual stress is small. The residual stress measured on the ground surface is correlated with the surface roughness both for parallel and perpendicular directions as shown in Fig. 8. The compressive residual stress increases with increasing surface roughness and with decreasing stock removal.

The distribution of the residual stress beneath the ground surface is shown in Fig.9. The depth of the compressive zone is about 20 µm for the case of a stock removal of 20 cm^3, and decreases with measured stock removal. It is about 5 µm for the case of a stock removal of 80 cm^3 and is independent of the depth of cut. In conclusion, after dressing, the cutting edge of a wheel is sharp and the cutting force is small. The ground surface is rough and has a large compressive zone near the surface.

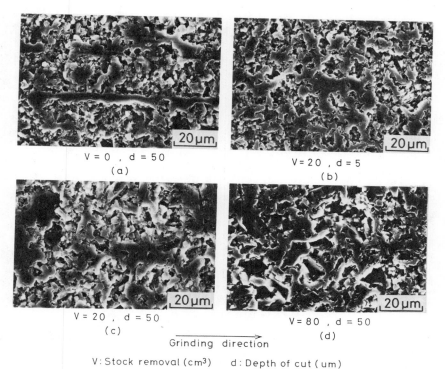

V = 0 , d = 50
(a)

V = 20 , d = 5
(b)

V = 20 , d = 50
(c)

V = 80 , d = 50
(d)

Grinding direction

V: Stock removal (cm^3)   d: Depth of cut (um)

Fig.5 Scanning electron micrographs of ground surfaces

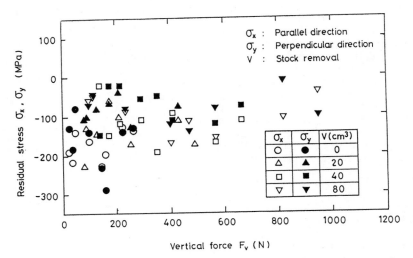

Fig.6 Relation between residual stress and vertical force

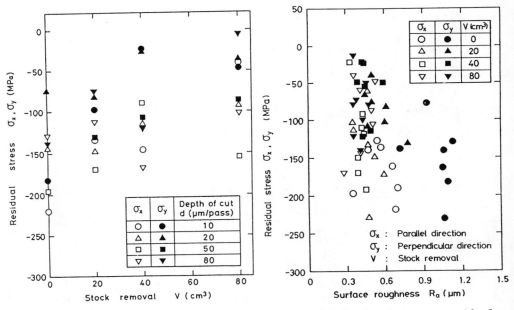

Fig.7 Relation between residual
stress and stock removal

Fig.8 Relation between residual
stress and surface roughness

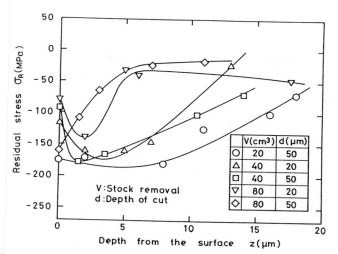

Fig.9 Distribution of residual stress beneath ground surface

CONCLUSIONS

The results obtained are as follows:

(1) Residual stresses measured on ground surfaces were all in compression. The compressive residual stress in the grinding direction was larger than that perpendicular to the grinding direction. The compressive residual stress was larger for rougher surfaces.

(2) The residual stress decreased with increasing amount of stock removal. It was rather insensitive to the depth of cut.

(3) The depth of the compressive zone was about 20 μm for the case of a stock removal of 20 cm³, and decreased as stock removal increased. It was about 5 μm for the case of a stock removal of 80 cm³ and was independent of the depth of cut.

ACKNOWLEDGMENT

The authors acknowledge Nikko Ltd. for providing us with experimental materials.

REFERENCES

1.   K. Tanaka, K. Suzuki and T. Kurimura, to be published in "Residual Stress in Science and Technology" (1989).
2.   T. Mishima, Y. Nanayama, Y. Hirose and K. Tanaka, X-Ray Fractography of Fracture Surface of Alumina Ceramics, in "Advances in X-Ray Analysis," Vol. 30, 545-552 (1987).
3.   T. Mishima, H. Yoshida, Y. Hirose and K. Tanaka, Pre-Cracking Technique and its Application to X-Ray Fractography of Alumina Ceramics, in "Advances in X-Ray Analysis," Vol. 31, 261-268 (1988).

RESIDUAL STRESSES NEAR SCC FRACTURE SURFACES OF

AISI 4340 STEEL

Zenjiro Yajima

Department of Mechanical Engineering, Kanazawa Institute of
Technology, 7-1 Oogigaoka, Nonoichi, Kanazawa 921, Japan

Masaaki Tsuda, Yukio Hirose

Department of Materials Science, Kanazawa University
1-1 Marunouchi, Kanazawa 920, Japan

and

Keisuke Tanaka

Department of Engineering Science, Kyoto University
Yoshida-hommachi, Sakyo-ku, Kyoto 606, Japan

INTRODUCTION

X-ray diffraction observation of the material beneath the fracture
surface provides failure analysists with useful information to judge the
mechanical condition of fracture.[1~13]

In the present paper, stress corrosion cracking (SCC) tests were
conducted by using the bluntly notched compact tension (CT) specimens of
200°C tempered AISI 4340 steel in a 3.5% NaCl solution environment.
The distribution of the residual stress beneath the fracture surface near
the root of the notch was measured with the X-ray diffraction technique.[4]
The effect of the notch root radius on crack nucleation with stress corrosion
was discussed on the bases of the results of X-ray observation.

EXPERIMENTAL PROCEDURE

Material

The material used in the experiments was AISI 4340 steel. The chemical
composition (wt%) of the material was as follows: 0.39C, 0.74Mn, 1.38Ni,
0.78Cr, 0.23Mo. The specimens were first normalized at 880°C for 1 hour.
After austenized at 850°C for 1 hour, they were quenched into oil and then
tempered at 200°C for 2 hours. Their mechanical properties are listed in
Table 1. The CT specimens with five different notch radii, ρ, of 0.12,
0.25, 0.5, 1.0 and 2.5 mm were machined from plates which were cut at 90° to
the axis of a hot rolled round bar of 100 mm diameter. The thickness of the

451

Table 1. Mechanical properties of test material

Yield stress σ_Y(MPa)	Tensile strength σ_B(MPa)	Elongation ε_l(%)	Fracture toughness K_{IC}(MPa√m)
1530	1880	4.5	54

specimen is 7 mm. After heat treatments, the specimen surfaces were ground off by 1.2 mm to remove the decarburized surface layers. The notches were carefully made by an electrical discharge machine, and then all the specimens were electro-polished.

Stress Corrosion Cracking Tests

Stress corrosion cracking tests were conducted in a simple lever arm tensile machine by keeping the stress intensity factor constant. The crack nucleation at the notch root was detected with the electrical potential method. A direct current of a constant value of 30A was maintained and the voltage difference around the notch mouth was measured. The tests were interrupted when the voltage difference started to increase. All the specimens were fractured in liquid nitrogen to observe the crack nucleation site with a scanning electron microscope. A circulating 3.5% NaCl solution was used as environment and the temperature of the solution was kept at 16±2°C.

X-Ray Observation

The residual stress near fracture surfaces was measured with the X-ray diffraction method. The method adopted was the standard $\sin^2\psi$ method by using the parallel beam of Cr-k_α X-rays as described in a previous paper.[10] The area irradiated by X-rays was of 2 mm in diameter at the middle of the specimen thickness on the fracture surface, touching the notch tip as indicated in Fig. 1. The conditions of X-ray observation are given in Table 2. The distribution of the residual stress beneath the fracture surface was measured by removing the surface layer successively by electro-polishing.

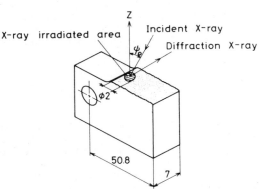

Fig. 1. Schematic illustration of X-ray irradiated area on the fracture surface (in mm).

Table 2. X-ray conditions for stress measurement

Characteristic X-ray	$Cr-K_\alpha$
Diffraction plane	(211)
Filter	V foil
Counter	Scintillation counter
Tube voltage	30 kV
Tube current	18 mA
Scanning speed	4 deg/min
Time constant	5 sec
Slit divergence angle	0.15 deg
Irradiated area	$\phi 2$ mm

EXPERIMENTAL RESULTS

Crack Nucleation during Stress Corrosion

Figure 2 shows the relation between the time to crack nucleation, t_n and the apparent stress intensity factor, K_ρ for each value of the notch radius. It is clear that the crack nucleation time, t_n increases with the notch radius. The figure also shows that there are threshold values of the apparent stress intensity factor, $(K_\rho)_{th}$ below which no crack nucleation occurs. Barsom et al.[14] examined the effect of notch radius on fatigue crack nucleation for high strength steel and suggested that crack nucleation life in notched specimens can be regulated by parameter $2K_\rho/\sqrt{\pi\rho}$. The data in Fig. 2 are replotted in Fig. 3 to see the relation between $2K_\rho/\sqrt{\pi\rho}$ and t_n. A linear relationship can be seen in the log-log diagram; this relationship

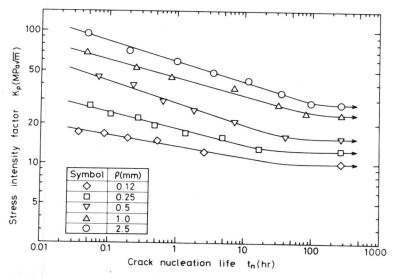

Symbol	ρ(mm)
◇	0.12
□	0.25
▽	0.5
△	1.0
○	2.5

Fig. 2. Relation between stress intensity factor and crack nucleation life.

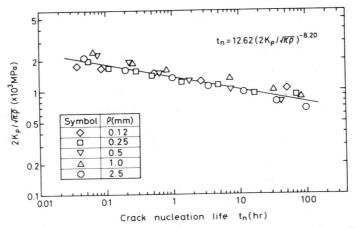

Fig. 3.   Relation between crack nucleation life and $2K_\rho/\sqrt{\pi\rho}$.

is as follows:

$$t_n = C \ (\ 2K_\rho/\sqrt{\pi\rho} \ )^m \tag{1}$$

where C is constant and m is equal to -8.20.

<u>X-Ray Observation of the SCC Fracture Surface</u>

Figure 4 shows the distribution of the residual stress on the SCC fracture surface measured by irradiating with an X-ray beam 2 mm in diameter. The residual stress in the vicinity of the fracture surface is tensile. It increases gradually and then decreases with increasing distance from the surface. The plastic zone size, ω_y, is defined as the distance at which the residual stress approaches to the initial value (49MPa).

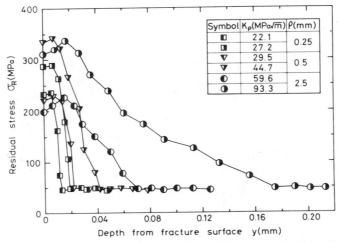

Fig. 4.   Residual stress distribution beneath fracture
          surface in stress corrosion cracking.

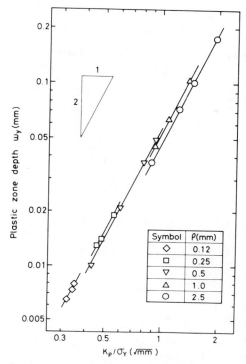

Fig. 5. Relation between plastic zone depth and apparent
stress intensity factor divided by yield stress.

The relation between ω_y and K_ρ/σ_Y is shown in Fig. 5, where σ_Y is the yield stress. It is noted that ω_y is proportional to $(K_\rho/\sigma_Y)^2$ for the case of SCC fracture. The relation is expressed as

$$\omega_y = \alpha \ (\ K_\rho/\sigma_Y \ )^2 \tag{2}$$

where α is a constant that is dependent upon the notch radius, as follows:

$\rho = 0.12$ mm	$\alpha = 0.063$
$\rho = 0.25$ mm	$\alpha = 0.061$
$\rho = 0.5$ mm	$\alpha = 0.057$
$\rho = 1.0$ mm	$\alpha = 0.054$
$\rho = 2.5$ mm	$\alpha = 0.047$

Thus α decreases as ρ increases.

DISCUSSION

In Table 3, the values of α obtained in a similar way for the cases of the fracture toughness test and SCC are summarized. The α value for the SCC test is smaller than that for the fracture toughness test. Levy et al.[15] derived $\alpha = 0.15$ on the basis of the elastic-plastic finite element method for elastic perfectly plastic material. The α value being different from 0.15 is now assumed to be caused by the difference of the yield stress in the plastic zone from that in the simple tension test. The yield stress in the plastic

Table 3. Values of $\alpha[= \omega_y/(K_\rho/\sigma_Y)^2]$ and yield stress in the plastic zone

	Tensile test	Fracture toughness test	SCC		
			$\rho=0.12$ mm	$\rho=0.5$ mm	$\rho=2.5$ mm
α	——	0.140	0.063	0.057	0.047
Yield stress $\sigma_Y{'}$ (MPa)	1530	1584	2361	2482	2733

zone, $\sigma_Y{'}$, is evaluated from the following equation:

$$\omega_y = 0.15 \; (K_\rho/\sigma_Y{'} \;)^2 = \alpha \; (K_\rho/\sigma_Y \;)^2 \tag{3}$$

$$\sigma_Y{'} = (\; 0.15/\alpha \;)^{\frac{1}{2}} \cdot \sigma_Y \tag{4}$$

The calculated values of $\sigma_Y{'}$ are given in Table 3.

From the previously published data of α measured for the fracture surface of various steels[10,12] and ceramics[16] the value of $\sigma_Y{'}$ was calculated by using equation (4) and correlated to σ_Y in Fig. 6. The following linear relation is obtained between $\sigma_Y{'}$ and σ_Y:

$$\sigma_Y{'} = -161.8 + 1.743\sigma_Y \tag{5}$$

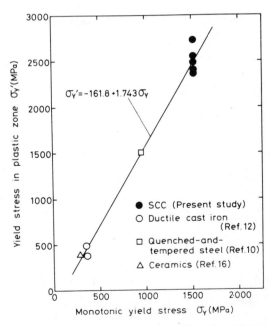

Fig. 6. Yield stress determined from plastic zone size.

The result of the present study agrees with this equation. From equations (4) and (5), the α value is given as a function of σ_Y.

$$\alpha = 0.15 \ [\ \sigma_Y/(\ -161.8 + 1.743\sigma_Y \ ) \ ]^2 \tag{6}$$

In the analysis of failure accidents, the apparent stress intensity factor can be determined from the measurement of the plastic zone by using α obtained from equation (6).

CONCLUSION

The main results obtained in the present study are summarized as follows:

(1) The depth of the plastic zone, ω_y, **was determined on the** basis of the distribution of the residual stress. It was related to the apparent stress intensity factor, K_ρ, divided by the yield stress, σ_Y, by

$$\omega_y = \alpha \ (\ K_\rho/\sigma_Y \ )^2$$

where α is a function of notch radius.

(2) The published data on the α-value determined for various kinds of steel show that α is related to the yield stress, σ_Y, by

$$\alpha = 0.15 \ [\ \sigma_Y/(\ -161.8 + 1.743\sigma_Y \ ) \ ]^2$$

REFERENCES

1.   S. Taira, and K. Tanaka, "Fracture Surface Analysis by X-Ray Diffraction Techniques," J. Iron and Steel Ins. Jap., 65:450 (1979).
2.   Committee on X-Ray Study on Mechanical Behavior of Materials, "X-Ray Fractography," J. Soci. Mat. Sci. Jap., 31:244 (1982).
3.   K. Tanaka, and N. Hatanaka, "Residual Stress Near Fatigue Fracture Surfaces of High Strength and Mild Steels Measured by X-Ray Method," J. Soci. Mat. Sci. Jap., 31:215 (1982).
4.   Z. Yajima, Y. Hirose, K. Tanaka, and H. Ogawa, "X-Ray Diffraction Study on Fracture Surface Made by Fracture Toughness Tests of Blunt Notched CT Specimen of High Strength Steel," J. Soci. Mat. Sci. Jap., 32:783 (1983).
5.   Z. Yajima, Y. Hirose, and K. Tanaka, "X-Ray Diffraction Observation of Fracture Surfaces of Ductile Cast Iron," Advances in X-Ray Analysis, 26:291 (1983).
6.   K. Tanaka, and N. Hatanaka, "X-Ray Fractography of Fatigue Fracture of Steels," J. Soci. Mat. Sci. Jap., 33:324 (1984).
7.   Y. Hirose, Z. Yajima, and K. Tanaka, "X-Ray Fractography on Stress Corrosion Cracking of High Strength Steel," Advances in X-Ray Analysis, 27:213 (1984).
8.   Y. Hirose, Z. Yajima, and K. Tanaka, "X-Ray Fractographic Approach to Fracture Toughness of AISI 4340 Steel," Advances in X-Ray Analysis, 28:289 (1985).
9.   Y. Hirose, and K. Tanaka, "X-Ray Measurement of Residual Stress Near Fatigue Fracture Surfaces of High Strength Steel," Advances in X-Ray Analysis, 29:265 (1986).
10. Z. Yajima, Y. Hirose, and K. Tanaka, "X-Ray Fractography of Fatigue Fracture of Low-Alloy Steel in Air and in 3.5% NaCl Solution," J. Soci. Mat. Sci. Jap., 35:725 (1986).

11. M. Tsuda, Y. Hirose, Z. Yajima, and K. Tanaka, "Residual Stress Near
     SCC Fracture Surface of AISI 4340 Steel Under Controlled Electrode
     Potential," J. Soci. Mat. Sci. Jap., 37:599 (1988).

12. Z. Yajima, Y. Simazu, K. Ishikawa, Y. Hirose, and K. Tanaka, "X-Ray
     Fractographic Study on Fracture Surface Made by Fatigue Crack
     Propagation Tests of Ductile Cast Iron," J. Jap. Soc. Strength and
     Fracture of Materials, 22:121 (1988).

13. M. Tsuda, Y. Hirose, Z. Yajima, and K. Tanaka, "X-Ray Fractography
     of Stress Corrosion Cracking in AISI 4340 Steel Under Controlled
     Electrode Potential," Advances in X-Ray Analysis, 31:269 (1988).

14. J. M. Barsom, and R. C. McNicol, "Effect of Stress Concentration on
     Fatigue Crack Initiation in HY-130 Steel," ASTM STP 559, 183 (1973).

15. N. Levy, P. V. Marcal, W. J. Ostengren, and J. R. Rice, "Small Scale
     Yielding Near a Crack in Plane Strain: A Finite Element Analysis,"
     Int. J. Frac., 7:143 (1971).

16. T. Mishima, H. Yoshida, Z. Yajima, Y. Hirose, and K. Tanaka, "Pre-
     Cracking Technique and It's X-Ray Fractographic Application on
     Fracture Surfaces of Alumina Ceramics," Proc. of The 24th Symposium on
     X-Ray Studies on Mechanical Behavior of Materials, The Society of
     Materials Science, Japan, 202 (1987).

RESIDUAL STRESS MEASUREMENT OF SILICON NITRIDE AND SILICON CARBIDE

BY X-RAY DIFFRACTION USING GAUSSIAN CURVE METHOD

Masanori Kurita, Ikuo Ihara

Nagaoka University of Technology
Kamitomioka, Nagaoka, 940-21 Japan

Nobuyuki Ono

Citizen Watch Co.
Shinjuku-ku, Tokyo, 163 Japan

INTRODUCTION

The residual stress induced by grinding or some thermal treatment has a large effect on the strength of ceramics. The X-ray technique can be used to nondestructively measure the residual stress in small areas on the surface of polycrystalline materials. The X-ray stress measurement is based on the continuum mechanics for macroscopically isotropic polycrystalline materials. In this method, the stress value is calculated selectively from strains of a particular diffraction plane in the grains which are favorably oriented for the diffraction. In general, however, the elastic constants of a single crystal depend on the plane of the lattice, since a single crystal is anisotropic. The behavior of the deformation of individual crystals in the aggregate of polycrystalline materials under applied stress has not yet been solved successfully. Therefore, the stress constant and elastic constants for a particular diffracting plane should be determined experimentally in order to determine the residual stress accurately by X-ray diffraction.

Gaussian and Cauchy functions are used for approximating an x-ray diffraction line profile[1]. The parabola method approximating a diffraction peak by a parabola is most widely used for determining the peak position of a diffraction line in x-ray stress measurement[2,3]. A Gaussian function can approximate a wider angular range of the diffraction peak than a parabola. In the previous papers[4,5], Kurita proposed the equations for rapidly calculating the peak position by the Gaussian curve method together with its standard deviation due to counting statistics.

The stress constant and x-ray elastic constants of beta silicon nitride and alpha silicon carbide specimens were determined by the Gaussian curve method. The standard deviations of these constants, which represent the variability due to counting statistics, were also calculated by using the equations proposed in the previous paper[6] in order to investigate the

reproducibility of the measured values. The residual stress of the silicon
nitride specimen induced by grinding and lapping were measured by using the
stress constant determined by x-rays.

PEAK POSITION AND ITS STANDARD DEVIATION

To obtain net diffracted x-ray counts, observed diffracted counts y
above background should be corrected for LPA (Lorentz-polarization and
absorption) factor. The corrected x-ray counts z are given by

$$z_i = l_i(y_i - y_{bi}) \tag{1}$$

In Eq.(1), y_{bi} are the background counts and l_i is the reciprocal LPA factor
given by

$$l_i = \frac{1 - \cos x_i}{(3 + \cos 2x_i)[1 - \tan\psi \cot(x_i/2)]}$$

where x_i is the diffraction angle and ψ is the angle between specimen
normal and diffraction plane normal. If the background counts y_{bi} in Eq.(1)
are determined for simplicity by connecting two end points of the diffraction line, (x_0, y_0) and (x_{n+1}, y_{n+1}), as shown in Fig.1, these are given by

$$y_{bi} = E_i y_0 + F_i y_{n+1}$$

where

$$E_i = (x_{n+1} - x_i)/(x_{n+1} - x_0)$$

$$F_i = (x_i - x_0)/(x_{n+1} - x_0)$$

However, the background can not be determined accurately for some
materials such as hardened steels with martensitic structure which have
very broad diffraction profile and some ceramics which have several
diffraction profiles being located closely. In addition, the omission of
the background subtraction is desirable for practical purposes in order to
reduce the measurement time. Fortunately, the diffraction peak can be
approximated by a Gaussian function even if the background subtraction is
omitted. In this case, the observed x-ray counts y are corrected for the LPA

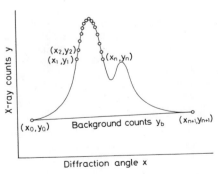

Fig.1 Schematic diffraction line.

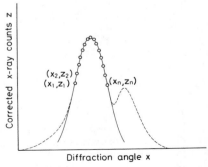

Fig.2 Fitting a Gaussian function
to $K\alpha_1$ diffraction peak.

factor as

$$z_i = l_i y_i \tag{2}$$

The peak position is determined by the Gaussian curve method as the main axis of the Gaussian function that is fitted to n data points of (x_i, z_i) around the peak of the diffraction line by using a least squares analysis as shown in Fig.2. Since the detailed derivation of the equation for calculating the peak position p is given in the previous papers[4,5], only the final result is given here as

$$p = \frac{x_1 + x_n}{2} - d\frac{\Sigma t_i w_i}{\Sigma T_i w_i} \tag{3}$$

where Σ denotes the summation from $i=1$ to n throughout this paper, c is an angular interval, and

$$d = 0.4c(n^2 - 4)$$
$$t_i = i - (n + 1)/2, \quad (i = 1, 2, \ldots, n)$$
$$T_i = 12t_i^2 - n^2 + 1$$
$$w_i = \ln z_i, \quad (\ln \text{ denotes the natural logarithm.})$$

The following well known relationship holds between x-ray counts y and the variance σ_y^2 of y due to counting statistics[7,8].

$$\sigma_y^2 = y \tag{4}$$

According to the statistical theory[9], the variance σ_p^2 of peak position p can be calculated from

$$\sigma_p^2 = \Sigma(\frac{\partial p}{\partial y_i})^2 \sigma_{y_i}^2 \tag{5}$$

Substituting Eqs.(3) and (4) into Eq.(5) and differentiating, we obtain the standard deviation σ_p of the peak position p. In case of the peak position using Eq.(2) without correction for background, σ_p is given by

$$\sigma_p = d\frac{\sqrt{\Sigma G_i^2/y_i}}{(\Sigma T_i \ln z_i)^2} \tag{6}$$

where

$$G_i = t_i \Sigma T_i \ln z_i - T_i \Sigma t_i \ln z_i$$

In the case of the peak position using Eq.(1) corrected for background, σ_p is given by

$$\sigma_p = d\frac{\sqrt{(\Sigma E_i H_i/v_i)^2 y_0 + \Sigma H_i^2 y_i/v_i^2 + (\Sigma F_i H_i/v_i)^2 y_{n+1}}}{(\Sigma T_i \ln z_i)^2} \tag{7}$$

where

$$v_i = y_i - y_{bi}$$
$$H_i = t_i \Sigma T_i \ln z_i - T_i \Sigma t_i \ln z_i$$

The standard deviations σ_p in Eqs.(6) and (7) represent the size of the variability in the peak position caused by counting statistics[10].

STANDARD DEVIATION OF STRESS

Let

$$u_i = \sin^2\psi_i$$

The stress S is to be determined from k peak positions for k inclinations of ψ. The slope M of the straight line fitted to k data points of (u_i, p_i) by using the least squares method as shown in Fig.3 is given by

$$M = \frac{\Sigma'(ku_i - \Sigma'u_i)p_i}{k\Sigma'u_i^2 - (\Sigma'u_i)^2} = \Sigma'a_i p_i$$

where Σ' denotes the summation from $i=1$ to k and

$$a_i = \frac{ku_i - \Sigma'u_i}{k\Sigma'u_i^2 - (\Sigma'u_i)^2}$$

The stress S is given by

$$S = KM = K\Sigma'a_i p_i \tag{8}$$

where K is the stress constant.

By using statistical theory[11], the variance σ_S^2 of stress can be calculated from Eq.(8) as

$$\sigma_S^2 = K^2\Sigma'a_i^2\sigma_{p_i}^2 \tag{9}$$

where the variance of peak position σ_p^2 is given by Eq.(6) or (7). The standard deviation σ_S of stress, the square root of the variance in Eq.(9), represents the size of the stress variability caused by counting statistics[10]. Since the values measured by X-ray diffraction such as the peak position p and the stress S have normal distributions[10], the 95% confidence limits of the measured value are given by

(measured value) $\pm$ 1.96σ \hfill (10)

where σ is the standard deviation of the measured value.

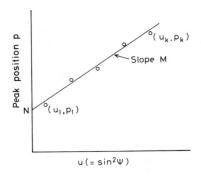

Fig.3

Fitting a straight line to k peak positions on a $\sin^2\psi$ diagram.

TEST PROCEDURES

Two kinds of materials, a hot-pressed beta silicon nitride Si_3N_4 and
sintered alpha silicon carbide SiC, were used to prepare two specimens.
The surfaces of both specimens were ground by about 50 μm with a diamond
grinding wheel of grain size of No.200. The surfaces for the x-ray irradiation were ground further with grinding wheels of No.400, 600 and 1000
succesively to remove the surface layer by about 100 μm. The direction of
the grinding is the longitudinal direction of the specimens. Finally, the
surfaces were finished by lapping with grains of No.1500. The final dimensions of the silicon nitride and silicon carbide specimens were 10.0 x 55.0
x 5.0 mm and 10.0 x 50.0 x 5.0 mm, respectively.

The specimen was set in a four points bending device and a load was
applied to produce a uniform bending stress on the surface of the specimen. The load applied to the specimen was measured with a load ring mounted
in the bending device. At the same time, the stress on the surface of the
specimen was measured with copper $K\alpha$ radiation by the Gaussian curve method
using an automated X-ray diffractometer developed in our laboratory[12]. A
slit for parallel beam optics was used. The residual stresses in the
silicon nitride specimen were also measured with copper and chromium radiations in order to investigate the X-ray penetration depth on the measured
stress value. Table 1 shows the conditions for X-ray stress measurement.

Table 1 Conditions for x-ray stress measurements

	Silicon carbide	Silicon nitride	
Characteristic x-rays Filter Diffraction plane Preset time Angular interval	Copper $K\alpha$ Nickel foil (2.0.15) 8.5 s 0.05°	Copper $K\alpha$ Nickel foil (323) 7 s 0.04°	Chromium $K\alpha$ Vanadium foil (212) 16 s 0.04°
Tube voltage Tube current Divergence angle of slit Irradiated area	40 kV 25 mA 0.5° 2 x 7 mm^2		

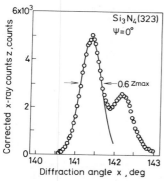

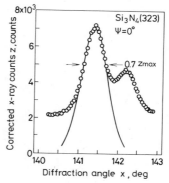

Fig.4 Gaussian function approximating $K\alpha_1$ diffraction profile
corrected for background for
silicon nitride.

Fig.5 Gaussian function approximating $K\alpha_1$ diffraction profile
uncorrected for background for
silicon nitride.

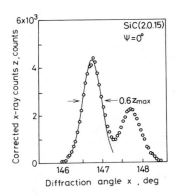

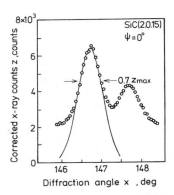

Fig.6 Gaussian function approxi-   Fig.7 Gaussian function approxi-
       mating Kα₁ diffraction profile   mating Kα₁ diffraction profile
       corrected for background for   uncorrected for background for
       silicon carbide.   silicon carbide.

TEST RESULTS

     Figure 4 shows the diffraction line of the silicon nitride specimen;
the x-ray counts were corrected for both background and LPA factor. The $K\alpha_1$
and $K\alpha_2$ profiles separate. The curve in Fig.4 is the Gaussian function
fitted to $K\alpha_1$ profile by a least squares analysis using the data points
above 60% of the maximum corrected counts z_{max}. Figure 4 shows that the $K\alpha_1$
single diffraction profile can be approximated by a Gaussian function fairly
well.

     Figure 5 shows a diffraction line corrected only for the LPA factor
without subtraction of background. The curve in Fig. 5 is the Gaussian
function calculated by a least squares analysis using the data points above
70% of the maximum counts z_{max} corrected only for LPA factor. Figure 5 shows
that the $K\alpha_1$ diffraction peak that is not corrected for background also can
be approximated by a Gaussian function. Similar results were obtained with
the silicon carbide as shown in Figs. 6 and 7. From the results of Figs.4 to
7, the peak positions were determined throughout this study by using corrected counts above 60 and 70% of the maximum counts z_{max} for the cases

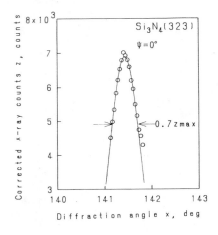

Fig.8

Peak position determination by
the Gaussian curve method with
an automated x-ray diffractometer
using a personal computer.

Table 2   Effect of background correction on measured stress value

Applied stress τ (MPa)	95% confidence limits of stress, $S \pm 1.96\sigma_S$ (MPa)	
	Uncorrected for background	Corrected for background
0	-6 ± 7	-6 ± 8
60	60 ± 7	63 ± 8
120	121 ± 7	119 ± 8
180	178 ± 7	179 ± 8
238	240 ± 7	240 ± 8
299	292 ± 7	289 ± 8

with and without correction for background, respectively. These data points fall almost in the same angular range.

Figure 8 shows an example of peak position determination, without correction for background, by using an automated diffractometer developed in our laboratory[12]. The peak position together with its standard deviation can be determined rapidly with this system.

EFFECT OF BACKGROUND SUBTRACTION ON MEASURED STRESS VALUE

Various bending stresses were applied with the bending device and the stresses on the surface of the specimen were measured by x-rays. The stresses were determined from both peak positions with and without correction for background and are shown in Table 2 together with their 95% confidence intervals. Table 2 shows that the background correction does not affect the stress value and its standard deviation, and that it can be omitted in determining the stress by the Gaussian curve method. The same result was also obtained with the silicon carbide specimen.

The background can not be determined accurately for hardened steels with martensitic structure which have very broad diffraction profile, or for some ceramics which have several diffraction profiles being located

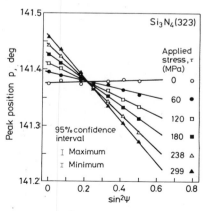

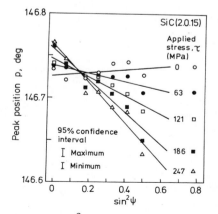

Fig.9   Sin$^2\psi$ diagram of silicon nitride for various applied stresses.

Fig.10   Sin$^2\psi$ diagram of silicon carbide for various applied stresses.

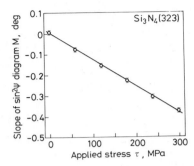

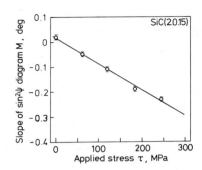

Fig.11 Variation of slope M and its 95% confidence interval with applied stress τ for silicon nitride.

Fig.12 Variation of slope M and its 95% confidence interval with applied stress τ for silicon carbide.

closely. Therefore, the Gaussian curve method which can determine the stress without subtraction of background is very useful for practical purposes. Since the background subtraction is found not to affect the stress value in the Gaussian curve method as shown in Table 2, the peak positions were determined by omitting the background subtraction hereafter.

X-RAY STRESS CONSTANT AND ELASTIC CONSTANTS

Figures 9 and 10 show the $\sin^2\psi$ diagrams for various applied stresses τ. The maximum and minimum 95% confidence intervals of peak positions were calculated from the equations given in the previous paper[6] and are shown in Figs.9 and 10. The straight lines in Figs.9 and 10 were determined from each seven data points by using the least squares method. The peak positions of the hot-pressed silicon nitride specimen shown in Fig.9 vary linearly with $\sin^2\psi$. However, the $\sin^2\psi$ diagram for the sintered silicon carbide specimen has poor linearity as shown in Fig.10. The nonlinearity of the silicon carbide specimen is considered to be caused by the texure induced in sintering.

Figures 11 and 12 show the change in the slopes M of the straight lines in Figs.9 and 10 with the applied stress τ. The 95% confidence intervals of

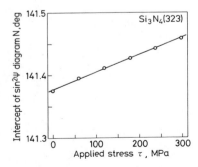

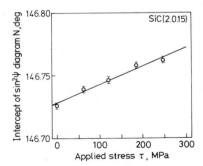

Fig.13 Variation of intercept N with applied stress τ for silicon nitride.

Fig.14 Variation of intercept N with applied stress τ for silicon carbide.

M shown in Figs.11 and 12 were calculated from the equations given in the previous paper[6]. The slope M decreases linearly with the applied stress τ as shown in Figs.11 and 12. Although the peak positions in Fig.10 for the silicon carbide specimen do not vary linearly with $\sin^2\psi$, the slope M of the straight lines fitted to the peak positions by the least squares method vary linearly with the applied stress τ as shown in Fig.12.

Figures 13 and 14 show the change in the intercept N of the $\sin^2\psi$ diagrams in Figs.9 and 10 with the applied stress τ. The 95% confidence intervals of N in Figs.13 and 14 were calculated from the equations given in the previous paper[6]; the circles in Fig.13 are drawn such that the diameters equal the 95% confidence intervals because the intervals are too small to be shown. The intercepts N of both specimens vary linearly with the applied stress τ as shown in Figs.13 and 14. Table 3 shows the 95% confidence limits of the x-ray stress and elastic constants. The limits were calculated from Eq.(10); the standard deviations of these constants were calculated from the equations given in the previous paper[6].

RESIDUAL STRESSES INDUCED BY GRINDING AND LAPPING

Table 4 shows the result of the residual stress measurement of the silicon nitride specimen ground and lapped by using the conditions described previously. The depth of the surface layer that diffracts 90% of the total diffracted x-rays is 25μm for the chromium Kα radiation for $\psi=0°$, but the depth for the copper Kα radiation is 80μm. The measured residual stress values are small as shown in Table 4 because of the gentle grinding conditions as described previously. The values measured with the chromium Kα radiation are larger than the values measured with the copper Kα radiation. This is because the copper Kα x-rays penetrate deeper through the grinding surface layer of less than 20μm thick.

Table 3   95% confidence limits of x-ray elastic constants

		Silicon nitride (323) plane	Silicon carbide (2.0.15) plane
Compliances $(10^4\text{GPa})^{-1}$	s_1	-8.5 ± 0.3	-3.7 ± 0.4
	$s_2/2$	39.2 ± 1.1	26.1 ± 1.3
Young's modulus E (GPa)		325 ± 9	446 ± 20
Poisson's ratio ν		0.28 ± 0.01	0.17 ± 0.01
Stress constant K (MPa/deg)		-779 ± 21	-971 ± 48

Table 4   95% confidence limits $S \pm 1.96\sigma_S$ (MPa) of residual stress in silicon nitride induced by grinding and lapping

Surface finish	Direction of stress	Copper Kα, (323) plane	Chromium Kα, (212) plane
Grinding	Ground direction	7 ± 10	0 ± 22
	Perpendicular to ground direction	-3 ± 11	-85 ± 23
Lapping	Ground direction	3 ± 7	-2 ± 20
	Perpendicular to ground direction	7 ± 8	35 ± 21

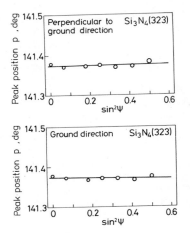

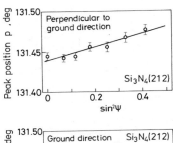

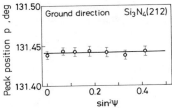

Fig.15 Sin$^2\psi$ diagram for ground
        surface of silicon nitride
        measured with copper Kα
        radiation.

Fig.16 Sin$^2\psi$ diagram for ground
        surface of silicon nitride
        measured with chromium Kα
        radiation.

Figures 15 and 16 show the sin$^2\psi$ diagram for the ground surface measured with the copper and chromium Kα radiations, respectively. The 95%
confidence intervals of the peak position are shown in Figs.15 and 16; the
intervals in Fig.15 are drawn so as to equal the diameters of the circles.
Almost horizontal straight lines in Fig.15 show that the specimen is stress
free, but the line in Fig.16 has a gentle slope showing that the specimen
has residual stress perpendicular to the grinding direction.

SUMMARY

The 95% confidence limits for the stress constant and X-ray elastic
constants of hot-pressed silicon nitride and sintered silicon carbide specimens were determined by the Gaussian curve method in order to investigate
the reproducibility of the measured constants. The diffraction lines of the
(323) plane of the silicon nitride and (2.0.15) plane of the silicon carbide
specimens were measured with copper Kα radiation. The Kα_1 and Kα_2 diffraction profiles are separated. The use of a sharp symmetric Kα_1 profile permits a precise determination of the stress by the Gaussian curve method. In
determining the peak positions, background subtraction can be omitted because
it does not affect the measured stress value.

The peak positions of the silicon nitride specimen vary linearly with
sin$^2\psi$, but those of the silicon carbide specimen give poor linearlity, which
is considered to be caused by the texture induced in sintering. However, the
slopes M and the intercept N of both specimens vary linearly with the applied
stress τ. The stress constant and x-ray elastic constants can be determined
from the slopes of the straight lines fitted to the M and N data points
against τ.

Residual stresses induced in the silicon nitride specimen by grinding
and lapping were measured with copper and chromium Kα radiations. The
chromium Kα radiation gave a larger residual stress value than the copper
Kα radiation because the former penetrates only the thin surface layer work
hardened by grinding, but the latter penetrates deeper through the work
hardened layer.

REFERENCES

1. Klug, H.P. and Alexander, L.E., "X-Ray Diffraction Procedures for Poly-
    crystalline and Amorphous Materials," John Wiley(1974), pp.635-637,
    661-665, 678, 679.
2. Hilley, M.E. ed., "Residual Stress Measurement by X-Ray Diffraction—
    SAE J784a," Society of Automotive Engineers, INC. (1980), pp.50— 60.
3. Kurita, M., Journal of Testing and Evaluation, Vol.9, No.2(1981), pp.
    133-140.
4. Kurita, M., Journal of Testing and Evaluation, Vol.9, No.5(1981), pp.
    285-291.
5. Kurita, M., "Role of Fracture Mechanics in Modern Technology," Sih,
    G.C., et al. ed., Elsevier Science Publishers B.V.(1987), pp.863-874.
6. Kurita, M., Advances in X-Ray Analysis, Vol.32(1989), Plenum Publishing
    Co., pp. 277-286.
7. Klug, H.P. and Alexander, L.E., "X-Ray Diffraction Procedures for Poly-
    crystalline and Amorphous Materials," John Wiley(1974), pp.360-364.
8. Kurita, M., Transactions of the Japan Society of Mechanical Engineers,
    Vol.43, No.368(1977), pp.1358-1360 (in Japanese).
9. Bowker, A.H. and Liebermann, G.J., "Engineering Statistics," Prentice-
    Hall(1959), p.62.
10. Kurita, M., Journal of Testing and Evaluation, Vol.11, No.2(1983), pp.
    143-149.
11. Bowker, A.H. and Liebermann, G.J., "Engineering Statistics," Prentice-
    Hall(1959), pp.48, 49.
12. Kurita, M., et al., Transactions of the Japan Society of Mechanical
    Engineers (A), Vol.54, No.500(1988), pp.854-860 (in Japanese).

RESIDUAL STRESSES IN Al$_2$O$_3$/SiC (WHISKER) COMPOSITES CONTAINING INTERFACIAL CARBON FILMS

Alias Abuhasan and Paul K. Predecki

Department of Engineering
University of Denver
Denver, CO 80208

ABSTRACT

Residual strains and stresses were measured in hot-pressed α-Al$_2$O$_3$ composites containing 25 wt% β(cubic) SiC whiskers with and without a 50-100Å thick carbon coating. X-ray diffraction methods were used including the separation of macrostresses from the microstresses in the Al$_2$O$_3$ and SiC phases by the method of Noyan and Cohen. The presence of the carbon coating on the whiskers reduced the magnitude of the residual microstresses in the whiskers by ∿13% and in the matrix by ∿10%. The measured residual stresses in the whiskers and the reduction due to the carbon film were in good agreement with calculations from recent models proposed by Hsueh, Becher and Angelini, and by Li and Bradt.

INTRODUCTION

In an earlier paper we reported the results of X-ray residual stress measurements on Al$_2$O$_3$/SiC (whisker) composites[1] including the separation of macro- and microstresses in each phase using the method of Noyan and Cohen[2,3]. In the present study these measurements were extended to composites in which the whiskers had been coated with a thin carbon film prior to incorporation into the composite. Such films are intended to reduce fiber/matrix interfacial bonding and thereby to increase toughness of the composite.

MATERIALS AND METHODS

Two samples were investigated: α Al$_2$O$_3$/25 wt% βSiC (uncoated) and α Al$_2$O$_3$/25 wt% β SiC (coated). In the second sample the whiskers were coated with a 50-100Å carbon film prior to processing[4]. Both samples were prepared by hot-pressing using essentially identical procedures and were obtained from Advanced Composite Materials Inc., Greer, SC. Samples obtained were rectangular (39 x 39 x ∿6 mm thick) with surfaces diamond ground to MIL Standard 1942-A (a procedure to minimize surface damage).

471

SEM examination revealed that the microstructures of the two samples were indistinguishable: whisker diameters were 0.6 - 0.9 μm, length-to-diameter ratios were 3-10 and whisker orientation was random in the plane normal to the hot-pressing direction (the plane exposed to X-rays in the diffraction experiments) for both samples. There was no evidence of micro-cracking in either sample.

Diffraction conditions were: CuK radiation, parafocussing optics, fixed slits and ψ tilts of the sample about the θ axis of the goniometer. Details are given elsewhere[1]. The coordinate system used for the measurements is defined in Fig. 1. Samples were X-rayed in the as-received condition. The hkl reflections utilized were:

SiC: 511,333 at $\sim 134° 2\theta$
Al_2O_3:   146 at $\sim 136°$ 2θ

Measurements were taken at $\phi = 0$ and $\phi = 90°$ at several ψ values between $+45°$ and $-45°$. The $\phi = 0$ direction was taken approximately parallel to the faint grinding marks on the samples. Diffraction measurements using the above reflections were also made on the pure SiC and Al_2O_3 starting powders to obtain the stress-free interplanar spacings.

RESULTS AND CALCULATIONS

The positions of all the diffraction peaks were determined by profile analysis using the Nicolet "Prof" program (R. A. Sparks' program). The peak positions were converted to residual strains for each phase in the directions

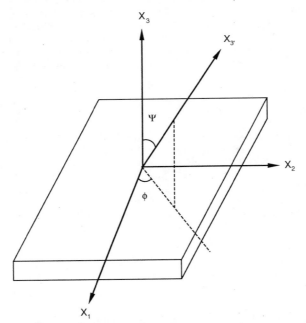

Fig. 1   Coordinate systems for specimen and measurements. Axes x_1, x_2 and x_3 are specimen axes. X-ray strain measurements are made along the $x_{3'}$ direction defined by angles ϕ and Ψ with respect to the specimen axes. The incident and diffracted beams (not shown) lie in the $x_3x_{3'}$ plane. The angle between these beams is bisected by $x_{3'}$.

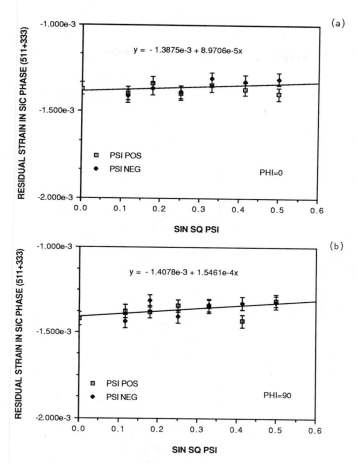

Fig. 2   Sin$^2\Psi$ plot for the 511+333 reflection from SiC in an Al$_2$O$_3$/25 wt. % SiC
(whisker) composite containing "coated" whiskers for (a): $\phi = 0$ and
(b): $\phi = 90°$. In figures 2-5, the equation of the linear least-squares fit to the
data is given at the top of each figure.

of the strain measurements as specified by the angles ϕ and ψ. Plots of
these strains against Sin$^2\psi$ are shown in Figs. 2 and 3 for the coated sample
and in Figs. 4 and 5 for the uncoated sample.

The Sin$^2\psi$ plots were approximately linear with little evidence of
ψ-splitting, consequently linear least-squares fits were made to the data in
Figs. 2-5 without regard to the sign of ψ. The total residual strains,
$\langle\varepsilon_{11}\rangle$, $\langle\varepsilon_{22}\rangle$ and $\langle\varepsilon_{33}\rangle$ in directions x_1, x_2 and x_3, respectively, (Fig. 1) in
each phase were then calculated from the slopes and intercepts in Figs. 2-5
as described previously[1]. The results are summarized in Table I.

From Table I it may be seen that for both samples, the residual strains
are compressive in the whiskers and tensile in the matrix. The magnitude of
the strains in the coated whiskers is smaller by ∿16% than in the uncoated
whiskers in these samples. (This is qualitatively evident by comparing Figs.

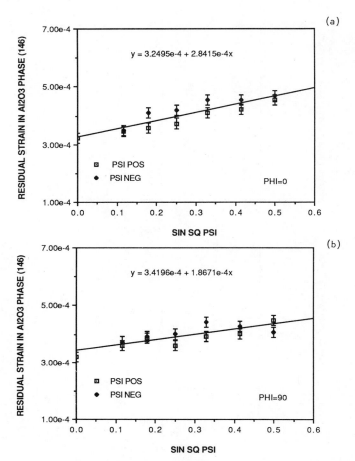

Fig. 3   Sin$^2 \Psi$ plot for the 146 reflection from Al$_2$O$_3$ in an Al$_2$O$_3$/25 wt. % SiC
(whisker) composite containing "coated" whiskers for (a): $\phi = 0$ and
(b): $\phi = 90°$.

Table I
TOTAL RESIDUAL STRAINS IN Al$_2$O$_3$/25wt%SiC (whisker) COMPOSITE
($\pm$ values are standard deviations)

Samples	Reflection	$\langle \epsilon_{11} \rangle \times 10^6$	$\langle \epsilon_{22} \rangle \times 10^6$	$\langle \epsilon_{33} \rangle \times 10^6$
Coated	Al$_2$O$_3$ (146)	609 ± 42	529 ± 43	333 ± 18
	SiC (511+333)	-1300 ± 72	-1250 ± 83	-1400 ± 37
Uncoated	Al$_2$O$_3$ (146)	557 ± 32	603 ± 30	364 ± 18
	SiC (511+333)	-1530 ± 79	-1550 ± 83	-1630 ± 37

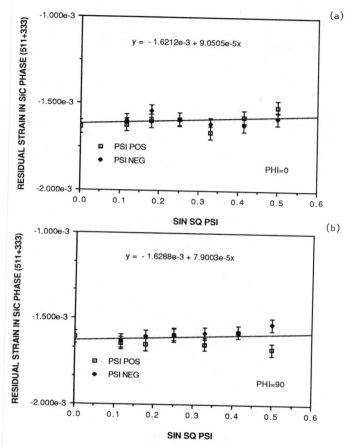

Fig. 4   Sin² Ψ plot for the 511+333 reflection from SiC in an Al₂O₃/25 wt. % SiC (whisker) composite with "uncoated" whiskers for (a): φ = 0 and (b): φ = 90°.

2 and 4.) The average strains in the matrix are smaller by ∿3% in the coated sample compared with the uncoated sample.

The total residual stresses in each phase were also obtained from the slopes and intercepts of Figs. 2-5. The macrostresses, $\langle^m\sigma_{ij}\rangle$, and the microstress components in each phase, $\langle^\mu\sigma_{ij}\rangle$, were then calculated as described elsewhere[1,2,3] using the following elastic constants:

α - Al₂O₃; ν = .23,   E = 407 GPa
β - SiC; ν = .21,   E = 352 GPa

where ν and E are Poisson's ratio and Young's modulus, respectively. The justification for the use of these constants was given previously[1]. The results of the stress calculations are given in Table II. Note that the macrostresses are defined to be the same in both phases and are assumed not to vary in the x_1, x_2 plane. As a consequence, $\langle^m\sigma_{33}\rangle$ is zero.

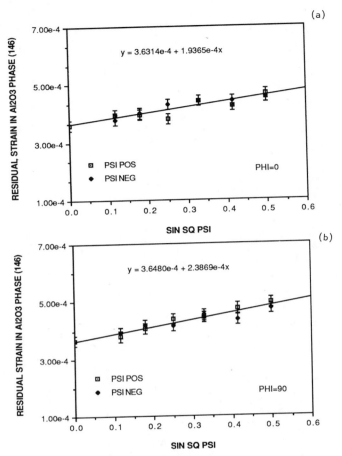

Fig. 5   Sin^2 Ψ plot for the 146 reflection from Al$_2$O$_3$ in an Al$_2$O$_3$/25 wt. % SiC
(whisker) composite with "uncoated" whiskers for (a): ϕ = 0 and
(b): ϕ = 90°.

Table II shows that the microstresses are nearly hydrostatic and are compressive in the whiskers and tensile in the matrix for both samples. The non-zero values for $\langle {}^\mu \sigma_{33} \rangle$ in both phases result from the fact that the X-ray penetration depth for 95% of the diffracted intensity (90 μm for ψ = 0 and 50 μm for ψ = ± 45°) is large compared with the whisker dimensions.

Table II also shows that the magnitudes of the microstresses in the coated whiskers are smaller by ∿13% than in the uncoated whiskers in these samples. The average microstresses in the matrix are smaller by ∿10% in the coated sample compared with the uncoated sample. Evidently the interfacial carbon film is reducing the residual microstresses by the above amounts.

The macrostresses in both samples are much smaller as expected for a gently ground ceramic surface. Differences in macrostress between the two samples are not considered significant.

Table II
RESIDUAL STRESSES (MPa) IN Al$_2$O$_3$/25wt%SiC (whisker) COMPOSITES
($\pm$ values are standard deviations)

	Microstresses - SiC whiskers		
	$\langle^\mu\sigma_{11}\rangle$	$\langle^\mu\sigma_{22}\rangle$	$\langle^\mu\sigma_{33}\rangle$
Coated	-855 $\pm$ 22	-819 $\pm$ 24	-822 $\pm$ 19
Uncoated	-949 $\pm$ 22	-962 $\pm$ 22	-968 $\pm$ 19
	Microstresses - Al$_2$O$_3$ matrix		
Coated	353 $\pm$ 9	338 $\pm$ 10	318 $\pm$ 12
Uncoated	391 $\pm$ 9	398 $\pm$ 9	335 $\pm$ 11
Samples	Macrostresses		
	$\langle^m\sigma_{11}\rangle$	$\langle^m\sigma_{22}\rangle$	$\langle^m\sigma_{33}\rangle$
Coated	59 $\pm$ 27	42 $\pm$ 29	0
Uncoated	8 $\pm$ 26	18 $\pm$ 53	0

COMPARISON WITH THEORY

The effect of interfacial films on the residual thermal stresses to be
expected in whisker-reinforced ceramic composites has recently been treated
by Hsueh et al.[5] They consider a single, long whisker having a coating
or film of uniform thickness in an infinite matrix. The whisker, the film
and the matrix are all assumed isotropic and displacement continuity is
assumed to be maintained at both interfaces during cool-down from the
processing temperature.

No Interfacial Film

In the absence of the film, the radial and axial stresses in the
whisker: σ_{ao} and σ_{zo}, respectively, are given by[5]:

$$\sigma_{ao} = \frac{(\alpha_m - \alpha_w)\,\Delta T}{[(1 - 2\nu_w)/E_w] + [(1 + \nu_m)/(E_m + \nu_w E_m)]} \tag{1}$$

$$\sigma_{zo} = 2\,\nu_w\,\sigma_{ao} + E_w\,(\alpha_m - \alpha_w)\,\Delta T \tag{2}$$

where α is the average thermal expansion coefficient and ΔT the temperature
difference between room temperature and the temperature at which residual
stresses begin to be frozen in. Subscripts w and m refer to whiskers and
matrix, respectively.

The stresses σ_{ao} and σ_{zo} for a whisker in the composite studied can be
calculated from eqs. (1) and (2) if we neglect whisker interactions and

Table III
THERMOELASTIC CONSTANTS USED

	E(GPa)	ν	α $(^{o}C^{-1})$
β - SiC	352[1]	.21[1]	4.45 x 10^{-6} [6]
α - Al$_2$O$_3$	407[1]	.23[1]	8.34[6]
Matrix (v_w = .3)	391	.224	7.173 x 10^{-6}
film*	15.45[7]	.175[8]	3.41 x 10^{-6} [7]

* Assumed to have the properties of randomly oriented bulk graphite

assume the "matrix" is isotropic having E, ν and α values in proportion to the volume fractions of the two phases present (rule-of-mixtures). E, ν and α values calculated for this "matrix" with a volume fraction of whiskers, v_w ≃.3 (equivalent to 25 wt%) are given in Table III. Using the values in Table III and ΔT = -1000°C, eqs. (1) and (2) yield σ_{ao} = -643 MPa and σ_{zo} = -1229 MPa. These values are quite comparable to values of -500 MPa and -1700 MPa, respectively, for the transverse and axial stresses in the whiskers calculated by Li and Bradt for the same composite and the same ΔT[6]. Li and Bradt used a more sophisticated model taking into account the elastic anisotropy of the whisker and applying an Eshelby approach to the micro-mechanics[6].

The average microstress in the whiskers (2/3 of the radial stress plus 1/3 of the axial stress) is -838 MPa for the Hsueh model and -900 MPa for the Li and Bradt model. These values can be approximately compared with the average of the measured microstresses in the uncoated whiskers in Table II, i.e., -960 MPa since the measured stresses are averaged over all orientations of the 511 and 333 planes in the whiskers. Considering the assumptions involved, the agreement is good.

Interfacial Film Present

With an interfacial film present, the equation of the Hsueh model for the radial stress in the whisker, σ_a, becomes more complex (see eq. 6(a) of reference 5) and is not reproduced here. The axial stress in the whisker, σ_{zw}, is still given by eq. (2) with σ_{ao} replaced by σ_a.

Before using the equations for the stresses in the whisker and the film (eqs. 6(a)-(d) of reference 5), their derivation was checked and the solutions confirmed using the MACSYMA computer algebra program (Symbolics Inc., Cambridge, MA). MACSYMA was also used to confirm that the equation for σ_a reduced to eq. (1) in the limit as the film thickness tends to zero and to obtain numerical values of σ_a and σ_{zw} using the data in Table III. (The data in Table III were obtained from references 1, 6, 7 and 8.)

Assuming that the film has the E, ν, and α properties of unoriented graphite (see Table III) and using a whisker diameter of 0.75 μm, a film thickness range of 50-100Å and ΔT = -1000°C, the radial and axial stresses in the whisker came out to be -567 to -505MPa and -1197 to -1171 MPa, respectively, for the Hsueh model. This gives an average stress in the whiskers of -777 to -727 MPa which is in reasonable agreement with the average of the measured whisker microstresses in Table II, i.e., -832 MPa.

The decrease in the average whisker stress due to the presence of the 50-100Å thick carbon film is therefore 7 to 13% for the Hsueh model, which compares favorably with the decrease of ∿13% measured by X-rays.

CONCLUSIONS

(1)   Residual microstresses in both phases of an $\alpha Al_2O_3/25$ wt% β SiC
       (whisker) composite are decreased by the presence of a 50-100Å thick
       interfacial carbon film on the whiskers.

(2)   The decrease in the whisker stresses due to the film is $\sim 13\%$ as
       measured by X-rays using the 511 + 333 SiC reflection and is readily
       detected. This decrease is in good agreement with the 7-13% decrease
       calculated for this composite from the model of Hsueh, Becher and
       Angelini[5].

(3)   The average whisker stresses measured in both the uncoated and coated
       samples: -960 and -832 MPa, respectively, are in good agreement with
       the average whisker stresses predicted by the models of Li and Bradt[6]
       (-900 MPa uncoated) and Hsueh et al[5] (-838 MPa uncoated and -777 to
       -727 MPa coated).

ACKNOWLEDGEMENTS

        This research was supported by the U.S. DOE on grant #DE-FG02-86ER45248
from the Division of Materials Sciences. The authors are grateful to Dr. J.
Rhodes of Advanced Composite Materials Inc., Greer, SC, for kindly providing
the samples, and to Drs. Z. Li and R. C. Bradt for sending preprints of their
latest research. The authors are indebted to Prof. Joel Cohen of the
Mathematics Dept. for his assistance with the MACSYMA program.

REFERENCES

(1)   P. Predecki, A. Abuhasan and C. S. Barrett, _Advances in X-Ray Analysis_
       31, 231-243 (1988).

(2)   I. C. Noyan and J. B. Cohen, _Mat. Sci. and Eng._ **75**, 179 (1985).

(3)   J. B. Cohen, _Powder Diffraction_ **1**, 15 (1986).

(4)   Personal Communication, Dr. J. Rhodes, Advanced Composite Matls. Inc.

(5)   C-H. Hsueh, P. F. Becker and P. Angelini, J. Am. Ceram. Soc., 71 [11],
       929-33 (1988).

(6)   Z. Li and R. C. Bradt, "Micromechanical Stresses in SiC Reinforced
       Al_2O_3 Composites," presented at the 90th Annual Mtg. of the American
       Ceramic Soc., Cincinnati, OH, May 1988. To be published in Jan. 1989
       issue of J. Am. Ceram. Soc.

(7)   J. Delmonte, "Technology of Carbon and Graphite Fiber Composites," Van
       Nostrand, 1981, p. 43.

(8)   Information provided by Pure Industries, Inc., St. Marys, PA.

PARALLEL BEAM AND FOCUSING X-RAY POWDER DIFFRACTOMETRY

W. Parrish
IBM Research Division
Almaden Research Center
650 Harry Road
San Jose, California 95120-6099

M. Hart
Department of Physics
The University
Manchester, U.K. M13 9PL

ABSTRACT

Comparison of results using synchrotron radiation and X-ray tubes requires a knowledge of the fundamentally different profile shapes inherent in the methods. The varying asymmetric shapes and peak shifts in focusing geometry limit the accuracy and applications of the method and their origins are reviewed. Most of the focusing aberrations such as specimen displacement, flat specimen and θ-2θ mis-setting do not occur in the parallel beam geometry. The X-ray optics used in synchrotron parallel beam methods produces narrow, symmetrical profiles which can be accurately fit with a pseudo-Voigt function. They have the same shape in the entire pattern. Only the width increases as $\tan\theta$ due to wavelength dispersion but with higher resolution systems dispersion can be eliminated. The constant instrument function contribution to the experimental profile shape is an important advantage in studies involving profile shapes, e.g., small particle sizes and microstrains, and accurate integrated intensities. The absence of systematic errors leads to more precise lattice parameter determinations.

INTRODUCTION

The increasing use of synchrotron radiation for powder diffraction has important advantages in particular situations but also raises the question as to how the data compare to those obtained with conventional X-ray tube focusing methods. This is important in selecting a method and also because of the large body of published data using conventional methods. The limited availability of storage ring sources makes it unlikely they will be used for the routine analytical applications required on a daily basis for materials characterization. The use of parallel beam geometry with X-ray tube sources results in a very large loss of intensity for powder specimens even when using a rotating anode tube. On the other hand, parallel beam grazing incidence experiments on thin films near the critical angle of total reflection have been successful.

This paper reviews some of the major differences in the data obtained with synchrotron parallel beam and X-ray tube focusing methods in terms of the profile shapes and systematic errors. The origins of the focusing aberrations are briefly reviewed and contrasted with the symmetrical constant instrument function obtained with the synchrotron parallel beam method. The major advantages of storage rings, namely the parallel beam, higher intensity, easy wavelength selectivity, and new freedom in geometrical design have been described (Parrish and Hart, 1987; Parrish, 1988) and are summarized at the end of this paper. Schematic drawings of the two geometries are shown in Fig. 1.

FOCUSING OPTICS, X-RAY TUBE

True focusing in the sense of optical lenses and mirrors is not realized in powder diffraction because of the limitations imposed by the polycrystalline specimen and the instrument geometry, and hence the term "parafocusing" has been used. The X-ray optics shown in Fig. 1(a) is the same as the original parafocusing diffractometer (Parrish, 1949) with the exception of the graphite monochromator M added to the diffracted beam. The specimen focusing circle SFC whose radius changes with θ, the diffractometer circle DC, and monochromator focusing circle MFC all intersect at the receiving slit RS in a properly aligned instrument.

The divergent beam from the narrow line focus F of the X-ray tube is limited by the entrance (or "divergence") slit ES to the specimen S length, and after reflection the rays converge at the receiving slit. The incident and diffracted beam apertures in the focusing

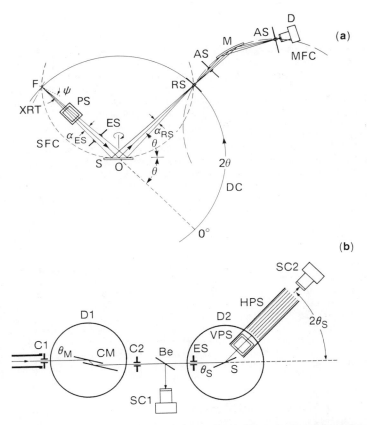

Fig. 1. X-ray optics of conventional focusing powder diffractometer (a), and parallel beam method used at Stanford Synchrotron Radiation Laboratory (b). Symbols described in text. Schematic drawings not to scale.

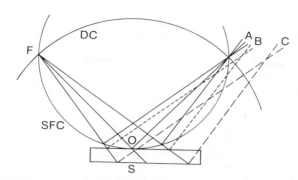

Fig. 2. Origin of displacement aberrations in reflection specimen θ-2θ scanning focusing geometry. Curved specimen matching focusing circle SFC produces reflections having the correct 2θ-position at A. The flat specimen causes shifts to B, and specimen surface displacement and transparency shifts peak to C.

plane are α_{ES} and α_{RS}, respectively. The distances F-O and O-RS are equal and determine the diffractometer radius. The image of the diffracted beam is formed at RS. If a crystal monochromator is used after the RS, it need not be a perfect crystal (e.g., Si). Therefore, a perfect crystal monochromator is not required because it would cause a large loss of intensity. Although pyrolytic graphite is imperfect and has a large rocking angle, it has a very high reflectivity for CuKα, does not change the shape of the profile, and is commonly used. Compact solid state detectors with Peltier cooling (instead of liquid nitrogen) and selectable single-channel output are being introduced (e.g., Kevex Corp.) to replace the monochromator and conventional detectors but it cannot handle very high count rates (see, for example, Bish and Chipera, 1988).

Parallel ("Soller") slits PS normal to the focusing plane are used in the incident beam to limit the axial divergence; they are not necessary in the diffracted beam because the monochromator also acts as a collimator and limits the accepted axial divergence. The focusing is maintained at all angles by scanning the detector at twice the speed of the specimen. Mis-setting of this θ-2θ relation causes the profile to broaden and the peak intensity to decrease; the effect decreases with increasing 2θ. The θ-2θ relation is not required in the parallel beam geometry and this freedom makes possible the development of new methods.

The intensity and resolution are mainly determined by the entrance and receiving slit widths. The intensity increases linearly with ES width but this also increases the length of specimen illuminated which leads to increasing flat specimen aberration. The intensity also increases rapidly with RS width and causes a symmetrical broadening but no peak shift (other than the overlapping Kα doublet components). The use of a flat specimen (rather than one with a continuously varying curvature to match the focusing circle radii) causes asymmetric broadening which increases with decreasing 2θ and causes peak shifts as illustrated schematically in Fig. 2. (The use of a flat specimen is required to rotate it rapidly around an axis normal to the surface to obtain better values of the intensities.) Specimen transparency causes a similar aberration with the maximum effect at 90° and falls to zero at 0° and 180°. Thus the intensity increases with increasing ES and RS widths but the resolution decreases. Specimen surface displacement, the most common source of error in 2θ causes peak shifts but no broadening; it varies with $\cos\theta$ in the reflection specimen mode, and is $\sin\theta$-dependent for the transmission specimen mode.

In the smaller diffraction angle region the flat specimen and axial divergence aberrations combine to cause large profile asymmetries and peak and centroid shifts. The specimen surface displacement adds to the shifts and altogether cause the systematic angular errors. The asymmetries also increase the complexity of the profile fitting function, and the varying separation of the $K\alpha_{1,2}$ components add a large asymmetry. Typical profiles of standard specimens are shown in Fig. 3. The profiles marked R were obtained with the specimen in the reflection mode using the geometry shown in Fig. 1(a). The dotted profiles T were recorded with the specimen in the transmission mode with a diffracted beam asymmetric quartz monochromator bent to approximate a logarithmic spiral.

The spectral distribution of $CuK\alpha_{1,2}$ is plotted on an angle scaled as dashed lines in Fig. 4. It shows the profile shape that would be obtained without the aberrations or contribution from the specimen. The differences between the solid line diffractometer profiles and the spectral profile are a measure of the instrumental broadening. At small 2θs the shape mainly results from the aberrations but the effect diminishes with increasing 2θ and at large 2θs they are nearly the same. The constantly varying broad asymmetric instrument function requires a more complicated profile fitting method (a sum of Lorentzians or Pearson VII functions) than the parallel beam method described below.

Is there an advantage in increasing the diffractometer radius? Even with a vacuum or helium path to eliminate air absorption a longer radius causes an intensity loss. The

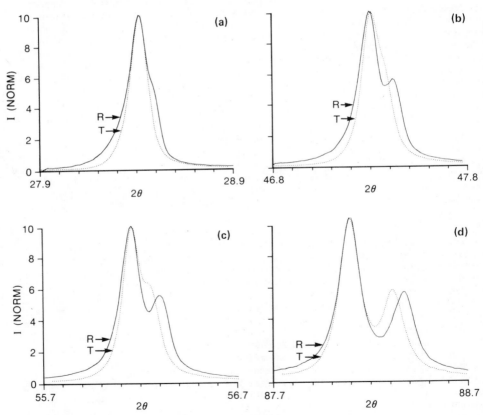

Fig. 3. Profile shapes obtained with θ-2θ scanning and focusing geometry. R reflection specimen (ES 1°, RS 0.09°), and T transmission specimen (ES 2°, with incident and receiving parallel slits to limit axial divergence). Silicon <10 μm, CuKα, (a) (111), (b) (220), (c) (311), (d) (422).

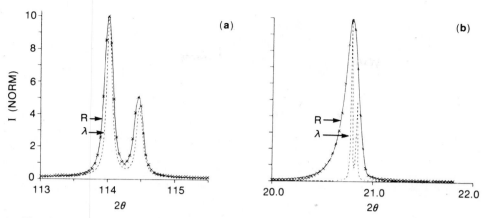

Fig. 4. Experiment profiles (crosses and solid line) from reflection specimen (R) and spectral profile of $CuK_{1,2}$ doublet (dashed line, λ) for (a) silicon (531) and (b) quartz (100) powder reflections.

entrance slit width would have to be decreased to maintain the same illuminated specimen length. The receiving slit width could be increased to maintain the same resolution (by an amount determined by the relative radii); it could also be decreased (but not narrower than the projected width of the X-ray tube focus) to increase the resolution. The axial divergence would be less because the same receiving slit length would intercept the same chord length of the larger diffraction cone; the profile asymmetry and also the intensity would then be reduced. The specimen surface displacement and transparency aberrations would be somewhat smaller since they vary inversely with R. There have been recent descriptions of the higher resolution obtained with a 250 mm radius and a FWHM of about 0.05° has been reported (see, for example, Munekawa and Toraya, 1988).

PARALLEL BEAM OPTICS, SYNCHROTRON RADIATION

The extraordinary storage ring X-ray source produces a high brightness parallel beam of white radiation. It requires a totally different diffractometer geometry from that used for focusing and has made possible the development of new methods. Figure 1(b) is a schematic drawing of the optics developed by the authors at the Stanford Synchrotron Radiation Laboratory. Unlike the narrow X-ray tube source, the primary beam can be any width (limited by the specimen size) and about 5 mm high (CI). We often use a 2.7 by 16 mm beam for powder diffraction. A silicon (111) channel monochromator CM is mounted on the first diffractometer. It is rotated to select a wavelength from the continuous radiation and reflects it in the same direction but offset by about twice the separation of the plates. A wide range of wavelengths from about 0.5 to 2Å can be used. This range straddles the power peak of the white spectrum and any wavelength can be selected in practice without realignment. $C2$ is part of the shielding and the scatter from a thin beryllium foil Be is used to monitor the incident beam intensity with scintillation counter SCI.

The diffracted beam from specimen S mounted on the second diffractometer $D2$ passes through a long parallel slit collimator HPS with 0.05° full aperture which determines the shape and resolution of the profiles. The axial divergence is limited by vertical parallel slits VPS with full aperture $\delta = 1.9$.

In this geometry the pattern can be recorded with θ-2θ scanning, or the specimen can be kept stationary and only the detector scanned without loss of profile symmetry as

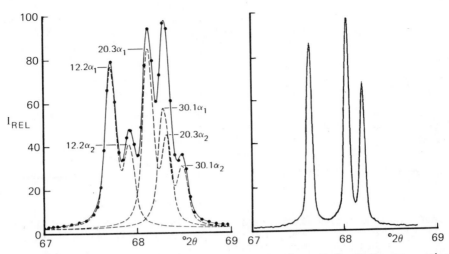

Fig. 5. The well-known overlapping three quartz reflections. Left: CuKα, conventional focusing method, dots are experimental points. Lorentzian profile fitted curves are dashed lines and solid line is their sum. Right: 1.54Å synchrotron X-ray profiles with higher resolution experimental data.

would occur in the focusing method. It is this important difference which makes it possible to use the method for high-pressure studies where the specimen must be fixed and for grazing incidence diffraction and other methods which require special angular relations between specimen and detector (Lim et al., 1987; Hart et al., 1987). Alternatively the channel monochromator can be step-scanned with the scintillation counter fixed to obtain energy dispersive diffraction data; this method has at least several orders of magnitude higher resolution than conventional white beam methods (Parrish and Hart, 1987).

The profile shapes are virtually symmetrical and have the same shape at all scattering angles. The FWHM is about 0.05° around 30° (2θ) and increases as $\tan\theta$ due to wavelength dispersion. At small angles the axial divergence causes a small asymmetric broadening. A pseudo-Voigt profile fitting function using the sum of 75% Gaussian and 25% Lorentzian components (the ratio can be varied) has been successfully used to determine integrated intensities and peak 2θs for crystal structure determination (Will et al., 1988). The constant shape, high resolution and single peak reflections make the instrument function easy to correct for in profile shape studies (Huang et al., 1987), and is an important advantage over focusing methods.

The flat specimen and specimen surface displacement aberrations which are the major sources of systematic angular errors in focusing geometry do not occur in the parallel beam geometry. (Note: if a receiving slit is used instead of the parallel slit collimator, specimen surface displacement shifts do occur, Parrish et al., 1986.) The difference in profile shapes obtained with the two methods is illustrated by the well-known quartz triplet cluster in Fig. 5.

Preliminary tests of the parallel beam method with an X-ray tube source showed a very large loss of intensity using standard powder specimens. We used a germanium (220) channel-cut monochromator to obtain a parallel beam and a 0.17° aperture parallel slit assembly to define the diffracted beam. A Rigaku rotating Cu anode tube with 0.5 by 10 mm line source and a 12° take-off angle was run at 50 kV, 250 mA. The intensity was about 100 times lower than the focusing geometry with a sealed-off Cu tube operated at 25 mA. If the channel monochromator is replaced by a narrow entrance slit having the same width as the projected width of the source, the intensity loss is smaller but the beam is not strictly parallel. It, therefore, appears to be impractical to try to duplicate the synchrotron radiation parallel beam method with an X-ray tube source for routine powder diffraction. However, there would be ample intensity for single crystal diffractometery. Grazing incidence reflectivity experiments near the critical angle of total reflection were also successful for studying multilayer films.

ADVANTAGES OF SYNCHROTRON RADIATION

Some of the advantages of the storage ring source are:

1. *Wavelength Tunability.* The easy wavelength selection makes it possible to obtain the highest possible peak-to-background by selecting a wavelength slightly longer than the absorption edges of the elements in the specimen (except for the low atomic number elements); see, for example, Parrish and Hart, 1987, Fig. 2). Short or long wavelengths may be used depending on the dispersion requirements. The patterns have a low uniform background out to the highest back-reflection angles so that many more reflection pairs can be selected for Fourier analysis. The wavelength tunability has also been applied to anomalous scattering studies (Will et al., 1987).

2. *Resolution and Profile Shape.* The higher resolution, single wavelength reflections and symmetrical shape provide a simple constant instrument function which can be used in studies involving profile shape analysis. The experimental profiles yield the specimen broadening contribution by deconvolving the instrument function. This is much more difficult and uncertain in the complex focusing profiles. Consequently, profile fitting to obtain the widths, integrated intensities, 2θs and the separation of overlapping peaks by profile fitting is far more precise than in the focusing case and more complex patterns can be analyzed.

3. *Systematic Errors.* The absence of systematic errors in 2θ due to the displacement aberrations leads to greater accuracy in the ds and lattice parameter determination. The relative peak intensities can be measured without the systematic errors caused by the overlapping $K\alpha_{1,2}$ doublet.

4. *Intensity.* The higher intensity leads to better counting statistics, and the possibility of doing rapid time-resolved experiments such as phase changes with temperature/pressure.

5. *Geometrical Advantages.* The specimen and detector do not have to maintain a θ-2θ relation. The profile shape is the same when the specimen is uncoupled from the detector and this makes it possible to do grazing incidence diffraction, quantitative preferred orientation analysis, high resolution energy dispersion diffraction and other studies not possible with the focusing method. The specimen can also be measured in the transmission mode without instrumental changes.

There are, of course, a number of disadvantages. The availability of storage ring sources is growing but is limited to certain time frames and is still not sufficient for all who wish to use it. The specimen preparation requires particles <10 μm to obtain precise values of the intensities (Parrish et al., 1986). The instrumentation is more complicated because all adjustments that must be made with the beam must be done by remotely controlled stepper motors.

ACKNOWLEDGMENTS

The authors are indebted to the staff of the Stanford Synchrotron Radiation Laboratory for making available the facilities and their support during our runs at Stanford University.

REFERENCES

D. L. Bish and S. J. Chipera, 1988, Comparison of a solid-state SI detector to a conventional scintillation detector-monochromator system in X-ray powder diffraction, *37th Denver Conf. on Applic. of X-ray Anal.*

M. Hart, W. Parrish and N. Masciocchi, 1987, Studies of texture in thin films using synchrotron radiation and energy dispersive diffraction, *Appl. Phys. Lett.* 50:897-899.

T. C. Huang, M. Hart, W. Parrish and N. Masciocchi, 1987, Line-broadening analysis of synchrotron X-ray diffraction data, *Jour. Appl. Phys.* 61:2813-2816.

G. Lim, W. Parrish, C. Ortiz, M. Bellotto and M. Hart, 1987, Grazing incidence synchrotron X-ray diffraction method for analyzing thin films, *Jour. Mater. Res.* 2:471-477.

S. Munekawa and H. Toraya, 1988, Development of a high resolution X-ray powder diffractometer and its evaluation, *37th Denver Conf. on Applic. of X-ray Anal.*

W. Parrish, 1949, X-ray powder diffraction analysis: film and Geiger counter techniques, *Science* 110:368-371.

W. Parrish, 1988, Advances in synchrotron X-ray polycrystalline diffraction, *Australian Jour. Phys.* 41:101-112.

W. Parrish and M. Hart, 1987, Advantages of synchrotron radiation for polycrystalline diffractometry, *Zeit. Krist.* 179:161-173.

W. Parrish, M. Hart and T. C. Huang, 1986. Synchrotron X-ray polycrystalline diffractometry, *Jour. Appl. Cryst.* 19:92-100.

G. Will, M. Bellotto, W. Parrish and M. Hart, 1988, Crystal structures of quartz and magnesium germanate by profile analysis of synchrotron-radiation high-resolution powder data, *Jour. Appl. Cryst.* 21:182-191.

G. Will, N. Masciocchi, M. Hart and W. Parrish, 1987, Ytterbium L_{III}-edge anomalous scattering measured with synchrotron radiation powder diffraction, *Acta Cryst.* A43:677-683.

CHEMICAL CONSTRAINTS IN QUANTITATIVE X-RAY POWDER DIFFRACTION FOR MINERAL ANALYSIS OF THE SAND/SILT FRACTIONS OF SEDIMENTARY ROCKS

D. K. Smith and G. G. Johnson, Jr.

Department of Geosciences and Materials Research Laboratory
The Pennsylvania State University, University Park, PA 16802

and

M. J. Kelton and C. A. Andersen

Advanced Analytical Services
Exploration and Production Research Laboratories
ARCO Oil and Gas Company, Plano, TX 75075

ABSTRACT

Quantitative X-ray powder diffraction using the complete digitized diffraction pattern has proved to be an effective approach to improving the accuracy of the analysis of complex mineral mixtures, provided representative reference patterns and accurate Reference Intensity Ratio (RIR) factors are available for each component phase. However, chemical and structural variability of common rock-forming minerals may complicate the pattern fitting approach. A method has been developed which utilizes X-ray fluorescence chemistry of an unknown and realistic compositional ranges for component phases as constraints on the quantitative XRD analysis without significant compromise of the pattern fit. This unique approach no only yields accurate weight fractions, but also provides indications of the specific compositions of each phase present in the mixture.

INTRODUCTION

Quantitative X-ray powder diffraction performed by comparing the full digitized diffraction pattern of an unknown to a data base of mineral diffraction patterns has proved to be an effective approach to improving the accuracy of the analysis of sand/silt fractions of sedimentary rocks and other complex mineral mixtures (Smith et al., 1987). The integration of this technique with chemical data obtained using X-ray fluorescence analysis of the unknown and actual compositional ranges for the diffraction reference minerals further improves the accuracy of the analysis without significant compromise of the XRD pattern fit. This unique method uses the chemical data as constraint conditions during the least squares refinement of the solution of mineral weight fractions determined from the XRD analysis, and therefore avoids pitfalls of most normative techniques. By maintaining a high quality XRD pattern fit, the method can easily determine relative abundances of minerals with similar compositions, such as pyrite/siderite or K-feldspar/muscovite. The technique provides a single, best-fit result which gives both accurate mineral weight fractions and information regarding the chemical compositions of each mineral phase present.

489

A master reference data base of diffraction patterns of the mineral phases present in the mixtures being analyzed is used to solve the composite pattern of the unknown by a least-squares method (Haskell and Hansen, 1978, 1979, and Hubbard et al., 1982). Each point of the unknown pattern is considered to be a weighted sum of the patterns of the component phases. Compositions of the data base minerals are included as a range for each element, where appropriate, or can be fixed if the chemical composition of a given phase is known. Constraint equations compare the weighted sum of each element contribution from each mineral to the chemical analysis of the unknown simultaneously, rather than iteratively, during the least-squares calculation. Iterations of the procedure allow the initial range of mineral compositions to be narrowed to a best fit chemistry simultaneously with the refinement of the diffraction data.

The effects of significant cation substitution or solid solution in feldspars and carbonate minerals pose problems in pattern fitting when using standard data base patterns. The use of chemistry has helped define these phases better than could be done only with diffraction information. Where pattern shifts or intensity variations are significant, multiple patterns are included in the data base, and the best fitting pattern is determined for each unknown.

THEORY

In order to employ constraint equations in the data matrix that is solved by the least-squares procedure, it is necessary to use the generalized formulation of the quantitative analysis equations derived by Karlak and Burnett (1966). The expression

$$U_{i,unk} = \sum_{j=1}^{n} W_j C_j I_{ij} \qquad (1)$$

(where $U_{i,unk}$ is the intensity of a given $\Delta 2\theta$ at $2\theta = i$ in the unknown pattern and I_{ij} is the intensity of the same $\Delta 2\theta$ interval at $2\theta = i$ in the reference phase j) from Smith et al. (1987) cannot be used because the intensities on the two sides of the equation are not on the same scale. All I_{ij}'s are normalized to $I_{max, j} = 1000$, and the U_i's depend on the count time used during data collection. Consequently, although the reference-intensity-ratio C_j is known, the weight factor W_j is related to the weight fraction x_j by some factor α.

This expression may be generalized by ratioing each intensity U_i to the strongest intensity U_k in the unknown pattern. The ratio is

$$R_{ik} = \frac{U_i}{U_k} = \frac{\sum_{j=1}^{n} W_j C_j I_{ij}}{\sum_{j=1}^{n} W_j C_j I_{kj}} = \frac{\sum_{j=1}^{n} x_j C_j I_{ij}}{\sum_{j=1}^{n} x_j C_j I_{kj}} \qquad (2)$$

and the α factor disappears. Collecting terms in x_j leads to a new set of intensity equations

$$\sum_{j=1}^{n} C_j(I_{ij}-I_{kj}R_{ik})x_j = 0. \tag{3}$$

Note that the expression for $i=k$ is identically equal to zero, which results in one less intensity equation in the data matrix than in set (1).

A problem with the equations (3) is that one solution is all x_j's = 0. In order to achieve a non-trivial solution, one additional equation must be added which is the relation among the weight fractions

$$\sum_{j=1}^{n} x_j = 1. \tag{4}$$

If the sample contains amorphous material(s), or if some of the phases are not included in the analysis, this sum would be less than unity. The weight fractions of crystalline components can still be determined by adding a known amount of an internal standard to the sample and determining each component's weight fraction relative to the weight fraction of the internal standard.

CHEMICAL CONSTRAINTS

Using the least-squares algorithm of Haskell and Hansen (1978, 1979), with the equations derived above, allows additional constraint equations to be included with the data matrix expressed as equality or inequality conditions. Using the chemical information as constraints is best done as inequalities because of the variable compositions of the reference phases and the errors in measuring the chemical compositions of the unknowns.

The data-base of reference patterns has been designed to allow the chemistry of the reference phases to be included as a compositional range. This range may be an analytical error, but in this project it is the solid-solution range over which the data-base reference pattern is considered to be valid. It may be small for minerals like quartz, or large for minerals like plagioclase. The elemental compositions of the selected end members of the solid-solution range are calculated and entered as the high and low values for each element in each reference mineral. The sum of the products of the elemental composition for each phase should equal the total value for the element in the unknown. This condition is the basis for the normative calculations usually done for mineral analyses based on chemistry. However, using this condition exclusively discounts the importance of the X-ray diffraction data. For example, normative calculations cannot determine relative abundances of polymorphs (such as pyrite/marcasite or calcite/aragonite) in a mixture, or minerals which differ only in light element chemistry when standard XRF techniques are employed to determine sample composition. If equations can be set up to allow all the variations to be taken into account, they can be used as constraint equations on the solution of the X-ray data, and should lead to a better quantitative solution than either of the data sets used separately.

Figure 1 illustrates the principal behind the equations which have been developed. For a given diffraction pattern solution of weight fractions, there is a minimum and maximum possible sum of elements based on the reference data. There is also a minimum and maximum analytical value for the same element in the unknown.   As shown in figure 1, the sum of the minimum contributions from the reference information must be less than or equal to the maximum value for that element in the analytical range.   Likewise, the sum of maximum contributions must be greater than or equal to the minimum analytical value. These conditions lead to the inequalities

$$\sum_{j=1}^{n} x_j e_{ij}(min) \le e_i + \Delta e_i \qquad (5)$$

and

$$\sum_{j=1}^{n} x_j e_{ij}(max) \ge e_i - \Delta e_i \qquad (6)$$

where e_{ij} is the concentration of element i in phase j, e_i is the total concentration of element i in the sample, and Δe_i is the known analytical error.

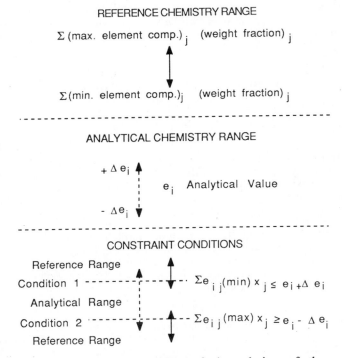

Figure 1.   Diagrammatical representation of the relation of the compositional ranges which lead to the inequality constraints.

TABLE 1

Reference-Intensity-Ratios for Selected Minerals

Quartz	1.0*
Microcline	0.360
Plagioclase	0.362
Calcite	0.752
Dolomite	0.696
Siderite	0.805
Pyrite	0.503
Kaolinite	0.359
Muscovite	0.205

* All RIR values are referenced to quartz

Once a solution has been found which satisfies both the X-ray diffraction data and the chemistry, it is possible to narrow the chemistry attributed to the reference phases and resolve the system of equations. In this way the chemistry of the minerals in the unknown may be more accurately defined. Some assumptions must be made, however, to allow this iteration. It is assumed that all the reference minerals containing a given element are adjusted proportionately. If specific information is known about any of the phases from other measurements, such as the Na/Ca ratio in plagioclase feldspar, this information may be used a priori to fix the approximate composition of that reference mineral. These chemical iterations are under development at this time.

EXPERIMENTAL

To set up this method of chemical constraints, it is necessary to prepare a data base of digitized diffraction patterns of reference minerals and to determine the appropriate reference-intensity-ratios (RIR's). These steps were accomplished in the manner of Smith et al. (1987). Improved RIR values were obtained by preparing 0.25/0.75 and 0.75/0.25 as well as 0.50/0.50 weight fraction mixtures and using the slope of the best-fit line in a plot of $x_Q I_{is} I^{rel}_{jQ}/I_{jQ} I^{rel}_{is}$ versus x_S, where Q is the internal standard (e.g. quartz) and S is the reference standard for which the RIR is being determined. Table 1 lists the values used in this study.

Mixtures of well characterized mineral specimens were used as test samples. All samples were ground in a McCrone micronizing mill to a 20 μm maximum particle size with the average particle size close to 10 μm. The samples were top-loaded into a cavity-type sample holder and compacted using a pellet press. Excess powder was removed using a razor blade to expose an interior surface flush with the front reference surface of the sample holder.

Chemical analyses were obtained from the mixtures using X-ray fluorescence. Only the analyses for Al, Si, Ca, Fe, K, and Mg were used for constraints. Their relative errors were established to be 3, 3, 5, 10, 10, and 50% respectively.

RESULTS

Six complex mineral mixtures (containing between five and eight common rock-forming mineral phases) were analyzed using the new approach. Four quantitative solutions were obtained for each data set. One solution was based only on the X-ray diffraction data. The others used chemical constraints. The first constrained solution used the analytical chemistry with the errors indicated, and variable chemistry for the reference minerals. The second constrained solution used the ideal expected sample chemistry and an assigned 1% relative analytical error. For comparison, a third constrained calculation used ideal sample chemistry and fixed compositions for the reference data base minerals. This result is equivalent to a normative calculation provided the program is using the constraints correctly. It will not be exact because the program uses a 1% error on the input data, and not all elements are used as constraints. The normative results prove that the chemical constraints are working properly.

Results of analyses of the six complex mixtures are presented in Tables 2-7. Analytical errors (reported as total absolute error for each analysis) are detailed in Table 8. Each of these mixtures generally show excellent results without chemical constraints (average total absolute error of 8.6%), which is not surprising because the minerals should be exact matches for those in the data base. In spite of this closeness, the inclusion of chemistry further improves the quality of the results in most cases, reducing the average total absolute error of the six analyses to 7.0% (see Table 8). The results shown in Table 4 illustrate the manner in which the use of chemical constraints eliminates the error resulting from the diffraction solution identifying a phase (in this case microcline) that is actually not present in the sample. The use of ideal chemistry improves all results (average total absolute error of 6.0%). All of these calculations are done with the same elements for the constraints except that the analytical error was only 1%. Using a modified reference data base with fixed mineral compositions further refines the results to produce the ideal normative solution. The normative result proves the validity of the approach. Note that because S, C, and O are not included in the chemistry, the calculated values are not exact (average total absolute error of 2.2%). The most obvious problem observed is the inability to precisely set the values for pyrite and siderite. Because Fe is the only controlled element in both phases, it is the diffraction information that fixes the weight ratio of these phases (see Tables 2, 3, 5, and 6)

TABLE 2   SAMPLE 1629

Mineral	No. Chem.	Anal. Chem.	Ideal Chem.	Fixed D.B.	Weight %
Quartz	34.6	32.5	31.4	30.4	30.0
Calcite	4.9	6.5	6.1	5.0	5.0
Dolomite	14.1	13.5	13.0	14.9	15.0
Pyrite	10.3	9.0	11.6	11.6	10.0
Siderite	14.2	12.3	13.4	13.3	15.0
Albite	6.8	10.3	9.1	9.9	10.0
Microcline	15.2	16.0	15.4	14.9	15.0
	R=0.25	R=0.27	R=0.26		

TABLE 3   SAMPLE 1630

Mineral	No. Chem.	Anal. Chem.	Ideal Chem.	Fixed D.B.	Weight %
Quartz	74.6	69.4	69.7	69.9	70.0
Dolomite	4.8	5.9	4.8	5.1	5.0
Pyrite	4.0	4.3	5.4	5.6	5.0
Siderite	3.8	3.8	4.5	4.6	5.0
Albite	3.9	5.2	5.0	5.2	5.0
Microcline	5.8	7.0	6.8	5.1	5.0
Kaolinite	3.2	4.5	3.9	4.7	5.0
	R=0.19	R=0.19	R=0.19		

TABLE 4   SAMPLE 1631

Mineral	No. Chem.	Anal. Chem.	Ideal Chem.	Fixed D.B.	Weight %
Quartz	11.1	10.4	10.4	10.0	10.0
Calcite	49.2	53.2	53.2	49.9	50.0
Dolomite	28.0	27.7	27.6	29.9	30.0
Pyrite	4.2	3.7	3.5	5.1	5.0
Albite	5.1	4.7	5.3	5.1	5.0
Microcline	2.5	0.3	0.1	0.1	0.0
	R=0.24	R=0.26	R=0.24		

TABLE 5   SAMPLE 1632

Mineral	No. Chem.	Anal. Chem.	Ideal Chem.	Fixed D.B.	Weight %
Quartz	25.9	26.2	25.3	25.9	25.0
Calcite	19.7	20.9	20.2	19.7	20.0
Dolomite	20.2	20.2	19.5	19.8	20.0
Pyrite	4.2	4.0	5.1	5.7	5.0
Siderite	9.5	8.4	8.9	9.3	10.0
Albite	7.3	4.2	7.5	5.1	5.0
Microcline	5.5	6.2	5.1	5.1	5.0
Kaolinite	7.8	10.0	8.5	9.5	10.0
	R=0.24	R=0.25	R=0.25		

TABLE 6   SAMPLE 1633

Mineral	No. Chem.	Anal. Chem.	Ideal Chem.	Fixed D.B.	Weight %
Quartz	64.1	59.1	60.2	59.8	60.0
Calcite	5.3	5.5	5.0	5.1	5.0
Dolomite	5.3	5.4	5.0	5.1	5.0
Pyrite	4.6	4.8	4.8	5.1	5.0
Albite	3.2	4.9	5.0	5.1	5.0
Microcline	9.7	10.4	10.6	10.4	10.1
Muscovite	7.9	10.0	9.6	9.5	10.0
	R=0.25	R=0.26	R=0.25		

TABLE 7   SAMPLE 1634

Mineral	No. Chem.	Anal. Chem.	Ideal Chem.	Fixed D.B.	Weight %
Quartz	50.2	49.8	48.6	49.8	50.0
Calcite	10.8	10.7	10.1	9.4	10.0
Dolomite	11.5	11.2	10.8	10.2	10.0
Pyrite	4.4	4.5	5.1	5.4	5.0
Siderite	4.2	4.1	4.5	4.7	5.0
Albite	7.0	8.4	8.3	10.1	10.0
Microcline	12.0	12.4	12.7	10.3	10.0
	R=0.19	R=0.19	R=0.19		

TABLE 8 TOTAL ABSOLUTE ERROR

Sample	No Chem.	Analy. Chem.	Ideal Chem.	Fixed D.B.
1629	10.2	10.5	8.9	3.9
1630	10.8	6.1	4.3	1.8
1631	4.7	7.6	7.8	0.3
1632	7.8	7.0	6.3	3.5
1633	9.4	2.5	1.4	1.5
1634	8.9	8.4	7.2	2.1
Average	8.6	7.0	6.0	2.2

The R-factor reported in the tables is defined as the sum of the absolute differences between the measured and calculated intensities divided by the sum of the calculated intensities, and is a good indication of the quality of the diffraction pattern fit. These factors show that the fit of the diffraction patterns is not compromised significantly with the addition of chemical constraints. Differences are only 2 percentage points, with approximately one half the magnitude of the R-factors being attributed to noise in the background. Because the diffraction data fit is not modified significantly while the quality of the analysis is improved, it is evident that an intermediate answer has been reached that "best-fits" both XRD and chemical analyses simultaneously. The results presented here indicate the quality of the analysis is dependent on: 1) high quality XRD data and accurate RIR values for the analyte phases, 2) accurate XRF sample chemistry, and 3) tight compositional ranges for data base minerals. It is clear from this study that accurate mineral compositions, as could be determined from microprobe or analytical electron microscopy, will aid greatly in producing superior quantitative analyses of sedimentary rocks and other complex mixtures when utilizing this new approach.

It is difficult to test this method with natural mixtures because the mineral components are not necessarily exact matches for data base minerals, and the true weight fractions are, of course, unknown. The same is true for normative methods of analysis. Older quantitative X-ray methods relied entirely on the diffraction data and physical calibration procedures. This combined method has the benefits of both approaches and the results should be closer than either method employed independently.

REFERENCES

Alexander, L. E. and Klug, H.P., (1948) Anal. Chem. 38, 196.

Haskell, K. H. and Hansen, R. J. (1978) Report SAND77-0052, Sandia National Laboratories, New Mexico.

Haskell, K. H. and Hansen, R. J. (1979) Report SAND78-1920, Sandia National Laboratories, New Mexico.

Hubbard, C. R., Robins, C. R. and Snyder, R. L., (1982) Adv. X-ray Anal. 26, 149.

Hubbard, C. R. and Smith, D. K., (1977) Adv. X-ray Anal. 20, 27-39.

Karlak, R. F. and Burnett, D. S., (1966) Anal. Chem. 30, 1741.

Smith, D. K., Johnson, G. G., Jr., Scheible, A., Wims, A. M., Johnson, J. L. and Ullman, G., (1987) Powder Diffraction 2, 73.

The Crystal Structures of the Cubic and Tetragonal Phases of $Y_1Ba_3Cu_2O_{6.5+\delta}$ and $Y_1Ba_4Cu_2O_{7.5+\delta}$

M. A. Rodriguez, J. J. Simmins, P. H. McCluskey, R. S. Zhou
and R. L. Snyder

Institute for Ceramic Superconductivity

New York State College of Ceramics at Alfred University
Alfred, NY 14802-1296

1 Introduction

The discovery of the superconducting material $Y_1Ba_2Cu_3O_{6+\delta}$ ("123" material) resulted in a world wide interest in the pseudo-ternary system BaO-YO-CuO [1,2]. A complete study of the phases present in this system was initiated to develop a better understanding and processing of the superconducting 123 material. The crystal structures were established for two of the three ternary compounds in this system immediately after the discovery of superconductivity [3]. One such phase was a green insulating compound $Y_2Ba_1Cu_1O_5$ ("211") which has the space group P_{bnm} [2,4]. The superconducting 123 compound was found to have the space group P_{mmm} and an ordered triple-celled perovskite structure [5,6].

Very little work has been published on the remaining ternary compound $Y_1Ba_4Cu_2O_{6.5+\delta}$ ("142"). This phase was first reported as $Y_1Ba_3Cu_2O_{6.5+\delta}$ ("132") by Frase, et. al [2]. Conflicting reports about the symmetry of the phase have come from Roth, et. al [7]. This study was undertaken to learn more about the 142 compound and examine its structural relation to the superconducting 123 phase with the hope of gaining insight into the requirements for superconductivity in this system.

2 Experimental Procedures

The 142 phase, and another point in its solid solution,132, were prepared by mixing stoichiometric proportions of yttrium oxide, barium carbonate and copper (II) oxide precursor materials in a mortar and pestle. The powder was calcined in an alumina crucible at 980°C for 12 hours in a static air environment at a heating rate of 10°C/min, and a cooling rate of 4°C/min. Powder diffraction showed the presence of barium carbonate, the 211 material along with other lines attributed to the 142 phase.

The sample powder was then ground and pressed into a 1.25 cm diameter cylindrical pellet and sintered in air at 980°C on an alumina setter. The heating rate was 10°C/min, soak time was 12 hours, and the cooling rate was 4°C/min. XRD revealed a good pattern for the 142 and 132 phases as well as a trace of the 211 green phase and barium carbonate. Si (NBS standard reference material 640a) was used as an internal standard with program INTCAL[8] to correct the d spacings for systematic errors.

Table 1. Diffraction Pattern for Tetragonal 142

d_{calc}(Å)	d_{obs}(Å)	I^{rel}	$hk\ell$	$2\theta_{obs}$	$2\theta_{calc}$	$\Delta 2\theta$
8.0413	8.0528	2	001	10.9779	10.9936	0.0158
5.7968	5.8063	1	100	15.2471	15.2722	0.0251
4.0989	4.1040	11	110	21.6358	21.6631	0.0272
4.0207	4.0212	4	002	22.0870	22.0901	0.0031
2.8984	2.8980	52	200	30.8284	30.8244	-0.0040
2.8703	2.8708	100	112	31.1285	31.1335	0.0051
2.3512	2.3522	23	202	38.2314	38.2483	0.0169
2.0495	2.0495	28	220	44.1521	44.1529	0.0009
2.0103	2.0106	12	004	45.0523	45.0593	0.0069
1.8331	1.8329	9	310	49.7015	49.6952	-0.0063
1.8259	1.8255	8	222	49.9158	49.9035	-0.0123
1.8049	1.8053	3	114	50.5148	50.5247	0.0099
1.6679	1.6678	31	312	55.0151	55.0090	-0.0061
1.6519	1.6517	13	204	55.5952	55.5896	-0.0056
1.4492	1.4492	16	400	64.2161	64.2169	0.0007
1.4352	1.4352	12	224	64.9187	64.9216	0.0029
1.2962	1.2961	18	420	72.9238	72.9199	-0.0039
1.2937	1.2935	17	332	73.0935	73.0863	-0.0072
1.2739	1.2738	19	116	74.4153	74.4122	-0.0031
1.2337	1.2336	4	422	77.2821	77.2733	-0.0089
1.2165	1.2164	2	206	78.5794	78.5750	-0.0044
1.1756	1.1755	4	404	81.8823	81.8745	-0.0077
1.0940	1.0939	4	512	89.5192	89.5179	-0.0013
1.0894	1.0897	4	424	89.9644	89.9953	0.0309
1.0819	1.0818	4	316	90.8046	90.7908	-0.0137

All patterns were collected using a Siemens D500 diffractometer with a diffracted beam graphite monochromator using Cu K_α radiation. The intensities of the diffraction lines were measured by using a side drifting sample mount and averaging over three independent runs. The resulting d's and I's for the 142 phase are given in table 1. A similar pattern was found for the tetragonal 132 phase. The resulting d's and I's are displayed in table 2.

A second phase of the 132 compound was obtained when approximately two grams of the tetragonal 132 material was calcined as a powder in a controlled atmosphere tube furnace at 1050°C in a flowing oxygen environment. The heating rate was approximately 5°C/min, the soak time was 16 hours, and the cooling rate was approximately 20°C/min. XRD analysis indicated the presence of a new phase along with trace quantities of the 211 material and the initial tetragonal 132 phase. The d's and I's were obtained as described above to get the pattern shown in table 3. The 142 phase material exposed to similar conditions did not display this cubic phase.

X-ray data for crystal structure refinement calculations for the tetragonal 142 and cubic 132 phases were collected from 10–120° 2Θ, with a count time of eight seconds and a step size of 0.050° 2Θ. The powder sample was mounted in a top loaded sample holder. Higher angles were not used since no significant peaks existed above 120° 2Θ. The conditions of the scan were room temperature and ambient atmosphere.

The diffraction patterns from the two phases were refined using our version[9] of the Wiles-Young Rietveld[10] program. Crystal structure plots were made using a local modification of the program ORTEP[11].

Table 2. Diffraction Pattern for Tetragonal 132

d_{obs}(Å)	I^{rel}	d_{obs}(Å)	I^{rel}	d_{obs}(Å)	I^{rel}
7.9981	1	1.9856	2	1.2700	4
5.808	1	1.8328	3	1.2325	3
4.0962	11	1.8240	2	1.1737	4
4.0053	4	1.8000	2	1.0931	8
3.6450	1	1.6664	39	1.0880	7
2.8970	53	1.6483	15	1.0792	6
2.8661	100	1.4486	7	1.0239	1
2.3481	20	1.4317	10	0.9642	4
2.0483	25	1.2949	7	0.9160	2
2.0048	8	1.2921	10		

3   Results

The d values shown in table 1 were analyzed using the indexing procedures of Louer [12,13], Visser [14], Taupin [15], and Goebel and Wilson [16]. All procedures suggested a tetragonal unit cell with dimensions of a = 4.097Å, and c = 7.996Å. Initial refinement of tetragonal 142 unit cell from the internal standard corrected powder pattern using the National Bureau of Standards version of the Appleman least squares gave a = 4.093(7)Å and c = 8.011(6)Å. However, the peak near 5.810Å did not index and could not be attributed to the two known low concentration impurity phases present. Since x-rays yield mostly information about the heavy atoms, a subtle arrangement of the oxygen sub-lattice would alter the structure and lattice parameters. The initial model that was refined shall be described. A slight alteration of this initial structure was needed to get to the actual cell. This subtle change is believed to be related to the oxygen sub-lattice.

The Crystal Data data base[17,18] was consulted via the software of Himes and Mighell[19] and of Harlow and Johnson[20] to see what materials may have similar unit cells and possibly be isostructural. Among the materials which were found by computer, single and double celled perovskites were prominent. The space group of the double celled perovskites in Crystal Data is $P_{\frac{4}{m}mm}$. Applying space group theory from Volume A of the International Tables for X-ray Crystallography[21] left little choice on how to place the atoms in the cell. The two barium atoms can be placed in the center of each perovskite half of the double

Table 3. Diffraction Pattern for Cubic 132

d_{calc}(Å)	d_{obs}(Å)	I^{rel}	hkℓ	$2\theta_{obs}$	$2\theta_{calc}$	$\Delta 2\theta$
5.7338	5.7412	1	110	15.4210	15.4409	0.0199
4.0544	4.0533	9	200	21.9098	21.9038	-0.0060
2.8669	2.8673	100	220	31.1676	31.1714	0.0037
2.3408	2.3407	8	222	38.4262	38.4239	-0.0023
2.0272	2.0272	21	400	44.6642	44.6636	-0.0006
1.8132	1.8130	3	420	50.2829	50.2784	-0.0045
1.6552	1.6549	23	422	55.4806	55.4678	-0.0128
1.4335	1.4340	8	440	64.9785	65.0079	0.0294
1.2821	1.2821	8	620	73.8559	73.8522	-0.0037
1.1704	1.1704	2	444	82.3142	82.3143	0.0001
1.0836	1.0836	7	642	90.6115	90.6095	-0.0019
0.9556	0.9556	4	660	107.4260	107.4211	-0.0050

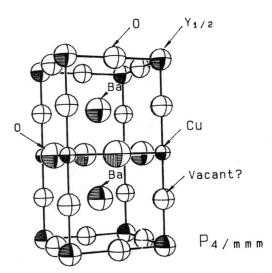

Figure 1. Initial Crystal Structure For Tetragonal 142

cell at $\frac{1}{2}, \frac{1}{2}, \frac{1}{4}$ and $\frac{1}{2}, \frac{1}{2}, \frac{3}{4}$. Keeping the analogy to the perovskite structure the Cu and Y atoms should go at the cell corners. The actual formula based on two barium atoms would be $Y_{\frac{1}{2}} Ba_2 Cu_1 O_{3.75+\delta}$. Therefore there is one copper and a half Y in this cell. If the copper atom is at $0, 0, \frac{1}{2}$, the half Yttrium atom must be placed at the origin. If the copper atoms are in the $+3$ oxidation state then there will be $4\frac{1}{4}$ oxygens in the cell. If all are $+2$ then the cell will contain $3\frac{3}{4}$ oxygens.

The $4\frac{1}{4}$ (or less) oxygen atoms can be distributed with two at $0, \frac{1}{2}, 0$ and $\frac{1}{2}, 0, 0$, two more at $0, \frac{1}{2}, \frac{1}{2}$ and $\frac{1}{2}, 0, \frac{1}{2}$, and the remaining disordered over the two sites at $0, 0, \frac{1}{4}$ and $0, 0, \frac{3}{4}$. This model is shown in figure 1.

For the 132 structure we can only fit two Ba atoms into the structure. The cell will contain two thirds of a $Y_1 Ba_3 Cu_2 O_{6.5+\delta}$ formula unit, thus the cell will contain $Y_{\frac{2}{3}} Ba_2 Cu_{\frac{4}{3}} O_{5-\delta}$. The presence of the unrefined peak at approximately 15° 2Θ led to suspicion about the structure. The peak could not be associated with an impurity phase or be ruled out as error. Thus, to consider this peak a new cell was proposed based on lattice parameters determined by Roth et. al [7]. To get the new cell four original tetragonal cells were taken together and placed side to side along the basal planes as shown in figure 2. Viewed from the top, the four cells were cut down the c-axis along the face diagonals indicated by dashed lines in the diagram. This new tetragonal cell is shown in figure 3. The difference between the initial and proposed structures must lie with the ordering of the oxygen atoms.

Least squares refinement using the new lattice parameters of the tetragonal unit cell gave a $= 5.796(8)$Å and $c = 8.041(3)$Å with a cell volume of 270.2Å^3 and a powder pattern figure of merit $F_N[22]$ of $F_{25} = 30.23$ ($\overline{|\Delta 2\theta|} = 0.0092$, # possible $= 90$).

To carry out a Rietveld refinement the background for the powder pattern was visually estimated and chosen points fit to a polynomial. The U, V, and W profile parameters were chosen by analyzing the profiles of a low and high order reflection and adjusting the pa-

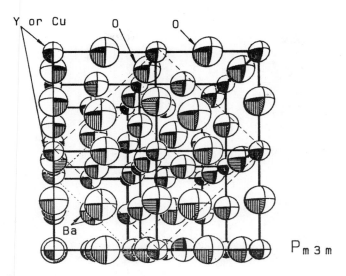

Figure 2. Determination of New Tetragonal Cell (Crystal Structure For Cubic 132)

rameters until the full width at half maximum values fit correctly. The asymmetry of the peaks in this pattern was sufficiently small not to justify invoking the full Parrish type of deconvolution [23] so a symmetric modified Lorentzian profile was used. The Rietveld for the tetragonal 142 phase converged to an R value of 8.79%. The expected error calculated for the x-ray pattern was 5.09%. The calculated and observed patterns are shown in figure 4. Much of the error can be attributed to the presence of $BaCO_3$. The carbonate peaks can be seen in the difference pattern of the refinement.

In the proposed structure the four Ba atoms occupy the $0, \frac{1}{2}, \frac{1}{4}$ sites. The one Y atom was placed at the origin of the lattice. This leaves the $\frac{1}{2}, \frac{1}{2}, 0$ $\frac{1}{2}, \frac{1}{2}, \frac{1}{2}$ $0, 0, \frac{1}{2}$ sites to place the two Cu atoms. Therefore a vacancy of one of the sites was predicted. One Cu was refined to the $\frac{1}{2}, \frac{1}{2}, 0$ site, but placement of the second Cu was more difficult. Both the $\frac{1}{2}, \frac{1}{2}, \frac{1}{2}$ and $0, 0, \frac{1}{2}$ sites refined as light in Cu but neither site was vacant. The best that can be determined from the x-ray data is that Cu is diminished over the two sites.

The oxygen atoms can be distributed with four at $\frac{1}{4}, \frac{1}{4}, 0$ and $\frac{1}{4}, \frac{1}{4}, \frac{1}{2}$, and two more at $0, 0, \frac{1}{4}$ and $\frac{1}{2}, \frac{1}{2}, \frac{1}{4}$. Analogous to the initially proposed structure, it seems probable that the $0, 0, \frac{1}{4}$ and $\frac{1}{2}, \frac{1}{2}, \frac{1}{4}$ sites are candidates for vacancies. Further work needs to be done to determine the nature of the oxygen sub-lattice with neutron diffraction.

The second phase of the 132 compound shown in table 3 immediately indexed[24] using a cubic procedure to give a cubic cell with a lattice parameter of a $=8.108(9)$Å with a cell volume of 533.1 Å^3 and $F_{12} = 26.22$ $(\overline{|\Delta 2\theta|} = 0.0075, \#$ possible $= 61)$. The tetragonal 132 described above would become cubic with space group P_{m3m} if the Y and Cu atoms were to become disordered. This model was input to the Rietveld and was refined in the same way described for the tetragonal structure. It converged to an R value of 11.91% with an expected error of 5.96%. Figure 2 is an illustration of the unit cell for this structure. The calculated and observed patterns as well as the difference pattern are shown in figure 5. The error in this refinement can be attributed to the high degree of disorder, presence of the 211 phase, and the initial tetragonal 132 phase material.

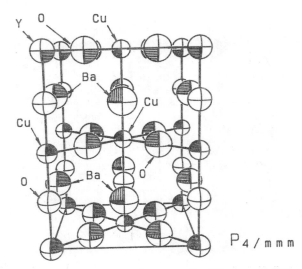

Figure 3. Proposed Crystal Structure For Tetragonal 142

4   Discussion of Results

The initial structure, illustrated in Figure 1, is similar to that of the tetragonal form of the superconducting triple celled perovskite 123 material. The 123 and 142 compounds both have a tetragonal multiple celled perovskite phase with copper-oxygen planes perpendicular to the c-crystallographic direction. In the superconducting 123 material, the yttrium atoms sit between the copper-oxygen planes on the central A site of the ABO_3 perovskite cell. In the 142 crystal structure, the yttrium atoms occupy the perovskite B site at the cell corners.

The x-ray powder diffraction patterns are primarily due to the scattering from the heavy atoms in the structure. Thus, refinement yields little information on the location of the oxygen atoms. Structurally it is reasonable to think of the edge oxygen between the copper and yttrium atoms in the initial tetragonal cell as the one most likely to be absent. This would produce the favored five fold coordination of copper. However, if only one of these oxygens is missing the cell symmetry must degrade. This fact, coupled with the presence of the peak at 15° 2Θ, led to the proposal of the new cell. A neutron diffraction study is underway with the proposed 5.8Å cell structure to establish the nature of the oxygen sub-lattice. It should be emphasized that the oxygen positions pictured in figures 1-3, although refined, are mostly speculative in the current study.

A sample of the tetragonal 132 phase sintered at 1000°C for 12 hours in oxygen showed a mixture of tetragonal and cubic phases, indicating partial phase transformation. The peaks for the cubic material lie between the split peaks of the tetragonal material. The tetragonal phase never appears to be completely converted to the cubic phase. This is illustrated in the raw data and the difference pattern for the cubic 132 shown in figure 5. Close exami-nation of the raw pattern shows small overlapping tetragonal peaks to the left of the cubic peaks near 22°, 45°, and 55° 2Θ. These small peaks and the associated displacements in the difference pattern indicate that the tetragonal phase was still present in small quantities.

Figure 2 shows the refined crystal structure for the cubic 132 material. The structure is similar to that of cubic barium titanate. In this structure the Y and Cu atoms in the cubic phase are fully disordered.

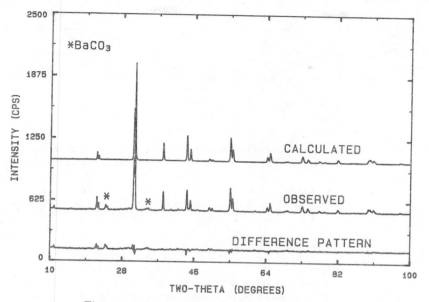

Figure 4. Rietveld Refinement For Tetragonal 142

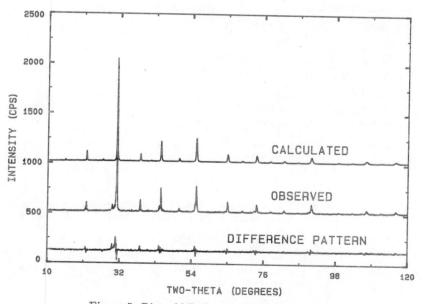

Figure 5. Rietveld Refinement For Cubic 132

Samples of the tetragonal 132 and 142 material were tested for diamagnetism and showed no superconducting behavior at liquid nitrogen temperatures. Further evidence of non-superconducting behavior was found with a sintered bar of a mixture of tetragonal and cubic phase 132 material. The bar was sintered at 1000°C in oxygen for 12 hours. The bar was electroded and tested for T_c at liquid nitrogen temperatures. It showed normal resistive behavior for a ceramic. The electrodes were removed and the bar was tested for Meissner effect at liquid nitrogen temperature. Results of this test proved negative as well. The presence of the copper-oxygen plane in the tetragonal phase is clearly not a sufficient condition for superconductivity above 77K. A lower temperature study on this material is currently underway.

5   Conclusions

$Y_1Ba_4Cu_2O_{7.5+\delta}$ and a similar compound $Y_1Ba_3Cu_2O_{6.5+\delta}$ are tetragonal with space group $P\frac{4}{m}mm$. The 132 structure has a disordered cubic phase with space group $Pm3m$. The tetragonal 142 and 132 phases have double perovskite cells which contain a copper-oxygen plane perpendicular to the c-axis similar to the 123 superconductor. The structure for the cubic phase has a single celled perovskite structure with the yttrium and copper atoms disordered on the B site. The preliminary investigation of superconductor properties at liquid nitrogen temperature showed no superconductivity behavior for the phases studied.

References

[1] C. W. Chu, P. H. Hor, R. L. Meng, L. Gao, Z. J Huang, Y. Q. Wang, M. K. Wu, J. R. Ashburn and C. Y. Torng, "Superconductivity at 93K in a New Mixed Phase Y-Ba-Cu-O Compound System at Ambient Pressure", Phys. Rev. Lett., 58, 908-910, March 1987.

[2] K. C. Frase, E. G. Liniger, D. R. Clarke, "Phase Compatibilities in the System Y_2O_3-BaO-CuO at 950C", J. Am. Ceram. Soc., 70 [9] C204-C205 (1987).

[3] A. L. Robinson, "An Oxygen Key to the New Superconductors", Science, 236 1063-1065 (1987).

[4] C. Michel and B. Raveau, "Les Oxydes A_2BaCuO_5 (A = Y,Sm,Gd,Dy,Ho,Er,Yb)", J. Sol. St. Chem., 43 73-80 (1982).

[5] W. J. Gallagher, R. L. Sandstrom, T.R. Dinger, T.M. Shaw and D.A. Chance, "Identification and Preparation Of Single Phase 90K Oxide Superconductor and Structural Determination by Lattice Imaging", Sol. St. Comm., 63 [2], 147-150, July 1987.

[6] M. A. Beno, L. Soderholm, D. W. Capone, II, D. G. Hinks, J. D. Jorgensen, J. D. Grace, Ivan K. Schuller, C. U. Segre and K. Zhang, "Structure Of The Single-Phase High-Temperature Superconductor $YBa_2Cu_3O_{7-\delta}$" Appl. Phys. Lett., 51 (1), 57-59 6 July 1987.

[7] R. S. Roth, K. L. Davis and J. R. Dennis, "Phase Equilibria and Crystal Chemistry in the System Ba-Y-Cu-O", Adv. Ceram. Mats. 2, 303-312 (1987).

[8] R. L. Snyder, "A Program to Apply Internal Standard Corrections", New York State College of Ceramics (1984).

[9] S. A. Howard and R. L. Snyder, "SHADOW - A Program for X-ray Powder Diffraction Profile Analysis", New York State College of Ceramics Report, 62 pages (1984).

[10] D. B. Wiles, R. A. Young, "A New Computer Program for Rietveld Analysis of X-Ray Powder Diffraction Patterns", J. Appl. Cryst., 14 149-151 (1981).

[11] C. K. Johnson, "ORTEP", ORNL-3794, Oak Ridge National Laboratory (1965).

[12] D. Louer and M. Louer, "Methode d'Essais et Erreurs pour l'Indexation Automatique des Diagrammes des Poudre", J. Appl. Cryst., 5 271-275 (1972).

[13] D. Louer, and R. Vargas, "Indexation Automatique des Diagrammes de Poudre par Dichotomies Succesives", J. Appl. Cryst., 15 542-545 (1982).

[14] J. W. Visser, "A Fully Automatic Program for Finding the Unit Cell from Powder Data", J. Appl. Cryst. 2, 89-95 (1969).

[15] D. Taupin, "A Powder-Diagram Automatic-Indexing Routine", J. Appl. Cryst. 6, 380-385 (1973).

[16] J. B. Goebel and A. S. Wilson, "INDEX, A Computer Program for Indexing X-ray Diffraction Powder Patterns", U.S. Atomic Energy Commission R&D Report (Batelle-Northwest Report BNWL-22), (1965).

[17] J. K. Stalick and A. D. Mighell, "Crystal Data", NBS Technical Note 1229 (1986).

[18] International Centre for Diffraction Data – JCPDS distributors, 1601 Park Lane, Swarthmore, PA 19081.

[19] V. L. Himes and A. D. Mighell, "NBS*LATTICE: A Program to Analyze Lattice Relationships", NBS Technical Note 1214 (1985).

[20] R. Harlow and G. G. Johnson Jr., Programs distributed by the JCPDS[18].

[21] T. Hahn, "International Tables for Crystallography", Vol A., D. Reidel Publishing Co. Dordrecht, Holland (1983).

[22] G. S. Smith and R. L. Snyder, "F_N: A Criterion for Rating Powder Diffraction Patterns and Evaluating the Reliability of Powder Pattern Indexing", J. Appl. Cryst., 12 60-65 (1979).

[23] S. A. Howard and R. L. Snyder, "Estimation of Particle Size and Strain using Direct Convolution Products in Profile and Pattern Fitting Algorithms", J. Appl. Cryst. in press (1988).

[24] R. L. Snyder, "Micro-Index A Package of Powder Pattern Indexing Programs", MDI Inc. PO Box 791, Livermore CA 94550.

USING X-RAY POWDER DIFFRACTION TO DETERMINE THE STRUCTURE OF VPI-5 -

A MOLECULAR SIEVE WITH THE LARGEST KNOWN PORES

Cyrus E. Crowder

Analytical Sciences, 1897 Bldg.
Dow Chemical Co., Midland, MI

Juan M. Garces

Central Research, 1776 Bldg.
Dow Chemical Co., Midland, MI

Mark E. Davis

Virginia Polytechnic Institute
Blacksburg, VA

INTRODUCTION

The synthesis of a new family of aluminophosphate-based molecular sieves containing pores defined by 18 tetrahedrally linked atoms has been described by Davis et al[1,2] at Virginia Polytechnic Institute (VPI). This development represents the first reported synthesis of a molecular sieve with pores larger than those defined by 12 tetrahedrally linked atoms[3,4]. This material has been shown to readily absorb triisopropyl benzene and exhibits a pore size distribution, determined from its argon adsorption isotherm, which suggests the existence of molecular sieve pores larger than those found in known molecular sieves[2]. Confirmation of a pore defined by 18 tetrahedrally linked aluminum and phosphorus atoms was accomplished by crystallographic structure determination using X-ray powder diffraction data. Single crystal methods were impossible due to an inability to synthesize crystals of suitable size. Details of the structure determination are reported in this work.

EXPERIMENTAL

Approximately 500 mg of washed, uncalcined VPI-5 was used for X-ray powder diffraction analysis. This sample contained approximately 3 to 5% silicon as determined by neutron activation experiments. For the diffraction experiment, the sample was placed on a thin cavity, zero-background sample mount. This mount utilizes a single crystal quartz support cut "off-axis" which results in virtually no interfering diffraction intensity from the mount. The thin cavity is approximately 1 mm deep and has the advantage of limiting depth-penetration effects that degrade diffraction peak resolution and symmetry. These effects are especially apparent in materials composed of lighter elements such as aluminum and phosphorus. Diffraction was performed using a Siemens D-500 diffractometer equipped with a long fine focus Co X-ray

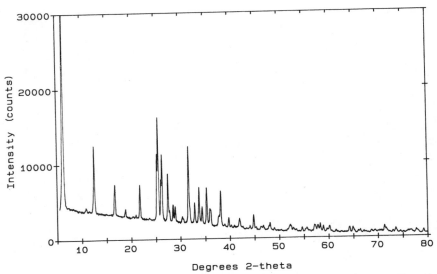

Figure 1. X-ray Powder Diffraction Data for VPI-5 using Co Kα_1
Radiation (λ = 1.788965Å).

tube operated at 50 kv and 45 ma. A primary beam quartz crystal
monochromator was used to provide a Co Kα_1 incident beam radiation (λ =
1.788965 Å). A linear position sensitive detector (PSD) was used to collect
the diffraction pattern. Data were taken over the range of 5.5 to 80° 2θ at
a rate of 0.1° /min utilizing a channel increment of 0.02° . Because a PSD was
used the total count time per channel was approximately 100 minutes. The
resulting diffraction pattern is shown in Figure 1.

ANALYSIS OF X-RAY DATA

. Precise peak centroids for 45 diffraction peaks between 5 and 50° 2θ
were obtained by profile fitting of the data. Profiles were based on a
split-Pearson VII function and obtained using an interactive computer
graphics routine employing a Newtonian fitting algorithm[5]. The resulting d-
spacings and intensities are given in Table I. Indexing of the pattern was
accomplished using the program DICVOL by Louër and Vargas[6]. A hexagonal cell
was obtained with a = 19.0 Å and c = 8.11 Å. Further refinement of these
parameters using Appleman's least squares program[7] gave a = 18.989 $\pm$ 0.001 A
and c = 8.113 $\pm$ 0.001 Å. Miller indices are given in Table II for the fitted
peaks. Transmission electron imaging using a Philips CM30 unit operating at
300 kv confirms the 16.5 Å spacing in the structure.

Inspection of the list of Miller indices from Table II shows that for
peaks of the h0l type, l is always even. This evidence indicates the
presence of a c-glide plane perpendicular to the major axes of the hexagonal
unit cell. Of the 27 possible hexagonal space groups[8], only five contain the
indicated c-glide plane. These are the P6cc, P6$_3$cm, P$\bar{6}$c2, P6/mcc, and the
P6$_3$/mcm. This field can easily be narrowed to three, since P6cc and P6/mcc
contain additional c-glide planes perpendicular to the secondary axes of the
hexagonal cell. If these additional c-glide planes were present, then
l must also be even for hh$l$ peaks as well. This is <u>not</u> the case for this
material as 111, 221, and 113 peaks are clearly evident.

TABLE I. Positions and Intensities for the X-ray Diffraction Pattern of VPI-5 (calculated intensities from DLS-refined structure).

2-Theta*	d-spacing	I(obs)	I(calc)
6.220	16.48	697	540
10.815	9.49	7	1
12.496	8.22	100	100
16.667	6.17	46	19
18.786	5.48	10	11
20.315	5.07	2	-
20.881	4.94	4	1
21.710	4.75	43	76
23.629	4.37	2	-
25.206	4.10	86	83
25.476	4.06	110	137
25.995	3.98	49	66
26.251	3.94	71	66
27.424	3.77	60	68
27.737	3.73	10	1
28.458	3.64	20	21
28.868	3.59	17	0
30.321	3.420	8	8
30.547	3.396	3	6
31.628	3.282	105	24
31.865	3.259	7	3
32.838	3.165	25	31
33.467	3.107	2	0
33.717	3.084	47	24
34.327	3.031	20	2
35.249	2.954	50	39
35.901	2.902	18	6
36.081	2.888	17	1
37.605	2.775	11	2
37.786	2.762	10	17
38.093	2.741	48	31
38.706	2.699	1	-
38.894	2.687	3	0
39.715	2.633	13	37
40.221	2.602	2	1
40.828	2.564	3	1
41.856	2.504	12	21
42.025	2.495	5	8
42.497	2.468	2	2
43.318	2.423	1	-
43.818	2.397	1	2
44.006	2.387	4	7
44.752	2.350	21	20
45.215	2.327	4	17
46.184	2.281	4	4
46.387	2.271	3	2
46.703	2.257	7	6
47.804	2.208	6	7
48.070	2.196	9	5
48.917	2.160	3	2

* Co Kα₁ radiation (λ = 1.788965 Å)

TABLE II. Hexagonal Indexing for VPI-5.

Miller indices	2-Theta* (obs)	2-Theta* (calc)	2-Theta* difference
1 0 0	6.220	6.236	-0.016
1 1 0	10.815	10.812	0.003
2 0 0	12.496	12.491	0.005
1 1 1, 2 1 0	16.667	16.678	-0.011
3 0 0	18.786	18.783	0.003
**	20.315		
2 1 1	20.881	20.890	-0.009
2 2 0	21.710	21.722	-0.012
**	23.629		
2 2 1, 4 0 0	25.206	25.220	-0.014
0 0 2	25.476	25.479	-0.003
3 1 1	25.995	26.004	-0.009
1 0 2	26.251	26.256	-0.005
3 2 0	27.424	27.430	-0.006
1 1 2	27.737	27.749	-0.012
2 0 2	28.458	28.469	-0.011
4 1 0	28.868	28.868	0.000
3 2 1	30.321	30.315	0.006
2 1 2	30.547	30.534	0.013
4 1 1, 5 0 0	31.628	31.633	-0.005
3 0 2	31.865	31.844	0.021
3 3 0	32.838	32.835	0.003
4 2 0	33.467	33.455	0.013
2 2 2	33.717	33.722	-0.005
3 1 2	34.327	34.328	-0.001
5 1 0, 3 3 1	35.249	35.257	-0.008
4 2 1	35.901	35.904	-0.003
4 0 2	36.081	36.092	-0.011
5 1 1	37.605	37.603	0.002
3 2 2	37.786	37.785	0.001
6 0 0	38.093	38.096	-0.003
**	38.706		
4 1 2	38.894	38.878	0.016
5 2 0	39.715	39.715	0.000
1 1 3	40.221	40.233	-0.012
4 3 1	40.828	40.822	0.006
5 2 1, 6 1 0	41.856	41.848	0.008
3 3 2	42.025	42.014	0.011
4 2 2	42.497	42.518	-0.021
**	43.318		
6 1 1	43.818	43.843	-0.025
5 1 2	44.006	44.002	0.004
2 2 3	44.752	44.751	0.001
5 3 0	44.752	44.760	-0.008
7 0 0	44.752	44.760	-0.008
3 1 3	45.215	45.231	-0.016
6 2 0, 4 4 1	46.184	46.187	-0.003
6 0 2	46.387	46.392	-0.005
5 3 1, 4 3 2	46.703	46.706	-0.003
5 2 2	47.804	47.780	0.024
6 2 1, 3 2 3	48.070	48.088	-0.018
4 1 3	48.917	48.935	-0.018

* Co Kα₁ radiation (λ = 1.788965 Å)
** Weak peaks due to minor impurity

The volume for the hexagonal unit cell given above is 2533 Å‡. From the reported overall density[2] of approximately 1.4 g/cc it was determined that there existed approximately 18 AlPO₄ formula units per unit cell. This eliminates the P6₃/mcm space group which requires a minimum of 24 formula units of AlPO₄ to meet symmetry requirements. Thus two hexagonal space groups are possible, namely P6₃cm and P6̄c2.

Absorption data, reported elsewhere[2], indicated 40% of the volume of the material was composed of pores readily accessible to triisopropylbenzene. This suggests a pore diameter greater than 10 Å. For a pore of this size to

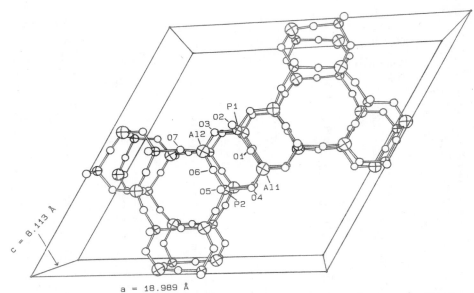

Figure 2.   Proposed Structure of VPI-5.   (Al = larged crossed spheres,
           P = smaller crossed spheres, O = open spheres).

fit within the crystallographic constraints of the given hexagonal cell, it
could only be oriented parallel to the c-axis of the cell. Furthermore, one
such pore, parallel to the c-axis, would occupy the indicated 40% of the
crystal volume, if its diameter were approximately 12 to 13 Å.

    Of the known aluminophosphate materials of this type, all show
tetrahedral coordination for both the aluminum and phosphorus atoms. Data
obtained from solid state NMR experiments[9] support the same coordination
scheme for VPI-5. Furthermore, an alternating arrangement of AlO_4 and PO_4

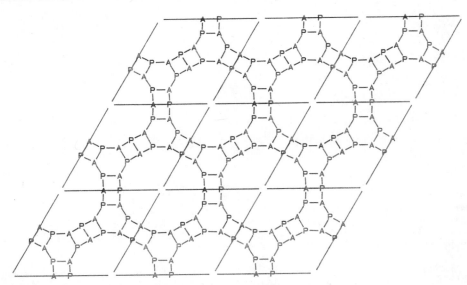

Figure 3.   VPI-5 Lattice Showing Arrangement of Pores

TABLE III.   DLS-Refined Atom Positions for the VPI-5 Structure.

Atom	Site	x	y	z
P-1	6c	0.4122	0.4122	0.2273
P-2	12d	0.4949	0.6646	0.2187
Al-1	6c	0.5755	0.5755	0.3329
Al-2	12d	0.3305	0.5006	0.3220
O-1	6c	0.4887	0.4887	0.2919
O-2	6c	0.5882	0.5882	0.5378
O-3	12d	0.3343	0.4141	0.2784
O-4	12d	0.5717	0.6573	0.2613
O-5	12d	0.3331	0.5054	0.5287
O-6	12d	0.4142	0.5871	0.2628
O-7	12d	0.2424	0.4956	0.2710

TABLE IV.   Bond Lengths and Angles in the DLS-Refined Structure.

Atoms	Distance (Å)	Atoms	Angle (°)
P1 - O1	1.544	O1 - P1 - O2	110.1
P1 - O2	1.537	O1 - P1 - O3	110.3
P1 - O3	1.554	O1 - P1 - O3'	110.3
P1 - O3'	1.554	O2 - P1 - O3	105.3
P2 - O4	1.570	O2 - P1 - O3'	105.3
P2 - O5	1.542	O3 - P1 - O3'	115.2
P2 - O6	1.545		
P2 - O7	1.538	O4 - P2 - O5	103.8
Al1 - O1	1.682	O4 - P2 - O6	112.8
Al1 - O2	1.680	O4 - P2 - O7	112.7
Al1 - O4	1.694	O5 - P2 - O6	104.0
Al1 - O4'	1.694	O5 - P2 - O7	104.3
Al2 - O3	1.716	O6 - P2 - O7	117.4
Al2 - O5	1.679		
Al2 - O6	1.687	O1 - Al1 - O2	109.7
Al2 - O7	1.679	O1 - Al1 - O4	111.2
		O1 - Al1 - O4'	111.2
		O2 - Al1 - O4	106.1
		O2 - Al1 - O4'	106.1
		O4 - Al1 - O4'	112.5
		O3 - Al2 - O5	104.1
		O3 - Al2 - O6	113.8
		O3 - Al2 - O7	114.4
		O5 - Al2 - O6	104.1
		O5 - Al2 - O7	104.5
		O6 - Al2 - O7	114.3
		Al1 - O1 - P1	171.6
		Al1 - O2 - P1'	171.5
		Al2 - O3 - P1	125.0
		Al1 - O4 - P2	128.6
		Al2 - O5 - P2'	176.7
		Al2 - O6 - P2	175.3
		Al2 - O7 - P2''	171.2

tetrahedra, similar to that found in ALPO-5, is suggested by the NMR experiments. These assumptions were then used in an exhaustive search via model building for arrangements of alternating PO_4 and AlO_4 tetrahedra that formed a 3-dimensional network defining a 12 Å pore parallel to the c-axis. This arrangement had to fit within the given hexagonal unit cell and also had to obey the symmetry constraints of either the P6c2 or P6₃cm space groups. Only one such arrangement could be obtained (see Figure 2) and its symmetry was that of the P6₃cm space group. Figure 3 shows the hexagonal net of pores formed using this unit cell. This configuration has the lattice origin centered in the large pore and matches the 81(1) net catalogued by Smith & Dytrych[10]. Based on the experimental data, this appeared to be a "best guess" for the structure of VPI-5.

The atom positions for the model were refined using a procedure known as DLS or Distance Least Squares refinement[11]. This procedure uses a least-squares technique to make small adjustments in atom positions within the constraints of the space group symmetry. This is done with the goal of optimizing all nearest neighbor and second-nearest neighbor distances in the structure. There are 11 unique atom sites which include 6-fold and 12-fold aluminum atom sites, 6-fold and 12-fold phosphorus atom sites, and two 6-fold plus five 12-fold oxygen sites. These are labelled in Figure 2. The optimal distance for P-O nearest neighbors was entered as 1.53 Å and that for Al-O as 1.70 Å. These values are in agreement with those reported by Bennett et al.[12] for other aluminophosphates. The resulting refined positions for the atoms in terms of the hexagonal cell edges are given in Table III. The resulting interatomic distances and bond angles are given in Table IV.

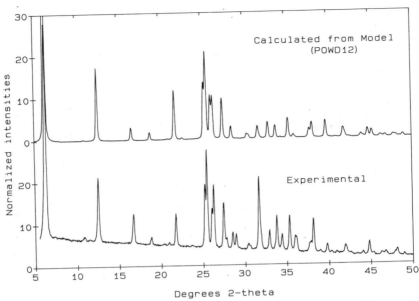

Figure 4.   Comparison of Experimental Powder Diffraction Pattern for
VPI-5 and that Calculated from the Proposed Structure.

A simulated X-ray powder diffraction pattern was computed for this
proposed structure after optimization by DLS refinement. This pattern is
shown in Figure 4 and is compared to the experimental pattern. This
positional match of the diffraction peaks is excellent, with calculated
positions being within 0.02° 2θ of those obtained experimentally. Calculated
intensities match less well but show a reasonably close correlation. These
are tabulated in Table I. A recent article by Bennett, et al[13] illustrates
the typical agreements between computed X-ray powder patterns for alumino-
phosphate molecular sieves and those measured experimentally. For all but
one case, the computed patterns in that report were derived from structures
determined by single crystal methods. The same sort of correlation is
indicated - good agreements for peak positions with a much poorer correlation
of peak intensities. In fact, for most of the patterns reported, the intensity agreement is poorer than that reported here for VPI-5. Another recent
article by Baerlocher[14] on zeolite structure refinements from powder data
states "The most informative way of judging the goodness of fit . . . is the
visual inspection of a plot of the intensity versus 2θ of the entire observed
and calculated patterns." Based on these observations, the agreement
obtained for the experimental and calculated X-ray powder patterns for VPI-5
indicated that the proposed structure model is essentially correct. If one
assumes a Van der Waals radius for oxygen to be 1.3 Å, then the pores in VPI-
5 have a free diameter of approximately 12.1 Å.

Further refinement of the proposed structure was attempted using the
Rietveld method[15]. This proved unsuccessful as intensity discrepancies in
the simulated pattern could not be resolved with the framework atoms alone.
The problem is believed to be due to absorbed water within the structure of
the material. It is known that the as synthesized material contains absorbed
water and that water will undoubtedly affect the experimental diffraction
pattern. Attempts have been made to model the configuration of the absorbed
water in the channels, but with no success in terms in providing a closer
match to the experimental X-ray powder diffraction pattern.

Another sample was obtained which had been vacuum dried for 48 hours. The X-ray diffraction pattern for this material shows a closer match to the calculated intensities for the model, however refinement of this pattern is not complete at this time.

CONCLUSIONS

From the excellent agreement between calculated and observed peak positions, precision lattice constants may be reported for the as synthesized VPI-5 as a=18.989(1) and c=8.113(1) Å for the hexagonal cell. While there are a few discrepancies among the peak intensities of the two patterns, it is believed that the basic aluminophosphate network reported here is correct, based on the X-ray, NMR, absorption, and density data. It is also again worth noting that this network corresponds to the 81(1) network derived by Smith and Dytrych[10] in their enumeration of possible molecular sieve frameworks. While they predicted that such a structure was theoretically possible, no evidence of its existence has been reported until now.

The reported atomic coordinates for the structure are the result of DLS[11] refinement of the proposed model. The agreement of the computed and experimental X-ray patterns indicates that the model is essentially correct, but it should be noted that the atomic coordinates have been obtained by DLS refinement and have not been refined directly from the experimental pattern.

Based on the described structure, VPI-5 is a molecular sieve with pores having a free diameter of approximately 12.1 Å Thus VPI-5 has pores larger than any molecular sieve known to date by at least 30%. This represents a significant advance in molecular sieve synthesis and has the potential for new absorption and catalytic applications.

The authors would like to gratefully acknowledge the support of the NSF for work carried out at the Virginia Polytechnic Institute and to thank Dr. Don Beaman for the transmission electron imaging work.

REFERENCES

1. M. E. Davis, C. Saldarriaga, C. Montes, J. Garces, and C. Crowder, "A Molecular Sieve with Eighteen-membered Rings", Nature 331:698 (1988).
2. M. E. Davis, C. Saldarriaga, C. Montes, J. Garces, and C. Crowder, "VPI-5: The First Molecular Sieve With Pores Larger Than Ten Angstroms", Zeolites 8:362 (1988).
3. D. W. Breck, "Structure of Zeolites", in: Zeolite Molecular Sieves Wiley, New York, (1974).
4. E. M. Flanigen, B. M. Lok, R. L. Patton, and S. T. Wilson, "Alumino-phosphate Molecular Sieves and the Periodic Table", Pure & Appl. Chem. 58:1351 (1986).
5. A. Brown and J. W. Edmonds, "The Fitting of Powder Diffraction Profiles to an Analytical Expression and the Influence of Line Broadening Factors", in: Advances in X-ray Analysis, 23:361, J. R. Rhodes ed., Plenum Press, New York, (1980).
6. D. Louër, and R. Vargas, "Indexation Automatique des Diagrammes de Poudre par Dichotomies Successives", J. Appl. Cryst. 15:542 (1982).
7. D. E. Appleman and H. T. Evans, "Indexing and Least-Squares Refinement of Powder Diffraction Data", USGS-GD-73-003, (1973).
8. N. F. M. Henry and K. Lonsdale, eds. "International Tables for X-ray Crystallography - Vol. I", Kynoch Press, Birmingham, England (1969).
9. M. E. Davis, C. Montes, P. E. Hathaway, J. P. Arhancet, D. Hasha, and J. Garces, "Physicochemical Properties of VPI-5", submitted to J. Amer. Chem. Soc.

10. J. V. Smith and W. J. Dytrych, "Nets with Channels of Unlimited
     Diameter", _Nature_ 309:607 (1984).
11. Ch. Baerlocher, A. Hepp, and W. M. Meier, "DLS-76 - A Program for
     the Simulation of Crystal Structures by Geometric Refinement",
     Institut für Kristallographie & Petrographie, ETH, Zurich, (1977).
12. J. M. Bennett, J. P. Cohen, E. M. Flanigen, J. J. Pluth, and
     J. V. Smith, "Crystal Structure of Tetrapropylammonium Hydroxide -
     Aluminum Phosphate Number 5", _in_: Intrazeolite Chemistry, ACS
     Symposium Series No. 218:109 (1983).
13. J. M. Bennett, W. J. Dytrych, J. J. Pluth, J. W. Richardson, Jr.
     and J. V. Smith, "Structural Features of Aluminophosphate Materials
     with Al/P = 1", _Zeolites_ 6:349 (1986).
14. Ch. Baerlocher, "Zeolite Structure Refinements Using Powder Data",
     Zeolites 6:325 (1986).
15. H. M. Rietveld, "A Profile Refinement Method for Nuclear and Magnetic
     Structures", _J. Appl. Cryst._ 2:65 (1969).

Optimizing the calculation
of standardless quantitative analysis

Lu Jinsheng, Xie Ronghou and Tan Xiaoqun

Central Iron and Steel Research Institute, Beijing, P.R. of China

C. Nieuwenhuizen

Philips Analytical, Almelo, The Netherlands

Introduction

A method for quantitative phase analysis without standards (QPAWS) has been published[1] in 1977 and has gained considerable interest, as the calibration for quantitative XRD may sometimes be difficult. Standards for quantitatative XRD are not generally available, and in many cases even the pure phases cannot be obtained.

The QPAWS method is based on *(i)* analysis of all phases present in the samples, *(ii)* foreknowledge of mass absorption coefficients (MAC) *(iii)* measuring samples which contain all phases in varying concentrations. In this method the number of samples used is equal to the number of phases to be analysed.

This latter requirement causes a limitation, because for a multi-component system it is often difficult to find samples which have a wide concentration range for each of the phases, in so few samples. And it is important to note that the analytical errors may become large if the selected samples are not sufficiently different in composition.

The least squares regression method was used by Chen [3] which avoids serious error propagation. It allows the use of more samples than the number of phases to be analysed, which leads to a better analytical accuracy, but it is based on the addition of an internal standard, it only can be applied to powder samples.

The method described in this paper uses the least-squares regression calculation for QPAWS analysis. In this way it is ensured that the widest possible composition range is included, which leads to improved accuracy. The method can be applied to both soild and powder samples. Experimental results and comparative calculations are included.

Equations for the calculation of concentrations

In the application of standardless quantitative analysis, some foreknowledge of the mass absorption coefficients is required. Here, three different situations can be distinguished: *(i)* mass absorption coefficients of the phases are known, *(ii)* mass absorption coefficients of the samples are known and *(iii)* mass absorption coefficient ratios are determined by mixing samples [1,2]. For each of these three situations a different calculation method is used.

1. MAC of the phases are known

A set of linear equations for QPAWS with known MAC of the phases has been previosly given by Zevin [1] as follows:

$$\begin{cases} \sum_{i=1}^{n}(1 - \dfrac{I_{ij}}{I_{is}})x_{is}.\mu_i = 0 \\ \\ \sum_{i=1}^{n} x_{is} = 1 \end{cases} \tag{1}$$

and $\quad x_{ns} = 1 - (x_{as} + x_{bs} + ... + x_{ms}) \quad , \quad x_{ms} = x_{ns} - 1$

I_{ij}, I_{is} Intensity of diffraction line of phase i in sample j or s

x_{is} Weight fraction of phase i in sample s

x_{ns} Weight fraction of phase n in sample s

μ_i Mass absorption coefficient of phase i

Equation (1) can be written as follows:

$$(1 - \frac{I_{aj}}{I_{as}})\mu_a x_a + (1 - \frac{I_{bj}}{I_{bs}})\mu_b x_b + ... + (1 - \frac{I_{mj}}{I_{ms}})\mu_m x_m$$
$$= -(1 - \frac{I_{nj}}{I_{ns}})\mu_n(1 - x_a - x_b - ... - x_m) \tag{2}$$

let: $\quad (1 - \dfrac{I_{aj}}{I_{as}})\mu_a = H_{aj} \ , \quad (1 - \dfrac{I_{bj}}{I_{bs}})\mu_b = H_{bj} \ , \quad (1 - \dfrac{I_{mj}}{I_{ms}})\mu_m = H_{mj} \ ,$

$$(1 - \frac{I_{nj}}{I_{ns}})\mu_n = H_{nj}$$

then equation (2) can be written as

$$(H_{aj} - H_{nj})x_a + (H_{bj} - H_{nj})x_b + ... + (H_{mj} - H_{nj})x_m = H_{nj} \tag{3}$$

Following the least squares regression approach proposed by Chen [3] the sum of the squares of the differences (Q) for p samples ($p > n$) and its derivative are

$$Q = \sum_{j=1}^{p} \left[(-H_{nj}) - \sum_{i=1}^{m} (H_{ij} - H_{nj}) x_{is} \right]^2 \qquad and \qquad \frac{\partial Q}{\partial x_i} = 0 \qquad (4)$$

let

$$\sum_{j=1}^{p} H_{aj} H_{nj} = A \quad , \qquad \sum_{j=1}^{p} H_{bj} H_{nj} = B \quad , \qquad \sum_{j=1}^{p} H_{mj} H_{nj} = M$$

then

$$(\sum_{j=1}^{p} H_{aj} H_{aj} - A) x_a + (\sum_{j=1}^{p} H_{aj} H_{bj} - A) x_b + ... + (\sum_{j=1}^{p} H_{aj} H_{mj} - A) x_m = -A$$

$$(\sum_{j=1}^{p} H_{bj} H_{aj} - B) x_a + (\sum_{j=1}^{p} H_{bj} H_{bj} - B) x_b + ... + (\sum_{j=1}^{p} H_{bj} H_{mj} - B) x_m = -B$$

$$(\sum_{j=1}^{p} H_{mj} H_{aj} - M) x_a + (\sum_{j=1}^{p} H_{mj} H_{bj} - M) x_b + ... + (\sum_{j=1}^{p} H_{mj} H_{mj} - M) x_m = -M \qquad (5)$$

or

$$\begin{bmatrix} -A \\ -B \\ ... \\ -M \end{bmatrix} = \begin{bmatrix} \sum_{j=1}^{p} H_{aj}H_{aj} - A & \sum_{j=1}^{p} H_{aj}H_{bj} - A & ... & \sum_{j=1}^{p} H_{aj}H_{mj} - A \\ \sum_{j=1}^{p} H_{bj}H_{aj} - B & \sum_{j=1}^{p} H_{bj}H_{bj} - B & ... & \sum_{j=1}^{p} H_{bj}H_{mj} - B \\ \sum_{j=1}^{p} H_{mj}H_{aj} - M & \sum_{j=1}^{p} H_{mj}H_{bj} - M & ... & \sum_{j=1}^{p} H_{mj}H_{mj} - M \end{bmatrix} \cdot \begin{bmatrix} x_a \\ x_b \\ ... \\ x_m \end{bmatrix}$$

Using these equations the weight fractions of the phases $a, b .. m$ in sample s can be calculated.

2. MAC of the samples are known

A set of linear equations for QPAWS with known MAC of the samples has been previously given by Zevin[1] as follows:

$$\begin{cases} \sum_{i=1}^{n} \frac{I_{ij}}{I_{is}} \frac{\mu_j}{\bar{\mu}_s} x_{is.} = 1 \\ \\ \sum_{i=1}^{n} x_{is} = 1 \end{cases} \qquad (6)$$

and $\qquad x_n = 1 - (x_a + x_b + ... + x_m) \quad , \qquad x_m = x_{n-1}$

$\bar{\mu}_j, \bar{\mu}_s$ Mass absorption coefficient of samples j or s

Equation (6) can be written as follows:

$$\frac{I_{aj} \bar{\mu}_j}{I_{as} \bar{\mu}_s} x_{as} + \frac{I_{bj} \bar{\mu}_j}{I_{bs} \bar{\mu}_s} x_{bs} + ... + \frac{I_{mj} \bar{\mu}_j}{I_{ms} \bar{\mu}_s} x_{ms} =$$

$$1 - \left[\frac{I_{nj} \mu_j}{I_{ns} \bar{\mu}_s} (1 - x_{as} - x_{bs} - ... - x_{ms}) \right] \qquad (7)$$

let

$$\frac{I_{aj}}{I_{as}}\frac{\bar{\mu}_j}{\bar{\mu}_s} = H_{aj} \; , \qquad \frac{I_{bj}}{I_{bs}}\frac{\bar{\mu}_j}{\bar{\mu}_s} = H_{bj} \; , \qquad \frac{I_{mj}}{I_{ms}}\frac{\bar{\mu}_j}{\bar{\mu}_s} = H_{mj} \; , \qquad \frac{I_{nj}}{I_{ns}}\frac{\bar{\mu}_j}{\bar{\mu}_s} = H_{nj} \qquad (8)$$

then equation(7) can be written as

$$(H_{aj} - H_{nj})x_a + (H_{bj} - H_{nj})x_b + \ldots + (H_{mj} - H_{nj})x_m = 1 - H_{nj} \qquad (9)$$

Following the least squares regression approach proposed by Chen[3] the sum of the squares of the differences(Q) for p samples ($p>n$) and its derivative are

$$Q = \sum_{j=1}^{p}\left[(1 - H_{nj}) - \sum_{i=1}^{m}(H_{ij} - H_{nj})x_{is}\right]^2 \qquad and \qquad \frac{\partial Q}{\partial x_i} = 0 \qquad (10)$$

let

$$\sum_{j=1}^{p} H_{aj}H_{nj} = A \quad , \qquad \sum_{j=1}^{p} H_{bj}H_{nj} = B \quad , \qquad \sum_{j=1}^{p} H_{mj}H_{nj} = M$$

then

$$
\begin{bmatrix}
\sum_{j=1}^{p} H_{aj} - A \\
\sum_{j=1}^{p} H_{bj} - B \\
\ldots \\
\sum_{j=1}^{p} H_{mj} - M
\end{bmatrix}
=
\begin{bmatrix}
\sum_{j=1}^{p} H_{aj}H_{aj} - A & \sum_{j=1}^{p} H_{aj}H_{bj} - A & \ldots & \sum_{j=1}^{p} H_{aj}H_{mj} - A \\
\sum_{j=1}^{p} H_{bj}H_{aj} - B & \sum_{j=1}^{p} H_{bj}H_{bj} - B & \ldots & \sum_{j=1}^{p} H_{bj}H_{mj} - B \\
\ldots & \ldots & \ldots & \ldots \\
\sum_{j=1}^{p} H_{mj}H_{aj} - M & \sum_{j=1}^{p} H_{mj}H_{bj} - M & \ldots & \sum_{j=1}^{p} H_{mj}H_{mj} - M
\end{bmatrix}
*
\begin{bmatrix}
x_a \\
x_b \\
\ldots \\
x_m
\end{bmatrix}
$$

Using these equations the weight fractions of the phases $a, b \ldots m$ in sample s can be calculated.

3. MAC of the samples and phases are unknown

If the mass absorption coefficients of the samples as well as the phases are unknown, the required foreknowledge of the mass absorption coefficients can be obtained from a separate experiment involving mixing of samples[1,2]. By mixing equal weights of samples s and j, and measuring the diffraction line intensity of phase i, the mass absorption coefficient ratios can be determined:

$$\frac{\bar{\mu}_j}{\bar{\mu}_s} = \frac{I_{is} - I_{i(j+s)}}{I_{i(j+s)} - I_{ij}} \qquad (11)$$

$I_{i(j+s)}$ Intensity of diffraction line of phase i
         in a 1:1 mixture of samples j and s

The mass absorption coefficient ratios determined in this way are input into equations (8) and the calculations follow the procedure of method 2 (MAC of the samples are known).

Experimental conditions

For a practical case study a three-phase system (ZnO, LiF and NiO) has been chosen and a set of four synthetic standards has been prepared. This allows the method described in this report to be demonstrated ($p > n$) and to perform comparative calculations with the QPAWS method by taking a subset of three out of four samples.

For the measurements a Philips automated x-ray diffractometer has been used, equiped with an automatic divergence slit, an Fe-tube and a graphite monochromator. The selected diffraction lines for measurement were ZnO (100), LiF (111) and NiO (200).

Four samples have been composed by weighing and mixing finely powdered pure phases. The compositions of the samples as determined by weighing are listed in table 1.

Table 1 . Concentrations of phases in the test samples.

Phase	Sample 1	Sample 2	Sample 3	Sample 4
ZnO	33.33	56.67	33.33	30.00
LiF	33.33	10.00	56.67	30.00
NiO	33.33	33.33	10.00	40.00

From each standard five specimen were prepared and each prepared specimen has been measured twice. The average intensity of these 10 measurements has been used for the calculations in this study.

Experimental results

1. Measured intensities

Average intensities as obtained on the prepared samples for the three phases present are given in table 2.

Table 2 . Average intensity of ZnO, LiF and NiO phases.

	I_{ZnO}	I_{LiF}	I_{NiO}
Sample 1	33111	18563	75633
Sample 2	44272	4508	56211
Sample 3	38573	38691	26071
Sample 4	29432	15808	87086

The mass absorption coefficients of the phases are $\mu_{ZnO}=98.4$, $\mu_{LiF}=23.85$ and $\mu_{NiO}=72.42$. The calculated average mass absorption coefficients for the samples are $\bar{\mu}_1=65.55$, $\bar{\mu}_2=82.94$, $\bar{\mu}_3=53.75$ and $\bar{\mu}_4=66.44$. Unit of MAC is cm^2 / g.

2. Calculated concentrations using known MAC of phases

Results of concentrations calculations by the method of standardless quantitative analysis presented in this report using known mass absorption coefficients of the phases are listed in table 3.

Table 3 . Concentrations using known MAC of phases.

Phase	Sample 1	Sample 2	Sample 3	Sample 4
ZnO	35.27	59.07	32.36	30.74
LiF	34.04	10.70	58.81	30.56
NiO	30.69	30.23	8.83	38.70

The maximum relative error in this table is 11.7%   ($\varepsilon_{max} = 11.7\%$).

$$\varepsilon = \frac{concentration\ found\ -\ known\ concentration}{known\ concentration} \cdot 100\%$$

For comparison, the results of concentrations calculations by the method of quantitative phase analysis without standards (QPAWS) using the same data are presented in table 4.

Table 4 . Concentrations by QPAWS using known MAC of phases.

Phase	Sample 1			Sample 2		
	1-3-4	1-2-4	1-2-3	2-1-4	2-3-4	2-1-3
ZnO	17.11	36.11	35.87	56.96	58.13	61.05
LiF	50.44	21.96	34.10	6.29	10.91	10.54
NiO	32.45	41.93	30.03	36.77	30.96	28.41

(Continued)

Phase	Sample 3			Sample 4		
	3-2-4	3-1-4	3-1-2	4-2-3	4-1-3	4-1-2
ZnO	31.94	14.63	33.91	30.95	15.92	32.40
LiF	59.01	77.16	57.69	30.63	44.97	18.87
NiO	9.05	8.21	8.40	38.42	39.11	48.73

In table 4 the identification 1-3-4 for sample 1 means that the concentrations for sample 1 have been calculated using the data of samples 1, 3 and 4. The maximum relative error in this table is 56.1%   ($\varepsilon_{max} = 56.1\%$).

3. Calculated concentrations using known MAC of samples

Results of concentrations calculations by the method of standardless quantitative analysis presented in this report using known mass absorption coefficients of the samples are listed in table 5.

Table 5 . Concentrations using known MAC of samples.

Phase	Sample 1	Sample 2	Sample 3	Sample 4
ZnO	37.61	58.73	33.27	29.75
LiF	33.68	10.33	57.48	29.00
NiO	28.71	30.94	9.25	41.25

The maximum relative error in this table is 13.9%   ($\varepsilon_{max} = 13.9\%$).

For comparison, the results of concentrations calculations by the method of quantitative phase analysis without standards (QPAWS) using the same data are presented in table 6.

Table 6 . Concentrations by QPAWS using known MAC of samples.

Phase	Sample 1			Sample 2		
	1-3-4	1-2-4	1-2-3	2-1-4	2-3-4	2-1-3
ZnO	11.65	31.22	36.00	52.84	57.10	60.90
LiF	44.79	27.64	33.31	8.50	10.43	10.23
NiO	43.56	41.14	30.69	38.66	32.47	28.87

(Continued)

Phase	Sample 3			Sample 4		
	3-2-4	3-1-4	3-1-2	4-2-3	4-1-3	4-1-2
ZnO	32.24	11.06	34.39	30.41	10.46	28.13
LiF	58.00	76.62	56.94	29.29	38.68	23.85
NiO	9.76	12.32	8.67	40.30	50.86	48.02

The maximum relative error in this table is 66.8% $(\varepsilon_{max} = 66.8\%)$.

Discussion and conclusions

The method of standardless quantitative analysis presented in this paper allows accurate results to be obtained. The average accuracy obtained in the experiments is 5 % relative. And even better accuracies are achievable for samples covering a wider range of concentrations. This method is different from the method of 'quantitative phase analysis without standards'[1] (QPAWS), in that it can use more samples than analysed phases, and uses least squares regression in the calulations of the concentrations.

The accuracy of quantitative analysis without standards depends on the width of the concentration ranges of the phases in the samples analysed. The analysis error may become significant in cases where only a limited concentration range is present in the samples.

This can be illustrated by the results obtained on sample 4, when using samples 1, 3 and 4 (4-1-3) for the analysis (table 4). The ZnO contents of the samples are 33.33%, 33.33% and 30.00% respectively, so a quite limited range indeed. The result of analysis by the QPAWS method is 15.92% as oposed to the 'known' value of 30.00% (which was determined by weighing in the preparation of the sample). This can be compared with the situation when samples 2, 3 and 4 (4-2-3) are used for the determination. The ZnO contents of the samples are 56.67%, 33.33% and 30.00%, a much wider range now. The analysis result for the same sample 4 becomes 30.95% now, which is much closer to the 'known' value of 30.00%.

The analysis result for the method proposed in this paper (which uses all four samples) is 30.74% for sample 4 (table 3), which is in good agreement with the 'known' value of 30.00%.

As the proposed method uses more standards than unknown phases, it is easier to fulfill the requirements of a sufficiently wide concentration range in a multi-component system, and the related problems of loss of accuracy are easily avoided.

The accuracies obtained in this study using the proposed method are generally better than those obtained by the QPAWS method. In the analysis of samples with known mass absorption coefficients for the phases present, the results have a maximum relative error of 11.7%. In the analysis of samples with known mass absorption coefficients for the samples, the maximum relative error is 13.9%.

The maximum errors for the QPAWS method are 56% and 67% respectively.

It is evident that the accuracy of the mass absorption coefficients used in these methods has a large influence on the analysis accuracy obtained, it is an important factor.

The method of standardless quantitative analysis as presented is quite attractive: it uses least squares regression, it allows more samples to be included in the calculations, which leads to a better analytical accuracy. The features of standardless quantitative phase analysis are fully maintained: standards are not required, the analysed phases do not have to be available as pure reference substances. The method can generally be applied to powder samples, and if the MAC of the samples or phases is known the method can also be applied to solid samples.

Acknowledgement :Authors wish to thank Mr.Chen Minghao for his helpful discussion on the least squares regression method .

References

1. L.S. Zevin, J. Appl. Cryst., **10**,147(1977)
2. L.S. Zevin, J. Appl. Cryst., **12**,582(1979)
3. Chen Minghao, Acta Metallurgica Sinica, **24**,B214(1988)

SHADOW: A SYSTEM FOR X-RAY POWDER DIFFRACTION PATTERN ANALYSIS

Scott A. Howard

Department of Ceramic Engineering
University of Missouri- Rolla
Rolla, Missouri 65401

INTRODUCTION

The SHADOW system consists of a set of programs and files, the relation of which is illustrated in Figures 1 and 2. Of the three programs in the system, SHADOW and XRDPLT are used routinely. The third, INSCAL, is used in the determination of various parameters characterizing the instrument. Two files, SHADOW.DFT and XRDPLT.DFT, contain default values for parameters controlling the reduction and display of data. A third file, WSGDAT, contains the instrument parameters.

The analytical capabilities of program SHADOW include background evaluation, peak searching, profile fitting, and the analysis of crystallite size and strain. Routine analysis of a pattern results in a hardcopy listing and a file containing data to be plotted. The plot file is used by program XRDPLT to produce either a screen plot, which may be manipulated by an interactive user, or a

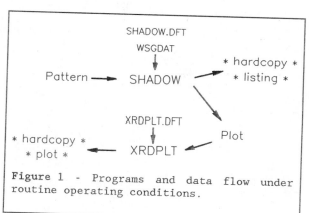

Figure 1 - Programs and data flow under routine operating conditions.

hardcopy on a plotter. The relationship between the programs and the files under routine operating conditions is illustrated in Figure 1.

In order to determine the instrument calibration curves, patterns must be obtained from defect-free specimens, i.e., specimens that do not exhibit line broadening due to small crystallite size or strain. After fitting profiles to the lines from these patterns, the refined parameters are written to a profile-data file. Program INSCAL reads this file, calculates the requisite parameters, and writes them to the file WSGDAT. Figure 2 shows the relationship between programs SHADOW and INSCAL and the file WSGDAT during the calibration process. Also shown is the profile-data file that carries the refined line information between the SHADOW and INSCAL programs. Program XRDPLT is also used for plotting the results from the various phases of this analysis.

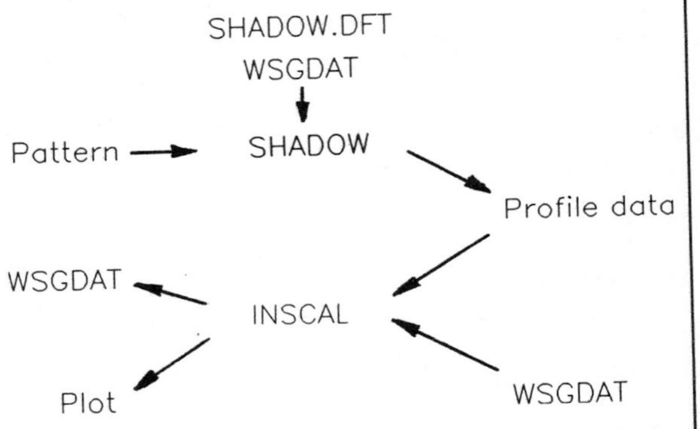

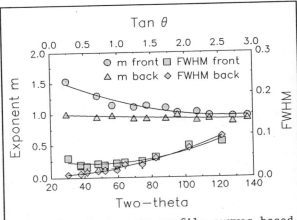

Figure 2 - Programs and data flow while generating the calibration curves for the instrument.

The goals in the design of the SHADOW system were: 1) the system was to be very flexible, and 2) the user's interactions with the program were to be kept to a minimum. Both of these goals were realized through the use of default files which are read by the programs upon start-up. The user sets the initial values for all parameters controlling the peak search, profile refinement, and display of data. Thus, the program can be configured to handle the routine problems in the laboratory while still maintaining the ability to address the most challenging of problems at the time the program is run.

WSGDAT: THE INSTRUMENT PARAMETER FILE

Among the data in the WSGDAT file are coefficients for a second-order polynomial describing the nominal split-Pearson VII line-shape for the instrument. These coefficients, derived from the dependence of the refined values of the exponent, m, on the angle 2θ and the FWHM values on $\tan\theta$, describe the low and high angle sides of the profile. Examples of the curves generated by these coefficients are shown in Figure 3. SHADOW uses the coefficients to establish the initial estimates for the profile parameters for refinement. These coefficients can also be used to constrain the shapes of the profiles when many overlapping reflections are simultaneously refined. The most significant role of the profile calibration curves is in generating a profile representative of instrumental contributions to line broadening. A second profile, modeling the specimen contributions to line broadening, is numerically convoluted with the instrument profile and the associated parameters adjusted until

Figure 3 - Instrument profile curves based on a Si profile standard.

the best fit to the observed line is obtained. Convolute profiles are typically used by program SHADOW in crystallite size and strain analysis.

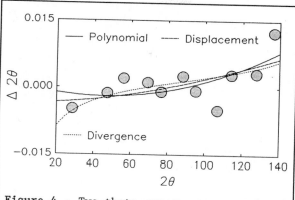

Figure 4 - Two-theta error curves produced by program INSCAL.

The instrument parameter file includes the coefficients for a second order polynomial used for correcting angular error. The 2θ correction curve, Figure 4, is applied to the search and profile-refinement results prior to printing. Program INSCAL also generates other curves which may aid in the determination of the nature of the angular error. The characteristic wavelengths and their relative intensities for the radiation source are also included in the file. The relative intensities are recorded since factors such as monochromator misalignment may affect the observed relative intensities.

PROGRAM SHADOW

The SHADOW program can run in three modes: <u>Interactive</u>, <u>Batch-Setup</u> and <u>Batch-Execute</u>. While in the interactive mode, the program will execute all commands as they are entered. In batch-setup mode, the program will record the interactive dialogue in a "control-file". Each time the program comes to the point where it would normally begin refinement, the user is instead reminded that the setup mode is active and that the commands have been recorded. This allows the user to describe complex and time consuming refinement problems, or the serial processing of one or more patterns, without waiting between steps. The recorded dialogue is fed into the program in the batch-execute mode at a later time. This feature also makes the SHADOW program ideal for repetitive analyses, such as those found in on-line quality control, since the control-file may be used to process other patterns.

Program SHADOW'S main menu, Table I, gives the type of analyses available. The <u>Search option</u>, S, is normally used to generate an initial set of line positions and intensities for profile refinement. If the <u>Refinement option</u>, R, is chosen, and a search has been performed on the

Table I
Program SHADOW's main menu

SHADOW performs the following functions:

S(S) - Search for peaks (and setup refinement regions)
R   - Refine peak parameters by profile fitting
A(S) - Auto setup for refinement (with auto peak search)
N   - Open new pattern for analysis
L   - Line list functions
P   - Read/write parameter file
E   - Exit program

Function ?

pattern, then the user is presented with the list of lines found during the search. The user may select the lines for refinement among these lines, and enter additional lines if necessary. In those cases where all lines in a pattern are to be refined, the user may direct the program to use the <u>Auto-setup for refinement with default search</u> option, AS. The program will search the pattern using the default search parameters, break the pattern down into small angular regions containing as few lines as practical, and write a control-file for SHADOW to use in batch-execute mode at a later time. If the pattern requires the use of search parameters different from the defaults, the user may run the search as a <u>Search with auto-setup of refinement regions</u>, SS, which will allow the search parameters to be specified. Following this search the user employs the <u>Auto-setup for refinement</u> option, A, and the program will write the control-file. SHADOW's control-file is written as a formatted file with comments so that it may be edited.

Whenever a peak search or profile refinement is performed, the results are stored internally by the program. The <u>Line List</u> option accesses two program functions which utilize the refined line list. This option is used to write the line parameters to a profile-data file for use by program INSCAL, and to calculate crystallite size and strain based on refined convolute-profile parameters. The <u>Parameter File</u> option allows the background, instrument parameters, search results, and refined line parameters to be written to a file. This obviates the need for repetitive

Table II
Profile-shape-function available in program SHADOW

##	Profile Function	Split	$\alpha_{2,3}$	# Parameters
1	Normal Lorentzian	no	no	3
2	Modified Lorentzian	no	no	3
3	Intermediate Lorentzian	no	no	3
4	Pearson VII	no	no	4
5	Gaussian	no	no	3
6	Voigt	no	no	4
7	Split Pearson VII	yes	no	6
8	Split Pearson VII Calibration based profile	yes	α_2	3
9	Split Pearson VII variable exponential factors variable FWHMs	yes	yes	10
10	Split Pearson VII equal exponential factors equal FWHMs	yes	yes	6
11	Split Pearson VII FWHMs and shape factors fixed according to calibration curves	yes	yes	2
12	Convolution product, Lorentzian*specimen profile	no	yes	3
13	Convolution product, Gaussian*specimen profile	no	yes	3
14	Convolution product, Lorentzian*Gaussian*specimen	no	yes	3

background analysis and peak searches when working with the pattern for an extended period of time. It also allows for documenting the instrument configuration at the time the pattern was collected.

The major function of program SHADOW is profile refinement. Since the diffraction line shape is dependent upon the physical characteristics of the radiation source, instrument and specimen, there must be a number of profile-shape-functions (PSF) available to the user. Table II contains a listing of the simple, compound, and convolution based PSF's available in program SHADOW. The simple PSF's, #1-7, consist of a single line and would be employed when a monochromatic radiation source is used, i.e., a synchrotron or neutron source. The compound PSF's, #8-11, are comprised of two or more lines: one for each of the characteristic lines in the radiation spectrum. For Cu K-α radiation, the compound profile would consist of two lines: one for the K-α_1 and the other for the K-α_2. The position and intensity of the K-α_2 line is determined with respect to the K-α_1 line by their wavelengths and their relative intensities. Thus, a compound profile requires fewer descriptive parameters than using two single-line PSF's. This reduces the problems associated with parameter correlations; particularly when there are a number of overlapping lines in a region (Howard and Snyder, 1983). The example shown in Figure 5 illustrates the advantage of using a compound profile in a case where the count rate is relatively low and the lines of interest overlap.

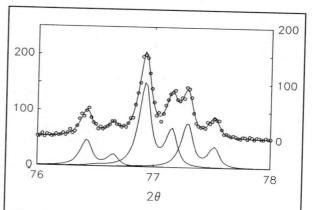

Figure 5 - Refined compound profiles, bottom, fit to a pattern from an Al_2O_3-Si mixture. The collected data and resultant pattern are offset for clarity. (Intensities in CPS.)

The asymmetry of the diffraction profile is accommodated through the use of a split profile. For example, line #7 in Table II uses two halves of a Pearson VII profile, each having a common peak position and intensity

Table III
Refinement options available in program SHADOW

Background known....... Y		Mono correction... 0.8005
A Profile............ 10	G	Asymmetry correction. N
B Size broadening...... N	H	Strain broadening.... N
C L/P correction....... Y	I	Rwp residual........ Y
D Use calc bkg as is... N	J	Ref calc bkg offset.. N
E Refine linear bkg.... Y	K	Amorphous profile.... N
F Use Gauss-Newton..... Y	L	Generate plot file... Y

(Brown and Edmonds, 1958). The FWHM and exponents for each half may be varied independently. Several of the compound profiles in Table II are based on split line functions but, to reduce the correlation between parameters, the shapes of the α_1 and α_2 lines are constrained in some fashion. The choice of PSF is generally based on the separation of the α_1 and α_2 lines as well as the degree of overlap of neighboring lines. In the most severely overlapped regions it may be necessary to constrain the lines to follow that of the instrument-profile-standard, #11, or to allow for a simple increase in the FWHM of these lines, #8.

Program SHADOW allows the user to tailor both the search and the refinement operations to a specific problem. Table III shows the option menu presented prior to refinement. The values of the options were taken from the default file SHADOW.DFT. The user has only to change the option(s) required for the special problem while the normal refinement will proceed with the user simply hitting the return key. A similar option menu is presented upon entry to the peak search section of the program.

CRYSTALLITE SIZE AND STRAIN ANALYSES CARRIED OUT WITH PROGRAM SHADOW

Program SHADOW may be used to determine crystallite size and strain in a specimen. The method of analysis is based on the angular dependence of the specimen-profile broadening. The profile-calibration curves, Figure 3, are used to generate the profile-broadening contribution due to the instrument. A second function, which may be either a Lorentzian, Gaussian, or convolution product of the two, may be used to represent the broadening contributions from the specimen. This function is convoluted with the instrumental function and the adjustable parameters refined until the best fit is obtained. Refinement of a single line will yield either a size or a strain value. When two or more lines are fit in a pattern, the angular dependence of the specimen profiles is used to determine the crystallite size and strain simultaneously. Crystallite size, τ, and strain, ϵ, are evaluated from:

$$\beta_\tau = \lambda / (\tau \cos\theta_k), \quad \beta_\epsilon = 4 \epsilon \tan\theta_k .$$

where β are the integral-breadths of the specimen lines (radians) and θ_k is the Bragg diffraction angle for the line. When both size and strain broadening is active, and both their contributions are modeled with a Lorentzian function, then the integral breadth of the specimen line is simply the sum of the two, i.e., $\beta=\beta_\tau+\beta_\epsilon$.

Use of the direct convolution products generated by program SHADOW in profile-fitting has yielded results of particular interest to those performing size and strain

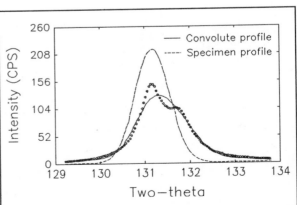

Figure 6 - The relatively poor fit obtained by fitting the W line with a convolute profile based on a Gaussian specimen function.

analysis. The line broadening contributions from small crystallite size is generally accepted to be Lorentzian in nature. That is, the specimen function can be best modeled by a Lorentzian function. Strain contributions, however, are generally believed to be Gaussian in nature. Work with the direct convolution products has indicated that, in many cases, the contributions from strain broadening can also be described by a Lorentzian contribution. Figures 6 and 7 illustrates the results obtained in fitting the lines from a W specimen that was highly strained. The relatively poor fit obtained

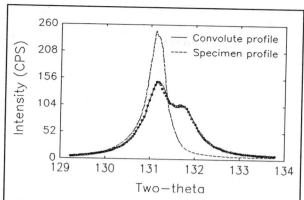

Figure 7 - The good quality of fit obtained using a Lorentzian based convolute profile in fitting the line from the strained W specimen is evident when this result is compared to figure 7.

when the specimen profile was modeled by a Gaussian function is evident from examination of Figure 6. Modeling the specimen contribution with a Lorentzian function, Figure 7, yielded the superior fit among the two convolute PSF's.

This result was not unique to this W specimen as similar results have been found in examining strained Si and Al metals (Howard and Snyder, 1988, Yau and Howard, 1988), as well as specimens in the $(Sr_{1-x}, La_x)TiO_{3\pm\delta}$ system (Howard, Yau and Anderson, 1988). This similarity in the broadening contributions indicates that the nature of the contribution, i.e., Lorentzian or Gaussian broadening, may not be reliably used in evaluating crystallite size and strain. However, the angular dependence of these contributions can be used to evaluate size and strain.

Figures 8 and 9 illustrate the concern with the nature of the specimen broadening. The integral-breadths of the specimen profiles are significantly larger when modeled by a Gaussian function than by a Lorentzian function. This translates directly to a difference in the calculated size and strain of the specimen. The nature of the broadening can best be illustrated in a Williamson-Hall style plot (1953), Figure 9. (All the integral-breadths values are from a Lorentzian specimen function.) Also

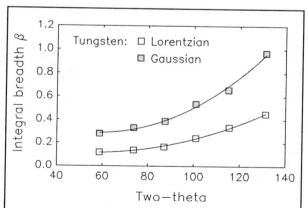

Figure 8 - Integral-breadth of the W lines based on the specimen function modeled by a Lorentzian and Gaussian profile. Size/strain estimates based on these curves would be significantly different.

illustrated in figure 9 are the results from the analysis of commercially available Linde A and C alumina specimens having a nominal crystallite size of 0.3 μm and 1.0 μm, respectively. The lack of slope in the fitted lines indicates the lack of strain in the Al_2O_3 specimens while the y-axis intercepts give a relative measure of the specimen's crystallite size.

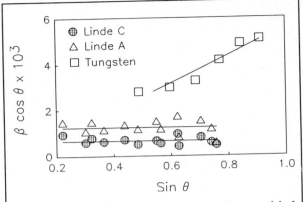

Figure 9 - Size and strain analysis provided by program SHADOW.

SUMMARY

The SHADOW system provides a means of characterizing a particular diffraction system in terms of its intrinsic diffraction profiles and the angular errors present in the measurement system. The variety of profile-shape-functions available allows for analysis of a wide range of specimens using different radiation sources. The compound PSF's will often allow refinement of overlapping lines that could not be achieved using single-line functions. SHADOW's ability to use direct-convolute PSFs based on different specimen functions will allow for a more detailed analysis of crystallite size and strain than those algorithms which make assumptions of the specimen line shape, e.g., Stokes or similar Fourier techniques.

The use of default parameter files, in conjunction with the ability to select the search and refinement parameters at run-time, provide for a flexible system requiring the minimum of user interaction. The batch-execute mode of operation allows for easy setup and editing of complex procedures. This mode of operation also makes repetitive analysis easy to perform.

REFERENCES

Brown A. & Edmonds J. W. (1980). Adv. in X-ray Anal., 23, 361-367.

Howard S. A. & Snyder R. L. (1985). Adv. in X-ray Anal., 26, 73-81.

Howard S. A. & Snyder R. L. (1988). Submitted, J. Appl. Cryst.

Howard S. A., Yau J. K., Anderson H. A. (1988). In press, J. Appl. Phys.

Williamson G. K. & Hall W. H. (1953). Acta. Met. 1, 22-31.

Yau J. K. & Howard S. A. (1988). Submitted, J. Appl. Cryst.

SPECIFIC DATA HANDLING TECHNIQUES AND NEW ENHANCEMENTS

IN A SEARCH/MATCH PROGRAM

P. Caussin, J. Nusinovici, and D. W. Beard*

SOCABIM, 9 bis, villa du Bel-Air
75012 Paris, France
* Siemens Analytical X-Ray Instruments, Inc.
P.O. Box 5477, Cherry Hill, NJ 08034, U.S.A.

ABSTRACT

A Search/Match program using digitized X-ray powder diffraction scans as inputs was presented in 1987 at the Denver Conference. This program has been rewritten to take full advantage of the newest capabilities of high end personal computers. Software improvements combined with the faster processing speed make it possible to search the 52,791 pattern data base in less than 16 seconds typically (COMPAQ DESKPRO 386/20 with 9 MB of RAM and 80387-20 numerical processor). This can be compared to the 90 seconds one year ago. A specific program allows the user to create and, if desired, customize the search data base directly from the CD-ROM PDF-2, making updates much easier. Practical examples are given to illustrate the operation of the proprietary pattern-recognition algorithm used in the search program.

INTRODUCTION

The use of sophisticated pattern recognition methods to search the complete JCPDS powder diffraction file (PDF1) on a personal computer has led to the use of more and more software optimization techniques in order to reduce the search time to less than one minute. The innermost loop of the program, which is executed more than 1 million times when the JCPDS-ICDD powder diffraction file (PDF) is searched, can fit entirely in the high-speed cache memory; while a compressed binary copy of the PDF is held in about 7 MB of RAM. This paper will focus on the data handling techniques used in the program.

INPUT DATA MAPPING

The continuous background is computed by the maximum concavity radius technique described in [1] and subtracted from the raw pattern by the graphics evaluation program (EVA). The search is then performed on the residual-intensity file.

The data is first reduced to a standard form which lends itself readily to the 32 bit fixed point computations used in the pattern-recognition

531

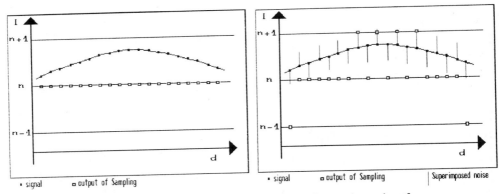

Figure 1.   Sampling a noiseless and a noisy signal

algorithm: 2 theta is mapped on a 4096/d scale ranging from 1 to 8192 (i.e. d=1.0 at the center of the scale). The intensity is mapped on a single byte per point, allowing reduced intensities from 0 to 255.

Quite surprisingly, and nearly by chance, an interesting effect was found in translating the scaling subroutine from Fortran to assembly language. The original program had been written to smooth the data while scaling it; and during the test of a simplified version (without smoothing) of the assembly language routine, it was found that the results of the search were significantly improved with respect to the original program although only the preprocessing routine had been changed.

A little bit of investigation showed that the noise naturally super-imposed on the signal statistically reduced the loss of information due to the quantization of the intensity into 256 discrete steps. This can easily be understood by examing Figure 1. If a slowly varying signal is periodically sampled, one has no information at all on the small variations of the signal as long as its value stays between two discrete steps. If a small random or deterministic noise is added to the signal (Figure 1), one sees that level n+1 will occur more frequently when the base level of the signal comes closer to the upper threshold. This technique of adding noise to a signal for improving the resolution of a quantization process is well known to signal experts[2]. Of course the improvement in resolution is at the expense of a reduced bandwidth as more samples are required to retain the basic accuracy.

As the critical part of the search process is an accurate decision as to what is background and what is diffracted intensity, the natural measurement noise helps by statistically improving the apparent intensity resolution by 2 to 4 (1 or 2 extra bits). This is true, unfortunately, for typical measuring parameters (e.g. step size = 0.03 degrees in 2 theta, measuring time = 1 second per step). The results cannot be expected to be better with poorer measuring conditions! There are two reasons for this: A little bit of noise helps, but more (significantly more than 2 or 3 units of quantization) does not, and the global bandwidth of the scaling process must remain consistent with the actual "bandwidth" of the measured data (closely related to the width of the narrowest peak in the pattern).

Practical aspects: The data should not be smoothed either before or after subtracting the background. The background subtracted scan can be displayed on the screen with the full Y-scale equal to the highest peak to produce an almost exact image of the search input data, because the intensity is mapped on 250 pixels on the screen compared to 256 bits for the search. If a Y-zoom discloses many weak peaks clearly, the square-root plotting scale can be used to write a file that enhances the weak peaks and that the search program will assume to be a linear one.

SORTING TECHNIQUES AND HARDWARE CONSIDERATIONS

The program uses the Johnson and Vand[3] strategy whereby one compares each "standard" pattern of the PDF1 (or of a subset) with the unknown diagram and assigns a score (the figure of merit or f.o.m.), depending on the likely presence of the corresponding phase, the list of the best possible matches being stored. The computation of the f.o.m. of each standard is carred out using almost exclusively <u>fixed</u> point 32 bit computations, whereas the ranking of each standard in the list is computed with conventional floating-point computations. This allows, in outline, for the <u>simultaneous</u> operation of the processor carrying out the evaluation of the next peak while the coprocessor updates the score of the standard.

The time saving benefits of fixed-point 32-bit computation techniques can be indicated by comparison of floating point and fixed-point 32-bit instruction times for the addition and division instructions which are the two most extensively used arithmetic operations in our program. The appropriate values are listed in Table 1.

The fine tuning of machine code which is possible through assembly-language programming makes it possible to ensure that the innermost loop of the program fits and stays resident in the cache memory of high-end personal computers (32KB of 35 nano sec SRAM for the COMPAQ DESK-PRO 386-20). The most frequently executed parts of the program are written to entail no subroutine calls or decision trees. Inline code and indexed decision tables are preferred.

The quality of the hard disk has a large influence on the duration of the search. Programs based on an index of standard lines are mainly influenced by the mean access time of the disk, whereas Johnson and Vand strategy programs, like the one described here, are mainly influenced by data throughput, because they have to read sequentially the complete data base. The typical PC-AT hard disk throughput is 5Mbit/sec and it may increase to 10Mbit/sec for some high end models. The availability of PCs with large memory (at least 8 Mbytes) also make it possible to store an image of the PDF1 (about 7 Mbytes) on a virtual disk. The use of this technique can speed up any Search/Match program.

Listed in Table 2 are the typical durations obtained when searching the complete PDF1 (set 1-37: 52,791 patterns) on various computers. An auxiliary

Table 1. Comparison between fixed-point 32-bit operations and floating-point ones (times are given in microseconds)

| | 8 MHz 16 bit uP | | 20 MHz 32 bit uP | |
	80286 fixed point	80287 floating p.	80386 fixed point	80387 floating p.
division	3.1	27.5	1.25	4.45
addition	0.6	13.1	0.10	1.40

Table 2. Typical search durations with different computer models (search of the same scan against the complete PDF1: 52.791 patterns)

Processor	Coprocessor	H.D. troughput	Memory cache	V Disk	sec.
80286-10	80287-10	High	No	No	58
80286-12	80287-8	Low	No	No	67
80386-16	80387-16	Low	No	No	55
80386-20	80387-20	Low	Yes	No	35
80386-20	80387-20	High	Yes	No	25
80386-20	80387-20		Yes	Yes	15

program is available for the on-site creation of the searchable data base from the CD-ROM version of PDF-2 issued by the JCPDS-ICDD.

SORTING PHILOSOPHY

The two main features of the sorting algorithm are:

- All d-I pairs of each standard, with I's not lower than 1%, are tested without using a preliminary step (usually called the "Search") based only on the strong lines. Any standard sharing at least one peak with the unknown goes through a procedure which is usually considered as the "Match" (and usually performed only on a selected set of standards).

- The positive search, where the f.o.m. of a candidate which shares lines with the unknown is improved; and the negative search, where the f.o.m. of candidates having lines in the unknown background areas is degraded, are weighted separately and are usually of comparable significance. (The best match will result in a minimal f.o.m.)

It follows that a reliable discrimination between sample background and weak peaks is of paramount importance for this Search/Match program. This is why the use of background subtracted scans instead of traditionally-used d-I data as input is strongly recommended.

The positive search is controlled by a "criterion" which is related to the large difference observed in the complexity of unknown scans, and in standard patterns as well. There is no theoretical reason to prefer a candidate sharing many lines with the unknown to a candidate having only one line in the d-field covered by the unknown. There are, however, many practical reasons to give priority to the first kind of candidate. This priority is set by the "criterion" value of 3, while it is also possible to encourage the selection of simple pattern standards (criterion = 1), or to select a neutral criterion (criterion = 2) between simple and complex pattern standards.

Practical aspects: The criterion for a first trial is selected according to the complexity of the unknown (e.g. start with criterion = 3 for complex diagram), but it is often helpful to use the other two criterion values on the same scan.

The negative search is controlled by the "penalty" factor for missing lines (penalty = 0-255). The increase of the penalty increases the discrimination power of the program but is limited by the quality of the scan and by possible preferred orientation in the sample. The penalty for a missing line also depends upon the relative height of that missing line, i.e. a 1% missing line scores almost nothing even with a large penalty value.

Practical aspects: Penalty = 8 (the default value) is suitable for a background-subtracted scan but is too high for traditional d-I lists (use 3-5). Values greater than 15 are used with high-quality scans. A very large penalty value gives priority to the negative search mode.

VERY SIMPLE CASE: PURE QUARTZ DIAGRAM

Any Search/Match program will find Quartz ranking first when using the data from pure quartz as input. This program also makes it possible to rank quartz first when using only a very restricted 2 theta scan. For example, with a scan that includes only one peak (d = 2.457 and I = 8 in the data base) using penalty = 255, it is possible to find Quartz ranking first from the complete PDF1 if a high quality diagram (step size = 0.02 degrees in 2 theta, measuring time = 10 seconds per step) is used as input; and ranking in the first ten with an average quality diagram (step size = 0.03 degrees in 2 theta, measuring time = 1 second per step).

Consider a   scan that includes the well known quartz 5 fingers (3 lines of less than 10% relative intensity with their $K\alpha_2$ components) with a very low background on both sides. A scan reduced to this field is adequate to find quartz ranking first from the complete data base with the high quality diagram and with the average quality one as well (using the default parameters: criterion = 3 and penalty = 8 after a background subtraction).

The use of criterion = 1 to enhance the selection of a simple pattern standard when evaluating the entire quartz diagram will make the standard pattern #36-1886 (Pigment blue 64) rank first above the two quartz patterns (33-1161 and 5-490). Pattern #36-1886 has only one peak in the scan field (18 to 70 degrees in 2 theta with Cu $K\alpha$). Its d value is exactly the same as the quartz 100% peak (d = 3.34 angstroms).

Table 3. Results of the search with criterion = 3 and penalty = 8

Sample: TB3
File : BTDENV.RAW   Data base: \JCP
Criterion: 3, penalty: 8 (Intensity file - no position window)
52791 phases processed, 52791 match chemistry/subfile, 52701 match intensity

Pattern - Formula / Name	mtc/nm	f.o.m.
31-0254 * CaB6 CALCIUM BORIDE	10/ 0	0.36
34-0427 * LaB6 LANTHANUM BORIDE	10/ 1	0.49
6-0401 D LaB6 LANTHANUM BORIDE	10/ 0	0.72
27-1122 C Co5Sm COBALT SAMARIUM	15/ 0	0.84
27-0214 * Eu3NbO6 EUROPIUM NIOBIUM OXIDE	10/ 1	0.85
37-0969 I Sr2EuNbO6 STRONTIUM EUROPIUM NIOBIUM OXIDE	13/ 3	0.86
27-1145 I Eu3TaO6 EUROPIUM TANTALUM OXIDE	9/ 2	0.89
20-0870 I KCrF3 POTASSIUM CHROMIUM FLUORIDE	10/ 0	0.94
25-1135 * BaCl2.2H2O BARIUM CHLORIDE HYDRATE	49/ 3	1.14
11-0137 D BaCl2.2H2O BARIUM CHLORIDE HYDRATE	42/ 2	1.33
37-1042 * ErRh3B ERBIUM RHODIUM BORIDE	8/ 0	1.37
17-0126 Ni5La LANTHANUM NICKEL	13/ 1	1.43
35-0739 * PbZrO3 LEAD ZIRCONIUM OXIDE	31/ 9	1.46
35-1400 I Co5Sm COBALT SAMARIUM	12/ 0	1.46
33-0518 I D.15LaNi5 LANTHANUM NICKEL DEUTERIDE	15/ 1	1.88

Elapsed time: 0 mn 24.11 s

Table 4. Results of the search with criterion = 1 and penalty = 8

```
Sample: TB3
File  : BTDENV.RAW                      Data base: \JCP
Criterion: 1, penalty:  8 (Intensity file - no position window)
52791 phases processed, 52791 match chemistry/subfile, 52701 match intensity
```

Pattern - Formula / Name	mtc/nm	f.o.m.
31-0254 * CaB6 CALCIUM BORIDE	10/ 0	3.58
34-0427 * LaB6 LANTHANUM BORIDE	10/ 1	4.88
6-0401 D LaB6 LANTHANUM BORIDE	10/ 0	7.24
27-1145 I Eu3Ta06 EUROPIUM TANTALUM OXIDE	9/ 2	8.05
27-0214 * Eu3Nb06 EUROPIUM NIOBIUM OXIDE	10/ 1	8.48
20-0870 I KCrF3 POTASSIUM CHROMIUM FLUORIDE	10/ 0	9.45
4-0850 * Ni NICKEL, SYN	3/ 0	9.82
37-0912 * Ag3As04 SILVER ARSENIC OXIDE	4/ 0	10.43
21-0264 Nd4Co3 COBALT NEODYMIUM	5/ 1	10.93
37-1042 * ErRh3B ERBIUM RHODIUM BORIDE	8/ 0	11.00
37-0969 I Sr2EuNb06 STRONTIUM EUROPIUM NIOBIUM OXIDE	13/ 3	11.20
21-0405 I Ni4In INDIUM NICKEL	4/ 0	11.96
17-0104 (Mo0C) MOLYBDENUM OXIDE CARBIDE	4/ 0	12.27
19-0558 HoRh HOLMIUM RHODIUM	3/ 0	12.50
27-1122 C Co5Sm COBALT SAMARIUM	15/ 0	12.58

Elapsed time: 0 mn 23.73 s

A MODERATELY COMPLEX SEARCH/MATCH PROBLEM

This problem was submitted for evaluation of different Search/Match programs. The sample was measured from 15 to 120 degrees in 2 theta with Cu Kα in about one hour. Measuring conditions were usual ones: step size = 0.03 degrees in 2 theta, measuring time = 1 second per step, and continuous step-scan to collect the data while moving. For the Search/Match the diagram was restricted to 15-80 degrees in order to avoid overemphasis of the few standards having recorded peaks above 80 degrees with Cu Kα.

The use of the default search parameters (criterion = 3, penalty = 8) after background subtraction led to a first list (see Table 3) of best potential matches. The graphic examination of the results with EVA (the list is directly interfaced with this program) made it possible to explain most of the unknown scan. CaB_6 (1st) is an isotype of LaB_6 (2nd and 3rd). Selection between them without some knowledge of the sample is a guess. The same is true for selecting between Co_5Sm (4th) and Ni_5La (12th) whereas $BaCl_2$ (9th and 10th) is an obvious match which has proved to be difficult to find by d-I list based programs. This program benefits from the fact that $BaCl_2$ has 50 lines which are in the non-background areas of the unknown pattern. Peak finding programs frequently have problems with such weak, heavily overlapped lines and therefore reduce the effectiveness of d-I list based search/match methods. The ranking of $BaCl_2$ can be improved by using a square-root scale for preparation of the file to be searched (moving $BaCl_2$ up to 5th and 7th).

At this step, two relatively high peaks of the unknown were still not explained. The use of criterion = 1 produced a list with Ni ranking 7th (see Table 4). Further trials to explain a few very weak peaks having failed, the solution of the problem was disclosed by the originator of the sample: Ni = 12.6%; $LaNi_5$ = 29.8%; LaB_6 = 26.1%; $BaCl_2$ = 31.5%, and Al in traces. The overlay of the Al pattern on the screen, Figure 2, makes it obvious that with the recorded scan it was not possible to identify or to reject Al (the 100% line of Al corresponds to 64 pulses on the diagram whereas the background is at 22). Knowing its presence and examing its pattern displayed as stick figures onto the scan of the unknown makes it easy to select the best parameters to identify Al.

Using a square-root intensity scale (to enhance the relative weight of weak Al peaks), criterion = 1 (to help find simple patterns), and penalty =

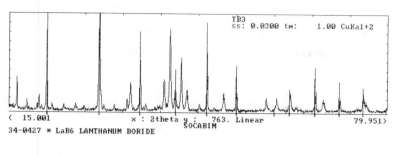

34-0427 * LaB6 LANTHANUM BORIDE

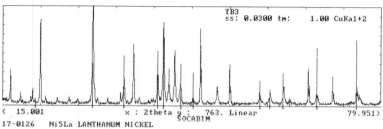

17-0126   Ni5La LANTHANUM NICKEL

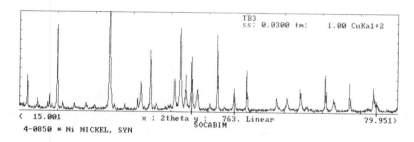

25-1135 * BaCl2.2H2O BARIUM CHLORIDE HYDRATE

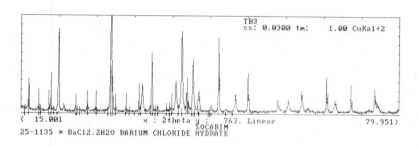

4-0850 * Ni NICKEL, SYN

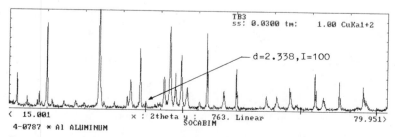

4-0787 * Al ALUMINUM

Figure 2.   Patterns of the actual components of the mixture
            overlaid on the raw data.

255 (to use a more negative search than positive one) causes Al to rise in rank to 46th out of the 52,791 candidates, which will not help in its identification, but is a remarkable illustration of the program's power.

CONCLUSION

Much effort has been devoted to speeding up the program because the search of the complete data base in about 20 sec. will open new possibilities for interaction with the user. Speeding the program is also required to prepare for the future: the search of a data base of digitized standard patterns, i.e. a data base of about 100 times more data for the same number of standards.

BIBLIOGRAPHY

1.   P. Caussin, J. Nusinovici, D. W. Beard, in Advances in X-Ray Analysis, Vol. 31, 1988, 423-430.

2.   B. A. Monitjo, Hewlett-Packard Journal, 39:70 (1988).

3.   G. G. Johnson, Jr., and V. Vand, Ind. Eng. Chem., 59:19 (1967).

USE OF THE CRYSTAL DATA FILE ON CD-ROM

Mark Holomany & Ron Jenkins

JCPDS-ICDD, Swarthmore, PA 19081

ABSTRACT

We have recently described the use of the Compact Disk Read Only Memory disk for the storage of data in the Powder Diffraction File (PDF). This work has now been extended to include the NBS (1987) Crystal Data File (CDF). The CDF contains 115,753 entries of which 59,613 are Inorganic materials and 56,140 organic. The data base can be accessed either by means of bit maps built on chemistry and subfile restrictions, or by means of a Boolean search system allowing combinations of search parameters including: chemistry; space group; cell volume; density and unit cell data.

USE OF THE CD-ROM FOR DATA STORAGE

In a previous paper [1] we described a search program "PC-PDF" which made use of Compact Disk Read Only Memory (CD-ROM) technology for the storage of the powder diffraction file PDF-2. We have demonstrated that use of the CD-ROM "laser disk" as a storage medium offers several advantages over hard disk systems. Traditional magnetic storage devices are limited in the density of information by the physical size of individual ferrite particles. On the other hand, the laser disk technology offers much higher track density by use of focused laser light onto an optically reflecting or non-reflecting surface. This gives first, a very narrow storage track - typically of the order of a micron or so, and second, of equal importance, allows the tracks to be very close together. A combination of these two factors gives a very large number of tracks - current technology allows about 20,000 tracks compared to about 2000 tracks on a typical hard disk, giving a total storage capacity of 550 Mbytes for a twelve centimeter platter. While CD-ROM access times are rather long (data transfer rate about 150 Kbytes/sec compared to about 625 Kbytes/sec for a hard disk) careful design of index files minimizes this problem and typical pattern seek times are less than 1 second.

Acceptance of this product has been very positive, both within the U.S. as well as overseas. The current revision of the PDF-2 Data Base requires 125 Mbytes of storage for the data files and 29 Mbytes for index files. During the course of the past year the International Centre has extended its use of the CD-ROM to now include the NBS Crystal Data File [2] and software for the accessing of this file is

Table 1   Speed comparison with different hardware

Problem #1 CDF - Search for inorganic phases containing only lead and oxygen

X/T	120 secs
A/T clone (8 MHz)	87 secs
Compaq10 (10 MHz)	63 secs

Problem #2 PDF - Search the mineral data base for phases containing silicon, and having a strong line between 3.32 and 3.36Å

X/T		310 secs
A/T clone (8 MHz)		280 secs
Compaq10 (10 MHz)	245 secs	

now in beta test. The current version of this data base requires 135 Mbytes for the CDF and 27 Mbytes for index files.

USE OF THE BOOLEAN SEARCH PROGRAM WITH THE PDF

The Boolean search program is based on the Query Plus program from Reference Technology Inc. [3]. While the initial version of this program was rather slow in execution, recent improvements have made it much more satisfactory. Table 1 compares execution times for two fairly complicated searches, one involving the PDF and the other, the CDF. In each case three different PC's have been timed to compare performances. The search output is shown in fig. 1 for the first test case described in table 1.

The search program works via a series of index files associated with the PDF-2 data base and uses a Boolean parsing expression consisting of a key range

```
 1)   060263   POTASSIUM ALUMINUM SILICATE HYDROXIDE            K AL2 ( SI3 AL ) 010*
 2)   070025   POTASSIUM ALUMINUM SILICATE HYDROXIDE            K AL2 SI3 AL 010 ( O*
 3)   070042   POTASSIUM ALUMINUM SILICATE HYDROXIDE            ( K , NA ) ( AL , MG*
 4)   090051   CALCIUM SILICATE HYDROXIDE                       CA2 SI O3 ( O H )2
 5)   090343   POTASSIUM IRON MAGNESIUM ALUMINUM SILICATE H     K0.5 ( AL , FE , MG *
 6)   090451   ALUMINUM SILICATE HYDROXIDE HYDRATE              AL2 SI2 05 ( O H )4 *
 7)   100483   LITHIUM POTASSIUM ALUMINUM SILICATE HYDROXID     K ( LI , AL )3 ( SI *
 8)   100484   LITHIUM POTASSIUM ALUMINUM FLUORIDE SILICATE     K ( LI , AL )3 ( SI *
 9)   100492   POTASSIUM MAGNESIUM ALUMINUM SILICATE HYDROX     K MG3 ( SI3 AL 010 )*
10)   100493   POTASSIUM MAGNESIUM ALUMINUM SILICATE HYDROX     K MG3 ( SI3 AL 010 )*
11)   100495   POTASSIUM MAGNESIUM ALUMINUM SILICATE HYDROX     K MG3 ( SI3 AL 010 )*
12)   100496   POTASSIUM ALUMINUM VANADIUM SILICATE HYDROXI     K AL V2 SI3 010 ( O *
13)   110046   ALUMINUM SILICATE                                AL2 SI 05
14)   110491   IRON MANGANESE BERYLLIUM SILICATE SULFIDE        ( FE , MN )4 BE3 SI3*
15)   130259   SODIUM MAGNESIUM ALUMINUM SILICATE HYDROXIDE     NA0.3 ( AL , MG )2 S*
16)   130535   BERYLLIUM CALCIUM ALUMINUM SILICATE HYDROXID     CA4 BE2 AL2 SI9 026 *
17)   160344   POTASSIUM MAGNESIUM ALUMINUM FLUORIDE SILICA     K MG3 ( SI3 AL 010 )*
18)   170548   IRON SILICATE                                    FE SI O3
19)   170754   SODIUM CALCIUM ALUMINUM SILICATE                 ( CA , NA )4 AL3 ( A*
20)   181171   SILICON OXIDE NITRIDE                            SI2 N2 O

2:"M  " & 6:"SI" & 7:332,336"
```

Fig.1   Search output from the test outlined in table 1.

Table 2   Index keys available for the retrieval of PDF data

1	PDF number	10	Inorganic chemical name fragments
2	Subfile	11	Mineral group codes
3	Inorganic chemical name	12	Reduced unit cell parameters
4	Mineral name	13	Principal author
5	Organic chemical element	14	Journal year
6	Inorganic chemical element	15	Journal coden
7	Three strongest lines	16	Color
8	CAS number	17	Density

specification, optional white space, parentheses, wild cards (*) and certain Boolean operators. These operators include AND (&); IOR (|) and ANDNOT (-). Retrieval from the PDF-2 data base is possible on eighteen index keys the specifications for which are given in table 2.

Fig. 2 gives an example of an experimental version of the program in which search parameters are selected by means of a menu table, and the output list can be displayed in terms of "d"; "2θ"; "$Sin^2\theta$" or "Q". In the example a search is made to examine the Inorganic Phases Subfile for phases containing the elements Eu, Mg and F. Fig. 3 shows the first page of the search output and fig.4 shows the diffraction pattern expressed in terms of values 2θ values.

USE OF THE BOOLEAN SEARCH PROGRAM WITH THE CDF

The NBS Crystal Data File (CDF) contains crystallographic unit cell parameters from both powder and single crystal sources. At the present time the CDF contains about 120,000 entries, each entry consisting of reduced cell and volume; crystal system; space group symbol and number; chemical name and formula; and literature reference. The primary use of the CDF is to match cells

```
6:Eu Mg F & 2:I

 * 0 MASTER INDEX                    10 INORGANIC CHEMICAL NAME FRAGMENTS
 * 1 PDF NUMBER                     *11 MINERAL GROUP CODES
   2 SUBFILE                        *12 REDUCED UNIT CELL PARAMETERS
 * 3 INORGANIC CHEMICAL NAME        *13 PRINCIPAL AUTHOR
 * 4 MINERAL NAME                   *14 JOURNAL YEAR
   5 ORGANIC CHEMICAL ELEMENT       *15 JOURNAL CODEN
   6 INORGANIC CHEMICAL ELEMENT     *16 COLOR
   7 THREE STRONGEST LINES          *17 DENSITY
 * 8 CAS NUMBER                     *18 REDUCED CELL VOLUME
 * 9 ORGANIC CHEMICAL NAME           19 EXIT

PLEASE SELECT AN OPERATION (AND,OR,NOT):


   I      INORGANIC FILE                   CP    COMMON PHASE
   O      ORGANIC FILE                     NBS   NBS PATTERN
   M      MINERAL FILE                     FOR   FORENSIC
   A      ALLOY,METAL,AND INTERMETALLIC FILE

PLEASE SELECT (I,O,M,A,CP,NBS,FOR):     I
 * WILL BE IMPLEMENTED IN THE FUTURE
```

Fig.2   Menu table for the set-up of a Boolean search of the PDF

32-378   JCPDS-ICDD COPYRIGHT 1987   QM=I

EuMgF
    4

EUROPIUM MAGNESIUM FLUORIDE

RAD: CuK 1   LAMBDA: 1.54056   FILTER: MONO.   D-SP:
CUTOFF: INT: DIFFRACTOMETER   I/ICOR:
REF: JENKINS, R., PH. D. THESIS, POLYTECHNIC INSTITUTE OF NEW YORK, NEW YORK,
USA., PRIVATE COMMUNICATION, (1980)

SYS: ORTHORHOMBIC   S.G.: CMCM (63)
A: 3.935   B: 14.43   C: 5.664   A: C:
A: B: C: Z: 4   MP:
REF: IBID.

Dx: 5.210   DM: 5.340   SS/FOM: F(30)=38.9(0.024,32)

Fig.3   Example of the search output in which the Inorganic subfiles has been searched for
         compounds containing the elements Eu, Mg and F.

32-378   JCPDS-ICDD COPYRIGHT 1987   QM=I
                    60 REFLECTIONS IN PATTERN. PAGE 1 OF 2. RADIATION= 1.5405

2-THETA	INT.	H K L	2-THETA	INT.	H K L
12.282	20	0 2 0	41.945	60	1 5 1
19.935	90	0 2 1	43.580	70	1 3 2
23.452	10	1 1 0	46.057	30	2 0 0
24.639	50	0 4 0	47.939	1	2 2 0
28.317	100	1 1 1	49.696	10	0 6 2
29.335	70	0 4 1	49.870	20	0 2 3
29.344	50		49.900	20	1 7 0
31.599	30	0 0 2	50.641	2	2 2 1
33.329	5	1 3 1	50.701	10	
33.993	5	0 2 2	50.731	5	1 5 2
37.359	30	0 6 0	52.614	4	1 7 1
38.677	<1	1 5 0	52.942	5	2 4 0
39.671	30	1 1 2	53.208	20	0 8 1
40.451	20	0 4 2	54.194	10	1 1 3
40.737	5	0 6 1	54.790	10	0 4 3

PRESS: PGDN   PGUP   HOME   END   ESC   F2(PRINT PDF CARD)   D21SQ TO CONVERT

Fig.4   Display of the powder pattern in terms of °2θ.

Table 3 Index keys available for the retrieval of CDF data

1 Space group symbol & number   7   Mineral name
2 Chemical elements   8   Pearson symbol
3 Journal CODEN   9   Unit cell parameters (reduced cell)
4 Cell volume (reduced cell)   10 CDIF ID number
5 Density   11 Organic chemical class
6 Subfile code   12 Journal year

```
BOOLEAN SEARCH - PDF-2 DATA BASE

EXPERIMENTAL VERSION  1.02X


ENTER QUERY USING AT MOST THE NEXT 3 LINES (PRESS ENTER ONLY ONCE) ...
2:"AG" & 3:"AMETAR*"
        65538           20          1865

  1)   707880  SILVER ZINC                        AG ZN
  2)   034835  LEAD SILVER (1^4)                   AG4 PB
  3)   706501  CADMIUM SILVER                      CD3 AG
  4)   707879  SILVER ZINC                         AG ZN

2:"AG" & 3:"AMETAR*"


CHOOSE ONE OF THE ABOVE BY LINE NUMBER (ENTER TO SKIP):
```

Fig.5 Example of the use of the Boolean search program with the Crystal Data File. In this example the journal *Acta Metall* has been sought for all patterns containing silver. Four hits have been found.

```
BOOLEAN SEARCH - PDF-2 DATA BASE

EXPERIMENTAL VERSION  1.02X


ENTER QUERY USING AT MOST THE NEXT 3 LINES (PRESS ENTER ONLY ONCE) ...
6:"M" & 5: 950, 960
        65537           19          7948

  1)   025988  LEAD OXIDE (1^2)                    PB O2
  2)   702937  LEAD OXIDE                          PB O2
  3)   120460  LEAD OXIDE (1^1)                    PB O
  4)   700071  OSMIUM SULFIDE                      OS S2

6:"M" & 5: 950, 960
```

Fig.6 An example of a search of the Crystal Data File for all minerals having an X-ray density between 9.5 and 9.6 g/cc.

determined from experimental data with those previously published. Searches can typically be made on cell parameters, chemistry or space group.

Since the data base format of the PDF and CDF were designed to be completely compatible [4], programs and techniques designed for searching the PDF are almost directly applicable to the CDF. The Boolean search program is no exception to this general rule. Table 3 lists the twelve index keys which are available for searching the CDF.

As a simple example of a search, suppose we wished to search for patterns containing silver, which have been published in the journal "Acta Metall.". The coden for this journal is AMETAR and by using this entry with key #3 and the element symbol AG with key #2, we find four hits, as shown in fig. 5. Similariy we could search the minerals file (search key #6 set to "M") for phases with an X-ray density between 9.5 and 9.6 (search key #5). Again four compounds are found matching these criteria, fig. 6.

OTHER SOFTWARE FOR SEARCHING THE CDF

In addition to the Boolean search software, other software programs are available which perform searches by alternative means. The most well known of these is the NBS*LATTICE program [5] which allows the identification of unknown materials, calculation of the reduced cell of the lattice, and the calculation and reduction of specified derivative supercells and/or subcells. This program has been available for a number of years running in a main frame environment, and has recently been ported down to a PC such that it will work with CD-ROM supported PC systems.

REFERENCES

[1]   Jenkins, R. and Holomany, M., "PC-PDF - A search/display system utilizing the CD-ROM and the complete Powder Diffraction File", Powder Diffraction, 4 215-219 (1987)

[2]   NBS Crystal Data File (1982), "A magnetic tape of crystallographic and chemical data compiled and evaluated by the NBS Crystal Data Center, National Bureau of Standards, Gaithersburg, MD 20899, USA; available from JCPDS-ICDD, 1601 Park Lane, Swarthmore, PA 19081, USA

[3]   Reference Technology Inc., Boulder, CO 80301, USA

[4]   Mighell, A.D., Hubbard, C.R and Stalick, J.K., "NBS*AIDS80. A Fortan Program for Crystallographic Data Evaluation", NBS Technical Note 1141 (1981), National Bureau of Standards, Gaithersburg, MD 20899,, USA

[5]   Himes, V.L. and Mighell, A.D., "NBS*LATTICE - A program to analyze lattice relationships", (1985) NBS Crystal Data Center, National Bureau of Standards, Gaithersburg, MD 20899, USA

A REFERENCE DATABASE RETRIEVAL SYSTEM:

INFORMATION AS A TOOL TO ASSIST IN XRD PHASE IDENTIFICATION

S.O. Alam, J.W. Edmonds, T. Hom,
J.A. Nicolosi and B. Scott

Philips Electronic Instruments, Co.
85 McKee Drive, Mahwah, N.J. 07430-2121

INTRODUCTION

A common analytical problem in a X-ray diffraction laboratory is the phase identification of an unknown sample. Presently, most phase identification is either performed manually or through computer search-match programs, as found on automated powder diffraction systems. Whether phase identification is performed manually or with computer search-match, often, the diffractionist wishes or needs to be able to make use of more sample information in the phase identification process than just d-I data. Additional information may be in the diffractionist's possession or can be readily obtained. The types of additional information which can be used in computer search-match programs are currently limited to information in the Powder Diffraction File-1 (PDF-1) [1], formerly Level I, database (ie., JCPDS sub-files, elemental and functional groups). The use of other information is not possible. Users of manual search procedures are not able to make optimum use of additional information.

The Powder Diffraction File-2 (PDF-2), formerly Level III, [1] database represents a dramatic increase in the volume of information available for the diffractionist in computer-readable form. The PDF-2 database contains additional information useful in phase identification. Its availability has provided us with an opportunity to develop a retrieval system and database, commercially available as PLUS37, which makes all the PDF-2 information accessible and searchable in stand-alone mode or integrated with Philips diffraction software. The ability to search and access PDF-2 information maximizes the use of this information, as will be illustrated in the examples.

Areas where improvements to the phase identification process may occur are 1) the search-match procedure, 2) creation of database sub-files prior to search-match, and 3) verification or validation of search-match results. In this paper, we will illustrate with examples how PLUS37 utilizes PDF-2 information in situation 2) and 3) for an expedient and thorough analysis.

The preparation of sub-files prior to search-match is a common practice to 1) shorten search-match times or 2) improve the quality of results [2]. Sub-files can be divided into two categories, JCPDS and user defined sub-files. To date, there are seven JCPDS sub-files in PDF-2. User defined sub-files are reference patterns selected based on the user's knowledge of the properties of the sample. As mentioned above, the types of sample information available for use in creating user defined sub-files is currently limited to elemental and functional group information. Providing the ability to search through the PDF-2 information gives the diffractionist an opportunity to create user defined sub-files based on, for example, crystal class, temperature during data collection, maximum value of the d-spacing of the pattern, geographic location for mineral searches, etc.

The availability of additional information is beneficial for the verification of search-match results. One area of particular difficulty in phase identification arises with reference patterns of solid solution and iso-structural materials. To resolve the ambiguity involving a single phase component, one can consider the use of sample information, such as, density, color, melting point, reference intensity ratio, etc. Alternately, further analysis may be necessary to resolve the ambiguity, as will be illustrated in the second example.

PLUS37

The PLUS37 database is built using Information Dimensions, Inc.'s database language, BASIS. The BASIS software is used to extract information from the PDF-2 database organized in NBS*AIDS83 format [3]. The PLUS37 database is produced for MS-DOS and Digital Micro-VAX computers on a Compact Disk Read Only Memory (CDROM) optical disk adhering to the industry-standard High Sierra logical file format. A CDROM is a compact, high density mass storage (650 Megabyte) medium.

The basic design of the PLUS37 database allows bi-directional accessing of information. The design permits either 1) the retrieval of all information relating a specified PDF number or 2) retrieval of the PDF numbers of all patterns whose contents satisfy a particular search criteria. The latter capability is made possible by the use of an inverted indexing scheme. The inverted indexes are built at the time of database creation and are read into memory during operation of the retrieval software, micro-BASIS, to obtain fast system response. Over 250 database fields (ie., color, melting point, temperature of measurement, etc.) have been indexed. 90 of the 220 Megabytes occupied by PLUS37 are used in the storage of indexing information.

The PLUS37 retrieval system provides an English-like command and query language to manipulate data, search the database, retrieve and display all or selected portions of the pattern information. Boolean and relational operations are supported. A hierarchical search can be performed. This allows the creation of a sub-file through the combination of search criteria entered on several input lines.

TWO EXAMPLES ILLUSTRATING THE METHOD

The first example illustrates the use of combining PLUS37 database information to assist in a manual Hanawalt search of a binary mixture of

D	I	D	I	D	I
4.53	17	2.544	18	1.730	13
3.702	100	2.372	12	1.599	19
3.660	61	2.268	11	1.510	13
3.470	17	2.142	34	1.400	11
3.203	19	2.097	19	1.372	20
2.642	20	2.090	22	1.230	11
2.614	32	2.013	22		
2.587	34	1.932	20		

```
1/   SET HIER ON
1/   FIND PDFNM=CARBONATE
*      620              1/ PDFNM=CARBONATE
2/   FIND DMAX LT 4.6
*      90              2/ DMAX LT 4.6        ** UNIVERSE= 1 **
3/   FIND D3=3.67/3.73
*       4              3/ D3=3.67/3.73       ** UNIVERSE= 2 **
```

Figure 1.   Computer-assisted Hanawalt Search.
Complete Unknown Pattern.

Witherite and Corundum. X-ray data from 10 to 80 degrees two-theta was obtained on a Philips diffractometer. The d-spacings and relative intensities are listed in the upper portion of Figure 1. The lower portion of Figure 1. represents the dialogue appearing on a computer screen between the user and PLUS37 retrieval software. The first command is to activate a hierarchical search. As indicated earlier, a hierarchical search allows the creation of a sub-file based on search criteria issued on multiple input lines. The second command uses information concerning the presence of carbonate functional group. This information may have been obtained by an geologist, applying a small amount of acid to his mineral sample and observing a chemical reaction characteristic of the presence of a "carbonate". The command results in a search of the PDFNM field for all PDF patterns containing "carbonate" in the PDF index name. Of the original 52,791 PDF patterns, 620 patterns satisfy the "carbonate" search criteria. The third command makes use of the fact that the d-spacing of first line of the pattern occurs at 4.53Å. Of the 460 patterns, 90 patterns remain after the issuing of the third command to search the DMAX field for all PDF patterns with d-spacing of line 1 less than 4.6 Å. The fourth command results in a range search of the D3 field using the strong intensity line of the pattern, 3.702 Å, and a one-percent error window for the d-value. D3 is a mapped field containing the three strongest Hanawalt lines. A mapped field contains multiple fields of the same type and index. A line-by-line examination between the measured pattern and the four remaining reference patterns shows pattern 5-374, Witherite, as having the best match.

Eliminating the identified lines from the pattern leaves the d-spacings as shown in the upper portion of Figure 2. The first two commands reset to the full database. There exists the possibility of line overlap of the phases. Therefore, the third command to search the DMAX field for all PDF patterns with d-spacing of line 1 less than 4.6 is used. This condition reduces the original set of 52,791 patterns to 11,450 patterns. Examination of the remaining lines of the residue pattern show that many of the lines are approximately equal intensity. This resulted in the decision to search the D5 field instead of D3 field.

```
    D       I       D       I       D       I

  3.470     17     2.372    12     1.599    19
  2.544     18     2.090    22     1.400    11

      4/  SET HIER OFF
      4/  SET HIER ON
      4/  FIND DMAX LT 4.6
  *      11450       4/ DMAX LT 4.6
      5/  FIND D5=2.08/2.12
  *       1531       5/ D5=2.08/2.12        ** UNIVERSE= 4 **
      6/  FIND D5=3.44/3.50
  *        103       6/ D5=3.44/3.50        ** UNIVERSE= 5 **
      7/  FIND D5=2.52/2.56
           3        7/ D5=2.52/2.56         ** UNIVERSE= 6 **
```

Figure 2. Computer-assisted Hanawalt Search.
Residual Unknown Pattern.

D5 is a mapped field containing the five strongest Hanawalt lines. Three range searches of the D5 field are performed using the most intense line (2.090 Å), the first (3.470 Å) and second (2.544 Å) lines of the residual pattern, respectively. A one percent window for the d-value was used in each case. Of the three remaining candidate patterns, Corundum, pattern 10-173, is the identified pattern.

Returning to the Witherite example, Table 1 illustrates the variety of possible combinations of types of information which can be used in arriving at the Witherite answer. In each case, Witherite appeared in the list of candidate phases. From an examination of TABLE 1, the following observations are emphasized:

1) A short manageable list of candidates can be obtained quickly with the use of information in addition to d-I data.
2) The minimum set of d-I data required for an identification is smaller when additional information is used.
3) As to be expected, a higher degree of confidence in the identification process can be achieved with the use of additional information.

The second example is two-phase mixture containing CoO and Co_3O_4. This example illustrates the difficulty in phase identification involving reference patterns of iso-structural and solid-solution materials because these materials give rise to very similar x-ray patterns.

X-ray powder data from 10 to 130 degrees was obtained on a Philips diffractometer. The computer search-match program properly identified CoO but it chose $(Cu_{0.3}Co_{0.7})Co_2O_4$, a Cu solid solution in Co_3O_4 for the second phase. Fe_3O_4, PDF# 26-1136; Co_3O_4, PDF# 9-418; and $MgAl_2O_4$, PDF# 21-1152 were listed as the next three highest scoring candidates and all are of the spinel-type structure.

A database search revealed that CoO and all the top candidates for the second phase except $MgAl_2O_4$ are black. The low relative intensities of the CoO pattern indicated that the second phase is the major component. $MgAl_2O_4$ is eliminated as a candidate, since $MgAl_2O_4$ is colorless and the sample is black (possibly, of course, from impurities).

TABLE 1.   Uses of Information in a Computer-assisted Hanawalt Search.
(Example: Witherite, PDF# 10-173)

JCPDS SUB-FILE	USER SUB-FILE	NUMBER OF MOST INTENSE LINES USED	NUMBER OF CANDIDATES FOUND
none	DMAX, carbonate	1	4
COMMON PHASES MINERAL	DMAX	1	2
MINERAL	DMAX	1,2,3	13,7,2
COMMON PHASES	DMAX	2*	4
none	DMAX	4	11
MINERAL	none	2*,3**	16,7
none	none	4	23

* most intense line used first, followed by third most intense line

** most intense line used first, followed by third and second most intense lines

As a further attempt to remove the ambiguity, it was decided to refine the lattice parameters of the second phase and perform a range search of the database using the value of the refined lattice parameters. An examination of the crystallographic information of the candidate phases shows that the literature lattice parameters (a) range from 8.074 to 8.0903 Å. Figure 3 shows the dialogue for retrieving information from PLUS37 for the refinement of the lattice parameters of the second phase. Using the lattice constants (8.074) for Copper Cobalt Oxide as a starting value and space group (Fd3m), the refined lattice parameter is 8.085 ± 0.001 Å. A range search of the database for patterns with lattice constants (a) between the values 8.083 and 8.087 Å yielded eleven patterns. Of the eleven patterns, only Co_3O_4 is listed as being black, five patterns had no color information and five patterns listed colors other than black. The pattern common to both the lists of eleven patterns and remaining second phase candidates is Co_3O_4.

During the development and packing of the PLUS37 database, the degree of completeness and accuracy of the information contained in PDF-2 was tested extensively. Not withstanding the fact that JCPDS is continuously improving the quality of PDF-2 database, errors do exist, especially, in non d-I data and this must be kept in mind when exercising search criteria.

CONCLUSION

The retrieval database system described above makes practical the usage of a large variety of PDF-2 information in phase identification. The reference database retrieval system proved useful in computer-assisting manual Hanawalt searches and providing on-line information necessary for further analysis when d-I diffraction data search-match results are

```
1/ 25-270

    1   PATTERN SAVED AS RESULT SET 1
2/ DISPLAY PDFNM,CF,CSC,A,SG

PATTERN 1

PDF index name              Copper Cobalt Oxide
Chemical Formula            (Cu_{0.30} Co_{0.70}) Co_2O_4
Crystal system code         C
                            Cubic
a0                          8.074 Angstroms
Author space group          Fd3m
```

Figure 3. Retrieval of Information for Lattice Parameter Refinement.

questionable or unsatisfactory. In computer-assisted manual Hanawalt searches the use of information in addition to d-I diffraction data produces a short manageable list of candidate patterns quickly and reduces the minimum set of d-I data needed for an identification.

REFERENCES

1. Jenkins, R. and Smith, D.K., "Powder Diffraction File" in "Crystallographic Databases", Data Commission of the International Union of Crystallography, 158-175, (1987).

2. McCarthy, G.J. and Johnson, Jr., G.G., "Identification of Multi-phase Unknowns by Computer Methods: Role of Chemical Information, The Quality of X-ray Powder Data and Sub-files", Adv. in X-ray Anal., 22, 109-120 (1979).

3. Mighell, A.D., Hubbard, C.R. and Stalick, J.K., 'NBS*AIDS80. A FORTRAN Program for Crystallographic Data Evaluation', National Bureau of Standards, Technical Note 1141 (1981), NBS, Washington, D.C.

ON THE SELECTION OF THE VALUE FOR THE EXPERIMENTAL WAVELENGTH IN POWDER DIFFRACTION MEASUREMENTS

Ron Jenkins

International Centre for Diffraction Data
Swarthmore, PA 19081

ABSTRACT

With the increasing use of automated powder diffractometers, more and more reliance is being placed on peak searching and α_2 stripping software to allocate a wavelength for the calculation of the d-value from the observed peak maxima. The complications in the correct allocation of the wavelength are well known and include problems due to the polychromatic nature of the diffracted X-ray beam, changes in the angular dispersion of the diffractometer over the angular range of the experimental data, and effects of instrument geometric aberrations on the profile shape. Further difficulties may accrue because of idiosyncrasy of the peak searching and α_2 stripping software.

INTRODUCTION

With the growing need for higher and higher quality X-ray powder diffraction data, one problem which is becoming increasingly more difficult to handle, is that of the selection of the "practical" wavelength in a powder diffraction experiment. Problems occur for several reasons. Firstly, because of the polychromatic nature of the diffracted beam, it is sometimes difficult to manually assess where the maximum of a peak occurs, especially in the range of angles from 30 to 60° 2θ. Where computer searching for peaks is employed, such methods are invariably more sensitive than manual techniques, meaning that a doublet is recognized as such at a lower angle. This in turn can lead to problems during phase identification because all of the observed lines may not match the correct standard pattern selection. This problem can be made even worse when α2 stripping, or profile fitting, techniques are employed on the experimental pattern, and have not been employed in the measurement of the reference pattern(s).

An additional problem is that, whether or not the correct maximum is established, an incorrect value of the wavelength may be employed in the calculation of the d-values. In a recent round robin [1] participants were asked to report the wavelength value for Copper Kα which they had employed in their reported work. These values varied from 1.5405 to 1.541Å!

551

FORM OF THE COPPER K SPECTRUM

For most practical purposes, in powder diffractometry the copper K spectrum is considered to be two pairs of lines, the $\alpha1/\alpha2$ doublet occurring from a 2p -> 1s transition; and the $\beta1/\beta3$ doublet from 3p -> 1s transition. In most experimental work the β-doublet intensity is typically reduced to less than a few percent of the α-doublet intensity, by use of filtration, or is removed by use of a diffracted beam monochromator or proportional detector. In each case, what remains is essentially bichromatic radiation. In applying this radiation for the measurement of interplanar spacings the main problems remain the angular dependant dispersion of the $\alpha1/\alpha2$ doublet, and differences in peak asymmetry between the two.

In point of fact the characteristic α radiation emission from copper is much more complex than the simple α-doublet and β-doublet model. The latest compilation of X-ray emission wavelengths by Cauchois and Senemaud [2] lists seven additional (satellite) lines for the copper $K\alpha$ series. Berger [3] has recently discussed a somewhat less complex model based on six α-lines which he denotes as $\alpha1$, $\alpha2$, $\alpha11$, $\alpha12$, $\alpha21$, and $\alpha22$, giving two triplets in place of the $\alpha1$ and $\alpha2$. These lines would be similar to the more familiar $\alpha3$ to $\alpha6$ satellite lines observed in low atomic number emission spectra such as aluminum. As in the case of aluminum these satellite lines arise from dual ionizations involving both K and L levels. Table 1 shows the wavelengths and relative intensities of the six lines.

Several points are of interest. First, it will be seen that the intensity of the weakest line in each triplet has roughly the same intensity relative to the $\alpha1$. However, the $\alpha2$ and $\alpha21$ lines are only about one third of the intensity of the $\alpha1$ and $\alpha11$ lines respectively. This accounts for the higher degree of asymmetry of the $\alpha2$ relative to the $\alpha1$ shown in figure 1. Secondly, the largest energy gap within any of the triplets is only about 2.5 eV. Since the absolute energy resolution of the powder diffractometer using Cu $K\alpha$ radiation ranges from about 200 eV at 10° 2θ to about 2.5 eV at 140° 2θ (see figure 2), the fine structure of the triplets is not revealed. The asymmetry of the $\alpha2$ will, however, start to become apparent at very high 2θ values. Even more important, where profile fitting methods are employed, the effective "fitting resolution" is probably only of the order of a few eV and here allowance must be

Table 1

Copper $K\alpha$ emission parameters (after Berger [3])

Line	Transition	Wavelength	Energy gap from $\alpha1$ (eV)	Relative Intensity
α_1	LIII -> K	1.540562	0	1.000
α_{11}	LIII(LII+) -> K	1.540560	<0.1	0.970
α_{12}	LIII(LI+) -> K	1.541022	2.41	0.095
α_2	LII -> K	1.544390	19.5	0.382
α_{21}	LII(LIII+) -> K	1.544374	19.9	0.331
α_{22}	LII(LI+) -> K	1.544685	21.5	0.093

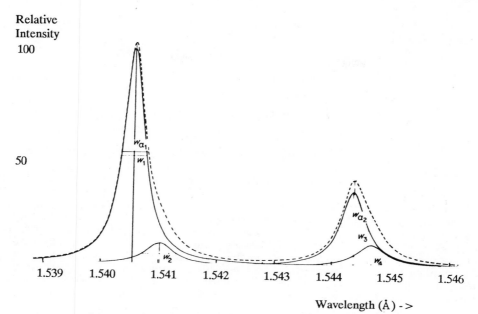

Fig.1 K- alpha emission spectrum of copper (after Berger [3])

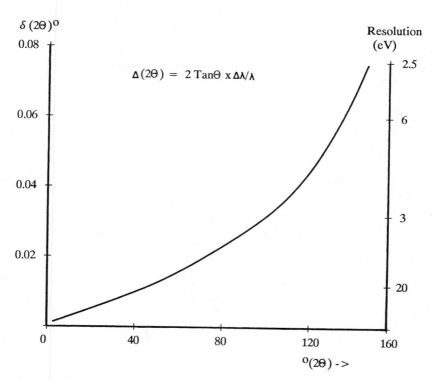

$$\Delta(2\Theta) = 2\,Tan\Theta \times \Delta\lambda/\lambda$$

Fig.2 Angular and energy resolution of the copper Kα1/Kα2 doublet measured
with the powder diffractometer

made for wavelengths other than the $\alpha 1$ and $\alpha 2$ if accurate ($<2\%$ or so) fitting is required [see e.g.4].

WHAT IS THE WAVELENGTH OF COPPER Kα?

The generally accepted value for the wavelengths of Cu K$\alpha 1$ and Cu K$\alpha 2$ are:

Cu Kα_1	1.54056Å
Cu Kα_2	1.54439Å

These values were reported by Bearden [5] and have now been recommended in the IUCr publication "International Critical Tables" [6]. The values are generally given in terms of a unit length in Angstrom units, based on the energy of the W K$\alpha 1$ line of 59.31821 KeV. There is an uncertainty which arises because of the conversion of X units to Angstrom units. Bearden uses a value of 1.002056 for this conversion factor. Because of this uncertainty, minor differences will be found in other tables. As an example, Cauchois and Senemaud list a value of 1537.400 Xu with a conversion factor of 1.0020802 to give 1.540598 Å. This value is also used by the National Bureau of Standards. The NBS value is based on the techniques used by Deslattes et al [7]. There is clearly some inconsistency between these values and Bearden et al have suggested [8] a new value of 1537.370 Xu for Cu K$\alpha 1$. These workers have also suggested that it be one of five new "secondary standards" for wavelength determination.

It follows from Bragg's law that an error in d-spacing is linearly related to an error in wavelength. For most qualitative phase identification, an accuracy of about 1 part in 1,000 is sufficient for $\Delta d/d$. Provided that the true $\alpha 1$ emission line is being used, i.e. either the diffracted beam is really monochromatic, or the contribution from the $\alpha 2$ is effectively removed from the diffraction profile, then the wavelength need also be known only to about 1 part in 1,000. For accurate lattice parameter determination, it is common practice to use an internal standard and, in these cases, the wavelength value is, to all intents and purposes, calibrated out.

MANIFESTATION OF THE PROBLEM

In "manual type" powder diffractometers, generally no attempt is made to modify the lines arising from characteristic X-ray tube wavelengths being diffracted by the specimen, other than to remove or reduce the β-component. In this instance, since a bichromatic $\alpha 1/\alpha 2$ wavelength is being used, problems occur in establishing the true maximum of a peak, because of the splitting of the α-doublet at angles in excess of about 45° 2θ. The resolving power $\Delta\lambda \backslash \lambda$ of a diffractometer is given by:

$$\Delta\lambda/\lambda = 2\,Tan\theta/\Delta 2\theta$$

Cullity [9] has suggested that, for two lines of similar height and width, this expression approximates to $Tan\theta/B$, where B is the peak breadth at half maximum intensity. Since the $\alpha 2$ is only one half the intensity of the $\alpha 1$, these two lines will start to resolve as soon as their peak are separated by an angle equivalent to about 1.16x the line width. As shown in fig. 2, for a line width of about 0.015 2θ, this situation will occur at about 55° 2θ. At angles lower than this, the two lines will be convolved with a shift in the observed 2θ maximum. To compensate for this, a weighted wavelength α^* of 1.54168Å is taken, and the peak of the convolved doublet is assumed to equal $(2\text{x}2\theta_{\alpha_1} + 2\theta\alpha_2)/3\text{x}2\theta\alpha_1$.

This is clearly an approximation and the error involved by assuming that either the $\alpha 1$ wavelength is always correct, or by assuming that the α^* is always correct, introduces an angular error of about $0.008^{\circ}/10^{\circ}$ 2θ. The corresponding error in $\delta\lambda/\lambda$ is about 0.7 parts per thousand. Post [10] has suggested that a sliding wavelength scale should be employed to compensate for this error

A second area of difficulty occurs where computers are used for such tasks as the estimation of peak position, the removal of the $\alpha2$ contribution and for profile fitting. Methods for $\alpha2$ elimination are generally based on either the Rachinger substitution correction technique [11], or on Fourier correction [12]. In practice, both methods are generally applied on a constant 2θ scale with a constant doublet separation and a similar profile shape for $\alpha1$ and $\alpha2$. The latter assumption can lead to oscillations on the high angle side of the residue $\alpha1$ peak. Ladell et al [13] have used the $\alpha1$ and $\alpha2$ components to effectively avoid this problem. Delhez and Mittemeijer [14] have discussed the need to use an angle dependent separation, similar to the sliding wavelength scale suggested by Post.

While these techniques are extremely useful, care must be taken in comparison of computer treated data and manual treated data. Such is frequently the case in the use of data from a modern Automated Powder Diffractometer (APD) and standard patterns form the ICDD Powder Diffraction File. A further complication is that most APD software systems now use a method of peak hunting based on the use of the second differential of the peak. This technique is extremely sensitive to partially resolved peaks and more peaks are generally found than using manual methods. As an example, a recent round robin test [1] revealed that the average number of peaks reported by workers using modern APD systems varied from 29 to 56. The actual number of peaks in the test case was 40!

A third area occurs with the increasing use of energy dispersive detectors in powder diffractometry. Such detectors offer several advantages over classical systems, one of which is that they have high inherent resolution thus obviating the use of a monochromator. Graphite monochromators are generally designed to give high quantum collection efficiency and as a result have rather poor energy resolution characteristics. A disadvantage of the Si(Li) detector is that the proportionality between the energy of the photopeak maximum (and, therefore, the estimated wavelength) is dependent upon the photon flux on the detector and on Compton Scatter. While these effects are controllable, generally by external calibration, their magnitude can be significant.

IS THERE A CASE FOR A WAVELENGTH CALIBRATION STANDARD?

In each of the cases mentioned above there are clearly many causes for experimental uncertainty in the estimation of the "practical" wavelength value. Since these uncertainties may differ from one instrument configuration to another, there seems to be a good case for the use of a "wavelength calibration standard". It is common practice to employ a 2θ error correction term to compensate for minor misalignments of the diffractometer, and even to remove inherent aberrations from the system. Since one has then to correct the observed 2θ value, then apply the wavelength to calculate the d-value, a process involving using the observed 2θ value and applying a corrected wavelength should be no more tedious.

While the major purpose of a wavelength standard would be to offer an independent means of establishing the value of the practical wavelength, it would also serve to establish what other lines from the X-ray tube (Cu $K\beta$, W $L\alpha$, etc.) might potentially give additional diffraction maxima. Single crystal materials such as Si(111) have been used for a number of years for measuring tube impurities [15] and while such an analyzer does not give data below 28° 2θ it does seem an obvious choice. The Instrument Data Collection Task Group of the International Centre is now testing such a standard and hopes to make a testing procedure generally available very soon.

REFERENCES

[1]   Schreiner, W.N and Jenkins, R., Adv. X-ray Anal., 32, (1988) in press

[2]   "International Tables of Selected Constants", Volume 18 - Wavelengths of X-ray Emission Lines and Absorption Edges, Cauchois & Senemaud, Pergamon press:Oxford (1978)

[3]   Berger, H., X-ray Spectrom., 15, 241-243 (1986)

[4]   Shreiner, W.N. and Jenkins, R., Adv. X-ray Anal, 26, 141-148 (1983)

[5]   Bearden, J.A., Rev. Mod. Phys., 39, 78 (1967)

[6]   "International Critical Tables for X-ray Crystallography", Volume IV - "revised and Supplementary Tables", Kynoch Press:Birmingham (1974)

[7]   Deslattes, R.D., Henins, A. and Kessler, E.G. Jr., "Accuracy in X-ray Wavelengths", in NBS Special Publication 567 "Accuracy in Powder Diffraction", U.S. Dept. of Commerce, NBS, Gaithersburg, MD., February 1980.

[8]   Bearden, J.A. et al, Phys. Rev.A (USA) 135, 899-910 (1964)

[9]   "Elements of X-ray Diffraction", Cullity,B.D., Addison-Wesley:New York (1956) p. 411

[10]   Post, B., private communication

[11]   Rachinger, W.A., J. Sci. Instrum., 25, 254-255 (1948)

[12]   Gangulee, A., J. Appl. Cryst., 3, 272-277, (1970)

[13]   Ladell, J., Zagofsky, A. and Pearlman, S., J. Appl. Cryst., 8, 499-506 (1975)

[14]   Delhez, R. and Mittemeijer, E.J., J. Appl. Cryst., 8, 609-611 (1975)

[15]   Parrish, W. and Kohler, T.R., Rev. Sci. Instrum., 27, 795-808 (1956)

RESULTS OF THE JCPDS-ICDD INTENSITY ROUND ROBIN

Walter N. Schreiner

Philips Laboratories, Briarcliff Manor, NY 10510

Ron Jenkins

JCPDS International Centre, Swarthmore, Pa 19081

INTRODUCTION

Last year at this conference we submitted a preliminary report on an X-ray powder diffraction round robin sponsored by the JCPDS-ICDD [1]. This round robin was designed primarily to study the intensities of diffraction lines found by users in routine work. At that time only a portion of the data had been analyzed, and we reported initial findings on the α-Al_2O_3 and $ZnO/CaCO_3$ samples. These included studies on counting statistics, resolution, and the effect of software on intensity precision. Since that time, all the data from the round robin has been entered into Lotus 1-2-3 (*) spread-sheets and numerous additional tests have been carried out. This paper discusses some of the more interesting findings. A complete paper on all of the tests performed is in preparation for submission to the "Methods and Practices Manual" published by the JCPDS-ICDD [2].

Briefly, the round robin had 22 participants mostly using commercial equipment from five major equipment manufacturers. Five units were hybrid and one data set was obtained with a synchrotron source. Four powder samples were sent to the participants with instructions to make specific runs; 177 data sets were received. Because of the previously observed effect of software on the intensities, most of the data presented here were obtained by manually reading computer printouts of the raw data files submitted. Only about two-thirds of the participants submitted raw data file printouts.

Between runs the sample was to be removed from the holder and reloaded. We were, therefore, able to test precision of data within each laboratory, including the contribution from specimen preparation. The single lab precision of the data obtained for Al_2O_3 is reported inFig.1. The standard deviation among the three runs was calculated for each lab for eleven different lines from 25.6° to 135.9° 2θ. The data are divided into two sets, those with fixed divergence slit (FDS) instruments, and those with variable divergence slit (VDS) instruments. The standard deviations are averaged across all laboratories in each set, and the resulting averages plotted as a function of 2θ. It is seen that at lower angles both instrument types obtain about the same precision (2.5%), however, at higher angles, the FDS data sets are nearly twice

---

* "Lotus" and "1-2-3" are registered trademarks of Lotus Development Corporation.

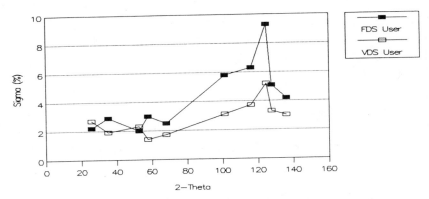

Fig. 1.   A1203 precision (variation of 3 runs);
            single person - single laboratory.

as bad as the VDS sets. This correlation should be expected if the data are counting
statistics limited, because the area of sample illuminated by VDS instruments is 2-3
times larger than in FDS instruments at high angles, hence the intensities are larger.

On the other hand, a quite unexpected result is obtained when the variation of
normalized intensities among laboratories is calculated, i.e. multi-laboratory precision
(reproducibility). The normalized intensity used for each laboratory was the average
of its 3 runs. Fig. 2 shows the multi-laboratory precision across 10 laboratories using
FDS instruments and 9 laboratories using VDS instruments. Note the vertical scale
here is 40% (vs 10% in Fig. 1) the average intensity value for the 3 runs. With the
exception of the line at 25.6°, multi-laboratory precision is the same at low angles, but
significantly worse at high angles for the VDS instruments. From the shape of these
curves, and by examination of the individual data sets, it is clear that the problem has
to do with alignment of the variable slit. The slit should be aligned so that the

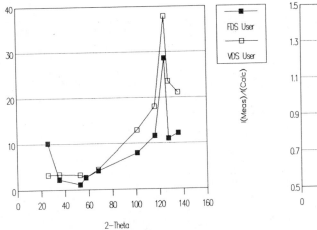

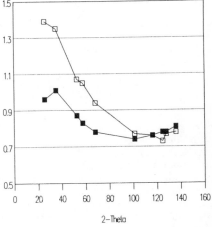

Fig. 2.   A1203 precision (average   Fig. 3.   Variation of intensity with
           of 3 runs); many people -   angle; measured-to-calculated
           many laboratories.   ratio.

illuminated area is constant at all angles, not just at low angles. Poor alignment results in non-constant areas with concomitant angular variations in the intensity. Since there is generally no direct divergence slit adjustment possible with an FDS system, this difference will not be present among these laboratories. Accurate methods of alignment of variable divergence slits have been reported in the literature [3].

An additional effect seen in Fig.2 is that FDS users show a much wider variation in the intensity for the line at 25.6°. It is not clear why this is the case. Examination of the data sets reveals a rather normal looking distribution, but with exceptionally large width.

The accuracy of measured intensities was studied by ratioing them to calculated intensity values. In Fig. 3 this ratio is shown for the Al$_2$O$_3$ data. The average intensities measured by the FDS users and that measured by the VDS users are plotted. The reference data, I(calc), was calculated with MICRO-POWD [4] for a fixed divergence slit system with a graphite monochromator. Calculated peak values were used. A significant variation with angle is seen by both FDS and VDS users. Because peak values were used, there is some question about the shape of these curves below 50°. A similar effect was reported in 1986 at this conference [5] for a VDS instrument where integrated intensities were measured.

A significant difference between the FDS and VDS data sets is also observed. The theoretical difference (proportional to $1/\sin\theta$) has been included in the I(calc) for the VDS data. To elucidate this difference, the ratio of the FDS and VDS data was computed for each line and compared to the theoretical difference. Fig. 4 shows this result for the Al$_2$O$_3$ sample and the β-Spodumene sample. Both samples exhibit a linear variation. The slopes are different; however, this is mostly traceable to the different normalization factors employed for the two data sets. It is likely this variation is due to an incorrect or incomplete description of the VDS instrument in the MICRO-POWD program. In the FDS case there must be additional systematic instrumental or sample-related effects not being accounted for in the calculations. Whatever these are, they appear to be highly reproducible and, therefore, lend themselves to correction via an experimentally determined angularly dependent, intensity calibration function.

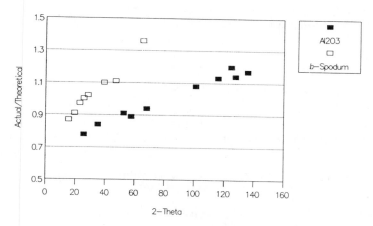

Fig. 4.   Variation of FDS/VDS with angle;
           actual-to-theoretical ratio.

CONCLUSIONS

Many additional tests were carried out in this round robin, including studies of peak shape, correlating peak hunting program performance with counting statistics, d-spacing accuracies and precision, orientation effects, peak to background ratios, and the like. As mentioned earlier, publication of these tables is in preparation. Other workers may find it of value to study the data themselves; in order to accommodate this, the Lotus spreadsheets containing all of the raw data are available through the Data Collection and Analysis Subcommittee of the JCPDS-ICDD. The spreadsheets are large, however, requiring in excess of 640Kb of memory to load into an IBM PC or equivalent. The major observations drawn from this round robin study to date are:

- The single laboratory precision of raw intensity data in the absence of orientation is essentially limited by counting statistics. This implies a significant potential for improved intensity data if other systematic errors can be modeled and removed from measured data.

- Systematic variations of intensity with 2θ angle are observed by all users. An intensity calibration standard would be of significant help in calibrating instruments to compensate for this effect.

- Significant intensity differences are apparent among users of variable divergence slit diffractometers. This problem can likely be corrected by proper alignment of the divergence slit.

- Peak profile shapes were independent of receiving slit.

- Peak finding efficiency is inversely correlated with Net Counting Error but is

- d-Spacing accuracy has not significantly improved with the advent of automated systems. External d-spacing calibration is not yet in routine use; in some cases it is used to compensate for exceptionally poor diffractometer alignment.

A new round robin is now in progress designed to explore diffractometer alignment and possibly to identify a material which could serve as an angularly dependent intensity calibration standard. It is hoped that round robins such as these will continue to elucidate common problems and result in the development of simple methods designed to address them. It is also clear that systematic errors in diffraction data remain abundant, but that, being systematic, these can be addressed in many instances by calibration procedures to result in significantly improved intensity data.

REFERENCES

[1]   "Preliminary Results from a Powder Diffraction Data Intensity Round-Robin", W.N. Schreiner, R. Jenkins, 1987 Denver Conference (in preparation).

[2]   Methods and Practices in X-ray Powder Diffraction, R.Jenkins, Ed., (1987), International centre for Diffraction Data, 1601 Park Lane, Swarthmore, PA 19081

[3]   "Towards Improved Alignment of Powder Diffractometers", W.N. Schreiner, Powder Diffr. Vol. 1, pp 25-33, (March 1986).

[4]   MICRO-POWD Version 1.01 for IBM PC, D.K. Smith, Marketed by Materials Data, Inc., Livermore, CA 94550.

[5]   "Observed and Calculated XRPD Intensities for Single Substance Specimens", W.N. Schreiner, G., Kimmel, Adv. X-Ray Anal., 30, 351-356, (1987).

ON THE PREPARATION OF GOOD QUALITY X-RAY POWDER PATTERNS

D. B. Sullenger
Monsanto Research Corporation–Mound*, Miamisburg, Ohio 45343

J. S. Cantrell, T. A. Beiter and D. W. Tomlin
Monsanto Research Corporation–Mound*, Miamisburg, Ohio 45343
and Chemistry Department, Miami University, Oxford, Ohio 45056

INTRODUCTION

For several years we have prepared and submitted a variety of quality powder x-ray diffraction patterns to the International Centre for Diffraction Data (ICDD) for inclusion as reference standards in their Powder Diffraction File (PDF).[1] Patterns submitted and/or currently under development include metal hydrides (inorganics), flavanoids and related compounds (organics), organic compounds involved in pollution (e.g., dioxins), explosives (organics and metal organics) and glass-ceramic phases (inorganics).

As a general procedure we have attempted to obtain the best possible experimental pattern, via automated diffractometric step-scans, from the substance in question with attention being given to specimen preparation, mounting, thickness, transparency and purity. When the structure has been known by whatever means, the powder pattern has also been calculated from its lattice and atomic positional parameters and a reconciliation sought between the two patterns. The experimental pattern is indexed, and the unit cell lattice parameter(s) are determined via a least squares "best fit" of the interplanar spacings derived from the peak positions. A figure of merit is computed for the pattern, to assess the peak discrimination and positional quality of the pattern. Finally, the data for the analyzed experimental pattern is assembled in PDF format and submitted to the ICDD.

EXPERIMENTAL AND CALCULATED PATTERNS

An attempt was made to obtain a pure, well crystallized sample of the substance under consideration. In general, it is much easier to obtain pure samples than well crystallized ones because purity is usually a concern of the synthesizing scientist, while crystallinity is not unless there is an evident need for it.

*MRC–Mound is operated by Monsanto Research Corporation for the U. S. Department of Energy under Contract No. DE-AC04-76-DP-00053.

Diffraction specimens of the samples were of two types. Reactive and/or radioactive substances were placed in special diffraction cells[2] that had flat polycrystalline beryllium windows for entry of the incident beam onto and exit of the diffracted x-ray beams from the specimen as it underwent its characterizing diffraction. Air-stable samples were free-fall sprinkled onto double-sided tape fastened to plain glass microscope slides for diffraction mounts. External standards were used with specimens contained in the Be-window cells, while internal standards were used with those mounted on microscope slides. In both cases the standard utilized was NBS Si (No. 640a).

Diffraction data collection was by means of automated step-scans. The instrument employed for this work was a Rigaku horizontal, wide angle automated diffractometer equipped with a 12 kW rotating anode x-ray generator. Initial scans were performed with the generator operating at 40 kV and 80 mA for a copper anode. A 1° dispersion slit, a 0.15 mm (~0.05°) receiving slit and a 1° scatter slit collectively defined the x-ray beams. A Ni foil filter was placed in the diffracted beam to reduce the Cu K_β radiation. Scans were performed in 0.05° steps with 0.4 sec. counts per step. The instrumental and step-scan parameters were adjusted as warranted by the quality of the pattern obtained vs the pattern quality desired. Maximum operating levels for the x-ray generator are 60 kV and 200 mA, while step-scan increments can be reduced to 0.001°, and data collection times can be increased several orders of magnitude if desired. The patterns under discussion have undergone this initial step-scan level of data collection and are in various stages of further experimental scrutiny as this article is written.

Pattern analysis was by means of Mound DEC PDP-11/44 computer software, viz., SPECPLOT[3] for display, peak location (peak-maximum or peak-centroid algorithms) and calculated pattern overlay, and LATTICE[4] for peak indexing and least squares lattice-pattern determination.

Powder diffraction patterns were calculated by POWD[5] and stored in a Mound local reference computer file as sets of d-space, I_{Rel} pairs for "stick-figure" overlay comparison with experimental patterns. To perform such calculations input lattice and atomic positional parameters are required. When the structure had been solved the known structural parameters were used for input. On one occasion single-crystal intensity data were converted to polycrystalline intensity data for comparative purposes.

SOME EXAMPLES

Beta-Uranium Hydride

Beta-uranium hydride, β-UH_3, is one form of a metal hydride of considerable interest for hydrogen storage applications. It crystallizes in the primitive cubic space group Pm3n, O_h^3 (No. 223) with eight formula units in a unit cell with a lattice parameter of 6.6444Å. There are two types of uraniums in the structure, one occupying (2a) positions and the other (6c) positions. The hydrogens are in (24k) positions. Rundle obtained the uranium positions via powder x-ray diffraction[6] and the hydrogen positions via powder neutron diffraction (of the deuteride)[7]. Accurate lattice parameters of both the α- and β- forms of UH_3, UD_3 and UT_3 were determined by Q. Johnson, et al[8]. A recent neutron diffraction study of β-UD_3

and β-UH$_3$ was made by Bartscher, et al[9]. There is a PDF pattern (No. 12-436) for this phase, but it has only ten visually estimated lines with d-spacings given to but two decimal places.

Through the good offices of M. C. Nichols of Sandia National Laboratories, Livermore, we obtained a well-annealed sample of β-UH$_3$ that had been prepared at Lawrence Livermore National Laboratory. A specimen of it was placed in a Be-window cell for collection of diffraction data. A comparative powder pattern was calculated using input lattice parameters from Johnson, et al[8]., and atomic positions from Bartscher, et al[9]. From the obviously high quality pattern obtained, 30 peaks (of which eleven displayed distinct K_{α_I}, $K_{\alpha_{II}}$ pairs) were detectable above background. All were satisfactorily indexed, and a cubic lattice parameter of 6.6431(3)Å was derived. The Smith-Snyder figure of merit[10] obtained for this pattern was 25.6. A comparison of the experimental pattern with the calculated pattern overlaid is shown in Figure 1. Interfering lines from the containing cell's face plate are indicated by adjacent asterisks.

Alpha-Barium Tritide

Alpha-barium tritide, α-BaT$_2$, is the room temperature form of the radioactive isotopic variety of barium hydride. It crystallizes[11] in the primitive orthorhombic space group Pbnm, D_{2h}^{16} (No. 62) with four formula units in the cell having lattice parameters of $a_o = 7.829$, $b_o = 6.788$, $c_o = 4.167$ Å. All atoms are in (4c) special positions. There is only one type of barium in the structure, but there are two types of hydrogen (tritium). Each type of atom has two variable positional parameters for a total of six in the structure.

Zintl and Harder[11] synthesized and performed both mono- and polycrystalline x-ray diffraction analyses on the isomorphous alkaline earth hydrides, CaH$_2$, SrH$_2$ and BaH$_2$. They determined the structure type for all of them, via the SrH$_2$ case, from Weissenberg photographic data, finding the metals to be nearly hexagonally close packed. The H's were then placed in plausible interstices of the metal array. Their Debye-Scherrer powder x-ray pattern for BaH$_2$ is included in the PDF as No. 28-137. No powder XRD data for BaT$_2$ is found there or in other scientific literature.

For our experimental pattern we used a powder sample synthesized by D. R. Grove at Mound. The diffraction specimen was contained in a Be-window cell for the automated

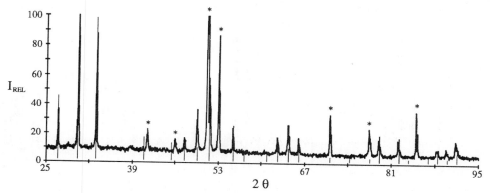

Figure 1 – Comparison of experimental and calculated β-UH$_3$ patterns.

diffractometeric step-scan. A comparative pattern was calculated, using the BaH_2 lattice parameters and the SrH_2 atomic positional parameters from Zintl and Harder.

Twenty-five peaks were read and indexed (eight overlaps). Of these 22 were included in the least squares' lattice parameter determination which found $a_o = 6.790(2)$, $b_o = 7.820(4)$ and $c_o = 4.161(2)$ Å. A figure of merit[9] of 18.5 was found for the experimental pattern. The overlay of the calculated BaH_2 pattern onto the experimental BaT_2 pattern (Fig. 2) indicates their agreement. Again, interfering cell face-plate lines are identified by adjacent asterisks.

Alpha-Lithium Disilicate

Alpha-lithium disilicate, $\alpha\text{-}Li_2Si_2O_5$, is the room temperature form of a common phase in the $Li_2O\text{-}SiO_2$ glass-ceramic system. It crystallizes[12] in the primitive monoclinic space group Cc, C_s^4 (No.9) and is pseudo-orthorhombic, since its beta- angle is approximately 90°. Four formula units are contained in its monoclinic cell having parameters of $a_o = 5.82$, $b_o = 14.66$, $c_o = 4.79$Å and $\beta = 90°$. There are nine atoms in the asymmetric unit, viz., two Li's, two Si's and five O's, all in (4a) positions. Thus there are 27 variable parameters for the structure.

Kracek[13] first thoroughly explored the phase relationships in the $Li_2O\text{-}Si_2O$ system, and Donnay and Donnay[14] studied the habits of the alkali silicates. Liebau[12] determined the crystal structure of $\alpha\text{-}Li_2Si_2O_5$ from single crystal data, and Glasser[15] described an experimental procedure for crystallizing alpha-lithium disilicate from Li_2O, SiO_2 glasses. The PDF contains a pattern for $\alpha\text{-}Li_2Si_2O_5$ (No. 17-447), but the relative intensities of its three strongest lines [viz., (130), (040) and (111)] never agreed with those of experimental patterns that we obtained during glass-ceramic investigations.

A sample of $\alpha\text{-}Li_2Si_2O_5$ was carefully prepared by D. P. Kramer at Mound by a modification of Glasser's method[15]. When finely pulverized and dusted through a sieve onto a patch of double-sided tape affixed to a plain glass microscope slide it yielded a diffraction pattern having much better intensity agreement with our previously obtained experimental patterns as well as with a pattern calculated from Liebau's structure[12]. For d-spacing determination in the experimental pattern no background subtraction was employed, but for the I_{Rel} evaluation a polynomial fit of the background was utilized. Twenty-nine peaks were

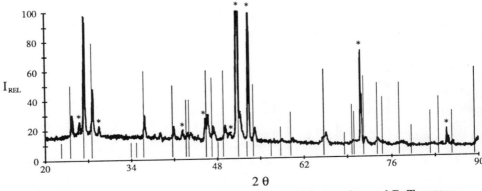

Figure 2 – Comparison of calculated BaH_2 pattern with experimental BaT_2 pattern.

observed and indexed in this pattern, and 23 of those were included in the least squares' parameter determination that found monoclinic cell edges of $a_o = 5.823(3)$, $b_o = 14.665(6)$, $c_o = 4.775(2)$ Å, and $\beta \cong 90°$. The Smith-Snyder figure of merit[10] for the first 29 lines of this pattern was approximately 30. Figure 3 displays contrasting overlays of PDF No.14-447 and of our calculated pattern on the (130), (040) and (111) triplet of the experimental pattern. Since in the POWD calculation[4] random orientation of the crystallites was forced, we believe that the PDF reference pattern has preferred orientation and that our experimental pattern is a viable candidate for its replacement in the PDF. Mention was made in most of the background publications of a strong propensity for preferred orientation in powdered samples of this phase.

2,2′,4,4′,6,6′– Hexanitroazobenzene–III

2,2′,4,4′,6,6′– hexanitroazobenzene–III (HNAB–III), is one of five known forms of an important secondary explosive. The crystal structures of HNAB–I and HNAB–II have been solved from single crystal x-ray diffraction data[16], but though a set of single crystal intensity data has been collected, a solution of its structure has not yet been achieved. However, its space group is known[17] to be end-centered orthorhombic C222$_1$, D$_2^5$ (No. 20), with lattice parameters of $a_o = 15.281$, $b_o = 40.925$, $c_o = 5.486$ Å. Its density has been measured as 1.751 g/cm³.

In order to obtain an experimental pattern of HNAB–III we prepared as thick a specimen as would adhere to a double-sided tape patch fastened to a plain glass microscope slide. The sample used for this work was prepared at Mound by D. W. Firsich and was carefully pulverized with a mortar and pestle to a fine powder. A thinner specimen was also prepared from this same finely ground sample because we had suspected possible specimen transparency difficulties with this low atomic number compound. The thick specimen exhibited 25 easily observable and indexable peaks. Fifteen of these reflections were included in a least squares' lattice parameter determination that yielded orthorhombic cell edges of $a_o = 15.365(2)$, $b_o = 41.370(4)$ and $c_o = 5.4955(2)$Å. The single crystal intensities were converted to powder intensities for comparison with the experimental pattern. The calculated pattern contained 149 reflections with relative intensities of one or greater, so no figure of merit has yet been computed for the experimental pattern. Instead we are currently

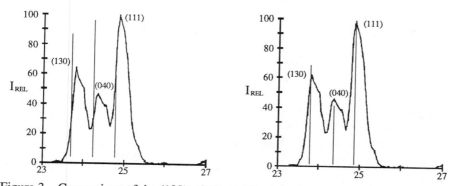

Figure 3 – Comparison of the (130), (040), (111) triplet from the experimental pattern of α-Li$_2$Si$_2$O$_5$ with PDF No.17-447 (left) and the calculated pattern (right).

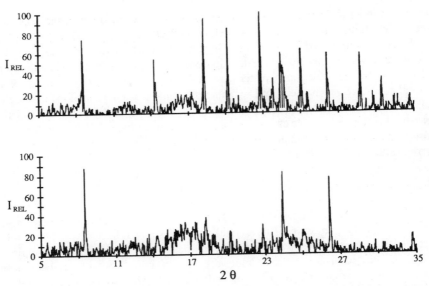

Figure 4 –   Comparison of a calculated powder pattern of HNAB–III with experimental patterns from thick (upper) and thin (lower) finely ground sample specimens.

performing higher power, longer count, finer step scans in an effort to bring more of the weaker peaks up from the background level. Figure 4 reveals the good agreement between the thick specimen and the calculated (overlaid) patterns and the transparency and, perhaps, preferred orientation problems of the thin specimen.

CONTINUING AND PROJECTED WORK

These four diffraction patterns are nearing a stage where they can be submitted to the ICDD for possible inclusion in the PDF as new or replacement reference patterns. We are also developing patterns from numerous other explosives and pyrotechnics, flavanoids and various isotopic varieties of BaT_2, UH_3 and other metal hydrides. In addition a new, improved cell, featuring an off-reflection cut single crystal of Be for a specimen face plate, is in an advanced stage of development. Initial experiments with it have indicated that a near zero diffraction background is obtainable, auguring well for obtaining diffraction patterns of much improved quality for air-sensitive and radioactive substances.

REFERENCES

1. "Powder Diffraction File," JCPDS – International Centre for Diffraction Data, 1601 Park Lane, Swarthmore, PA 19801.
2. Design of cells was by M. C. Nichols and associates, Sandia National Laboratories, Livermore, CA; fabrication was at MRC–Mound.

3. R. P. Goehner, SPECPLOT – An Interactive Data Reduction and Display Program for Spectral Data, in: "Advances in X-ray Analyses," J. R. Rhodes, C. S. Barrett, D. E. Leyden, J. B. Newkirk, P. K. Predecki and C. O. Ruud, eds, Plenum Pub. Co, New York, (1980), vol. 23, 305–311.

4. Program, LATTICE , adapted by E. F. Jendrek (MRC–Mound) from a program originally written by D. E. Appleman and H. T. Evans, USGS-GD-73-003, U. S. Geological Survey Report, Washington, D. C., (1973).

5. D. K. Smith, M. C. Nichols and M. E. Zolinsky, "POWD, A FORTRAN IV Program for Calculating X-Ray Powder Diffraction Patterns – Version 10", Pennsylvania State University, University Park, PA (March, 1983). MRC–Mound version adapted by E. F. Jendrek.

6. R. E. Rundle, The Structure of Uranium Hydride and Deuteride, J. Am. Chem. Soc. 69:1719 (1947).

7. R. E. Rundle, The Hydrogen Positions in Uranium Hydride by Neutron Diffraction, J. Am. Chem. Soc. 73: 4172 (1951).

8. Q. Johnson, T. J. Biel and H. R. Leider, Isotopic Shifts of the Unit Cell Constants of the α and β Trihydrides of Uranium: UH_3, UD_3, UT_3, J. Nucl. Mater. 60:23 (1976).

9. W. Bartscher, A. Boeuf, R. Caciuffo, J. M. Fournier, W. F. Kuhs, J. Rebizant, and F. Rustichelli, Neutron Diffraction Study of β-UD_3 and β-UH_3, Solid State Commun. 53:423 (1985).

10. G. S. Smith and R. L. Snyder, F_N: A Criterion for Powder Diffraction Patterns and Evaluating the Reliability of Powder-Pattern Indexing, J. Appl. Cryst. 12: 60 (1979).

11. E. Zintl and A. Harder, Constitution of Alkaline Earth Hydrides, Z. Elektrochem. 41:33 (1935).

12. F. Liebau, Investigations of Layer Silicates of the Formula Type $A_m(Si_2O_5)_n$, I. The Crystal Structure of the Room Temperature Form of $Li_2Si_2O_5$, Acta Cryst. 14: 389 (1961).

13. F. C. Kracek, The Binary System Li_2O-SiO_2, J. Phys. Chem. 34:2641 (1930).

14. G. Donnay and J. D. H. Donnay, Crystal Geometry of Some Alkali Silicates, Am. Mineralogist 38: 163 (1953).

15. F. P. Glasser, Crystallization of Lithium Disilicate from Li_2O SiO_2 Glasses, Phys. Chem. Glasses, 8: 224 (1967).

16. E. J. Graeber and B. Morosin, The Crystal Structures of 2,2',4,4',6,6'– Hexanitroazobenzene (HNAB), Forms I and II, Acta Cryst. B30: 310 (1974).

17. P. H. Swepston, Molecular Structure Corp., College Station, TX, Private Communication (November 24, 1987).

SEMI-QUANTITATIVE XRD ANALYSIS OF FLY ASH USING RUTILE AS AN INTERNAL STANDARD

A. Thedchanamoorthy and G.J. McCarthy
Department of Chemistry
North Dakota State University
Fargo, North Dakota 58105

ABSTRACT

XRD analysis of fly ash was quantitated using the Reference Intensity Ratio (RIR) method and rutile (TiO_2) as an internal standard. Rutile RIR's for 15 of the crystalline phases commonly observed in North American fly ash were determined. Error analysis on the various steps in quantitation indicated that precision ranged from $\pm 10\%$ of the amount present for phases that diffract x-rays strongly to $\pm 21\%$ for weakly diffracting phases. Limit of detection in the mostly glassy fly ashes ranged from 0.2% for lime, the most strongly diffracting phase, to 3.5% for weakly diffracting mullite. Accuracy evaluated with a simulated fly ash was within the limits established by precision, but in actual fly ash samples, accuracy will be a function of the match between the crystallinity and composition of the analyte and the analyte standard. Overlaps among peaks of some of the important phases require intensity proportioning; for this reason, the method is best described as semi-quantitative.

INTRODUCTION

Fly ash is a very fine, predominantly glassy, material removed from flues of fossil-fuel power plants. Characterization of its crystalline fraction by x-ray diffraction (XRD) is necessary for many studies of fly ash utilization and disposal. In reference (1), we described the XRD mineralogy of fly ash. Use of an internal standard for quantitative comparisons of the amounts of crystalline phases among ashes was also presented. The rutile form of TiO_2 was chosen as the internal standard because it had few overlaps with the key peak of each fly ash phase, moderate linear absorption coefficient, long term stability in air and during grinding, and low cost. A fixed amount of rutile is added to every ash sample and intensities are ratioed to the strongest rutile intensity. These ratios are being obtained for up to 1000 fly ashes, and they will be incorporated into a database of fly ash chemistry, mineralogy and physical properties (2,3). To date, the mineralogy of more than 400 samples has been determined.

The objective of the present study was to quantitate fly ash XRD mineralogy at minimum cost using a procedure reproducible among numerous student analysts. This quantitation is being done by the RIR method (4-6) using rutile as an internal standard. In this paper we present the rutile reference intensity ratios (RIR's) for the crystalline phases described in (1) and for calcite and portlandite (see Table 1 for compositions), which are also occasionally observed in fly ashes. For the case of solid solution phases, rutile RIR's

Table 1

CRYSTALLINE PHASES IDENTIFIED IN FLY ASH

Code	Name	Nominal Composition
Ah	Anhydrite	$CaSO_4$
C_2S	Dicalcium silicate (Bredigite)	$Ca_{1.75}Mg_{0.25}SiO_4$
Cc	Calcite	$CaCO_3$
C_3A	Tricalcium aluminate	$Ca_3Al_2O_6$
Hm	Hematite	Fe_2O_3
Lm	Lime	CaO
Ml	Melilite (ss)	$Ca_2Mg_{0.5}AlSi_{1.5}O_7$
	Akermanite	$Ca_2MgSi_2O_7$
	Gehlenite	$Ca_2Al_2SiO_7$
Mu	Mullite	$Al_6Si_4O_{13}$
Mw	Merwinite	$Ca_3Mg(SiO_4)_2$
Pc	Periclase	MgO
Pl	Portlandite	$Ca(OH)_2$
Qz	Quartz	SiO_2
$C_4A_3\bar{S}$	Sodalite structure	$Ca_3Al_6O_{12} \cdot CaSO_4$
Sp	Ferrite Spinel (ss)	$(Fe,Mg)(Fe,Al)_2O_4$
	Magnetite	Fe_3O_4
	Magnesioferrite	$MgFe_2O_4$
Tn	Thenardite	Na_2SO_4

were determined for several compositions. An error analysis was performed, and a test was performed with a simulated fly ash of known mineralogy. Additional details of the application of the protocol described here to the various classes of fly ash, and results for the four National Institute of Standards and Materials fly ash Standard Reference Materials, are given elsewhere (7).

MATERIALS AND EXPERIMENTAL PROCEDURES

Table 1 lists the codes, names and nominal compositions of 15 fly ash phases identified to date in the ~400 fly ash samples. The dicalcium silicate (C_2S) solid solution phase observed in several fly ashes had a pattern similar to that of bredigite and was modeled with the specific composition given in Table 1. Melilite and spinel are also solid solution phases. Gehlenite and akermanite are end members of the melilite solid solution series, and magnetite and magnesioferrite of the spinel series.

Most of the standards were synthesized or heat treated at 1200-1300°C for at least two hours to simulate the generally high crystallinity of the refractory fly ash phases resulting from their formation at high temperatures. Anhydrite was made by firing gypsum. Calcite and thenardite, which form at low temperatures in fly ash, were laboratory reagents. Portlandite, which forms from lime in some fly ashes on exposure to moist air, was made by exposing "soft-burned" lime to ambient moisture. Natural hematite, magnetite and quartz were used without further heat treatment. Lime was prepared by firing calcite. Reagent hematite and periclase were heat treated at 1300°C. Bredigite, C_3A, the melilites, magnesioferrite, merwinite, mullite and the sodalite phase were synthesized from chemically mixed starting batches. The general procedure was to combine stoichiometric amounts of solutions of metal nitrates and ammonium hydroxide, dry the solution/gel at 110°C, heat treat the residue at 500°C to remove volatiles, pelletize the product and fire the pellets to 1200-1350°C for a minimum of 6 h, using intermediate regrindings and refirings until a phase pure product was obtained.

Two Philips diffractometers were employed for data collection. Both had a Cu tube operated at 50 kV and 23 mA, θ-compensating divergence slits, graphite diffracted beam monochromators, scintillation detectors and NIM counting electronics. Intensity data on the manual instrument were obtained using the scaler-timer. On the automated diffractometer (APD) data were collected in 0.02° steps for 2-5 s count times. The resident software on the APD was used for data smoothing; α_2 stripping was not performed. Peak positions and heights are read manually using the x-y cursors on the graphics terminal. There is no provision for integrated intensities in the available version of the APD software.

Reagent grade rutile was fired at 1200°C for 2 h. The fired rutile and the analyte standard or fly ash were each ground to <10 μm (verified by SEM) in a McCrone Micronizing corundum mill for 5 min using an EtOH medium. For RIR determination, a 1:1 mix was ground by hand for about 20 min. Three or more side-drifted specimen mounts were used. Data acquisition parameters were adjusted so that at least 10^4 counts were collected for each phase. The ratio of the intensity of the key peak to that of the I = 100 27.5° rutile peak was measured. For general analyses of fly ash, peak instead of integrated intensities are used because of severe overlapping of peaks in the complex assemblages of up to a dozen phases. Because integrated intensities can be used for several phases (e.g. lime), integrated as well a peak RIR's were determined for the common fly ash phases using the manual diffractometer. Intensities were scaled using the correction factor $1/\sin\theta$ in order to convert θ-compensating divergence slit intensities into equivalent fixed-slit intensities.

The Reference Intensity Ratio method equation is:

$$x_{analyte} = x_{standard}/RIR \cdot I_{analyte}/I_{standard} \qquad [1]$$

where x = weight fraction. To evaluate the quantitation, analyses were made of a simulated fly ash made up of six crystalline fly ash phases and silica gel to simulate glass. Specimens were prepared with 10 wt% rutile and hand ground in an agate mortar for 20 min. In this mix, the RIR equation, expressed in weight percent, becomes:

$$wt\% \text{ of analyte} = 10/RIR \cdot I_{analyte}/I_{rutile} \cdot 1.11 \qquad [2]$$

The factor 1.11 (100%/90%) corrects the result to wt% in a rutile-free sample.

RESULTS AND DISCUSSION

Rutile RIR Values

Rutile RIR values, peak and integrated for the manual diffractometer and peak for the APD, along with the hkl and 2θ (CuKα) position of the I = 100 analyte peak, are listed in Table 2. Values for both θ-compensating slit (θS) and fixed slit (FS) are given. The number in parenthesis is one standard deviation for nine measurements with three specimens, each moved and scanned three times. A complete set of rutile RIR's was obtained only for the APD because this instrument is used in the database project (1). Peak and integrated manual diffractometer values were determined for the more common fly ash phases. In five cases where the most intense peaks frequently have overlap problems, RIR values for less intense alternate peaks are given. These less intense peak RIR's can be used when the phase is abundant enough to have statistically significant numbers of counts for these peaks.

The relative standard deviations in RIR values were generally below 5%. With greater than 10^4 counts collected for each peak and low backgrounds, counting statistics should not

Table 2

Rutile Reference Intensity Ratios for Fly Ash Crystalline Phases

Phase	2θ/hkl	Manual Integrated		Manual Peak		APD Peak	
		θS^a	FS^b	θS	FS	θS	FS
Anhydrite	25.5/002	1.25(4)	1.34	1.05(6)	1.12	1.11(4)	1.19
Bredigite (C_2S)	32.8/222	ND^c		ND		0.16(2)	0.14
Calcite	29.4/104	1.39(4)	1.30	1.13(2)	1.06	1.12(1)	1.05
C_3A	33.1/224	1.18(5)	0.99	1.18(3)	0.99	1.17(4)	0.99
Hematite	33.2/104	0.81(3)	0.67	0.58(3)	0.48	0.61(2)	0.51
	54.1/116					0.35(2)	0.18
Hematite (mineral)	33.2/104	0.86(3)	0.71	0.70(2)	0.58	0.74(4)	0.61
Lime (1300°C/36h)	37.4/100	1.78(3)	1.32	1.82(5)	1.35	1.83(4)	1.35
Melilite (ss)	31.1/211	0.71(3)	0.63	0.56(2)	0.50	0.59(1)	0.52
Akermanite	31.1/211	0.46(1)	0.49	0.34(1)	0.30	0.36(2)	0.32
Gehlenite	31.1/211	0.69(2)	0.69	0.53(1)	0.47	0.53(1)	0.47
Magnesioferrite	35.5/311	(d)		0.80(1)	0.62	0.82(2)	0.64
	30.2/220					0.20(1)	0.18
Magnetite	35.5/311	(d)		0.82(2)	0.64	0.80(3)	0.62
Merwinite	33.3/013	(d)		0.49(1)	0.41	0.51(1)	0.42
	33.8/020					0.22(2)	0.18
Mullite	26.3/210	(e)		0.22(1)	0.23	0.23(1)	0.24
Mullite (refractory)	26.3/210	(e)		0.26(1)	0.27	0.25(1)	0.26
	16.4/110					0.07(1)	0.12
Periclase	42.9/111	1.44(4)	0.93	1.30(1)	0.84	1.35(2)	0.87
Portlandite	34.1/101	ND		0.40(2)	0.32	0.41(2)	0.33
Quartz	26.7/101	1.42(3)	1.46	1.44(2)	1.48	1.48(4)	1.52
	20.9/100					0.24(1)	0.31
Sodalite ($C_4A_3\bar{S}$)	23.7/211	ND		ND		0.56(2)	1.17
Thenardite	32.2/113	ND		ND		0.54(1)	0.45
Lime (700°C/12h)	37.4/100	1.69(5)		1.56(7)			
Lime (900°C/12h)	37.4/100			1.78(4)			
Lime (1100°C/12h)	37.4/100			1.84(3)			
Lime (1300°C/12h)	37.4/100	1.76(4)		1.81(1)			

a. Value for theta-compensating divergence slit
b. Value for fixed divergence slit.
c. ND = Not Determined (Not a common phase in fly ash).
d. Not determined. Tail of peak overlapped with a rutile peak.
e. Not determined. Overlapping of two strong reflections in phase.

contribute more than 1% to the error (4). The principal causes of error in the RIR measurements should be lack of complete homogeneity in 1:1 mix and preferred orientation.

There was good agreement between peak intensity RIR values determined on the manual and automated diffractometers; the values were almost always within each other's standard deviations. For a few compositionally simple, highly crystalline phases (quartz, lime, C_3A, periclase), agreement between integrated and peak intensity RIR was good. With the other common fly ash phases, agreement was not as good; in some cases it was only ±25%. The integrated RIR's were usually greater than peak RIR's, as would be expected for less than fully crystalline phases. This points out the need to use the appropriate RIR (peak or integrated) for quantitation, and if peak RIR's are used, to look for a similar level of crystallinity by measuring FWHM and/or α_1/α_2 resolution.

It is noteworthy that the RIR values for the 50:50 melilite solid solution phase do not fall between those of the end members. McCarthy et al. (8) have demonstrated that RIR variations in solid solutions can be distinctly nonlinear or even show maxima and minima.

Ferrite spinel solid solution has as its principal components $FeFe_2O_4$ (magnetite) and $MgFe_2O_4$ (magnesioferrite), with minor substitutions of Al and Ti (1-3). There is no significant difference between the RIR's of these principal end members, so the ferrite spinel RIR can be approximated by a single RIR value.

Lime is an important analyte in fly ash and has a profound effect on its pozzolanic and cementitious properties. Experience in our laboratory (2,3) has shown that lime can react with water at various rates, which are thought to be dependent on how "hard-burned" is the lime. Hard-burned lime is less water reactive. To provide RIR values for limes of varying crystallinity, five heat treatments were made. The four entries at the bottom of Table 2, and the fifth lime entry, show the variation of RIR of lime with firing temperatures, and they clearly show the effect of crystallinity on RIR. The integrated intensity is nearly constant for firing between 700 and $1300^\circ C$, but the peak intensity increases with firing temperature to reach the integrated value at $1300^\circ C$. Even with 12 h of firing at $1300^\circ C$, the lime sample cannot be stored and handled because it reacts with moist air to form portlandite. The 36 h firing at $1300^\circ C$ apparently provides additional sintering sufficient to give a non-reactive standard.

Error Analysis

Precision. Analyst and instrument reproducibility, and particle statistics, preferred orientation and homogeneity of mixing of analyte and standard, are the principal factors affecting precision of quantitation by the protocol described here. From equation [2] above, these errors can be divided into those associated with the RIR term and with the $I_{analyte}/I_{rutile}$ quotient. Error in the determination of analyte to rutile intensity, discussed by in detail by McCarthy et al. (1), was found to be $\pm10\%$ and $\pm16\%$ for weak peaks and strong peaks respectively. These high values are due to the fact that fly ash is typically 60-90% glass (which produces high background) and count rates for analytes are low. Thus, in an experiment of reasonable duration, the analyte counts are often less than 10^3 and counting statistics become a more important factor. Combining this level of precision with that found for the RIR determinations (Table 2) gives the total error (root mean square of the individual errors) of the quantitation. For example, considering lime, with the strongest Bragg diffraction (largest RIR) and bredigite with the weakest diffraction, we get for the best and worst cases:

	I_a/I_r Term*	RIR Term (N=9)**	Total Error
Best case	10%	2%	$(104)^{1/2} = 10\%$
Worst case	16%	13%	$(425)^{1/2} = 21\%$

In our laboratory, the estimated total error of quantitation is prorated between these limits according to the RIR of the analyte under consideration.

Accuracy. The best way to assess accuracy of the quantitation would be use standard materials certified for mineralogical content. As there are no such standards, an simulated fly ash was made by combining six of the common crystalline phases with silica gel in the proportions shown in Table 3. Three specimens were made from the mix. For all six crystalline phases, analytical recovery (agreement between measured and known concentrations) was within 10% of the known values (Table 3), better than the $\pm10\%$ discussed above for precision in the best cases. The excellent recovery in the amorphous phase (by

*reference (1)
**APD peak RIR

Table 3

Analysis of a Synthetic Fly Ash

Phases	wt% in mix	wt% Recovered	
		$\bar{x}$	s
Anhydrite	5	4.8	0.5
C$_3$A	6	5.8	0.3
Mullite	5	4.7	0.7
Periclase	5	5.4	0.4
Quartz	5	4.9	0.2
Spinel	6	6.3	0.4
Silica gel	58	58.3*	
Rutile	10	10	

*by difference

Table 4

Limits of Detection and Rutile RIR's of the Crystalline Phases in Fly Ash

	Ah	Mu	Qz	Ml	Hm	C$_3$A	Mw	Sp	Lm	Pc	So	Pl
LOD[a]	0.5	3.5	0.3	1.0	1.0	0.5	1.1	0.7	0.2	0.4	1.0	1.4
RIR[b]	1.11	0.25	1.48	0.59	0.61	1.17	0.51	0.82	1.83	1.35	0.56	0.41

a. in wt%
b. APD peak

difference) is a reflection of the balance of positive and negative errors in recovery for the crystalline phases.

It must be cautioned that the crystallinity and composition of analyte and analyte standard were identical in this test, and substantial RIR differences could arise when these parameters differ in real fly ashes (8-10). The inherently inhomogeneous nature of fly ash is not consistent with a unique analyte standard for many fly ash phases. For solid solution phases, there is no single intermediate composition, but instead one finds a range of compositions in and between fly ash grains (i.e., intra- and intergrain inhomogeneity). Thus, an analyte standard can, at best, only approximate the average composition of a fly ash solid solution phase.

Limit of Detection

The lower limit of detection (LOD) is the minimum wt% of a phase in a fly ash sample to give a measurable peak. LOD was determined empirically by noting the wt% derived from the smallest peak of each phase observable in the ~400 ashes studied to date. These results are shown in Table 4, along with the rutile RIR's for these phases. The values varied from a low of 0.2 wt% for lime to a high of 3.5% for mullite. There is a correlation between the diffracting power of the phase, as indicated by a large RIR, and the LOD. Counting statistics in the low count rate, high background, largely glassy fly ashes affect LOD. Davis (10) also reports that weakly diffracting substances have the highest limits of detection.

Overlapping Peak Problems and Recommendations for Improvements in the Protocol

On application of the protocol to actual fly ashes, problems with overlapping peaks require proportioning of intensity at certain 2θ angles. A detailed discussion of proportioning procedures is given elsewhere (7). The phases affected are hematite, C_3A, merwinite and ferrite spinel (magnetite). The necessity to do this for up to four important phases in some fly ashes, combined with the impossibility of exactly matching analyte and analyte standard, are the reasons that this protocol is termed "semi-quantitative."

The intensity proportioning procedures would be more accurate if integrated rather than peak intensities were used, as they are only completely correct when peaks coincide rather than overlap. An obvious improvement on this protocol would result from computer deconvolution of overlapping peaks that would permit use of integrated rather than peak intensities. Deconvolution was not possible with the equipment and software employed.

Another improvement to the protocol would be use of the full digital diffractogram. One option for quantitation using computer data reduction and analysis developed by Smith and coworkers (11,12), is to modify the RIR method so that full profile RIR's, measured experimentally from the synthesized pure phases or calculated from crystal structure data, are summed to give the convoluted diffractogram of the crystalline phases in an unknown. This method has not yet been applied to a material as complex as fly ash.

In another full diffraction profile option, the Rietveld method (13) is used to refine the crystal structures of each crystalline phase in the fly ash, and results in a scaling factor of each phase's contribution to the full diffraction profile (14-17). Intensity in the profile can thereby be apportioned to crystalline and noncrystalline phases. An advantage of the Rietveld method would be that solid solution composition could be factored into the analysis as parameters to be refined, and that the full diffracted intensity of poorly crystallized phases and trace phases would be accounted for. This method has been applied to quantitation of the mineralogy of zeolite tuffs, materials that share mineralogical complexity and the presence of noncrystalline phases with fly ash, by Bish and Chipera (17). A potential problem with the Rietveld method is the paucity of Bragg diffraction from largely glassy fly ash that could make refinement of the hundreds of chemical and structural parameters from a dozen or so fly ash crystalline phases nearly impossible.

CONCLUSIONS

Rutile is an almost ideal internal standard material for quantitative XRD analysis of fly ash. Rutile RIR values for fly ash analyte standards have been determined. Where appropriate, these standards were fired to match the crystallinity of fly ash analytes or, as was the case with portlandite, the standard was synthesized under conditions that approximate its formation in the ash. Where the analytes were solid solutions, RIR's for several representative solid solution members were determined. The relative standard deviation in the RIR values was generally below 5%. There was good agreement in peak intensity RIR's determined manually and with the APD. Integrated intensities RIR's were generally greater than those from peak intensities. The variation in lime's peak RIR with firing temperature, contrasted to the minimal variation in its integrated RIR, demonstrated the importance of approximating crystallinity in the analyte and the analyte standard (by measuring FWHM and/or α_1/α_2 resolution) in cases, such as fly ash analyses, where peak intensities must be used in a quantitation.

An analysis of overall precision in the steps in quantitation of fly ash mineralogy by the protocol described in reference (1) and here ranged from $\pm10\%$ for the most strongly diffracting (high RIR) phases to $\pm 21\%$ for the most weakly diffracting phase.

A test of the protocol with a simulated fly ash made up of a dominant amorphous phase and six crystalline phases gave very good recoveries, and the relative standard deviation for the three measurements was better than the $\pm 10\%$ estimated for precision in the best cases. Such good recoveries are not to be expected in actual fly ashes where the analyte standard will not be an exact match to the analyte in crystallinity and composition.

The limit of detection for 12 of the most common fly ash phases was found to vary from 0.2% for lime, the most strongly diffracting phase, to 3.5% for mullite, a weakly diffracting phase.

Problems with overlapping peaks in many actual fly ashes require proportioning of intensities. Because of the additional uncertainty inherent in this procedure, the protocol is best termed semi-quantitative.

ACKNOWLEDGMENTS

This research was supported by the sponsors of the Western Fly Ash Research, Development and Data Center (Northern States Power, Nebraska Ash Company, National Minerals Corporation, Cooperative Power Association, Otter Tail Power).

REFERENCES

1. G.J. McCarthy, D.M. Johansen, S.J. Steinwand, A. Thedchanamoorthy, Adv. X-Ray Anal. 31:331-342 (1988).
2. G.J. McCarthy, O.E. Manz, R.J. Stevenson, D.J. Hassett and G.H. Groenewold, in Fly Ash and Coal Conversion By-Products: Characterization, Utilization and Disposal II, Mat. Res. Soc. Symp. Proc. Vol. 65, Materials Research Society, Pittsburgh, pp. 165-166 (1986).
3. G.J. McCarthy, O.E. Manz, D.M. Johansen S.J. Steinwand, R.J. Stevenson, and D.J. Hassett, in Fly Ash and Coal Conversion By-Products: Characterization, Utilization and Disposal III, Mat. Res. Soc. Symp. Proc. Vol. 86, Materials Research Society, Pittsburgh, pp. 109-112 (1987).
4. H.P. Klug and L.E. Alexander, X-Ray Diffraction Procedures for Crystalline and Amorphous Materials, Wiley Interscience, New York (1974).
5. F.H. Chung, J. Appl. Cryst. 7:519-526 (1974).
6. B.L. Davis, Reference Intensity Method of Quantitative X-ray Diffraction Analysis, 2nd Edition, South Dakota School of Mines, Rapid City (1988).
7. G.J. McCarthy and A. Thedchanamoorthy, in Fly Ash and Coal Conversion By-Products: Characterization, Utilization and Disposal V, Mat. Res. Soc. Symp. Proc. Vol. 136, Materials Research Society, Pittsburgh, in press.
8. G.J. McCarthy, R.C. Gehringer, D.K. Smith, V.M. Injaian, D.E. Pfoertsch and R.L. Kabel, Adv. X-Ray Anal. 24:253-264 (1981).
9. R.C. Gehringer, G.J. McCarthy and R.G. Garvey, Adv. X-Ray Anal. 26: 119-128 (1981).
10. B.L. Davis, Adv. X-Ray Anal. 31:317-323 (1988).
11. D.K. Smith, G.G. Johnson, Jr, A. Scheible, A.M. Wims, J.L. Johnson and G. Ulmann, Powder Diffraction, 2: 73-77 (1988).
12. D.K. Smith, G.G. Johnson, Jr. and A.M. Wims, Aust. J. Phys. 41: 311-321 (1988).
13. H.M. Rietveld, J. Appl. Cryst. 2: 65 (1969).
14. R.J. Hill and C.J. Howard, J. Appl. Cryst. 20: 467-474 (1987).
15. B. Jordan and B.H. O'Connor, in Proc. AXAA-88 Conf., Perth, Australia, August 1988, pp. 309-318.
16. B.H. O'Connor and M.D. Raven, Powder Diffraction, 3: 2-6 (1988).
17. D.L. Bish and J. Chipera, Adv. X-Ray Anal. 31: 295-308 (1988).

MECHANICALLY-INDUCED PHASE TRANSFORMATIONS IN PLUTONIUM ALLOYS

P. L. Wallace, W. L. Wien, and R. P. Goehner*

Lawrence Livermore National Laboratory
*Nicolet Instrument Corporation

ABSTRACT

In this article, we show that mechanically-induced phase transformations can be readily achieved in two Pu-alloy systems. We have observed mechanically-induced phase transformations in both Ti-stabilized β-Pu and Ga-stabilized δ-Pu. In both of these alloys, the parent phase has been largely transformed to α-Pu, and the cause of these transformations was mechanical strain introduced by the metallographic sample preparation. For the Ga alloys, x-ray diffraction (XRD) patterns were taken at about 1.5-μm steps down to the undisturbed material in order to develop depth profiles of the surface damage. The total depth of the disturbed material in these alloys is estimated to be about 7.6 μm, but this depth was not measured for the Ti alloys. The proportions of α-Pu and δ-Pu in the Ga alloys have been estimated using (a) a new quantitative phase analysis program (SPECQUAN) that uses multiple peaks of each phase in order to minimize the effects of preferred orientation and (b) an older manual technique (i.e., hand calculations). The results from these techniques are compared. SPECQUAN was developed to use the SPECPLOT data file structure directly, thus reducing our data processing. The program is written in Fortran 77 and employs an external intensity ratio quantification procedure to obtain its results. XRD calibration has been done independently by means of accurate density measurements on a reference Ga alloy.

INTRODUCTION

It has been observed by many authors[1-4] that Pu and its alloys can be susceptible to mechanically-induced phase transformations. In this work, we extend this knowledge to an alloy system (Pu-Ti) for which the effect had not been documented. We also report on how the amount of transformed material varies with depth in a second alloy system (Ga-Pu).

To perform this work:

- We developed an electropolishing technique to control the amount of material removed in each step.

- We used a profilometer to determine (a) whether or not electropolishing removed layers uniformly and (b) the average amount of material removed.

- We wrote a computer program to produce the required quantitative analyses.

All of these techniques could be adapted readily to problems involving non-Pu systems.

ANALYTICAL PROCEDURES

We prepared the samples metallographically using techniques that had been successfully applied in the past.[5] These included grinding with silicon carbide papers and polishing with diamond paste on polishing cloths. A relatively recent addition to these techniques was the use of a vibratory polisher for the final polishing step.

When we determined that a precise measurement of the depth of surface damage was not required, we used routine electropolishing in the metallographic laboratory. This was the process for the Pu-Ti alloy and some of the Ga-Pu alloys.

For those Ga-Pu samples requiring a precise determination of the depth of damage, a controlled electropolishing procedure was used. First, we used a commercial vinyl tape to mask off surface areas that were not to be electropolished. Then we used either the epoxy resin of the sample mount or a non-conductive lacquer to cover the sides and backs of the samples, leaving room only for the electrical contact.

The electropolished sample area was 75 square mm. This size had been determined previously to be appropriate for the desired sensitivity of XRD analysis. In order to be sure that we were removing a similar amount of material in each electropolishing step, the voltage, electropolishing time and the exposed area were held constant. The electropolishing parameters are shown in Table 1.

We performed profilometer studies using a modified commercial instrument attached to a glove box.[6] We made these measurements only after completing all required electropolishing steps, rather than after each step. This was necessary because of (a) the long setup time for each profilometer measurement and (b) the very small amount (estimated at 1.5 μm) removed in each electropolishing step.

The XRD data were taken with an automated diffractometer[7] that had been adapted for metallographic sample mounts. The experimental conditions are shown in Table 2. Each sample was wrapped in commercial plastic wrap prior to removal from a chemical hood used for sample handling. The procedure is based on one described in Wick.[1]

We first analyzed the resultant XRD spectra using SPECPLOT[8] to determine the phases present. For the Pu-Ti alloy, that was all that was done. If phase quantification was desired (as for some Ga-Pu alloys), the SPECPLOT treatment was followed by analysis by either hand calculation or the use of the new program SPECQUAN.[9]

Table 1. Experimental conditions for controlled electropolishing

Sample area (square mm)	75
Electropolishing voltage (volts)	10
Electropolishing time (seconds)	30
Alloy	Ga-Pu
Electrolyte	10% HNO3, 90% C2H5OH

Table 2. Experimental conditions for x-ray diffraction

Parameters for routine analyses:

Diffractometer	Automated[7]
Radiation	Cu Kα
Tube settings	50 kV and 30 mA
Divergence slit	1 deg, fixed slit
Monochromator	Diffracted-beam, pyrolytic graphite
Scanning range	Usually 24 to 72 degrees 2-theta
Step size	0.04 degrees 2-theta
Count time	4 Sec

Additional parameters for quantitative analysis:

α-Pu peaks	211 + 020 (peaks overlap)
δ-Pu peaks	111, 220 and 311
Integration widths	0.2 degrees 2-theta
Background points	29.9, 35.6, 40.9, 53.6, 61.5, and 69.5 degrees 2-theta

---

A reference Ga-Pu alloy was used as a calibrant for the quantitative phase analyses. This alloy had been cast and slow-cooled to produce a mixture of α-Pu and δ-Pu that would be useful in our analyses. Density determinations on this alloy were done using an immersion technique that is very similar to those described in Wick.[1] The density determination was performed by two different operators on two different days in order to gain a measure of the reproducibility of the process. Based on these two determinations, the density was established at 16.48 ± 0.04 g/cc.

Calculation of the fractions of α-Pu and δ-Pu is based on two assumptions:

(1)   α-Pu and δ-Pu are the only phases present.

(2)   Observed densities (rather than crystallographic ones) are used for the calculation. The δ-Pu density used was 15.75 g/cc; the α-Pu density was 19.65 g/cc. The latter figure is lower than theoretical because of micro-cracks in the α-Pu.

Based on the absence (or, in some cases, the extremely low intensity) of impurity phases, these assumptions have proven to be acceptable. Given the above assumptions, the reference alloy was estimated to contain 18.0 ± 0.3 vol% α-Pu.

XRD intensities determined from the reference alloy were used as input into both our manual method and SPECQUAN.

SPECQUAN, a Computer Program for Quantitative Phase Analysis

SPECQUAN[9] is a recently-developed phase analysis program designed to use the SPECPLOT[8] data file structure. The program also supports three additional file structures, and others can be easily added by changing the parameter and data subroutines. The program is written in Fortran 77 and is operational under both RSX-11M and VMS operating systems. The program could be modified for other computer systems.

The program uses an external intensity-ratio quantification procedure.[10] By using sample peaks that are relatively close in 2-theta, many instrumental and sample errors are reduced. Of particular importance to us, the program uses many peaks for each phase to minimize preferred-orientation problems. Such problems are almost always present when handling metallographic samples; this is particularly true for δ-Pu, a soft phase that readily takes on a crystallographic texture.

RESULTS

Important in all the results that follow is the concept of the infinite depth for copper radiation in Pu-based alloys. For these alloys, the calculated depth is about 1 μm.

Pu-Ti Alloy

The composition of this alloy was about 2.45 at.% Ti, as determined by the relative amounts of Pu and Ti that were alloyed. Heat treatment of the cast alloy was done in a fashion so as to maximize the amount of β-Pu produced.[11] The heat treatment at relatively high temperature (425 °C) followed by quenching in silicone oil was designed to take advantage of the relatively sluggish transformation behavior of these alloys. This treatment resulted in an alloy that was nearly all β-Pu, as confirmed by subsequent density XRD measurements.

However, XRD initially indicated that the alloy was predominantly α-Pu with some β-Pu also present; see lowest curve in Fig. 1. After electropolishing, the surface concentration of α-Pu had largely

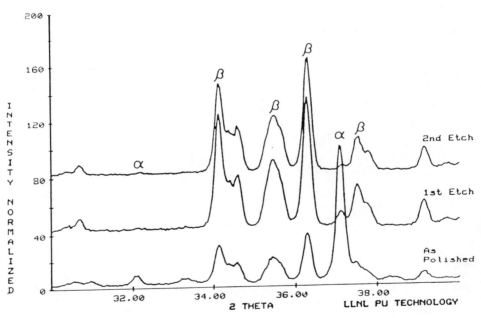

Fig. 1.   XRD patterns for a Pu-2.45 at.% Ti alloy. The bottom pattern reflects the as-polished sample With its damaged surface layer. The middle and upper patterns are of the same sample after one and two electropolishing steps, respectively. The major α-Pu lines are marked with arrows. Note that all but the strongest line disappear after the electropolishing, confirming that the bulk of the α-Pu observed in the bottom pattern is a surface phenomenon.

disappeared, and the alloy was indeed mostly β-Pu. This can be seen in the two upper curves of Fig. 1 and points directly at the sample preparation as being the source of the damaged surface layer in this metastable alloy.

Ga-Pu Alloys

Surface transformations were first seen in these alloys when a sample with an extremely low microhardness (less than 30 diamond pyramid hardness number, dphn, at a 25-g load) was prepared. This 3.3 at.% Ga alloy had been heat-treated a very long time at 440 °C. On three different occasions, this alloy was prepared; each time the phase analysis was the same: predominantly α-Pu with some δ-Pu (lower three curves of Fig. 2.). Two subsequent electropolishings removed nearly all the transformed surface layer and produced a diffraction pattern (upper curve, Fig. 2) that agreed very well with metallographic observations.

Note that, although the surface had been damaged by the original sample preparation, the underlying δ-Pu metallurgical structure could still be seen microstructurally. Additionally, the microhardness reflected the soft δ-hardness, not that of the relatively hard (270+

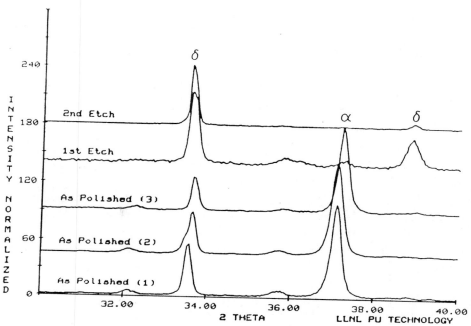

Fig. 2. XRD patterns for a Ga-Pu alloy where the problems with metallographic preparation were first noted. This ultra-soft alloy brought to light the problems associated with a new vibratory polisher. Each of the lower three patterns represents a slightly different sample preparation; the patterns show α-Pu as the primary phase present. But the softness of the alloy and the morphology seen in microscopic analysis led to further studies (upper two patterns) to see if we were seeing a surface effect. We were, since the latter patterns taken after electropolishing are mainly δ-Pu.

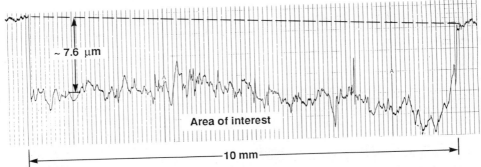

Fig. 3.   A typical profilometer trace of the type used to determine both
the total depth of material (7.6 μm) removed by electropolishing
(EP) and the average amount removed per step (1.5 μm).   The
upper curve shows a cross-sectional profile of a sample after
EP.   At left and right are the original surface levels, and the
middle portion is where EP was done.   The step height is
equivalent to about 7.6 μm, and the distance across the
electropolished area is about 10 mm.   Instrument drift is
indicated by the slope of the line connecting the two reference
points.

dphn) α-Pu overlayer.   We believe that the extreme softness of this alloy
made it possible to readily transform its surface from δ-Pu to α-Pu.

Determination of the Depth of Damage in Ga-Pu Alloys

This determination was done on a cast 3.3 at.% Ga alloy with a
metallographically polished surface,[12] the last polishing stage being
done with 1-μm diamond paste.   Figure 3 shows the profilometer trace that
overall removal of 7.6 μm.   Figure 4 shows the as-polished spectrum (at
bottom) and the successive patterns as Pu is removed at about 1.5 μm per
step.   Since this step size exceeds the infinite depth of penetration for
copper radiation, there can be little (if any) overlap in the metal
volume examined in adjacent alloy layers.   The final pattern (Fig. 4 top)
shows the pattern that best represents the actual phase mixture.

These data were quantified using both SPECQUAN and our previous
calculation method.   We found no significant difference between the two
methods.   Figure 5 gives the quantitative data as a function of depth.

CONCLUSIONS

Phase transformations are relatively easy to induce in the Ga-Pu and
Pu-Ti alloys discussed here.   The combination of electropolishing step-
wise, profilometry, and quantitative XRD permits the successful
evaluation of the depth of the surface damage.

While the evidence is circumstantial, we believe that vibratory
polishing does contribute to surface phase transformations in Pu alloys.
Hence, our polishing procedures have been modified to allow for
additional electropolishing (and for supplemental XRD analysis in some
cases)..   The authors conclude (as have others[5]) that at least a light
electropolishing is mandatory if good XRD analyses are to be obtained on
any metallographically-prepared Pu alloy.

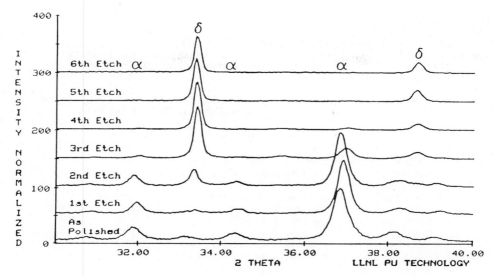

Fig. 4.   XRD patterns for a Pu-3.3 at.% Ga alloy used to determine the
depth of surface damage caused by metallographic polishing.   The
lowest pattern is the as-polished sample, while the subsequent
patterns indicate the change in the surface as successive layers
are removed.

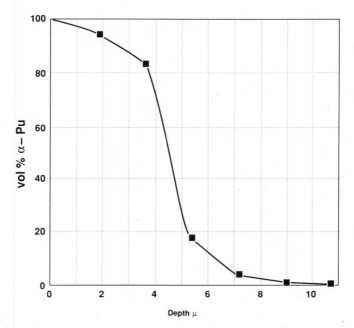

Fig. 5.   The depth profile of the surface damage as determined from the
XRD patterns shown in Figure 4..   This profile is typical of
those seen on several samples.

ACKNOWLEDGMENTS

Metallography was performed by John Bergin, Harold Clark, David Haggerty and Bobby Vallier. Charles Peters did the profilometry, and Peter Biltoft and Gilbert Gallegos provided the alloys and performed the requisite heat-treatments.

Work performed under the auspices of the U.S. Department of Energy by the Lawrence Livermore National Laboratory under Contract W-7405-Eng-48.

REFERENCES

1. O. J. Wick, ed., "Plutonium Handbook, a Guide to the Technology," The American Nuclear Society, La Grange, Ill. (1980).
2. J. R. Morgan, New Pressure-Temperature Phase Diagram of Plutonium, in: "Plutonium 1970 and other Actinides, Part II," W. N. Miner, ed., The Metallurgical Society of AIME, New York (1971).
3. A. Goldberg and T. B. Massalski, Phase Transformations in the Actinides, in: "Plutonium 1970 and other Actinides, Part II," W. N. Miner, ed., The Metallurgical Society of AIME, New York (1971).
4. A. Goldberg and J. C. Shyne, Thermodynamic Analysis of Stress and Pressure Effects on the $\delta \to \alpha$ Transformation in Plutonium Alloys, J. Nucl. Mat. 60:137 (1976).
5. J. B. Bergin, private communication (1982).
6. C. M. Peters, private communication (1988).
7. P. L. Wallace, F. Y. Shimamoto, and T. M. Quick, "Large-Scale Automation of the Lawrence Livermore Laboratory X-ray Analytical Facilities," Lawrence Livermore National Laboratory, Livermore, Calif., UCRL-52953 (1980).
8. R. P. Goehner, SPECPLOT—An Interactive Data Reduction and Display Program for Spectra Data, in: "Advances in X-ray Analysis, Vol. 23," pp. 305-311, Plenum Publishing Corp., New York (1980).
9. R. P. Goehner, private communication (1986).
10. R. P. Goehner, X-ray Diffraction Quantitative Analysis Using Intensity Ratios and External Standards, in "Advances in X-ray Analysis, Vol. 25," pp. 309-313, Plenum Publishing Corp., New York (1982).
11. P. J. Biltoft, private communication (1986).
12. G. F. Gallegos, private communication (1986).

THE DETERMINATION OF α-CRISTOBALITE IN AIRBORNE DUST

BY X-RAY DIFFRACTION - THEORY AND PRACTICE

M. Jeyaratnam and N.G. West

Occupational Medicine and Hygiene Laboratories
Health and Safety Executive
403, Edgware Road, London NW2 6LN, UK

INTRODUCTION

One of the earliest occupational diseases to be recognised was silicosis, a pathological condition of the lungs caused by the inhalation of dust containing free crystalline silica. During the industrial revolution in the UK the incidence of silicosis rose dramatically; for example in the Sheffield cutlery industry, where dry grinding of knives and forks was practised, the life expectancy of workers was reduced on average by twenty five years[1]. At the present day there is a wide spectrum of industry in which exposure to crystalline silica may occur; in the UK alone it is estimated that more than 100,000 workers are potentially at risk. The great majority of this exposure is to quartz but cristobalite is also commonly encountered where silica rich materials are subject to very high temperature e.g. ceramics, refractories, brick manufacture. Cristobalite is also found in certain natural deposits including bentonite clay and chert.

To prevent the development of silicosis under modern working conditions, the crystalline silica content of respirable dust samples is routinely monitored. Samples of dust are collected on a filter medium in the worker's breathing zone, analysed, and the results compared with published exposure standards. These are referred to as Threshold Limit Values (TLV) in the USA [2] and Occupational Exposure Limits (OEL) in the UK[3]. These limits refer to airborne concentrations of substances and at present the limits for quartz and cristobalite, in both the USA and the UK, are 0.1 and 0.05mg m-3 respectively. The lower limit for cristobalite reflects the fact that there is evidence it is more toxic than quartz[4].

The determination of quartz in airborne dust deposits on filters has been a routine procedure for some years. Recommended methods have been published by various governmental agencies including NIOSH (Method 7500) [5] in the USA and HSE (MDHS 51) [6] in the UK. There are important differences in the analytical approaches used in these two methods in particular NIOSH Method 7500 involves re-deposition of the dust deposit from the sampling filter onto a silver membrane for analysis, whereas in HSE MDHS 51 the sampling filter is placed directly in the diffractometer. Each method has advantages and disadvantages but the major problem with both is the presence of interfering phases. The only practical and effective way around this problem is to measure a number of quartz diffraction lines and select an interference free line for quantification. In MDHS 51 four lines are routinely measured, the (100), (101), (112) and (211); all four have sufficiently sensitivity to measure down to half the exposure limit for a 500 litre air sample (25 µg quartz on the filter). [7]

In principle, the same analytical approach should be applicable to cristobalite determination, but in practice a number of factors combine to make the analysis much more demanding. Firstly, the lower exposure limit for cristobalite means that it is necessary to measure down to ~12µg mass on the filter, a considerable challenge for conventional X-ray powder diffractometry. Secondly, as figure 1 illustrates, although there is a strong primary line (101) for cristobalite, the secondary lines are relatively weak and only two, the (102) and (200/(211), are useful analytically.

Thirdly, and most importantly, experience has shown that the term 'cristobalite' covers a range of crystalline structure. This is in contrast to quartz which exhibits remarkably little structural or chemical variation. 'Cristobalites' from different sources show marked variation in the positions, shapes and relative intensities of the diffraction peaks, these differences being related to the degree of disorder and the presence of impurities in the crystal. [8] [9] Examples of structural properties of three different 'cristobalites' are given in Table 1.

The high temperature (HSE) cristobalite was formed by conversion of quartz at 1550°C. The diffraction peaks are sharp and intense, closely matching the reference pattern for the mineral in the JCPDS

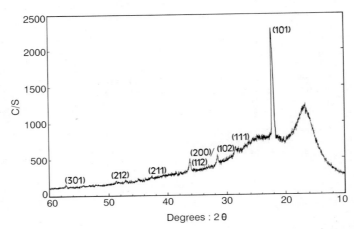

Fig. 1.   X-ray diffraction pattern of an α-cristobalite dust deposit (~ 200 µg) on a Gelman DM800 filter substrate. 1° divergence slit.

Table 1. X-ray powder diffraction data on 'cristobalite' from different sources (1° divergence slit)

| Source | d-spacing (Å) | | | Full width half maxima (101) °2θ | Relative intensity (peak height)% | |
	(101)	(102)	(200)/(112)		$\frac{(102)}{(101)}$	$\frac{(200)/(112)}{(101)}$
High temperature cristobalite (HSE)	4.05	2.845	2.485	0.30	11	16
Calcined diatomaceous earth (Manville Research)	4.07	2.854	2.492	0.40	5	15
Natural cristobalite from India (Steetley Minerals Ltd)	4.12	2.858	2.498	0.90	3	20
α-cristobalite JCPDS file 11-696	4.05	2.841	2.485	-	13	20

file. The cristobalite formed by calcination of diatomaceous earth
has a similar pattern but there is a systematic small increase in the
'd' spacings, the peaks are broader and the relative intensity of the
(102) line is much reduced. The third example is a clay deposit
containing cristobalite from India. There is a further shift in the
'd' spacings from the reference pattern and marked broadening of the
primary line. The structure closely resembles that assigned to
Opal-CT in naturally occurring disordered free silica [10] and is closer
to β-cristobalite (JCPDS 4-359) than α-cristobalite. A phase having a
diffraction pattern very similar to β-cristobalite has also been
observed to form during the devitrification of alumino-silicate fibres
at 1150°C. [11]

From an occupational health viewpoint there is no evidence to suggest that
one source of 'cristobalite' is any more or less toxic than any other.
To control exposure the aim must be to measure all forms of cristobalite
on a common analytical basis using the same reference standard.

ANALYTICAL METHODOLOGY

Standard reference cristobalite

From the preceeding discussion it is clear that the most critical decision
regarding the method is the choice of standard cristobalite. There are a
number of reasons why the high temperature, ordered cristobalite is much
preferred, notably that it accords with the established cristobalite
structure [9] and also that it is the form most commonly encountered in
industry. However, experience has shown that the apparently simple
procedure of preparing 'standard' cristobalite by heating free silica is
beset by practical difficulties, which make for large inter-laboratory
variation in the X-ray response. [12] The recent availability of the NBS
respirable cristobalite standard (SRM 1879) represents a significant
advance and this SRM should be used as the basis for all calibrations.
Unfortunately, the present work was completed before its appearance and
makes use of the HSE high temperature cristobalite (Table 1) for
calibration. Preliminary data suggest that the response of SRM 1879
is significantly higher (∿20%) than the HSE standard.

Preparation of standard filter samples

On-filter standards were prepared by dispersing the HSE high temperature
cristobalite in a dust chamber and sampling the aerosol through cylcone
elutriators onto pre-weighed filters (Gelman DM800). By this means a
series of on-filter standards were made, covering the weight range
20-2000 µg, and having a similar particle size distribution and deposition
pattern to workplace samples. [6]

Instrumentation and sample presentation

A Philips PW1050 diffractometer, fitted with a PW1170 automatic sample
changer and linked on-line to a DEC PDP 11/44 computer system, was used
for the analyses. The operating parameters were as follows:

> 2.7 kw broad focus Cu X-ray tube run at 45 kV 45mA
> 1° divergence slit, 0.3mm receiving slit,
> 1° scatter slit
> Graphite diffracted beam monochromator

The filters used for sampling were presented direct to the diffractometer [6]; the standard holders were machined to provide a recess 0.1mm deep in which the filter was secured with a trace of flurocarbon grease (Voltalef 90).

Calibration of X-ray diffractometer

The diffractometer was calibrated on the three strongest lines for cristobalite by measuring the net peak areas on each of the filter standards. The net peak areas were determined by step scanning over the peaks, integrating to obtain the gross peak areas, and then making a simple linear background correction. A step size of 0.05° 2θ with a count time of 30 secs per step was used for optimum analytical performance. Any further increase in counting time could not be justified in terms of improved detection limits for routine analysis. Linear calibrations were obtained over the range 0 to 2000 μg cristobalite with the predicted large difference in sensitivity between the primary and two secondary lines (Figure 2).

RESULTS AND DISCUSSION

Detection Limits

The detection limits for the three diffraction lines, based on simple counting statistics are given in Table 2. The count time on each line is governed by the width of the base of the peak; in all three cases this was approximately 1° 2θ, making a count time of 10 minutes. For the less ordered forms of cristobalite it is necessary to count over a larger angle which leads to poorer detection limits. Given the requirement to determine cristobalite down to 12 μg, it is evident that only the primary line has sufficient sensitivity. Although the detection limit for the (200)/(112) line is just below 12 μg, the instrumental precision at this level would be very poor ($\sigma\% > 50$).

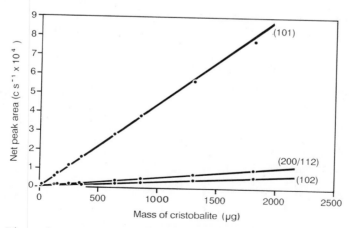

Fig. 2. Calibration curves for the (101), (102) and (200)/(112) diffraction lines.

Table 2.   Detection limits (µg) for α-cristobalite

X-ray diffraction line		
(101)	(102)	(200)/(112)
3	20	10

Detection limit*

* Detection limit = $\dfrac{3\sqrt{N}}{m}$

N = Integrated background area

m = Sensitivity (net peak area/µg)

Analytical precision

Figures for instrumental precision are listed in Table 3.   Since
the method is direct and involves no sample preparation, these figures
closely approximate the overall analytical precision.   Twelve repeat
determinations were made over a period of a week, on two samples, one
with a nominal cristobalite loading of 48 µg and the other 270 µg.   For
the (101) line the instrumental precision is very good for trace analysis
of this type.   As predicted, however, the precision on the secondary
peaks at the lower loading is poor.   For the (102) line the loading
was little more than twice the detection limit.   At the higher loading
the instrumental precision on the secondaries is much improved but
this is of marginal practical significance since very few workplace
samples approach this level.

Interferences

As in the determination of quartz in airborne dust[6], the analysis of
cristobalite can suffer from serious interference.   Unfortunately,
unlike quartz, it is seldom possible to avoid line overlaps by
selecting an alternative line for measurement since the secondary lines
have insufficient sensitivity.   Hence, in practice, the method is very

Table 3.   Instrumental precison for α-cristobalite
           determination

		X-ray diffraction line		
		(101)	(102)	(200)/(112)
Nominal	σ	2.09	16.2	8.8
48 µg	σ%	4.2	37.9	17.2
cristobalite	Mean	49	43	51
Nominal	σ	9.5	15.3	12.5
270 µg	σ%	3.2	5.6	4.3
cristobalite	Mean	292	273	292

Table 4. Potential interfering phases

α-cristobalite line		(101)		(102)		(200)/(112)	
Phase	JCPDS Ref. No.	°2θ	RI	°2θ	RI	°2θ	RI
α cristobalite	11-695	21.93	100	31.46	13	36.11	20
β cristobalite	27-605	21.61	100	–	–	35.63	12
quartz	5-490	20.83	34	–	–	36.52	12
tridymite	18-1170	21.62	100	–	–	35.90	16
albite (low)	20-554	22.03	60	31.43	3	36.05	14
						35.94	2
anorthite	12-301	22.00	60	31.64	20	36.77	9
						35.90	25
orthoclase	19-931	21.03	70	30.80	30	36.24	2
calcite	5-586	–	–	31.42	3	36.18	4
cordierite	9-472	21.60	80	–	–	35.97	14
corundum	10-173	–	–	–	–	36.64	10
kaolinite	14-164	21.20	44	–	–	35.13	90
		21.45	34			35.97	44
mullite	15-776	–	–	–	–	35.13	90
muscovite	7-42	22.95	10	30.99	16	35.89	12
						36.55	8
talc	3-0881	21.60	5	–	–	36.04	100
zircon	6-0266	–	–	–	–	35.63	44

RI - nominal relative intensity from JCPDS file
Cu X-ray tube

dependent on the primary (101) line alone. Table 4 lists some of the more common interfering phases, the most important of which are certain of the feldspar minerals which have lines almost coincident with the primary cristobalite line. It is possible to make a simple correction for the feldspar interference by measuring the intensity of an adjacent interference free feldspar line and hence making the appropriate adjustment to the cristobalite (101) line intensity.

CONCLUSIONS

The determination of α-cristobalite in airborne dust deposits is considerably more challenging than the determination of quartz, for three main reasons. Firstly, there is a requirement to measure down to lower levels due to the lower exposure limits, secondly there is an absence of strong secondary diffraction lines and thirdly the term 'cristobalite' covers a range of crystalline structure.

In practice it is necessary to set up the method using a high temperature, ordered cristobalite reference material for calibration and to make most measurements on the primary line alone. The two strongest secondary lines can only be used to confirm the primary results when the cristobalite loading is relatively high, certainly greater than 20 µg. Since the method relies so heavily on the primary line, it is vital to take account of possible interferences on this line.

Consideration should be given to the use of an alternative analytical technique such as infra-red spectrophotometry, to provide confirmation in difficult cases.

ACKNOWLEDGEMENTS

Thanks are due to Mr Peter Salt (E.C.C. International) for valuable discussions providing new insight into the problems of cristobalite determination.

REFERENCES

1.   D. Hunter, 'The Diseases of Occupations' Hodder and Stoughton, London, 1978.

2.   American Conference of Governmental Industrial Hygienists, 'TLVs, Threshold Limit Values and Biological Exposure Indices for 1986-1987.' ACGIH, Cincinnati, 1986.

3.   Health and Safety Executive, 'Occupational Exposure Limits 1988, Guidance Note EH40', HMSO, London, 1988.

4.   American Conference of Governmental industrial Hygienists, 'Documentation of the Threshold Limit Values, Fourth Edition, 1980.' ACGIH, Cincinnati, 1980

5.   National Institute for Occupational Safety and Health, 'NIOSH Manual of Analytical Methods, Third Edition, Volume 2' NIOSH, Cincinnati, 1984.

6.   Health and Safety Executive, 'Quartz in Respirable Airborne Dusts, MDHS 51/2,' HSE, London, 1988.

7.   K.J. Pickard, R.F. Walker and N.G. West, 'A Comparison of X-ray Diffraction and Infra-red Spectrophotometric Methods for the Analysis of α-Quartz in Airborne Dusts,' Ann. Occup. Hyg., <u>29</u>; 149 (1985).

8.   G.W. Brindley and G. Brown, 'Crystal Structures of Clay Minerals and their X-ray Identification' Mineralogical Society, London, 1980.

9.   W.A. Deer, R.A. Howie and J. Zussman, 'An Introduction to the Rock Forming Minerals' Longman, London, 1966.

10.   J B Jones and E.R. Segnit, 'The Nature of Opal. I. Nomenclature and Constituent Phases, Journal Geol Soc. Australia, <u>18</u>; 57 (1971).

11.   G. Vine, J. Young and I.W. Nowell, 'Health Hazards Associated with Aluminosilicate Fibre Products,' Ann. Occup. Hyg., <u>28</u>; 356 (1984).

12.   P.D.E. Biggins, 'The Selection of a Suitable Cristobalite Standard to be Used in the Analysis of Airborne Dust' ECSC Research Project 7257-67, 1986.

Automatic Computer Measurement of Selected Area Electron Diffraction Patterns from Asbestos Minerals

J. C. Russ, Materials Sci. & Eng. Dept., North Car. State Univ., Raleigh, NC
T. Taguchi, Hitachi Scientific Instruments, Rockville, MD
P. M. Peters, Div. Industrial Safety & Health, State of Wash., Olympia, WA
E. Chatfield, Chatfield Technical Consulting, Mississauga, Ontario, Canada
J. C. Russ, Biomedical Engineering Dept., University of Texas, Austin, TX
W. D. Stewart, Dapple Systems, Sunnyvale, CA

Conventional selected area diffraction patterns as obtained in the TEM present difficulties for identification of materials such as asbestiform minerals, although diffraction data is considered to be one of the preferred methods for making this identification[1]. The preferred orientation of the fibers in each field of measurement, and the spotty patterns that are obtained, do not readily lend themselves to measurement of the integrated intensity values for each d-spacing, and even the d-spacings may be hard to determine precisely because the true center location for the broken rings requires estimation. To overcome these problems, we have implemented an automatic method for diffraction pattern measurement. It automatically locates the center of patterns with high precision, measures the radius of each ring of spots in the pattern, and integrates the density of spots in that ring. The resulting spectrum of intensity vs. radius is then used just as a conventional X-ray diffractometer scan would be, to locate peaks and produce a list of d,I values suitable for search/match comparison to known or expected phases. The method is tolerant of incomplete or spotty rings, and sensitive to rings with very few and very weak diffraction spots. It is implemented on a personal computer (Macintosh][) that may also be useful for other image processing applications in the TEM laboratory.

Electron diffraction patterns such as those in Figure 1 were obtained from a variety of asbestiform minerals using a camera length of 0.4 m., and the negatives subsequently digitized into the computer system using a solid state video camera and light box. It would have been equally suitable to directly acquire the patterns (or other images) using a camera mounted on the TEM. Each diffraction pattern image was then brightness levelled by subtracting an image in which each pixel had the maximum brightness in a 7x7 pixel neighborhood[2]. This removes the overall variation in brightness common in electron diffraction patterns without affecting the spot and ring patterns, because the width of the spots and rings is much less than 7 pixels.

A circular Hough transform was applied to the resulting grey scale image, producing a three-dimensional image space whose coordinates are the X,Y location of the center and the radius of each ring. The brightness of each pixel in this space is the sum of brightnesses of all pixels in the original two-dimensional image which lie on a circle defined by the location of the pixel in Hough space.

The most common form of the Hough algorithm[3,4] is the linear transform, in which each point in the original image is mapped onto all of the points in a 2-dimensional space (whose coordinate system is values of M and B in the equation y=Mx+B) which would define a line

593

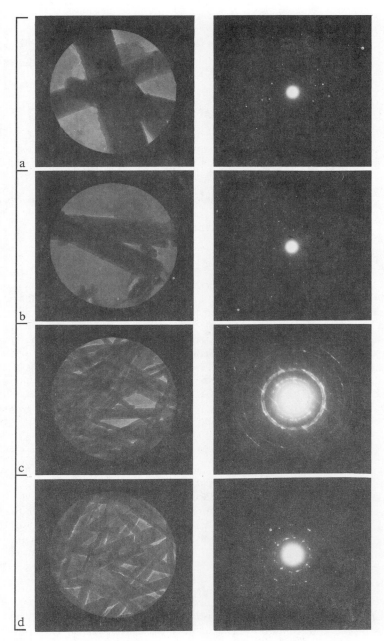

Figure 1. Images of SAD region (40,000x) and resulting ED patterns for
a) Amosite, b) Crocidolite, c) Chrysotile, d) Sepiolote.

that could pass through the object. Once this has been done for all points in the image, the total value from all superimposed mappings gives brightness peaks in Hough space corresponding to the linear structures present in the original image, and these can be found by discrimination and the corresponding lines plotted on the original.

The same method can be used for non-linear shapes[5,6]. In particular for circles, Hough space is three-dimensional (the X,Y coordinates of the center and the radius)[7]. Each point in the image can be mapped into this space in a rather simple way, because it generates a cone whose apex (in the plane radius = 0) is just the point itself.

Figure 2 shows the application of this method to electron diffraction patterns. The pattern of spots or rings from a single- or polycrystalline sample is the radius=0 plane, and the cones that form behind it intersect and superimpose (in proportion to the brightness of the original points) to form high density regions in Hough space as indicated schematically in the figure. The central line formed by these high density regions locates the actual center of the pattern. Locating the center is accomplished by finding the brightest point (interpolating to the nearest 0.1 pixel dimension) in each constant radius plane in the Hough space, and then using all of these points to perform a weighted least squares fit (using the brightness of the point as the weighting factor) to determine the center point.

Along this central line down through the Hough space, the brightness values produce a spectrum plot of integrated intensity in each circular ring as a function of the radius of the ring (and hence, the d-spacing since d is proportional to 1/radius). Figure 3 shows the diffraction pattern from the gold standard, with complete and uniform rings. If this entire image is subjected to the circular Hough transform, and the intensity along the center line plotted (Figure 4), the result shows the peaks that correspond to the rings, and the overall rise in brightness near the center.

By using the background levelling method described above, the intensity spectrum from the asbestos samples show a zero background intensity between peaks, and are also faster to map into Hough space since points in the spatial image with a zero brightness can be skipped. Figure 5 shows an example of a portion of an intensity vs. radius spectrum. The variation in the shape of the peaks reflects differences in the density and uniformity of the spots at various radii. Symmetric or continuous spot patterns produce the most symmetric peak shapes in the spectrum.

This spectrum was processed by software normally used for handling X-ray diffractometer scans[8], which locates spectral peaks and interpolates using a parabolic fit to obtain precise peak positions, and reports the location and net intensity of each peak. Figure 6 shows the peak positions, with the intensities normalized to assign a value of 100 to the brightest line. In this case, the calibration of the d-spacing values was established by the similar automatic measurement of the diffraction pattern from a gold foil (Figures 3 and 4), after which the calibration constant was stored for use with subsequent patterns measured with the same camera length. The calibration for the camera length used was obtained from a least squares fit to 8 well defined rings in the gold pattern, giving d [Ångstroms] = 1.4114 / radius [cm] as shown in Figure 7.

The resulting list of d,I values is then available for search/match identification. This can either be done using patterns measured on standard specimens, or using data from tables such as those from the Joint Committee on Powder Diffraction Standards[9]. The latter are available for more than 40,000 compounds, but the relative intensities from electron diffraction are not identical to those for conventional X-ray diffraction, and the asbestos minerals are poorly represented in the files because of the presence of many versions of the phases with slightly different compositions and different mineral names. (For instance, amosite is found listed as grunerite or cummingtonite, and crocidolite as reibeckite or magnesioreibeckite, on several cards with somewhat differing values of d-spacing and intensity). Accordingly, we elected to establish our own database in this instance.

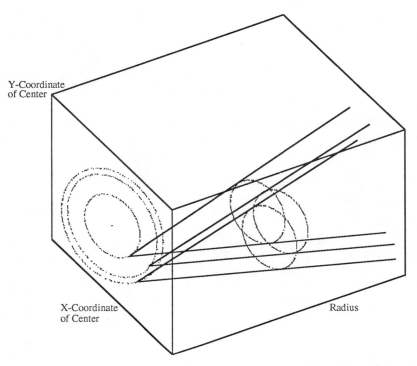

Figure 2. Principle of the Three-Dimensional Circular Hough Transform
for Points on Electron Diffraction Pattern.

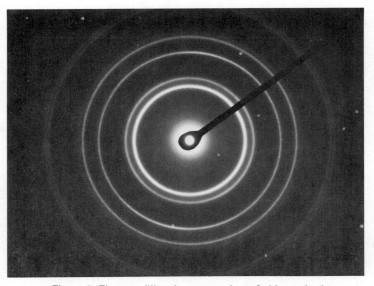

Figure 3. Electron diffraction pattern from Gold standard

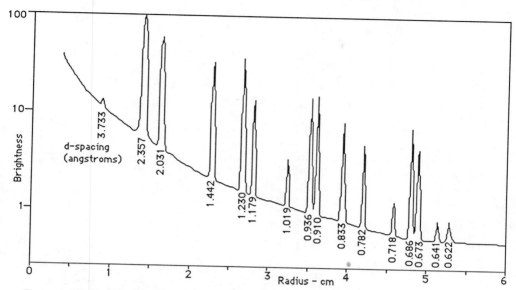

Figure 4. Integrated (Circular) Density along the central axis in Hough space for the image in Fig. 3, with the d-spacing of the peaks calculated from the measured radius marked.

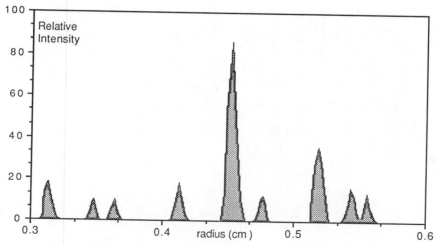

Figure 5. Expanded detail of intensity vs. radius spectrum (crocidolite).

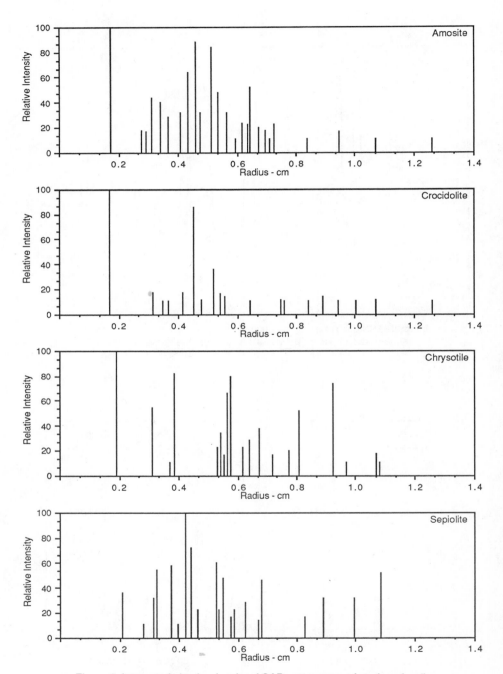

Figure 6. Integrated circular density of SAD patterns as a function of radius.

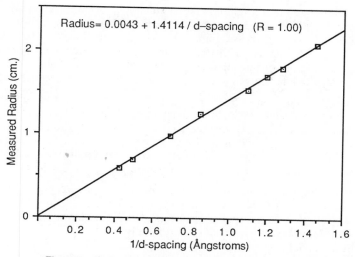

Figure 7. Camera Calibration plot from Gold ring pattern

Table 1. Intensity and d-spacing values measured for asbestos minerals using a circular Hough transform to process the selected area diffraction pattern.

Amosite		Crocidolite		Chrysotile		Sepiolite	
d-spac	inten	d-spac	inten	d-spac	inten	d-spac	inten
8.32	100	8.43	100	7.36	100	6.73	36
5.12	18	4.50	18	4.56	54	5.06	10
4.84	16	4.06	10	3.82	10	4.52	32
4.55	44	3.88	10	3.66	82	4.32	54
4.14	40	3.42	18	2.66	22	3.76	58
3.86	28	3.12	86	2.60	34	3.54	10
3.45	32	2.96	12	2.55	16	3.36	100
3.26	64	2.72	36	2.50	66	3.20	72
3.07	88	2.6	16	2.45	80	3.05	22
2.99	32	2.54	14	2.28	22	2.69	60
2.76	84	2.20	10	2.21	28	2.63	22
2.63	48	1.89	12	2.09	38	2.57	48
2.50	32	1.86	10	1.97	16	2.45	16
2.37	10	1.68	10	1.83	20	2.41	22
2.29	24	1.59	14	1.75	52	2.26	28
2.22	22	1.50	10	1.53	74	2.12	14
2.19	52	1.41	10	1.46	10	2.07	46
2.09	20	1.32	12	1.32	18	1.71	16
2.04	18	1.12	10	1.31	10	1.59	32
1.99	10					1.42	32
1.95	22					1.30	52
1.69	10						
1.49	16						
1.32	10						
1.12	10						

Data files for standard Amosite, Crocidolite, Chrysotile and Sepioloite were created directly from measured patterns. Table 1 summarizes the d,I data for these phases. The interpolation of the maximum brightness radius in Hough space corresponds generally to a precision of about 2 significant digits past the decimal point for d-spacings. Subsequently, these standard patterns were used to automatically identify unknowns (patterns from different specimens of the same minerals), with 100% reliability. The search/match programs[10] allow user setting of the number of lines that must match, and the permitted misalignment of d-values and mismatch of intensity values. They will also function for mixtures of phases.

This approach to electron diffraction pattern analysis is novel, and quite different from methods that locate each spot for single crystal patterns[11] or that involve matching observed patterns to synthetic ones[12]. Instead, measurement data for d-spacing and intensity are directly obtained, much like the results from an X-ray diffractometer scan on an unknown sample, except that far less material is required for selected area electron diffraction, and EM imaging can be used to select and examine the fibers or particles.

The method completely overcomes the problems associated with manual measurement of electron diffraction patterns. These are plagued by difficulty in locating the center of rings (especially spotty ones produced by oriented fibers). Small shifts in ring position (for instance due to minor changes in lattice parameter) are accurately measured, and minor rings which are often missed by manual methods but may be important to identify the presence of minor phases in the sample are reported with accurate intensity values, because the method automatically provides an integrated brightness value for each ring. In addition to application to asbestiform minerals, it is expected that this method will be important for multiphase materials of many kinds, including complex superconductors.

References

1. K. F. J. Heinrich, ed. (1980) Characterization of Particles National Bureau of Standards Special Publication 533
2. J. C. Russ (1988) The Analysis and Measurement of Images, Engineering Extension Service, North Carolina State University, Raleigh NC
3. P.V.C.Hough (1982) *Method and Means for Recognizing Complex Patterns* U. S. Patent 3069654, Dec. 18, 1962
4. R. O. Duda, P. E. Hart (1972) *Use of the Hough Transformation to Detect Lines and Curves in Pictures* Comm. Assoc. Comput. Mach. 15 11-15
5. D. H. Ballard (1981) *Generalizing the Hough Transform to Detect Arbitrary Shapes* Pattern Recognition 13 #2, 111-122
6. R. C. Gonzalez, P. Wintz (1987) Digital Image Processing, Addison Wesley, Reading MA
7. J. C. Russ (1988) *Automatic Methods for the measurement of curvature of lines, features and feature alignment in images* Journal of Computer-Assisted Microscopy (in press)
8. Thetaplus+, Dapple Systems, 355 W. Olive Ave., Sunnyvale, CA 94086
9. JCPDS, International Centre for Diffraction Data, 1601 Park Lane, Swarthmore, PA 19081
10. J. C. Russ, T. M. Hare, M. J. Lanzo *X-ray Diffraction Phase Analysis using Microcomputers*, in Advances in X-ray Analysis vol. 25 (J. C. Russ et. al., ed.) Plenum Press, 1982
11. D. S. Bright, E. B. Steel *Automatic Extraction of Regular Arrays of Spots from Electron Diffraction Images,* J. of Microscopy, in press
12. M. J. Carr *Automation of Electron Diffraction Analysis in an Analytical EM*, in Analytical Electron Microscopy 1981 (R. H. Geiss, ed.) San Francisco Press, p. 139-146

COMPARISON OF EXPERIMENTAL TECHNIQUES TO IMPROVE PEAK TO BACKGROUND RATIOS IN X-RAY POWDER DIFFRACTOMETRY

William K. Istone,[1] John C. Russ,[2]
and William D. Stewart[3]

[1] Champion International Corp., West Nyack, NY
[2] North Carolina State University, Raleigh, NC
[3] Dapple Systems, Sunnyvale, CA

ABSTRACT

High peak to background ratios are especially important in powder diffractometry when attempting to identify minor phases in a sample or improving the limit of detection in quantitative determinations. Instrumental techniques to improve peak to background generally involve the employment of monochromatic or partially monochromatic radiation through the use of filters, crystal monochrometers, or pulse height discriminators.

In this study, a digital pulse height discriminator, configured as a card in a personal computer (Apple IIe) with appropriate software, is used in conjunction with a scintillation detector to improve peak to background ratios. The software allows the pulse height distribution to be scanned and the optimum pulse height window to be set for a given set of sample and instrumental conditions. Results obtained by this technique are directly compared with results obtained using a pyrolytic graphite monochrometer and beta filters. Examples cited include qualitative phase identification in both fluorescent and non-fluorescent samples and semi-micro quantitative analysis (determination of airborne silica).

INTRODUCTION

Most diffraction techniques require X-radiation that is essentially monochromatic. The source of such radiation is usually the K-alpha doublet from the tube target. Other target lines (especially the K-beta) and sample fluorescence must be minimized in order to eliminate additional peaks in the pattern and to reduce background. While it is possible to obtain pure monochromatic

601

radiation, this is only at the expense of a considerable loss in intensity and hence some form of partial monochromatisation is utilized.   There are three important methods of partial monochromatisation: beta-filters, crystal monochrometers, and pulse height discriminators.

Beta-filters are an effective means of eliminating the major interfering tube line (K-beta), but are not usually effective in dealing with sample fluorescence problems. Crystal monochrometers are effective at dealing with both problems but result in a significant loss in intensity. Pulse height discriminators are effective at dealing with sample fluorescence problems, but usually ineffective at dealing with secondary tube lines.

In this study, we have used a digital pulse height discriminator in conjunction with a scintillation detector and have compared its performance to that of a curved-crystal monochrometer. Pulse height discriminators have been commonly used in conjunction with lithium drifted silicon detectors (Si(Li)) and gas proportional detectors, but are usually not used in conjunction with scintillation detectors due to the broad pulse distributions associated with this type of detector. In this study, we demonstrate that under proper conditions a pulse height discriminator with a scintillation detector can be used to provide partially monochromatic radiation in diffration studies and obtain results equivalent to or better than those obtained with a curved crystal monochrometer. The lower cost and increased ease of operation further justify this approach.

EXPERIMENTAL

For the purposes of this study, three model compounds were chosen: alpha-alumina as an example of a highly crystalline, non-fluorescent compound, magnetite as an example of a highly fluorescent sample, and alpha-quartz (100 micrograms on a silver membrane) as an example of a semi-micro quantitative analysis. All three compounds were analyzed using both the pulse height discriminator and the curved crystal monochrometer with instrument conditions optimized for each. Beta filters were used where appropriate. Data was obtained using a Philips vertical diffractometer fitted with a Dapple Systems "Thetaplus+" automation package. The software was modified to allow for manually setting the pulse height discriminator which was borrowed from a Dapple Systems "Dataplus+" XRF system. All measurements were made with a long fine focus copper target X-ray tube, scintillation detector, and a sample spinner (to minimize preferred orientation effects). The monochrometer used was a AMR 3-202 focusing monochrometer with a pyrolytic graphite crystal.

The alpha-alumina sample was prepared from NBS SRM-674 and used as received. The magnetite sample was prepared from Alfa Chemicals magnetite (14439), also used as received. The 100 microgram alpha-quartz sample was prepared by depositing NBS SRM-1878

TABLE 1

PARAMETER	FOR PHA	FOR MONOCHROMETER
POWER (keV/mA)	35/25 (40/30)	35/25 (40/30)
DIVERGENCE SLIT	0.5° (1.0°)	0.5° (1.0°)
RECEIVING SLIT	0.2° (0.2°)	0.1° (0.1°)
ANTI-SCATTER SLIT	1.0° (1.0°)	NONE (NONE)
BETA-FILTER	Ni (Ni)	NONE (NONE)
STEP-SIZE	0.05° (0.02°)	0.05° (0.02°)
DWELL TIME	1 sec (5 sec)	1 sec (5 sec)

Same conditions for Alpha Alumina and magnetite. Conditions for Alpha-Quarts are in parentheses.

(respirable alpha-quartz) from isopropanol onto a Hytrex 0.45 micron silver filter using vacuum. The alumina and magnetite samples were prepared as loosely packed powders, front packed into the sample cups for the sample spinner. The silica sample filter was cemented to an appropriate sample holder. Table 1, summarize the optimum operating conditions that were experimentally determined for alumina, magnetite, and silica (alpha-quartz).

RESULTS AND DISCUSSION

In all cases the pulse height discriminator (PHA) was in the circuit, but when the experiments involving the monochrometer or beta-filters alone were conducted the upper limit for the PHA was set to its maximum value and the lower limit was set to a predetermined minimal value so that the PHA functioned as an electronic noise filter.

The software employed in the Dapple "Thetaplus+/ Dataplus+" system controls the PHA card which is inserted into the computer bus (in an Apple IIe computer in this case). The upper and lower discriminator values are set in the PHA by 8 bit DAC's (digital to analog converters). Writing a numeric value from 0 to 255 to a memory-mapped address for each threshold is all that is needed to set the corresponding voltage value (0-10 V.)

The optimal setting is achieved by scanning the PHA window to collect a curve of throughput, from which the best operating points are selected. In XRF work (for which this PHA was originally designed), as the system drives the spectrometer to the two-theta angle for a given element, the wavelength corresponding to the angle through the Bragg equation (for the crystal in use) is converted to energy, and the appropriate energy window set on the PHA card to minimize background. At present, the software uses a fixed energy width for the window, which is scanned in proportion to the sine of the angle (replacing the need for a sine-potentiometer on the pulse amplifier). However, other algorithms in which window width is adjusted can be easily substituted.

For each of the model compounds, diffraction patterns were obtained utilizing the procedures and operating parameters described above. In each experiment the intensities of the major peak and at least two background points either side of the major peak were measured. The peaks used were 2.09D for alpha-alumina, 2.53D for magnetite, and 3.38D for alpha-quartz. From this data figures of merit were calculated for each experimental condition. "Figures of Merit" are designed to emphasize many different aspects of system performance. The P^2/B ratio (where P and B are peak and background intensities respectively) is often used because it combines the criterion of sensitivity to changes in concentration or amount of an element or phase, with the need for total signal in order to obtain good statistical precision. The P^2/B ratio also appears in the usual expressions for minimum detection limit in XRD and XRF. These are generally of the form $C_{mdl} = n*B^{0.5}/P*t^{0.5}$ where n is a number of standard deviations, B and P are background and peak counting rates, and t is the analysis time.

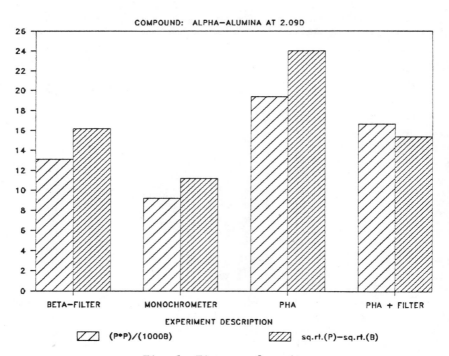

COMPOUND: ALPHA-ALUMINA AT 2.09D

EXPERIMENT DESCRIPTION

$\boxed{//}$ (P*P)/(1000B) $\boxed{//}$ sq.rt.(P)-sq.rt.(B)

Fig. 1. Figures of merit.

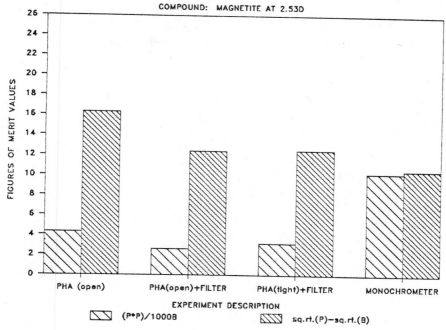

Fig. 2. Figures of merit.

Another commonly used figure of merit is derived from the equation for counting error: $E\% = 100/t^{0.5} * 1/(P^{0.5}-B^{0.5})$ where the abbreviations are as above. In this case the error will be at a minimum when the differences between the square roots of the peak and background intensities are at a maximum; therefore this value can be taken as a figure or merit. Finally, in the case of quantitative analysis such as airborne silica where peak area are plotted against concentration, peak area may be taken as a figure of merit.

Figures one through three show the figures of merit values for the three model compounds plotted as bar charts. Scaling factors were included in each calculation so that the values could be plotted on the same scale. For alpha-alumina, four sets of values are shown: PHA, Beta-filter, monochrometer, and PHA plus beta-filter. For both figures of merit the PHA exhibits the best performance (although beta peaks are not removed) and the monochrometer the poorest. The combination of the PHA plus beta-filter gives good figures of merit values while at the same time removing the beta peaks. For magnetite, the conclusion is not as straight forward as the PHA appears to be better according to the $P^{0.5}-B^{0.5}$ calculation, but the monochrometer appears better according to the P^2/B calculation. In the case of alpha-quartz, the PHA yields the best results according to all three calculations of figures of merit. Open and tight are used in Figures 2 and 3 to

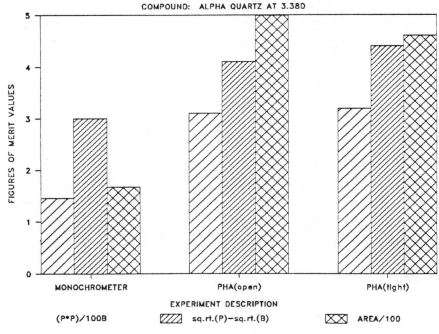

Fig. 3. Figures of merit.

refer to whether the pHA is set for a large window (open) or a narrow window (tight) in an attempt to eliminate some source of secondary radiation such as iron or copper K-beta. The monochrometer is more effective at reducing background than a pulse height discriminator, but does so with a large loss in peak intensity. Hence, in cases where background intensity is not abnormal (alumina and quartz), the PHA out performs the monochrometer; but in cases where the background is extremely high (fluorescent samples such as magnetite), the results are mixed.

SUMMARY

For a set of model compounds, diffraction patterns were obtained utilizing a pulse height discriminator, graphite monochrometer, and beta-filters in combination with a scintillation detector and a copper target tube. Figures of merit were calculated to show which experimental technique or combination of techniques was most effective at optimizing peak to background. The experimental data indicates that for non-fluorescent samples the combination of a pulse height discriminator with a beta-filter produces the best peak to background. For fluorescent samples, no firm conclusion can be drawn as different figures of merit calculations lead to different conclusions.

ACKNOWLEDGMENTS

The authors would like to acknowledge Champion International
Corporation for having made available the materials and their West
Nyack, New York facilities for conducting the experiments reported
in this study.

REFERENCES

1. H.P. Klug and L.E. Alexander, "X-Ray Diffraction Procedures,"
John Wiley and Sons, Inc., New York (1954)

2. E.P. Bertin, "Principles and Practice of X-Ray spectrometric
Analysis," Plenum Press, New York (1975)

3. R. Jenkins and J.L. de Vries, "X-Ray Powder Diffractometry,"
N.V. Philips, Eindhoven (1970)

4. T. Moori et al, Yogyo Kyokai Shi, 78(904), pp 396-400, (1970)

5. I.J. Delaney and M.E. Standen, Inst. Mining Met., Trans., Sect
C, 79 (Sept), pp 234-237, (1970)

X-RAY DIFFRACTION STUDIES OF SOLID SOLUTIONS

OF PENTAGLYCERINE-NEOPENTYLGLYCOL

D. Chandra*, C. S. Barrett** and D. K. Benson***

*   University of Nevada, Reno, NV
**   University of Denver, Denver, CO
***   Solar Energy Research Institute, Golden, CO

ABSTRACT

     An array of molecules that is anisotropic in the extreme has been discovered in certain thermal-energy storage materials and is reported here: neopentylglycol (NPG) and NPG-rich solid solutions with pentaglycerine (PG) have a crystal structure, stable at room temperature, that consists of bimolecular chains of molecules that are all unidirectionally aligned throughout a crystal. There are hydrogen bonds between every molecule in one chain and its neighbors in that chain, but none between molecules of one chain and any molecules of the neighboring parallel chains. Thus there are strong intermolecular bonds along each chain and only weaker bonds between the chains. The structure has been determined by using modern single crystal techniques with 529 independent reflections from a crystal of NPG ($C_5H_{12}O_2$). The structure is monoclinic with space group $P2_1/c$ - C_{2h}^5. This anisotropic structure transforms to a cubic structure at higher temperatures.

     A phase diagram for the binary PG-NPG system is presented that is based on differential scanning calorimetric (DSC) results combined with X-ray diffraction data. The transformation temperatures of various solid-solutions and the extent of single-phase and two-phase regions are shown and the reason for the broadening of the DSC endotherms are determined. The high temperature cubic phase, γ, transforms either to a PG-rich tetragonal phase, α, an NPG-rich monoclinic phase, β, or a two phase mixture, $(\alpha+\beta)$, depending on sample composition; earlier investigators seem to have found transformation to a single phase, they called "II," only. The solid-solid transformation occurs at 89°C in pure PG, at 40°C in pure NPG and at a minimum of 24°C in samples of 75 mol% NPG in PG-NPG.

INTRODUCTION

     Crystalline organic molecular materials such as polyalcohols are potential candidates for thermal storage devices[1]. These polyalcohols, which have tetrahedral globular molecules, are classified as "Plastic Crystals"[2]. These crystals have a large entropy of solid-solid transition and very low entropy of fusion. The low temperature structures of pure polyalcohols are generally layered and the high temperature structures are isotropic. The energy storage is in the isotropic high-temperature phase. The transformation temperatures of pure polyalcohols are fixed and generally higher than required for passive thermal energy storage applications but solid-solutions of two or three polyalcohols in different proportions can allow adjustment of these temperatures.

Differential scanning calorimetry (DSC) performed on solid solutions of pentaglycerine (PG) and neopentylglycol (NPG) showed broadening of the endotherms. To explain the broadening of these DSC endotherms, X-ray powder diffraction (XRPD) patterns have been made on solid solutions of various compositions. Combining the DSC results with the XRPD results, a binary phase diagram has been proposed for the PG-NPG system; single-phase and two-phase regions have been identified and delimited. The solid solutions in the composition range from approximately 60% to 75% NPG are of great interest for solar thermal energy storage because of their low transformation temperature. The understanding of the structural behavior of solid solution polyalcohols is of prime importance in determining the mechanisms by which they accept, store and release thermal energy.

The crystal structure of PG has already been determined[3], however previously reported determinations[4, 5, 6] of the structure of NPG at room temperature were controversial, and there were uncertainties in the indexing of its powder patterns. A new crystal structure determination of NPG has been made in this project by modern single crystal instrumentation and methods, and was used as the basis for interpreting XRPD patterns of the NPG-rich solutions in the low temperature region, as well as for gaining a better understanding of the molecular array and intermolecular forces in the crystal.

EXPERIMENTAL

Single crystals of NPG were grown at the University of Nevada-Reno for the crystal structure determination, which was conducted at the Crystalytics Company in Lincoln, Nebraska. A computer-controlled four-circle Nicolet autodiffractometer was used for this purpose, with copper Kα radiation. The structure of the crystal was determined by direct methods using SHELXTL programs. Hydrogen atoms were located by difference Fourier synthesis and refined as independent isotropic atoms. A powder pattern was generated from the single crystal results and this simulated pattern was used for indexing of the XRPD peaks.

The solid solution samples of PG-NPG were made by melting mixtures in capped tubes to prevent possible sublimation of the components. Several sets of these samples were made to check for reproducibility. These solutions were cycled above and below the transformation temperatures, then slowly cooled to room temperature, and then ground in a moisture-free environment for use in the X-ray diffraction experiments. The compositions used for XRPD were 12.5, 25, 40, 50, 60, 75, 80 and 87.5 mol% NPG in the PG-NPG series.

The phases present in these pure and solid solution powders were determined using an automated Norelco diffractometer with Bragg-Brentano focussing. This diffractometer was interfaced with a DEC microvax computer, and Nicolet software was used for the analyses. A Philips APD 3250 diffractometer equipped with a diffracted-beam graphite monochromator was also used for some of the experiments; copper Kα radiation was used, with the X-ray tube operating at 35 kV and 20 mA for these. In general, the data was acquired at 0.02° 2θ, 2-second steps; and once the X-ray patterns were

examined, some important 2θ ranges with overlapped or unresolved peaks were
scanned at slower speeds and smaller intervals to obtain better peak fitting.
Sample transparency was a problem but admixing the NBS silicon standard SRM
640b powder lessened the beam penetration and also served as an internal
standard.   The X-ray diffraction data from low-temperature phases having
compositions ranging from 50 to 87.5 mol% NPG, whose transformation
temperatures were near room temperature, were conducted at 19°C by cooling
the entire room down to this temperature.   The high temperature phase was
also obtained in a similar manner by increasing the room temperature.

RESULTS AND DISCUSSION

        The structure of NPG, as determined in this project, is monoclinic,
with space group $P2_1/c$ - C_{2h}^5 (No. 14).   The lattice parameters from this
single crystal study are a = 5.979 Å, b = 10.876 Å, c = 10.099 Å, β = 99.78°
at 20 $\pm$ 1°C.   The intensities of 887 independent reflections were measured
and processed; refinement cycles involving 529 selected independent
reflections and 79 parameters reduced the R value to 0.043.   The coordinates
of each atom in the four molecules of $C_5H_{12}O_2$ in the unit cell, thermal
parameters, bond lengths and angles, and details of the determination are to
be published elsewhere.

        Important features of the crystal structure of NPG can be pointed out
by using a projection of the structure perpendicularly onto the a-c plane,
i.e., a plot of the structure as viewed along the b direction.   The computer-
generated projection of atom coordinates is shown in Fig. 1 for one of the
alternate settings of the space group $P2_1/c$, in which the unit cell is
outlined in this figure.   There are symmetry centers at the corners of the
cell and at the position 1/2, 1/2, 1/2, as well as at the centers of the a
and c edges of the cell, and there are 2-fold screw axes normal to the paper
at the positions S in the figure.   At the upper left of Fig. 1, outside the
cell corner labeled R_1, a molecule is shown with its interatomic bonds drawn
with heavy lines.   This molecule is connected to one with its center inside
the cell at the upper left that is related to the first by the symmetry
center at R_1 and that also is drawn with heavy bond lines.   These two
molecules can be viewed as bonded together by strong hydrogen bonds, O-H...O,
shown as heavy dashed lines.

        Identical pairs of molecules, also drawn with heavy lines, occur at all
the other corners of the cell.   The pair at 0, 0, 0 and the one at 1, 0, 0 at
the upper right hand corner are bonded together by H-bonds, also shown by
heavy dashed lines, that are near the label R_2.   Along the x direction there
is thus a row of pairs of molecules with each molecular pair joined by strong
O-H...O bonds.   Each pair is joined to the next pair in the row by similar
strong O-H...O bonds and this row can be thought of as a chain with links
centered at R_1 and R_2.   A similar chain that is parallel to the above
mentioned one is located in the middle of the unit cell and is shown with
lighter lines.   It, too, is an H-bonded chain, but it is not connected to the
first mentioned chain by any H-bonds.   The second chain passes through point
R_3, which is at 1/2, 1/2, 1/2 and which is related to R_1 by the screw axis at
S.   There are four O-H...O bonds surrounding the point R_2 which is a symmetry
center; these four bonds lie in a plane, but that plane is not parallel to
the x-z plane, a fact that is determined from the coordinates of the oxygen
atoms at the ends of the bonds.

        The structure of NPG can thus be seen as an array of strong H-bonded
chains of molecules along the x direction with each chain only weakly van der

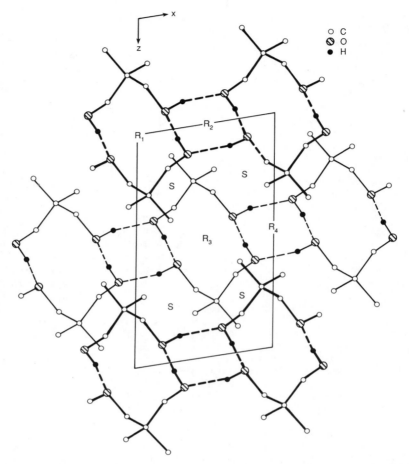

Figure 1. The newly determined chrystal structure of monoclinic NPG ($C_5H_{12}O_2$)
as viewed down the b axis, i.e., as projected perpendicularly onto
the a-c plane. Smallest circles = H, midsize = C, largest
(containing stripes) = O. Hydrogen bonds link each H atom to two
neighboring molecules, as indicated by a dashed and a solid line
extending out from each H atom.

Waals bonded to its neighboring parallel chains. This unidirectional array
of the low-temperature structure becomes isotropic when heated above its
phase transformation temperature and becomes the high temperature cubic γ
phase.

Considering NPG as a host structure, PG molecules may be expected to
enter the lattice interstitially to fit between the H-bonded chains, straining the van der Waals bonds but not the H-bonds, in preference to entering
substitutionally by replacing molecules of the chains and thus straining the
H-bond array. Thus the formation of a β-phase solid solution of the interstitial type would be expected, associated with little or no alteration of
the unit cell dimension a, but with alteration of the b and c dimensions. On
the other hand, for a dopant to form a substitutional solid solution with
NPG, displacing some molecules of the chains would be required which would,

therefore, alter the average length of the chains and consequently the a dimension of the unit cell. Whatever the nature of the solid solution of PG in NPG, an interesting anisotropic straining of the lattice could be expected. The XRPD patterns obtained thus far, however, result in shifts of the NPG peaks so small, when PG is dissolved within the limits of the β phase, that determination of the unit cell dimensions and distortions that result would only be disclosed by more thorough analysis of the patterns than has yet been done, and with more complete X-ray data than is available at present.

The X-ray diffraction analyses have shown that PG and NPG have limited solubility in the low temperature region as compared to complete solubility in the high temperature, γ phase, region. The DSC scans showed that the solid- solid phase transition endotherms gave generally broad, ill-defined peaks associated with low enthalpies. But the X-ray diffraction data clearly showed coexistance of two phases, (tetragonal)α and (monoclinic) β, at certain compositions. Coupling the X-ray and DSC data we are proposing a phase diagram for the PG-NPG binary system shown in Fig. 2. The DSC scans are also included in this figure to indicate the onset temperatures which describe the boundaries of two-phase regions. The enthalpies of solid-solid transitions and the mean transformation temperatures for the various solid solutions are reported in reference 7. It can be seen from the proposed phase diagram that in the solid-solid transformations of the PG-NPG solutions there is a significant change in composition of the cubic and low temperature phases when heated through the ($\beta+\gamma$) and ($\alpha+\gamma$) phase regions. It is these changes in the compositions, during the heating through the two-phase regions, that broaden the endotherms in the DSC scans. Font, et. al,[8] reported thermal studies on PG-NPG 67 mol% and 30 mol% NPG samples, but no structural studies leading to identification of two-phase regions were mentioned. A solidus curve is shown at the top right in the phase diagram of Fig. 2 to indicate the onset of melting.

The composition range 50 to 87.5 mol% NPG is important for the construction of the phase diagram; several sets of X-ray data were taken for 50, 60, 75 and 87.5 mol% NPG samples made from different batches of PG-NPG solid-solutions to determine reproducibility of the stable phases at various temperatures. A few solid solutions that had transitions near room temperature suffered undercooling, typically 3 to 6°C. Therefore, the temperature at which the data were acquired was very important in these cases. The room temperature thermostat was set so that a mercury thermometer near the sample holder of the diffractometer read 19°C for the low temperature runs. There were problems encountered in obtaining X-ray data for the low-temperature phases in spite of the fact that the transformation temperature was above 19°C; the high temperature phase, γ, was present along with the low temperature α and β phases due to undercooling. This under-cooling was eliminated by freezing the samples and then allowing them to equilibrate at room temperature.

It was observed that the solubility of NPG in the PG structure extended up to approximately 60 mol% of NPG as indicated by the X-ray diffraction analyses. The X-ray intensities of the reflection from the (101) plane of the α phase varied with the percent of NPG. As the concentration of NPG was further increased, a miscibility gap was found and lines of the monoclinic phase β appeared in the diffractions patterns. One of the peaks at approximately 12° 2θ was chosen as a reference peak for this β phase and, of course, the intensity of this peak increased with increasing percent NPG, while the intensities of X-ray peaks from the α phase decreased. As the NPG

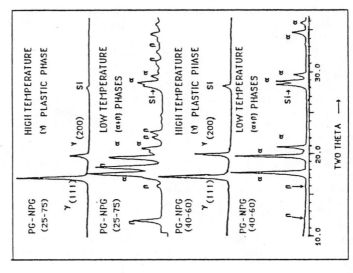

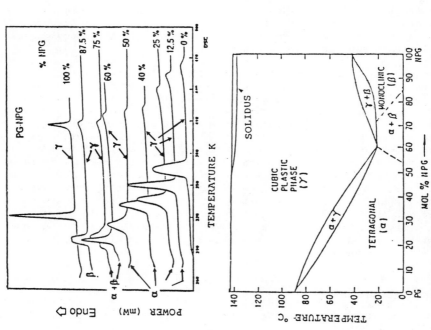

Figure 2. Differential Scanning Calorimetric (DSC) scans of PG-NPG series for various compositions are reproduced at the upper left. Typical XRPD patterns of two sets, compositions, PG-NPG (40-60) and (25-75). The new phase diagram for the system, (at the lower left), is based on combining the DSC and XRPD results; it delimits the phase regions for the tetragonal α, monoclinic β and the cubic γ. Approximation have been made to draw the phase boundaries, for the (α+β), (α+γ) and (β+γ) regions.

concentration increased beyond the limit of the two-phase region, only β phase peaks were present.

When solid solutions in the two-phase region were heated through their transformation temperatures, the single phase region, γ, appeared. It is interesting to note that every solid solution in the γ phase region yielded only two diffraction peaks; these were indexced as 111 and 200 of a face centered cubic structure (see Fig. 2). No other reflections appeared, presumably because of disorder in the plastic phase due to increased thermal vibration and molecular rotation[9,10], which will be studied in the future. The lattice parameters of the cubic γ phase at 60 and 75 mol% NPG at a temperature just above the phase-change temperature were found, with the 111 and 200 indexing, to be 8.814 A and 8.898 A, respectively. A table for the lattice parameters vs. composition for the α and β phases is not yet complete at this time. Differences in the XRPD patterns of α and β phases for the PG-NPG mol fractions 40-60 and 25-75 are shown in Fig. 2. The 12° 2θ peak of the monoclinic phase, β, in the PG-NPG 40-60 sample is of very low intensisty because of the small amount of this β phase present. Several runs have been made that confirm the presence of these low intensity monoclinic β-phase peaks. We have often observed preferred orientation problems associated with the 12° 2θ peak, which is the 011 reflection from the β phase.

To summarize, a new structure of NPG, stable at low temperatures, has been determined. It consists of a highly anisotropic bimolecular chain-like unidirectional array. Powder patterns generated from single crystal data were used in the analyses of the XRPD patterns obtained from the PG-NPG β solid solutions. Based on the X-ray diffraction and DSC results a phase diagram is proposed for the binary PG-NPG system: at 19°C, phase α extends from pure PG to approximately 60 mol% NPG; then a miscibility gap extends up to approximately 75 mol% NPG; from 75 to 100 mol% NPG there is only the single monoclinic phase β; at higher temperatures only the cubic phase, γ, is stable.

ACKNOWLEDGEMENT

The support of this research by the U.S. Department of Energy under contract No. DE-AC03-87SF16873 is acknowledged. The authors also acknowledge the work on the determination of the structure of NPG, (details to be published elsewhere), by Dr. Cynthia A. S. Day, with advanced single-crystal methods at Crystalytics Co., Lincoln, Nebraska.

REFERENCES

1.   D. K. Benson, J. D. Webb, R. W. Burrows, J. D. McFadden and
      C. Christensen, SERI Report, Task Nos. 1275.00 and 1464.00WPA304
      (available through NTIS. SERI/TR255-1828, Category 62e) March (1985).
2.   J. Timmermann, Solid State Phase Transitions in Pentaerytritol and
      related Polyhydric Compounds, J. Phys. Chem. Solids, Pergamon Press,
      18:8 (1961).
3.   P. Eilerman, E. Lipman and R. Rudman, Acta Cryst., B39:263, (1983).
4.   R. Zannetti, Unit Cell and Space Group of 2.2 Dimethyl-1,3 Propanediol,
      Acta Cryst., 14:203 (1961).
5.   E. Nakano, K. Hirotsu and A. Shimada, the Crystal Structure of
      Pentaglycerol and Neopentylglycol, Bull. Chem. Soc. of Japan, 42:3367
      (1969).

6.   H. T. Frank, K. Krzemicki and H. Voellenkle, Phase transition in
      Neopentylglycol, <u>Chem-Zig</u>, 97(4):16, (1973).

7.   D. Chandra, Crystal Structure Changes in the Solid-Solutions of
      Pentaglycerine-Neopentylglycol, Final Report to Solar Energy Research
      Institute, Midwest Research Institute, Golden, CO, Contract No:
      CA-5-00491-01, Sept. 10, (1986).

8.   J. Font, J. Muntasell, J. Navarro and J. L. Tarmaitt, Time and
      Temperature Dependence of the Exchanged Energy on Solid-Solid
      Transitions of Pentaglycerine/Neopentylglycol Mixtures, <u>Solar Energy
      Materials</u>, Elsevier Publishers BV., 15:403-412 (1987).

9.   D. Chandra and C. S. Barrett, Effect of Interstitial Dopants on the
      Thermophysical Properties of Solid State Phase Change Materials. Draft
      of Final Report DOE Contract No. DE-AC03-84SF-12205, Jan. 30 (1986).

10.   D. Chandra, C. S. Barrett, and D. K. Benson, Adjustment of Solid-Solid
      Phase Transition Temperatures of Polyalcohols by the use of Dopants,
      "Advances in X-Ray Analysis," ed. C. S. Barrett, J. B. Cohen, J. T.
      Faber, D. E. Leyden, J. C. Russ and P. K. Predecki, Plenum Publishing,
      29:305-313 (1986).

SIMULTANEOUS THERMAL AND STRUCTURAL MEASUREMENTS OF

ORIENTED POLYMERS BY DSC/XRD USING AN AREA DETECTOR

Steven T. Correale and N. Sanjeeva Murthy

Corporate Technology
Allied-Signal, Inc.
Morristown, N.J. 07960

INTRODUCTION

Differential scanning calorimetry (DSC) and variable temperature x-ray diffraction (XRD) are two complementary techniques which provide thermal and structural information. By using DSC and XRD simultaneously, one can directly correlate the results from the two techniques[1]. The simultaneous measurements eliminate problems which might arise due to sample inhomogeneity and instrumental differences. In addition, the DSC provides precise control over temperature, and heating and cooling rates.

XRD analysis of fibers and oriented films usually requires measurement of meridional and off-meridional reflections in addition to equatorial reflections. XRD patterns of oriented polymers could be obtained using film cameras and the photographs analyzed on a densitometer. Alternatively, several radial θ-2θ scans can be obtained at various azimuthal angles on a transmission goniometer. Both these techniques are time consuming. An area detector (2-dimensional positional sensitive detector or 2D-PSD) permits fast data acquisition (1 or 2 min exposures) from oriented polymers.

By combining the DSC with an area detector, one can study phase transitions, and changes in morphology (percent crystallinity, crystallite size and degree of orientation) as a function of temperature.

EXPERIMENTAL

A modified Mettler FP84 microscopy DSC cell, and a Nicolet area detector were mounted on a two-circle Huber goniometer in transmission geometry. The x-ray source was Ni filtered copper radiation from a Philips XRG-3000 generator. A 1 mm collimator was used. A schematic of the equipment layout is shown in Figure 1.

The DSC cell can be heated to 400°C and cooled to -60°C. The cell was modified[2] for use with x-rays as follows. The exit window for the diffracted x-rays in the cell was enlarged to 5 mm in diameter so as to be able to collect data up to 50° 2θ. The insulating glass slides were

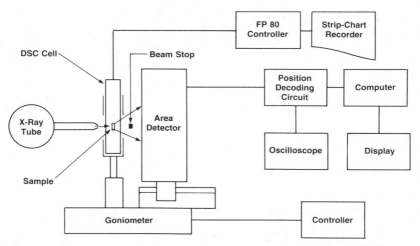

Fig. 1.   Schematic of relevant components of the DSC and the XRD equipment.

removed and a 0.25 mm thick Be foil was cemented over the exit window.   The
DSC cell was held vertically on the goniometer in a holder.   This holder
allowed us to mount or dismount the cell without disturbing the alignment
of the goniometer.   The modifications made to the DSC cell, and the absence
of a reference crucible (DSC being mounted vertically) affected the base-
line of the thermograms.   Therefore, we used the thermograms to determine
the temperatures of the thermal events and not to calculate such integral
parameters as heat of fusion.   The DSC cell was controlled by a Mettler
FP80 central processor and the thermograms were recorded on a strip-chart
recorder.

   The area detector, a multi-wire sealed proportional counter, has a spatial resolution of ~0.1 mm. A real-time image of the XRD pattern displayed
on a oscilloscope was used in aligning the DSC cell, and in monitoring the
experiments. A Cadmus computer was used for collecting, storing and analyzing the data.

   Polymer powders were packed into a 2 mm thick, 6 mm I.D. Teflon® ring,
and then wrapped in a 15 μm thick Al foil.   The fiber samples were wound on
a 0.25 mm thick Be foil and wrapped in Al foil.   The Al foil, which was used
to protect the DSC cell from the molten polymer, did not significantly
attenuate the x-rays and its diffraction pattern did not overlap that of
the polymer.   The thermocouples and their leads inside the DSC cell were
insulated from the Al foil by using a 0.25 mm thick Be foil coated with 5
to 10 μm layer of polyparaxylene (PPX)[3].   PPX melts at ~420°C, is a good
insulator, and can be polymerized by vacuum deposition at temperatures
below 30°C, resulting in a pinhole free coating[4].   This layer of PPX did
not produce any detectable diffraction pattern nor did it affect the thermograms.

   The XRD patterns were obtained as the samples were being heated or
cooled at a controlled rate.   Sometimes, the heating or the cooling cycle
was interrupted to keep the specimen at a given temperature as an XRD pattern was being recorded.   The DSC signal was lost during such interrup-
tions, and there was some delay before the signal returned to its normal
value after the heating or cooling was resumed.   The x-ray generator was

usually operated at 25 kV and 20 mA; this low setting, especially the voltage, was required to reduce air scattering. The sample to detector distance was between 6 to 8 cm and the goniometer was set with θ-2θ = 0.0°. XRD patterns were typically collected for 1 or 2 minutes.

RESULTS AND DISCUSSION

Unoriented Nylon 6

Thermograms from powdered nylon 6 (an Allied-Signal product, primarily in the γ form) heated from 30°C to 240°C at 5°C/min, and then cooled from the melt (240°C) to 40°C at 5°C/min are shown in Figures 2a and 2b, respectively. The labels on the XRD patterns (Figs. 2A-F) correspond to the labels on the thermograms. Plots of the radially integrated relative intensities are superimposed on each XRD pattern.

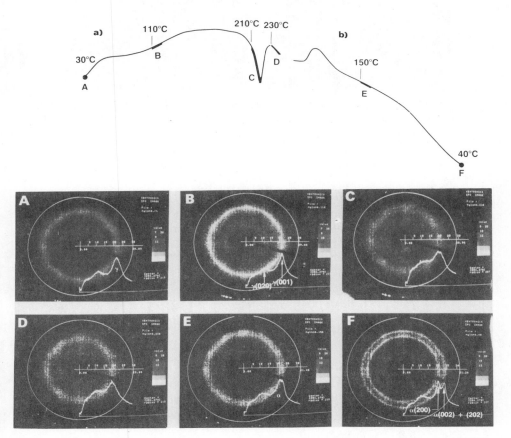

Fig. 2. Thermograms of unoriented γ nylon 6 during heating (Fig. a, 30°C to 240°C at 5°C/min) and cooling (Fig. b, from the melt at 240° to 40°C at 5°C/min). The x-ray diffraction patterns (A-F) are 2 min exposures. In this and the following figures, the labels on the XRD photographs correspond to the labels on the thermograms. The plots superimposed on the XRD patterns are radially integrated intensity vs. 2θ.

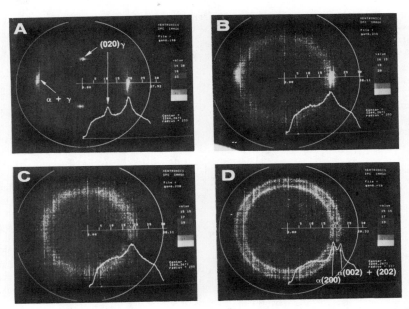

Fig. 3. The XRD patterns (2 min exposures) obtained at 190°C (A), 215°C (B), 219°C (C) and at 22°C after cooling from the melt (D). The fiber axis is vertical.

The XRD patterns (Figs. 2A and 2B) obtained at 30°C and 110°C during the heating cycle show the strong (001) and the weak (020) reflections from the γ phase at 21° and 11° 2θ, respectively; the latter is noticeable only in the radial integrations. The decrease in the half-width of the (001) peak in the pattern at 110°C (Fig. 2B), represents an increase in the crystallite size during heating. At 210°C, as nylon 6 begins to melt, the (001) reflection is barely visible above the amorphous halo in the radial integrated plot of the XRD pattern (Fig. 2C). The XRD pattern at 230°C (Fig. 2D) shows only an amorphous halo indicating that the sample has indeed melted. Upon cooling from the melt, the sample goes through an exotherm at ~197°C, indicating the onset of crystallization. At 150°C the XRD pattern (Fig. 2E) shows the emergence of two α crystalline peaks. As the sample is cooled, the XRD pattern at 40°C (Fig. 2F) clearly shows the two α crystalline peaks of nylon 6.

Nylon 6 Fiber

XRD data from a nylon 6 fiber heated at 1°C/min is shown in Figure 3. The fiber is approximately 50% crystalline, mostly in the γ phase with a small amount of α. The XRD pattern at 190°C (Fig. 3A) shows strong equatorial reflections, (001) due to γ, and (200) and (002)+(202) due to α, and a meridional (020) reflection due to γ[5]. Since the meridional reflection is due only to the γ phase, the changes in γ phase as a function of temperature can be monitored independent of the α phase.

The XRD pattern in Fig. 3B shows that as the nylon 6 fiber is heated through the γ endotherm (214°C), the equatorial reflection is still present even though the γ meridional reflection is no longer visible. This is accompanied by a considerable increase in unoriented diffuse scattering at

~20° 2θ. The increase in diffuse scattering is probably due to the melting
of the γ phase. As the sample is heated beyond the α endotherm (219°C),
the XRD pattern shows only an amorphous halo (Fig. 3C). The XRD pattern in
Fig. 3D, obtained after the sample is cooled to room temperature, is from
unoriented α nylon 6. The azimuthal intensity distributions of the
equatorial reflection, superimposed on the XRD patterns obtained during
heating at 150°C and 215°C (Fig. 4), do not show any change in full-width-
half-maximum, i.e., in the degree of crystalline orientation, due to the
melting (or conversion) of the γ phase. The higher background at 215°C can
be associated with onset of melting, which also reduces the intensity of
the equatorial reflection.

Polyethylene Fiber

We studied extended-chain polyethylene (PE) gel-spun fibers (Spectra®)
to understand the structural origins of the endotherms in these fibers.
Pennings and coworkers, in their studies of PE fibers made by surface-
growth technique, observed three endotherms between 136°C and 164°C. They
attributed the first endotherm at ~140°C to the melting of the unconstrained fibrils, the second at ~150°C to orthogonal to hexagonal lattice
transformation, and the last at ~160°C (which sometimes is split into two
peaks) to the melting of the hexagonal phase. A thermogram obtained from a
Spectra® PE fiber is shown in Fig. 5. The XRD patterns obtained at 30°C,
110°C and 135°C (Figs. 5A, B and C respectively) show the strong (110) and
the weak (200) orthorhombic reflections. These patterns show that as the
temperature is increased, the (200) reflection shifts to lower 2θ values.
The XRD pattern obtained after the endotherm at 150°C (Fig. 5D), does not
show any structural or phase change, and follows the same trend seen at
lower temperatures, i.e., the expansion of the unit cell with increase in
temperature. The difference between these results and the published data[6]
is likely due to experimental details (history of the fiber, sample preparation, heating rates, etc.). Further work is in progress to understand
the nature of these endotherms.

Halar® Fiber

A thermogram from a Halar® fiber (an ethylene/chlorotrifluoroethylene
copolymer from Ausimont Co.) heated from room temperature to 260°C at

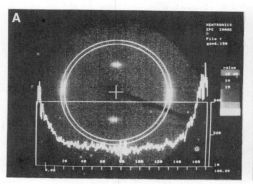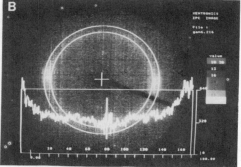

Fig. 4. Azimuthal integration of the equatorial reflections of nylon 6
    (α + γ) fiber during initial heat-up at 150°C (A), and at 215°C (B).

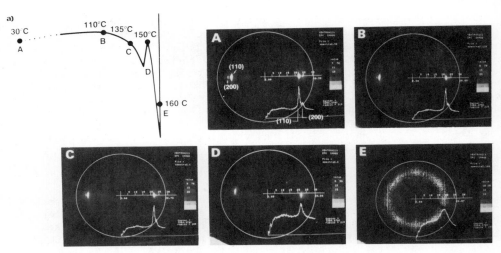

Fig. 5. A thermogram of extended chain polyethylene fiber (Spectra®)
heated at 5°C/min is shown in Fig. a. The XRD patterns in
Figs. A–E are 1 min exposures with the fiber axis vertical.

10°C/min and then cooled from the melt at 10°C/min is shown in Fig. 6. As
the fiber is heated, a second–order phase transition occurs near 150°C[7].
Comparison of the XRD patterns obtained at 22°C and 150°C (Figs. 6A and
6B), show a significant increase in the crystallite size at 150°C as indicated by the sharpening of the (11$\overline{2}$0) equatorial reflection. Upon
further heating up to the melt temperature, the width of the (11$\overline{2}$0) reflec-

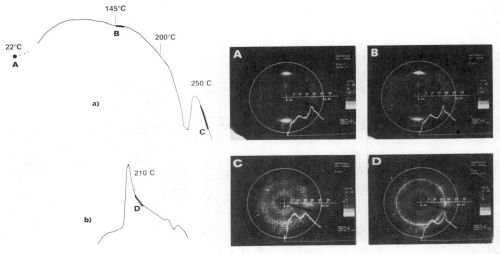

Fig. 6. Thermograms of Halar® (ethylene/chlorotrifluoroethylene copolymer)
fiber heated from room temperature at 10°C/min to above 260°C (Fig.
a) and then cooled from the melt at 10°C/min (Fig. b). The XRD
patterns (A–D) are 1 min exposures, and the fiber axis is horizontal.

tion continues to decrease. The XRD pattern obtained after the melting endotherm at 235°C (Fig. 6C) shows only the amorphous halo from the molten polymer. As the fiber is cooled through the endotherm (222°C), the XRD pattern at 210°C (Fig. 6D) shows that the fiber has crystallized and has become unoriented.

CONCLUSION

The installation of an incident beam focusing monochromator will significantly improve the resolution of the XRD patterns. A helium path is being designed to reduce the background scattering. Additional information can be obtained from the XRD patterns using appropriate image processing software. Preliminary measurements of the percent crystallinity of fibers have been done, but accurate and reliable measurements require further refinements of the technique.

The simultaneous DSC/XRD has proved to be a useful technique in eva-luating and understanding materials. The area detector is becoming indispensable in analyzing fibers and oriented films. The examples in this report illustrate the advantages of combining simultaneous DSC/XRD with an area detector, especially in the study of fibers, e.g., in studying phase transitions and following the changes in crystallinity, degree of orien-tation, off-axis reflections, as a function of temperature.

ACKNOWLEDGEMENT

We would like to thank Dr. S. Kavesh and Dr. S. Chandrasekaran for pro-viding the fibers used in these studies and for their permission to publish the results, and to Mrs. A. C. Reimschuessel and Dr. A. J. Signorelli for reviewing the manuscript.

REFERENCES

1. T. G. Fawcett, C. E. Crowder, L. F. Whiting, J. C. Tou, W. F. Scott, R. A. Newman, W. C. Harris, F. J. Knoll, V. J. Caldecourt, "The Rapid Simultaneous Measurement of Thermal and Structural Data by a Novel DSC/XRD Instrument", Adv. in X-ray Analysis, 28:227 (1985).

2. J. T. Koberstein, University of Connecticut, personal communication.

3. Paratronix, Attleboro, Massachusetts.

4. M. Szwarc, Polym., Eng. Sci., 16, 473 (1976).

5. N. S. Murthy, S. M. Aharoni, A. B. Szollosi, J. Polym. Sci., 23, 2549 (1985).

6. A. Zwijnenburg, "Longitudinal Growth, Morphology and Physical Properties of Fibrillar Polyethylene Crystals", Thesis, University of Gronigen (1978).

7. J. P. Sibilia, L. G. Roldan, S. Chandrasekaran, J. Polym. Sci., 10, 549 (1972).

VACUUM FREE-FALL METHOD FOR PREPARATION OF RANDOMLY ORIENTED XRD SAMPLES

Jim Ludlam, Brad Jacobs and Paul K. Predecki

Department of Engineering
University of Denver
Denver, CO   80208

INTRODUCTION

Quantitative phase analysis by XRD requires the attainment of (1) random crystallite orientation and (2) homogeneous and intimate mixing of the constituent phases in the samples. These two requirements must also be met in the preparation of random composite materials, particularly those containing randomly-oriented fibers or whiskers.

The two most common methods for producing random crystallite orientation are spray drying[1] and the air suspension method.[2,3] In the latter, an air suspension of the crystallites is rapidly collected onto a glass fiber filter. The crystallites then assume the random orientation of the filter fibers on which they are deposited. Neither of these methods is particularly suitable for preparing bulk random mixtures of phases which would be suitable for fabrication of composites.

It is well known from the experiments of Galileo that particles subjected to a constant gravitational force will fall in vacuum with the same acceleration independent of their shape, size or density. If while falling the particles are given sufficient rotational motion in an uncorrelated manner, their orientations in space should become random. The vacuum prevents both orientational effects and particle segregation caused by aerodynamic forces. The orientational effects on plate-like particles can readily be appreciated by observing the motion of a sheet of paper allowed to free-fall in air.

The purpose of this study was to test if a vacuum free-fall procedure would produce samples which meet the first requirement: random crystallite orientation.

APPARATUS AND EXPERIMENTAL PROCEDURES

A simple device (Fig. 1) was constructed to impart the desired random particle motion in vacuum and to collect the sample. A small amount of sample (0.1 gm) was loaded onto the top of the sample chutes (1). The device resting on a metal plate was then covered with a bell jar and evacuated with a fore pump to 10^{-3} torr. The vibrators (2), consisting of a

625

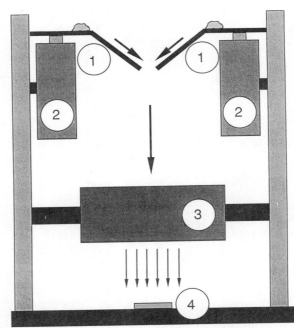

Fig. 1.   Schematic illustration of sample deposition device showing
sample chutes (1), vibrators (2), diffusing screen (3) and
sample substrate (4). Particle motion is indicated by
arrows. The vertical distance from (1) to (4) is ∿18 cms.

pair of 60 Hz vibratory engravers (Burgess Vibrocrafters Inc., Grayslake, IL)
that were mechanically attached to the sample chutes and the diffusing screen
(3), were then turned on. The amplitude of the vibrators and the chute
angles were such as to cause the sample particles to "dance" down the chutes
rather than sliding down. Particles falling from the chutes then passed
through one or two metal diffusing screens (3) and were collected on the
sample substrate (4). The purpose of the screens was to disperse the
particles more uniformly over the sample substrate and to impart additional
random motion rather than to differentiate the particles with respect to
size. Consequently the screen size (0.5 mm) was somewhat larger than the
largest particles used. The sample substrate was a glass slide (low
scattering glass) coated with a 1-10 μm thick film of petroleum jelly
prepared by dipping the slide into a hexane solution of the jelly. After
deposition was completed (∿10 sec) the vacuum was released and the sample
substrates were removed and X-rayed.

The material selected for testing was MoO_3 since its platy habit is
conducive to severe preferred orientation effects and because it has been
used in other preferred orientation studies.[1,3] The MoO_3 powder (E.M.
Industries, Cherry Hill, NJ, 99.5% pure) was used in the as-received
condition and had an average size of ∿30 μm (largest platelet dimension).
Samples were prepared with and without vacuum present during deposition.

X-ray diffraction patterns were run on a Philips type 42267 diffractometer using Cu K alpha radiation, a fixed divergence slit and a graphite,
diffracted-beam monochromator under the following conditions: tube voltage:
35 kv, current: 25 mA, step time: 1 sec, step interval: .04°.

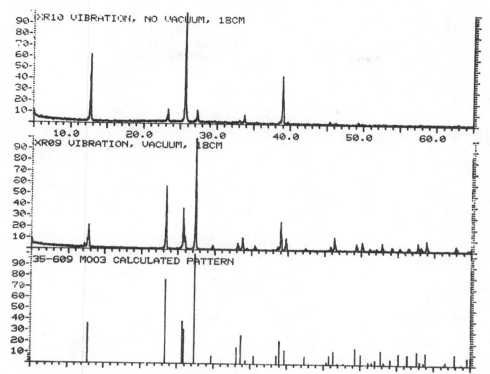

Fig. 2.   MoO$_3$ diffraction patterns for samples obtained from the
            deposition device in air (top) and in vacuum (middle). The
            PDF calculated pattern (#35-609) is shown at the bottom.

Table I.   Relative Intensities (Peak Heights) of MoO$_3$ Samples
           (The intensity of the 021 reflections is taken as 100.)

hkl	2θ(deg)	35-609	V-FF	FF	SD	AS	Calc
020	12.75	36	25	452	-	30	30
110	23.36	77	64	92	72	67	80
040	25.73	38	42	781	39	42	36
120	25.89	31	13	0	-		21
021	27.39	100	100	100	100	100	100
130	29.75	7	5	0	-	6	3
101	33.16	15	8	18	-	13	10
111	33.83	26	15	50	26	23	27
140	34.42	3	1	0	-	1	6
060	39.02	21	31	395	19	21	29
200	45.85	8	5	0	9	7	8
002	49.35	15	7	13	15	15	16

35-609 = JCPDS, PDF number for calculated pattern of MoO$_3$
V-FF   = vacuum free-fall sample
FF   = free-fall sample in air at 1 atmosphere
SD   = spray dried sample (ref. 1)
AS   = air suspension/filtration sample (ref. 3)
Calc   = calculated pattern (ref. 4)

RESULTS AND CONCLUSIONS

The patterns obtained are shown in Fig. 2 together with the calculated pattern (35-609) from the powder diffraction file. Relative peak-height intensities were obtained from the patterns using the Nicolet, second-derivative rapid peak-picking program. These are given in Table I together with the peak-height intensities from other studies.

From Fig. 2 and Table I it is evident that the sample prepared with air in the deposition device (sample FF) shows strong (0h0) orientation. With vacuum in the device (sample V-FF) the intensities are much closer to the values from either of the calculated patterns and are also comparable to the intensities obtained by spray drying (SD) and air suspension/filtration (AS). It should be noted that the intensities for samples V-FF and FF were not corrected for non-infinite sample thickness nor were additional efforts made (longer count times, profile analysis, higher resolution optics) to obtain better intensity data.

We conclude that air resistance causes pronounced orientation effects with plate-like particles. Removal of the air and use of vibration as employed in the deposition device results in close to random crystallite orientation.

ACKNOWLEDGEMENTS

This work was supported by U.S. DOE Grant #DE-FG02-86ER45248 from the Division of Materials Sciences. We are grateful to D. K. Smith of Penn State University for providing the calculated patterns "calc" in Table I.

REFERENCES

(1)   L. D. Calvert, A. F. Sirianni, G. J. Gainsford and C. R. Hubbard, Adv. X-Ray Anal 26, 105-110 (1983)

(2)   B. L. Davis and N-K. Cho, Atmos. Environ. 11, 73-85 (1977)

(3)   B. L. Davis, Adv. X-Ray Anal 27, 339-348 (1984)

(4)   Pattern provided by D. K. Smith, Pennsylvania State Univ.

APPLICATIONS OF DUAL-ENERGY X-RAY COMPUTED TOMOGRAPHY TO STRUCTURAL CERAMICS*

W. A. Ellingson and M. W. Vannier**

Materials and Components Technology Division
Argonne National Laboratory
Argonne, Illinois 60439

ABSTRACT

Advanced structural ceramics (Si_3N_4, SiC, Al_2O_3, ZrO_2) are rapidly being developed with sufficient fracture toughness to be considered for engineering applications such as internal combustion engine components, rotating turbine engine components, and heat recovery systems. X-ray computed tomography (CT) is a promising nondestructive evaluation method for these ceramics, but beam hardening presents a serious problem in the interpretation of CT images generated with polychromatic X-ray sources by creating artifacts. Dual-energy X-ray techniques have the potential to eliminate these problems. In addition, in theory, dual energy allows generation of quasimonochromatic equivalent images, which should allow verification of theoretically determined optimum energies. In using dual-energy methods, the high- and low-energy images are nonlinearly transformed to generate two energy-independent images characterizing the integrated Compton/photoelectric attenuation components. Characteristic linear combinations of these two "basis" images can serve to identify unknown materials and generate synthesized monoenergetic images.

The dual-energy method has been used to study structural ceramics as well as liquids that are close to ceramic materials in atomic number and mass density. The work was done on a Siemens DR-H CT machine with 85- and 125-kVp energy levels. Test samples included Si_3N_4 cylinders ranging from 10 to 50 mm in diameter, liquid Freon TF, and densified SiC.

---

*Work supported by the U.S. Department of Energy, Assistant Secretary for Conservation and Renewable Energy, Office of Transportation Systems, as part of the Ceramic Technology for Advanced Heat Engines Project of the Advanced Materials Development Program (Contract ACK-85234).
**Mallinckrodt Institute of Radiology/Washington University, St. Louis, MO.

INTRODUCTION

Computed tomography (CT) is a well-established imaging modality used in medical diagnostic radiology.[1-2] Industrial applications have been described in the nondestructive testing literature for the past 10 years.[3-7] The technology, especially for ceramic materials, is still in its infancy, and CT is thought of primarily as a research technique, but with important potential for future improvements.

CT scanning is, fundamentally, a method for producing spatial maps of the local X-ray attenuation within a slice through an object. Generally speaking, slice imaging is synonymous with tomography. It is possible to form slice images of an object by holding the object stationary and moving a source and detector about a point in the object which lies between them. Part of the object will be in focus on the resulting image and part will be blurred. If one-dimensional detectors are used and sufficient projections are available, the influence of blurring from over- and underlying structures may be removed mathematically.

The mathematical basis for CT scanning is reconstruction from projections. In much the same fashion that the temperature distribution within a flame or the distribution of stars within a galaxy may be computed from external projections, the spatial distribution of X-ray attenuation within an object may be reconstructed by obtaining consecutive line integrals through an object at different source-detector orientations (1-dimensional projections) followed by a suitable computational step.

One of the applications for which X-ray CT imaging holds great promise is materials characterization. Theoretically, CT should be able to provide very accurate information on X-ray absorption at points within an object. These X-ray absorption values should be useful in quantitative characterization of inclusions, cracks, short- and long-range density gradients, and other local inhomogeneities. This type of characterization should be possible for internal structures that occupy less than 1 mm^3. In addition, when X-ray CT is used for parts with complex shapes, one does not encounter mode conversion problems, as one does when using elastic-wave modalities such as ultrasonic nondestructive evaluation (NDE).

The potential for detailed materials characterization by CT imaging has been difficult to realize for objects of arbitrary complexity when conventional X-ray sources are used. Problems occur because the X-ray sources used in most CT scanners are polychromatic (i.e., they produce a spectrum of photon energies), and X-ray absorption by materials is energy-dependent. Correction for the polychromaticity of X-ray heads used in CT imaging requires knowledge of the source spectrum and the interaction of X-rays in the relevant energy range with the compounds that make up ceramic materials.

CT System Architecture

It is not possible or appropriate in this paper to give a comprehensive review of X-ray CT. Several excellent reviews[1-7] already exist, and the reader is referred to them.

Figure 1 shows a schematic diagram of an X-ray CT system. The major components are the X-ray source, the detector, the computer, and the image display system. Each component has been thoroughly discussed elsewhere and has been the subject of complete conferences. Therefore, the following discussion will be limited to some features of these components which are important for application to ceramics.

The X-Ray Source. For ceramic materials, high spatial resolution is needed, since it is desirable to be able to accurately determine defect sizes of <100 μm. Therefore, the X-ray focal spot should be small to minimize geometric unfocusing due to divergence between the source and detector.

The Detector. Although Fig. 1 shows a one-dimensional detector array, the detector can be a two-dimensional detector such as film or photostimulable phosphor that are subsequently digitized, an image intensifier with digitizing camera, or a two-dimensional detector array, as used in some recent experiments with a synchrotron as the X-ray source.[8-9]

The Computer. In general, the computing elements in a CT system consist of a data acquisition system, a control computer, and an array processor. The control computer maintains the data files, stores and

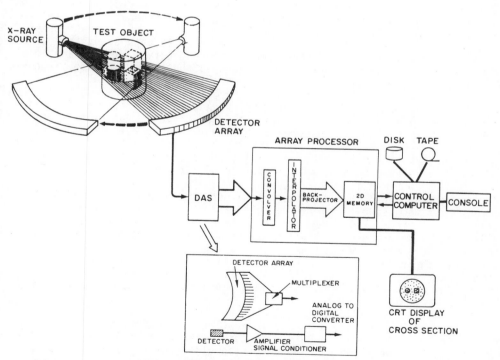

Fig. 1.   Schematic Diagram of X-Ray Computed Tomography System. DAS = data acquisition system.

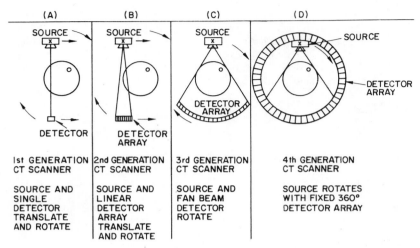

Fig. 2.   Source-Detector Configurations for CT Scanners.

retrieves programs, performs user interface and archival functions as
well as networking, and generally acts as the "assistant gatekeeper."
The array processor is the site of the computational steps needed to
perform the image reconstruction from projections. Generally speaking,
these array processors are not general-purpose devices but are optimized
to perform the convolution and back-projection (addition) operations that
are required for this reconstruction. The speed of data acquisition,
data "crunching" in the array processor, and production of the final
reconstructed image is totally dependent upon the speed and effectiveness
of the control computer.

     The Display System. High-resolution display systems of up to 1024 x
1024 pixels with at least 8-bit-deep (256 gray levels) information are
now commonly available. This means that over the longest dimension of an
object, 1024 segments can be individually displayed. For example, if the
detection system can be shown to resolve 25 μm, then a 25-mm object can
be fully displayed at a pixel-to-pixel resolution of 25 μm.

     The detector-source configurations used in various CT scanners are
classified into several so-called "generations," as shown schematically
in Fig. 2. Figure 2(A) shows the original "first-generation" system
developed and introduced commercially in 1974 by EMI, Ltd. Such systems
employ a "translate-rotate" motion. The X-ray tube and a single detector
traverse the object in a straight line while a pencil beam scans across
the object. The source-detector assembly then rotates about the object
by a small angular increment and the process is repeated until a rotation
of 180° or 360° is completed.

     Figure 2(B) shows the second-generation configuration introduced in
the late 1970s to speed up data collection without losing the advantages
of the first-generation multidetector translate-rotate geometry. In this
configuration, the source-detector system still undergoes translate-
rotate motion, but an array of detectors is used. Thus, for each linear
scan, data are collected for many sets of parallel rays (typically 20 to
60). The X-ray beam is in a fan configuration, allowing greater source
efficiency and faster scanning.

Figure 2(C) shows a third-generation configuration, which employs a "fan beam." This architecture involves only one motion (rotation), and it represents the geometry most commonly used in medical scanners. The array of detectors is made large enough to enclose the whole reconstruction area, and the source-multidetector combination rotates around the object. For each CT "slice," the data collection process can be completed in a few seconds.

Figure 2(D) shows another system architecture, designated fourth-generation geometry. This system employs a stationary circular array of detectors and a rotating X-ray source. It is a suitable architecture for fast data acquisition and reduction of motion artifacts. However, such a system requires a very large number of detectors, significantly more than third-generation designs, and the detectors must be very thin since each cell has to detect X-rays over a wide range of angular directions.

X-Ray-induced Artifacts

Beam hardening. Beam hardening (BH)[10-13] is most responsible for the inaccuracy of X-ray attenuation values obtained with CT. It occurs when a polychromatic X-ray source, such as tungsten, is used. Such a source produces X-rays with a spectrum of energies, extending from relatively low values (a few keV) to the accelerating potential of the X-ray tube. The lower energy X-rays contribute little to the formation of images, since they are almost totally attenuated by the materials of interest.

CONSIDERATIONS FOR CERAMIC MATERIALS

The sensitivity of ceramics to flaws necessitates carefully controlled processing and finishing operations to improve reliability.[14] The sizes of individual flaws that affect load-bearing capabilities are dependent upon the stress levels in the part, but are frequently on the order of 10-100 μm. The performance of ceramic parts is also affected by density gradients, binder/plasticizer (B/P) distribution in the green state, and porosity distributions. Since ceramic materials compete with metal materials for many market applications, the need to hold down fabrication costs becomes a driving force for the development of effective NDE methods.

Density gradients are of fundamental importance to ceramic processing. They are responsible for many of the problems with "shrinkage" cracking, warping, and deviations from near-net shape that have traditionally frustrated the ceramic engineer. X-ray CT offers a method of nondestructively mapping density gradients in both green (undensified) and densified ceramic parts within the limits of the contrast of the CT image.

CORRECTION FOR BEAM HARDENING

The aim of CT is to assign to every point inside an object a number that is specific to the material located at that point. A suitable candidate for this number is the X-ray attenuation coefficient of the material. As discussed earlier, a difficulty arises because the X-ray beam typically used in CT consists of photons at different energies (polychromatic X-rays). Since the attenuation at a given point is generally greater for photons of lower energy, the energy distribution or spectrum of the X-ray beam changes ("hardens") as it passes through the object. A possible solution is to assign to the point the attenuation coefficient of photons at a particular energy. If we used monoenergetic X-rays, beams from different directions should be attenuated in the same way at a given point. Reconstruction of such attenuation coefficients is a well-defined aim of CT. Generally, we are given the total attenuation of polychromatic X-ray beams through an object, and we wish to produce estimates of the total attenuation of monoenergetic X-ray beams.

Photons interact with matter by means of (1) interactions with atomic electrons, (2) interactions with nucleons, (3) interactions with the electric field surrounding nuclei or electrons, and (4) interactions with the meson field surrounding nucleons. The effect of any of these interactions can result in (1) complete absorption of the incident photon, (2) elastic scattering of the incident photon, or (3) inelastic scattering of the incident photon. Thus, theoretically, there are 12 different processes by which electromagnetic radiation can be absorbed or scattered by matter. In the energy range from a few keV to many MeV, however, all but a small number of minor effects are explainable in terms of just three of the twelve processes. These are the photoelectric effect, the Compton effect, and the pair production effect.

An important additional simplification occurs when the energy of interest is either well below or well above 1 MeV. That is, for each energy range, i.e., either <1 MeV or >1 MeV, the physics of the attenuation process is further simplified by the fact that attenuation is dominated by just two of the three principle mechanisms. In the former case, the attenuation process can be described, to an excellent approximation, as the result of photoelectric and Compton interactions; in the latter case, as the result of Compton and pair production interactions. Of the two cases, only the low-energy situation has been well studied, and it is this case which is applicable to ceramics.

The energy-dependent mass attenuation coefficient $(\mu/\rho)(E)$ of materials can thus be expressed with sufficient accuracy as a linear combination of the Compton and photoelectric coefficients. The mass attenuation coefficient can be expressed as a sum of two linearly independent basis vectors:

$$(\mu/\rho)(E) = \sum_{i=1}^{\infty} a\, f_i(E_i) \tag{1}$$

It can be shown[15-17] that the mass attenuation coefficient of any material can be expressed as a linear combination of the coefficients of two so-called basis or calibration materials:

$$(\mu/\rho)(E) = a_1 \cdot (\mu/\rho)_1(E) + a_2 \cdot (\mu/\rho)_2(E) \quad , \tag{2}$$

where subscripts 1 and 2 refer to reference material 1 and 2, respectively. Since any two linearly independent sums of two basis functions (the Compton and photoelectric components) span the space, they are also adequate basis functions. It follows then that any material ξ can be expressed as a linear combination of any other two materials, α and β, which are designated the basis-set materials.

For photoelectric and Compton scattering, we may write

$$\mu(E) = a_1 \frac{1}{E^3} + a_2\, f_{kn}(E) \quad , \tag{3}$$

where

$$f_{kn}(\alpha) = \frac{1+\alpha}{\alpha^2} \left[\frac{2(1+\alpha)}{1+2\alpha} - \frac{1}{\alpha} \ell n \ (1+2\alpha) \right]$$

and

$$+ \frac{1}{2\alpha} \ell n \ (1 + 2\alpha) - \frac{1 + 3\alpha}{(1 + 2\alpha)^2} \quad , \qquad \alpha = E/511 \ keV$$

$$a_1 \approx k_1 \frac{\rho}{A} z^n \ , \quad n \approx 4$$

$$a_2 \approx k_2 \frac{\rho}{A} z \ ,$$

where ρ = mass density, A = atomic weight, and z = effective atomic number. Now note that, for most elements, z/A is constant. Thus, a_1 is proportional to ρz^3, a_2 is proportional to ρ, and a_1/a_2 is proportional to z^3. Then, for an image generated in x,y space,

$$\mu(x,y:E) = a_1(x,y) \frac{1}{E^3} + a_2(x,y) \ f_{kn}(E)$$

and

$$\oint \mu(x,y:E)ds = A_1 \frac{1}{E^3} + A_2 \ f_{kn}(E)$$

$$A_1 = \int a_1(x,y)ds, \quad A_2 = \int a_2(x,y)ds$$

This implies that to measure $a_1(x,y)$, $a_2(x,y)$, and the integrals of A_1, A_2 we need to make intensity measurements with two energy spectra and we need to know the X-ray spectra at these two energies.

EXPERIMENTAL RESULTS

We present here results from two applications: (1) verification of optimum energy and (2) application to turbocharger rotors. A Siemens Somatom Model DR-H CT scanner, located at the Mallinckrodt Institute of Radiology, was used for these studies. This polychromatic-source scanner is in everyday use for medical examinations, and no special modifications were made for the work described here.

Verification of Optimum Energy

It was noted earlier that dual-energy software could be used to obtain quasi-monochromatic image data sets. Optimum energies can be determined from these data sets through statistical analysis of noise in the resulting images. We applied dual-energy BH correction methods to sets of densified and green-state Si_3N_4 specimens made especially for us by the Norton Advanced Ceramics Company. In addition, we experimented with Freon TF as a calibration fluid.

The green-state, cold-pressed cylinders had diameters of 64, 51, 38, 26, and 13 mm, and heights of 46, 37, 28, 20, and 9 mm, respectively; the densified specimens were of the same sizes. Theoretical optimum-energy calculations[18] have shown that for a 50-mm-diam green-state Si_3N_4 specimen, the optimum incident photon energy for maximum sensitivity to thickness change should be near 55 keV, as shown in Fig. 3.

A set of equivalent 50- to 125-keV monoenergetic-photon images of a 105-mm-diam polyethylene bottle filled with Freon TF were obtained. Previous work had established that Freon TF had a mass attenuation coef-

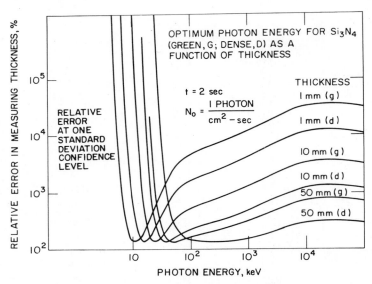

Fig. 3.   Theoretical Optimum Incident Photon Energy as a Function of
           Energy for Green and Dense Si_3N_4 Specimens of Various
           Thicknesses.

ficient close to that of green-state Si_3N_4. A region within each of the
images was analyzed statistically for noise. Thus, a statistical noise
level as a function of equivalent photon energy could be obtained.

    Figure 4 shows such a plot for the 51-mm-diam Si_3N_4 test cylinder.
The theoretical optimum energy is about 55 keV for the Si_3N_4 and about 70
keV for 105 mm of Freon TF. The experimental data show values of about
75 and 80 keV, respectively. This level of agreement (within about

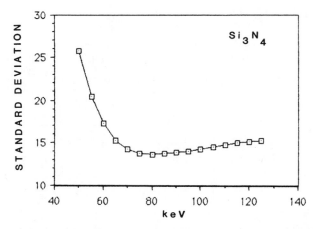

Fig. 4.   Experimentally Obtained Optimum Incident Photon Energy for a
           51-mm-diam Green-State Si_3N_4 Sample.

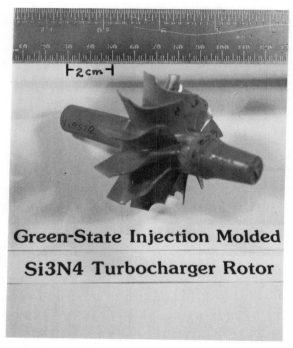

Fig. 5.   Green-State Turbocharger Rotor Used in Experimental Dual-Energy
           CT Scan Tests.

20 keV for the Si_3N_4 and 10 keV for the Freon TF) clearly shows the
potential of the dual-energy approach for quantitative densitometry
studies and significant reduction of BH.

Turbocharger Rotors

The dual-energy software package was implemented on the Somatom DR-H
to image both liquid Freon TF and a green-state injection-molded Si_3N_4
turbocharger rotor (see Fig. 5). An axial CT section of the entire rotor
was taken, with a slice thickness of 2 mm and the dual-energy BH correction. Figure 6 shows the image produced from the "Hi-kV" image
reconstruction data file.

Transaxial CT sections were also obtained. Figure 7, a section
through the vanes, shows edge artifacts (the star pattern in the hub) but
also shows an apparent low-density region near the core of the shaft.

SUMMARY AND CONCLUSIONS

We have established that the BH problem must be solved when applying
polychromatic X-ray CT to advanced ceramics. The most promising solution
appears to be the use of dual-energy approaches with careful attention to
the two energies selected for calibration. Medical CT scanners have tremendous potential for application to ceramics. The main difficulties are
related to the need for access to proprietary software and the limitation
in spatial resolution caused by focal-spot size and slice thickness.

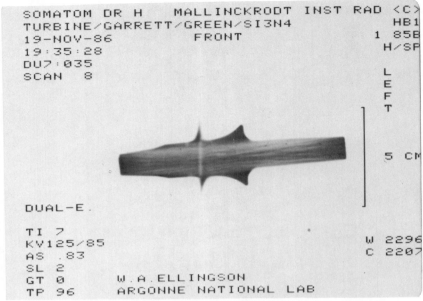

Fig. 6.   Axial CT Image of Rotor Shown in Fig. 5, Obtained with the
Dual-Energy Beam-hardening Correction.

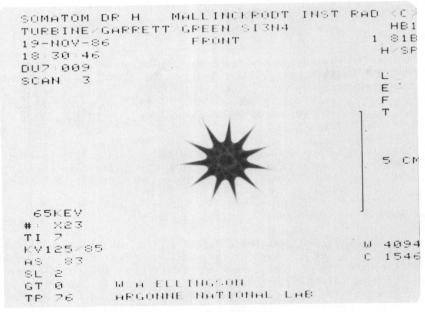

Fig. 7.   Transaxial CT Image of Same Rotor, Taken Through the Vanes.

Green-state parts up to 6 in. in size can be handled with existing
125-kVp X-ray heads. However, dense pieces of the same size cannot be
examined at low energies with desirable S/N ratios.

We have conducted many experiments with ceramic test pieces of
various types, sizes, and shapes. Some problems still exist with so-
called "streaming" or edge artifacts, even with dual energy. It is also
quite apparent that calibration fluids will be necessary. Freon TF
$z \approx 14$, $\rho = 1.8$) may be a suitable calibration fluid for green ceramics
and metholyzed iodide ($\rho \sim 3.39$ g/cm^3) may be suitable for densified
ceramics that are close to it in density.

ACKNOWLEDGMENTS

The authors would like to acknowledge the contributions of several
individuals who have been instrumental in this work. Mr. Bob Knapp and
Ms. Carolyn Offutt of the Mallinckrodt Institute of Radiology helped
acquire the CT data. Dr. Kamal Amin of the Norton High Performance
Ceramic Company, Northboro, MA, supplied many of the test samples used in
this work, and Dr. Harry Yeh of the Garrett Ceramic Components Division
of the Allied-Signal Aerospace Company, Torrance, CA, provided ceramic
test parts.

REFERENCES

1. Brooks, R. A., DiChiro, G., "Principles of Computer Assisted
    Tomography (CAT) in Radiographic and Radioisotopic Imaging," Phys.
    Med. Biol. 21, 689-732 (1976).
2. Brooks, R. A., DiChiro, G., "Theory of Image Reconstruction in
    Computed Tomography," Radiology 117, 561-572 (1975).
3. Reimers, P., Gilboy, W. B., Goebbels, J., "Recent Developments in
    the Industrial Application of Computerized Tomography with Ionizing
    Radiation," NDT Int. 17, 197-207 (1984).
4. Reimers, P., Goebbels, J., "New Possibilities of Nondestructive
    Evaluation by X-Ray Computed Tomography," Mater. Eval. 41, 732-737
    (1983).
5. Persson, S., Ostman, E., The Use of Computed Tomography in
    Nondestructive Testing of Polymeric Materials, Aluminum and
    Concrete, SKEGA AB, Ersmark, Sweden, 1985.
6. Segal, E., Notea, A., Segal, Y., "Dimensional Information Through
    Industrial Computerized Tomography," Mater. Eval. 40, 1268-1279
    (1982).
7. Gilboy, W. B., Foster, J., "Industrial Applications of Computerized
    Tomography with X- and Gamma-Radiation," in Research Techniques in
    Nondestructive Testing, ed. R. S. Sharpe, Vol. 6, Academic Press,
    New York, 1982, pp. 255-287.
8. Flannery, B. P., Deckman, H. W., Roberge, W. G., D'Amico, K. L.,
    "Three-dimensional Microtomography," Science 237, 1389-1544
    (1987).
9. Kinney, J. H., Johnson, Q. C., Bunse, V., Nichols, M. C., Saroyan,
    R. A., Nusshart, R., Paul, R., Brase, J. M., "Three-dimensional
    X-Ray Computed Tomography in Materials Science," Mater. Res. Soc.
    Bull. XIII(1), 13-17 (1988).
10. Segal, E., Ellingson, W. A., Segal, Y., Zmora, I., "A Linearization
    Beam-Hardening Correction Method for X-Ray Computed Tomographic
    Imaging of Structural Ceramics," in Review of Progress in

Quantitative NDE (Proc. Conf., Univ. of California, San Diego, August 3-8, 1986), ed. D. O. Thompson and D. E. Chimenti, Vol. 6A, Plenum Press, New York, 1987, pp. 411-417.

11. Segal, E., Ellingson, W. A., "Beam-Hardening Correction Methods for Polychromatic X-Ray CT Scanners Used to Characterize Structural Ceramics," in Nondestructive Characterization of Materials II (Proc. 2nd Intl. Symp., Montreal, Canada, July 21-23, 1986), ed. J. F. Bussiere et al., Plenum Press, New York, 1987, pp. 169-178.

12. Sawicka, B. D., Ellingson, W. A., Photon CT Scanning of Advanced Ceramic Materials, Atomic Energy of Canada, Ltd., Chalk River Nuclear Laboratory, Report AECL-9384, 1987.

13. Ellingson, W. A., Segal, E., Vannier, M. W., X-Ray Computed Tomography for Structural Ceramic Applications: Beam Hardening Corrections, Argonne National Laboratory, Report ANL-87-24, May 1987.

14. U.S. Congress, Office of Technology Assessment, New Structural Materials Technologies: Opportunities for the Use of Advanced Ceramics and Composites - A Technical Memorandum, OTA-TM-E-32, U.S. Government Printing Office, Washington, DC, September 1986.

15. Kalender, W. A., Perman, W. H., Vetter, J. R., Klotz, E., "Evaluation of a Prototype Dual-Energy Computed Tomographic Apparatus: I. Phantom Studies," Med. Phys. 13, 334-339 (1986).

16. Lehmann, L. A., Alvarez, R. E., Macovski, A., Brody, W. R., Pelc, N. J., Riederer, S. J., Hall, A. L., "Generalized Image Combination in Dual kVp Digital Radiography," Med. Phys. 8, 659-667 (1981).

17. Alvarez, R. E., Macovski, A., "Energy-Selective Reconstructions in X-Ray Computerized Tomography," Phys. Med. Biol. 21, 733-744 (1976).

18. Allemand, R., "Basic Technological Aspects and Optimization Problems in X-Ray Computed Tomography," in International Advanced Course on Physics and Engineering of Medical Imaging, Maratea, Italy, 1984, p. 11.

MICROTOMOGRAPHY DETECTOR DESIGN: IT'S NOT JUST RESOLUTION

H. W. Deckman, K. L. D'Amico, J. H. Dunsmuir,
B. P. Flannery and Sol M. Gruner[†]

Exxon Research and Engineering Company
Clinton Township, Route 22 East
Annandale, NJ   08801

INTRODUCTION

The design of an x-ray detector suitable for use in tomography must be optimized for the intended application. Recently[1], we have developed microtomography applications that require resolution of ~1 micron in three spatial dimensions and ~1% statistical accuracy in the reconstruction of attenuation coefficients for each cubic micron volume element in a ~.1 cubic mm specimen. X-ray detector design for these applications must take into account much more than just the demanding micron spatial resolution requirement. The detector must be optimized to take into account the physical properties of the specimen to be measured, the characteristics of the x-ray beam available to probe the specimen, signal to noise ratios needed in the reconstructed image and requirements of the data processing algorithm. In addition, the detector design should be sufficiently flexible to allow significant variation in the kinds of specimens that can be examined.

To exploit the potential identified[2] for creating a three dimensional x-ray microscope using intense synchrotron based x-ray sources, a panoramic detector was developed. Detector performance satisfied both resolution and statistical accuracy requirements for microtomography, and scan times as short as thirty minutes have been achieved using intense monochromatic x-rays from Exxon's x10A beamline at the National Synchrotron Light Source. The instrument functions not only as a three dimensional imaging x-ray microscope, but also allows spatially resolved elemental mapping. The ability to create spatially resolved elemental maps comes from tunability of this synchrotron beam line over a broad energy range. Details of the high resolution detector construction will be described elsewhere[3] and this paper will focus on design constraints imposed by alignment, speed and accuracy requirements. These requirements are derived primarily from x-ray source characteristics, and desired signal to noise ratios in reconstructed images.

[†]Department of Physics, Princeton University, Princeton, NJ   08544

In principle, a wide variety of methods exist for recording panoramic 5-20 keV x-ray images with micron spatial resolution. For instance, micron resolution can be obtained by magnifying the x-ray image with x-ray optics, sampling the image with a pinhole array, or by converting the x-ray image to either an electron or optical image which is magnified and detected. Many technical difficulties exist for each of these approaches, and available solutions are further constrained by the need to have digitally recorded data suitable for tomographic inversion. This paper will explore the nature and consequences of these constraints. In particular, detector design and operation for microtomography are constrained by characteristics of the x-ray source and desired signal to noise ratios in the reconstructed image. Constraints imposed by x-ray source characteristics lead to specific protocols (modalities) for image acquisition, while desired signal to noise ratios dictate observational strategies and impose stringent performance requirements on individual detector components. Before describing these design and operational constraints in detail, the panoramic detector we constructed will be described.

HIGH RESOLUTION PANORAMIC DETECTORS

The nature of the previously mentioned image acquisition and operational constraints led us to construct a high resolution electro-optic detector. Electro-optic[4] x-ray detectors can be broadly defined as panoramic position sensitive detectors, which utilize components developed for optical image recording such as vidicons, isocons, charge coupled devices (CCD), and charge injection devices (CID). Besides the readout device, electro-optic detectors can contain an energy converter (either a phosphor or photocathode), a gain element, and a format altering device to magnify or demagnify the image. The function of the energy converter is to distribute the energy of an x-ray photon amongst numerous, more easily handled quanta. Often the number of quanta produced in the energy converter is still too few to be effectively recorded in the readout device and a gain element (such as a magnetically or electrostatically focussed image intensifier) is interposed. A format alteration device is almost always required, because the x-ray image differs in size from the readout device. Only a few of the tremendous number of configurations for the different converter, gain, format alteration and readout devices available are suitable for tomography. The combination of these elements that we found most suitable for microtomography is shown in Figure 1.

The electro-optic detector contains no gain element and uses a phosphor as a energy converter, a microscope objective as a format altering device and a CCD readout device. To obtain the desired resolution we had to fabricate a new type of phosphor, which efficiently converts x-rays to visible light while preserving micron resolution in the image. Spreading of light within flat phosphor screens limits attainable resolutions to 2-4 times the phosphor thickness. To overcome this problem we lithographically[3] fabricated cellular phosphor screens in which the phosphor screen is divided into a honeycombed array with the phosphor material in each cell acting as its own optical waveguide. A schematic diagram of the high resolution cellular phosphor screen is shown in Figure 2. High resolution is achieved because light emission is localized within the dimensions of a single phosphor cell, which can be made smaller than one micron.

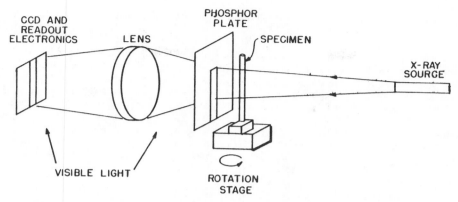

Figure 1.   Schematic of imaging x-ray detector components used for
            microtomography.

Format of the high resolution image produced by the cellular
phosphor screen is altered with a lens system, which relays the image
onto a CCD. CCD sensors have evolved to become the preeminent electro-
optical sensor technology[4]. The most important aspect of CCD sensors
for optical image detection is that they exhibit a readout and dark
noise that is at least an order of magnitude lower than vidicons,
isocons, orthocons and other electro-optic sensors. Commercially
available CCD sensors have a readout noise of less than 50
electrons/pixel and a dark noise of less than 5 electrons/minute-pixel
when operated at temperatures below -75°C. These low noise figures
result in substantial improvements in detective quantum efficiency of
electro-optic x-ray detectors constructed with CCD sensors. CCD sensors
also exhibit the largest dynamic range (saturation signal/r.m.s. readout
noise) of all electro-optic sensors. Saturation signals on many CCD
sensors approach $\sim 10^6$ electrons/pixel yielding a dynamic range for
signal detection of $\sim 10^5$. For signal levels below saturation, CCD

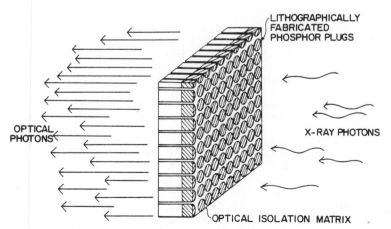

Figure 2.   Schematic of cellular phosphor. Each cell is an optical
            waveguide of about one micron diameter.

sensors exhibit linearity of response with signal intensity that is
independent of counting rate.

Lack of any count rate limitations make CCD sensors suitable as
area detectors for intense synchrotron radiation. The lack of count
rate limitations is due to integration of charge produced from separation of electron-hole pairs generated by the signal in each pixel until
the device is read out. Accumulated charge is read out from CCD sensors
by passing charge from pixel to pixel in a bucket brigade fashion to an
on-board charge sensitive preamplifier. Noise in read out of the
on-board preamplifier is substantially reduced through use of a double
correlated sampling technique, which cannot be used with other electro-
optical sensor technologies.

Performance and operational requirements for the individual
elements in the detector are imposed by data acquisition protocols and
signal to noise considerations.

PROTOCOLS FOR DATA ACQUISITION

Characteristics of the intense tunable synchrotron x-ray sources
dictate protocols for data acquisition. To generate accurate
tomographic images, low noise data must be obtained along a sufficient
number of coplanar paths through the target. In medical tomography, a
point x-ray source illuminates the patient with a fan beam of radiation
and rotates about a ring to illuminate a set of detectors opposite the
source. This protocol for data acquisition is referred to as fan beam
collimation. With highly collimated synchrotron radiation a different
data acquisition protocol, plane parallel, is used. In this mode,
collimated x-rays illuminate the target along a two dimensional set of
plane parallel paths, and radiographs of the sample are recorded at an
equally spaced number of view angles separated by an angle $\Delta\phi$, which
must satisfy the inequality,

$$\Delta\phi \leq \frac{\pi}{\sqrt{2}} \frac{\Delta t}{D} \tag{1}$$

where D is the sample size and Δt is the size of an individual resolution element imaged by the detector. Panoramic x-ray detectors used for
plane parallel data acquisition must have pixels that accurately map the
x-ray image onto a linear Cartesian grid. Geometric distortions from
true positional linearity that are greater than ~.5 pixels can produce
artifacts in the reconstructed image. These requirements are easily met
with the combination of a lens coupled format alteration and a CCD
readout. Most other format alteration techniques[5] (such as fiber-optic
expansion bundles) coupled with other readout devices do not meet this
requirement. Microscope objective lenses operating with micron resolution can relay distortion free images that meet positional linearity
requirements across the approximately one centimeter field of view of
the CCD. Pixels in CCD's are 5-50 microns in size and are lithographically laid out on a Cartesian grid. By choosing these as the format
alteration and readout device of the electro-optic detector, geometrical
distortion effects are virtually eliminated in microtomography
applications.

The plane parallel data acquisition protocol imposed by x-ray
source characteristics also creates several operational constraints.

Accurate reconstruction of plane parallel data requires that the imaging electro-optic detector must be aligned so that the axes of the Cartesian grid of pixels are parallel and perpendicular to the axis of rotation of the sample. Angular misorientation between these axes or wobble of the sample axis of rotation with respect to the detector results not only in loss of resolution in the reconstructed image but also can result in artifacts. Axes between detector and sample have to be aligned so that

$$\alpha_1 \leq \frac{1}{N} \tag{2}$$

where α_1 is the angular misorientation between the Cartesian grid of pixels and sample axis of rotation, and N is the number of pixels across the detector. Mechanical stages located below the sample rotation stage allow us to routinely reduce this angular misorientation to less than .1 milliradians. The magnitude of the angular misorientation is determined by using the photometric accuracy of the electro-optic x-ray detector to compare centers of mass of fiducial targets before and after rotation by 180°. By repeating this procedure at several different angles, the magnitude of the wobble for the sample axis rotation stage can be measured. For accurate reconstruction, the maximum spatial wobble should be less than ~.5 pixel, which corresponds to approximately ~.5 micron in our microtomography system.

The plane parallel imaging modality also sets a limit on the maximum sample size that can be scanned at a given resolution. The reconstruction algorithms for plane parallel collimation require that x-rays pass through only a single column of pixels in the sample. This requires that the sample size, D, satisfy the inequality,

$$D \leq \frac{2\Delta t}{\Delta \theta_1} \tag{3}$$

where $\Delta \theta_1$ is the maximum divergence of two principle rays passing through different points in the sample and Δt is the size of an individual resolution element imaged by the detector. A further restriction between the maximum allowable sample to detector distance, S, comes from considerations of resolution degradation by penumbral blurring. To prevent penumbral blurring degrading the resolution, the sum of sample size, D and maximum allowable separation between sample and detector, S, must satisfy the following inequality,

$$S + D \leq \frac{\Delta t}{\Delta \theta_2} \tag{4}$$

where $\Delta \theta_2$ is the maximum divergence angle between two x-rays passing through the same point in the sample. For synchrotron radiation collected from a bending magnet, horizontal beam divergences of two milliradians can be achieved. For micron resolution, Eqs. 2 and 3 restrict the maximum sample size to less than .5 mm and the maximum gap between sample and detector to be less than ~.1 mm. Source collimation sufficient to achieve ~10 micron resolution can be achieved with millimeter sized samples using conventional x-ray generators, although observation times become 10-100 times longer.

SIGNAL TO NOISE AND IMAGE QUALITY CONSIDERATIONS

Signal to noise considerations place strict limits on several aspects of detector performance as well as ways detectors can be used for data acquisition. The goal of tomography is to reconstruct a spatially resolved map of the sample's x-ray attenuation coefficient, F. This map is reconstructed from measurements of incident, I_o, and transmitted, I_t beam intensities, which are related by

$$I_t(\phi,t_1) = I_o e^{- \int F(t_1,t_2)dt_2} \qquad (5)$$

where t_1 and t_2 are x-ray impact parameters (coordinates) within the target and ϕ is the view angle. The data used in algorithms for tomographic inversion is the optical depth or projection, P (ϕ,t_1) defined as

$$P(\phi,t_1) = \ln[I_o(\phi,t_1)/I_t(\phi,t_1)] \qquad (6)$$

In our system, projections are derived from sequential measurements of I_o and I_t. Both systematic and statistical errors are introduced into the measurement of I_o and I_t as well as the reconstructed map of attenuation coefficients.

Statistical noise in reconstructions of attenuation coefficient maps arises from two sources: (1) noise in the projection data and (2) noise amplification introduced by the inversion method. Noise amplification (ω) is inherent in the tomographic reconstruction process and has been shown[6] to scale as

$$\omega^2 = N \geq \frac{D}{\Delta t} \qquad (7)$$

where N is the number of pixels per side in the reconstructed image. This amplification factor multiplicatively increases statistical errors in the projection data and is approximately a factor of 20 in our system. To obtain acceptable image quality in the reconstruction with this degree of noise amplification, statistical errors in the projection data have to be less than ~.4%. This means that statistical uncertainties must be less than .25% in individual measurements of I_o and I_t. Statistical uncertainties in observations of I_o and I_t come from x-ray counting statistics as well as noise introduced by the detector. A measurement of the noise introduced by the detector relative to an ideal detector is the detected quantum efficiency, ρ, which is

$$\rho = \left(\frac{S_D}{\sigma_D}\right)^2 \Bigg/ \left(\frac{S_I}{\sigma_I}\right)^2 \qquad (8)$$

where S means integrated signal, σ = the RMS integrated noise and subscripts D and I refer to the real and ideal detectors respectively. For an ideal detector ($\rho = 1$), obtaining observational uncertainties of I_o or I_t below .25% requires recording more than 160,000 x-ray photons. Many types of electro-optic x-ray detectors saturate before even 1,000 x-ray photons can be recorded. The maximum number of x-ray photons per pixel η_{max} that can be recorded in a single exposure of a lens coupled CCD detector before saturation occurs is

$$\eta_{max} = \frac{W}{\epsilon_p L_e \epsilon_c} \qquad (9)$$

where ϵ_p is the number of visible photons emitted per absorbed x-ray, L_e is the light gathering efficiency of the CCD, $\epsilon_c \approx .8$ is the quantum efficiency of the CCD, and W is the number of electrons that can be accumulated in the well before saturation. The maximum number of x-rays that can be accumulated per pixel is only a fraction of the number of electrons that can be accumulated in each well of the CCD. For effi-cient data collection, the CCD well depth must be large enough to record at least 160,000 x-rays per pixel in a single CCD exposure.

By recording at least 160,000 x-rays per pixel, all data for a single view can be acquired with a single exposure. Acquiring data with a single exposure minimizes observation time by eliminating readout time delays from multiple exposures. Readout rate is limited by the require-ment that the efficiency of transferring charge from pixel to pixel be greater than 99.99%. For most chips, readout rates faster than 1-10 microseconds/pixel degrade the charge transfer efficiency. Since defect free CCD sensors containing 5×10^5 pixels have been manufactured and chips containing 5×10^7 pixels are being developed, readout time for CCD chips can vary from .1 -10 sec. In our current system CCD readout time accounts for approximately 50% of the time required to scan a sample. Significant and undesirable increases in sample observation time occur if multiple exposures have to be summed to accumulate the necessary number of x-ray photons per pixel.

In our detector (which uses a scientific grade RCA SID-501 CCD), ~800,000 electrons can be accumulated in a single exposure before each well saturates. This means that to record 160,000 x-rays per pixel with a single exposure the phosphor and lens can produce and relay ($\epsilon_p L_e$) a maximum of five optical photons per detected x-ray. Energy conversion efficiency in the phosphor approached 2% and typically more than 100 optical photons are generated per detected x-ray (i.e. $\epsilon_p >100$). Light gathering efficiency, L_e, of the microscope objective is given by

$$L_e = (A)^2 \qquad (10)$$

where A is the numerical aperture of the objective lens. Microscope objective lenses that were used magnify the image 10-60 times and have light gathering efficiencies greater than 5% (A>.25). The number of optical photons that arrive at the CCD per detected optical photons is then $\epsilon_p L_e$, which is slightly larger than the maximum allowable number (5) computed from CCD storage capacity (well depth) considerations. As such, no gain element is needed in the electro-optic detector and the number of optical photons arriving at the CCD must usually be reduced by decreasing the lens efficiency with an optical stop. Even though a maximum of five optical photons can be relayed to the CCD for each detected x-ray, the detected quantum efficiency can be quite large due to the low readout noise of CCD detectors and the large number of x-rays being accumulated. The detected quantum efficiency, ρ, for lens coupled CCD x-ray detectors has been shown to be[5]

$$\rho = \cfrac{1}{\alpha\left[\cfrac{1}{T\alpha} + \cfrac{1}{T\alpha\epsilon_p} + \cfrac{1-L_e}{L_e}\cfrac{1}{T\alpha\epsilon_p} + \cfrac{1}{T\alpha\epsilon_p L_e \epsilon_c} + \left(\cfrac{Q_n + Q_d\tau}{T\alpha\epsilon_p L_e \epsilon_c}\right)^2\right]} \tag{11}$$

where T equals the x-ray absorption probability in the phosphor screen, L equals the number of x-rays per pixel incident, ϵ_c equals the quantum efficiency of the CCD, Q_n equals the readout noise of the CCD, Q_d equals the dark counting rate of the CCD and τ is the exposure time for a single image. Because our exposure time is on the order of one second and the RCA CCD dark counting rate is less than one per second, the last term in Eq. 11 can be ignored. Since microtomography requires recording a large number of x-ray photons, Eq. 11 can be reduced to

$$\rho = .56T^2 \tag{12}$$

where T is the absorption probability of a photon by the phosphor. Eq. 12 emphasizes the importance of making the phosphor layer thick enough so that the x-ray absorption probability approaches unity. This requirement means that for detection of 10 keV x-rays, the aspect ratio of micron sized cells in the cellular phosphor must be greater than ten to one.

Besides placing strict limits of performance on detector components, statistical noise considerations dictate an optimal sample observation strategy. The number of x-ray photons per pixel that must be incident on a sample can be estimated from the required statistical uncertainty (or detected accuracy), σ_F/F, of the reconstructed attenuation coefficient map, which is

$$(\sigma_F/F)^2 = \cfrac{e^{-\bar{F}D}}{\rho\bar{F}DN_0} \tag{13}$$

where N_0 is the number of x-rays incident per pixel, and $\bar{F}$ is the averaged linear x-ray attenuation coefficient. It is readily shown that the observational uncertainty is a minimum when the optical depth through the target $(\bar{F}D)$ equals two. For large optical depth, $\bar{F}D>2$, observational uncertainties increase because few photons are transmitted, while for small optical depth $\bar{F}D<2$ observational uncertainties increase because few photon interact with the sample. To achieve optimal observation conditions either the sample size D must be adjusted or the averaged linear attenuation coefficient of the sample tuned by changing x-ray energy. With synchrotron based tomography systems, the averaged linear attenuation can be readily tuned by changing beam energy.

Systematic errors as well as statistical errors occur in data collected by electro-optic detectors. Uncorrected systematic errors can introduce a variety of artifacts into reconstructed maps of x-ray attenuation coefficients. An example of the type of artifact introduced by systematic errors are rings that appear in reconstructed images. Persistent errors in response at a fixed position on the detector introduce rings, which overlay the reconstructed image. In general, systematic errors can be reduced to acceptable levels by a combination of detector design and calibration procedures. Major sources of system—

atic errors are signal dependent backgrounds, spatial inhomogeneities in detector response and offsets in the linearity of detector response. A flow chart of procedures used to correct these errors in measurements of $I_p(t_1)$ and $I_0(t_1)$ is shown in Figure 3.

To map spatial variation in the incident flux, $I_0(t_1)$, the sample is withdrawn from the beam and a calibration exposure recorded every 5-10 view angles. Recorded images of $I_0(t_1)$ and $I_t(t_1)$ contain ~250,000 data points digitized with 16 bit accuracy. These images are first processed to remove zero-offset effects. Zero-offset effects are

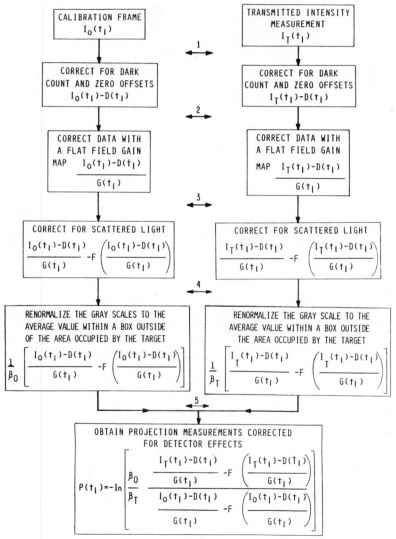

Figure 3.   Flow chart of calculation procedure used to remove systematic error from image.

removed (Figure 3, Step 1) using an offset map, which makes a corrected signal vanish when the incident x-ray flux = 0. To acquire this zero offset map, $D(t_1)$ an image is recorded with no incident x-ray flux. Following the zero offset correction, some of the effects of spatially inhomogeneous detector response are removed (Fig. 3, Step 2) by correcting the data with a flat field gain map. The gain map is designed to correct spatial response of all elements in the detector except the energy conversion device. Gain maps are obtained by removing the phosphor plate and recording the response of the detector to a spatially uniform optical beam. The next correction (Fig. 3, Step 3) removes effects of signal dependent background that are unique to electro-optic detector. Signal dependent backgrounds exist in all electro-optic x-ray detectors due to scattering of optical radiation between the energy conversion and readout devices. Scattered optical radiation produces a signal dependent background, which is spatially correlated with the signal as well as a component, which is spatially uncorrelated with the original signal.

To remove scattered light effects, a spatial filter function, F, must be designed. This filter function is designed from measurements of the recorded images of simple, well-characterized fiducial targets. Stripping scattered light effects out of recorded images can be difficult as well as time-consuming. Correction for scattered light effects is not needed for most targets because our detector was designed to significantly reduce scattered light. Scattered light effects were reduced by constructing the detector with a back-illuminated RCA CCD, which has a flat active surface that can be anti-reflection coated. Before projection measurements can be obtained from the data, effects due to time dependent x-ray source intensity variation must be removed (Fig. 3, Step 4). Effects of time dependent source variations are removed by normalizing the data to the average value in a region not obscured by the sample. Finally (Fig. 3, Step 5), fully corrected projection measurements are obtained.

REFERENCES

1. B. P. Flannery, Three-Dimensional X-Ray Microtomography, Science, 237:1439 (1987).
2. L. Grodzins, Optimum Energies for X-Ray Transmission Tomography of Small Samples - Applications of Synchrotron Radiation to Computerized Tomography, Nucl. Instrum. Methods, 206:541 (1983).
3. H. W. Deckman, J. H. Dunsmuir and S. M. Gruner, to be published.
4. S. M. Gruner, J. R. Milch and G. T. Reynolds, Survey of Two-Dimensional Electro-optical X-Ray Detectors, Nucl. Instrum. Methods, 195:287 (1982).
5. H. W. Deckman and S. M. Gruner, Format Alterations in CCD Based Electro-optic X-Ray Detectors, Nucl. Instrum. Methods, 246:527 (1986).
6. W. G. Roberge and B. P. Flannery, Computed Tomography by Direct Fourier Inversion II. Noise Amplification and Spatial Resolution, to be published.

REQUIRED CORRECTIONS FOR ANALYSIS OF INDUSTRIAL SAMPLES WITH MEDICAL CT SCANNERS

P. Engler, P. K. Hunt, E. E. Armstrong, and W. D. Friedman

BP America Research and Development
4440 Warrensville Center Road
Cleveland, Ohio 44128

INTRODUCTION

As a technique for non-destructive materials analysis, computed tomography (CT) has been especially useful for studying the dependence of the structure of ceramics on manufacturing processes.[1,2] CT also has been used for characterizing the lithology of reservoir cores while they are still contained in preservation material or a core barrel.[3,4]

The parameter measured by CT is the X-ray attenuation coefficient, which is a function of both material density and material composition. In order to simplify data handling, the X-ray attenuation coefficients are converted to units known as CT numbers (in units known as Hounsfield Units) by normalizing the measured attenuation coefficient to that of water:

$$CT = 1000(\mu - \mu_w)/\mu_w \qquad (1)$$

Here μ_w is the attenuation coefficient of water and μ the attenuation coefficient of the sample. In this way, the CT number of water is 0 and the CT number of air, which serves as absolute 0, is -1000 HU. The image is a matrix of these numbers.

Due to the large investment (>$400,000) required for scanners designed specifically for industrial use, industrial laboratories frequently purchase relatively inexpensive, used medical scanners. Used medical scanners can be obtained for less than $20,000. Although a lot is sacrificed in terms of spatial and contrast resolution by going this route, these instruments can provide very useful images for certain applications. Alternatively, acquisition of scanner time at nearby medical facilities can provide a low cost means of gaining access to this technology.

Usage of medical instruments can create several problems due to the fact that medical scanners and their corresponding reconstruction software are optimized to human samples whose size, shape, density and composition vary significantly from industrial samples. The linear attenuation coefficients of anatomical features fall in a rather narrow range compared to industrial materials. The highest CT number of a human body component (bone) is approximately +1000 HU, whereas sintered ceramics have CT numbers extending above +3000 HU.

For example, images of industrial or geological samples obtained on medical instruments frequently show artifacts due to a phenomenon known as beam hardening. Beam hardening occurs when polychromatic X-ray sources are used, as is the case for medical CT scanners. It is due to a preferential absorption of the lower energy X-rays as they pass through an object. Beam hardening induced artifacts include an apparent high density perimeter known as cupping and measured attenuation coefficients that are lower than the true values. These are problematic artifacts because their presence is not always obvious. Medical instrument manufacturers address this problem by using compensating algorithms, which adequately correct for beam hardening in samples with X-ray attenuating properties similar to those found in the human body. The algorithms succeed because medical systems are calibrated to water, the major component of the human body. However, industrial samples that attenuate X-rays more strongly require additional beam hardening correction. This paper describes correction procedures that can be applied when using medical scanners.

Also, the paper describes image enhancement and analysis software that can be implemented off-line to allow the use of CT as a quantitative NDE technique. With such development, CT could replace more costly, destructive analyses since it is fast and requires minimal sample preparation.

EXPERIMENTAL

Two CT scanners were used in this study: Data for beam hardening corrections were obtained on BP America's Deltascan 100. This is a 2nd generation instrument consisting of a sealed W X-ray source, operated at 120 kV and 25 mA, and three $Bi_4Ge_3O_{12}$ detectors. The scanner measures transmitted X-ray intensity during linear translations of the source and detectors at 3° increments for a 186° rotation. The scan takes 2 min, while image reconstruction takes an additional 10 s per slice. The image used as an example of data enhancement and analysis was obtained on a Technicare HPS 1440 (North Coast Imaging, Solon, OH). This is a 4th generation instrument consisting of a W rotating anode source that revolves around the sample in 4 s and a fixed ring of 1440 solid state detectors. Reconstructed images were transferred via magnetic tape to BP America's mainframe VAX cluster.

RESULTS AND DISCUSSION

Beam Hardening Artifact Reduction

The CT number of each pixel in an image of a cylinder that is homogeneous in terms of density and composition should be identical within statistical limits. A profile of CT numbers of a row of pixels through the sample should yield a flat line as shown in Figure 1 (left) for an image of a cylinder of water obtained on the Deltascan 100. The system software adequately corrected for beam hardening in this sample because the manufacturer calibrated with water, the major component of the human body. However, a line profile across the image of a cylinder of fused quartz, obtained on the same instrument produced a U-shaped profile, symptomatic of cupping (Figure 1, right). In this case, the manufacturer's pre-programmed parameters for the algorithm did not correct adequately for beam hardening.

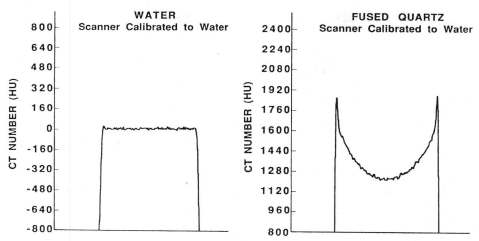

Figure 1. Line profiles of CT numbers through cylinders of water
           and fused quartz obtained on a Deltascan 100 calibrated
           to water.

It is not the high density itself that causes the artifact, but the
differential between the sample and the surrounding medium. These artifacts
can be minimized by reducing this differential. This can be accomplished by
immersing the sample in a medium whose CT number is close to that of the
sample. Now the large differential will no longer be at the sample edge,
but instead at the interface between the surrounding medium and the low
density plastic container wall. Cupping will now appear at the perimeter of
the container instead of the perimeter of the sample.

The objective of immersion is to come up with reproducible conditions
for producing an artifact free image whose CT numbers can be quantified for
comparison to other images or bulk properties. Figure 2 compares CT number
profiles across a 2" diameter sandstone scanned in air, water and in 3%
aqueous KI solution. In water, the severity of cupping has been reduced but
not eliminated. This figure now shows the profile of water as well as the
sample. At 3% KI, cupping in the image of the sample has been eliminated as

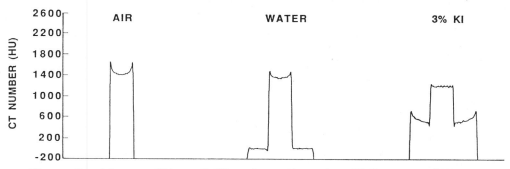

Figure 2. Line profiles of CT numbers through a 2" Berea sandstone
           core immersed in air, water and 3% aqueous KI solution
           obtained on a Deltascan 100 calibrated to water.

evidenced by the flat profile. The cupping in the image of the KI solution is expected and acceptable. Increasing the KI concentration above 3% produces a noisier image, since the KI solution hardens the beam and reduces the number of photons reaching the detectors. Thus, for this sample, the 3% KI solution, which has a CT number approximately 2/3 of the CT number of the sample, provides the optimal conditions for eliminating the cupping artifact from the image of the sample. (Absolute zero for CT numbers in Hounsfield units is -1000.) Analysis of other samples revealed that adjustment of the KI concentration in order to maintain a 2/3 ratio provided the best conditions for eliminating beam hardening artifacts.

In order to prevent KI solution from being imbibed into the porous sandstone, the surface had to be coated with silicone. Alternative immersion media are sand or silica beads, which meet the 2/3 rule for sandstones and green ceramics. The draw back of sand and glass beads is that it is difficult to get these materials into recessed areas. Also, they tend to pack unevenly. Consequently, the CT numbers in an image produced by immersing a part in sand will not be as reproducible as the numbers obtained when the part is immersed in aqueous KI solution. This can be an important point when trying to do quantitative work.

If one owns their own instrument, an alternative means of removing the cupping artifact is to recalibrate the system with a phantom made of a material having an X-ray attenuation similar to the samples to be scanned. In the recalibration process, a material phantom of homogeneous composition and circular cross section is substituted for the water phantom. The objective is to modify the parameters of the beam hardening correction software such that cupping is eliminated from the perimeter of the phantom. With recalibration, μ_w in Equation 1 is replaced by μ_c -- the attenuation coefficient of the calibration material. The CT values of the phantom become zero, and they will not vary across the image. Subsequent samples having attenuation coefficients similar to the new phantom will not exhibit cupping and will have CT numbers close to zero.

Figure 3 compares the line profiles of CT numbers across the diameter of an image of a 3" diameter fused quartz phantom for a system recalibrated

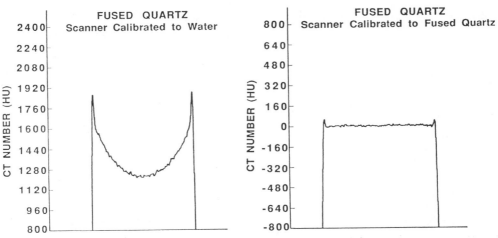

Figure 3. Line profiles of CT numbers through a fused quartz cylinder imaged on a Deltascan 100 calibrated to water and to fused quartz.

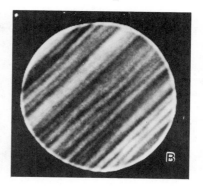

Figure 4.   CT images of a 3" Berea sandstone core obtained on a
Deltascan 100 calibrated to a)water and b)fused quartz.

with the quartz phantom to a system calibrated with water. The profile
becomes flat, except at the edges, indicating that cupping has been mostly
eliminated. The CT value of quartz is now zero with air still having a
value of -1000. Figure 4 compares a cross sectional slice of a 3" Berea
sandstone with the instrument calibrated to (a) water and (b) fused quartz.
In the image obtained with water calibration, a halo blurs the image of the
structure. In contrast, the laminae in the image after recalibration are
much more clearly defined. These results show that the optimal material for
recalibrating the scanner is a homogeneous cylinder with an X-ray
attenuation and diameter similar to the sample, e.g., fused quartz for
sandstone cores and aluminum for sintered SiC.

Image Enhancement and Analysis

Figure 5a shows an image of a slice containing the axis of a die-
compacted SiC, right cylinder. Although μ depends on both density and
composition, it correlates directly with density for regions of homogeneous
composition. On the gray scale in this image, white indicates the denser
portions of the material. There is an overall higher density at the two
lower corners and at the center of the top. Superimposed over this gradient
are a number of small localized regions of apparent high and low density

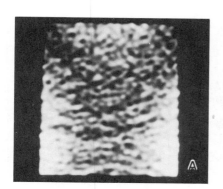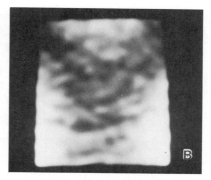

Figure 5.   Unsmoothed (a) and smoothed (b) CT images of die-
compacted, SiC, right cylinder.

that make it difficult to determine the more generalized density pattern. To resolve the density gradient, a 9x9 moving average smoothing algorithm was applied (Figure 5b). Now, the overall density gradients are readily apparent. To a ceramist, this means that during sintering the upper section having lower density will shrink more than the lower portion creating a non-uniform width. With this information and the product specification, ceramists can decide if they have to take steps to either insure more uniform density or to redesign the dimensions of the mold with a compensating shape so that the green part will have a shape that will sinter to the desired shape.

In order to quantify the density variations in the part, a calibration curve was established between the independently measured bulk density of a number of parts and the mean CT number of corresponding images. The mean CT number that can be obtained by using the region of interest marker on a commercial CT was not used, since the value obtained would have been incorrect. Instead, the CT value of each pixel was first weighted by its distance from the center line. The reason for the weighting factor is that the CT image represents a plane parallel to the long axis of a cylinder. If one assumes circular symmetry, then the value of a pixel at a distance r from the center line represents the density of all of the volume elements on a circle having a radius of $2\pi r$. Since there are more volume elements on a circle of larger radius, pixels further away from the center line have to be give proportionately more weight in calculating an average that is representative of a bulk parameter. The resulting calibration curve had a correlation coefficient of 0.98 for eight specimens having bulk densities ranging from 1.7 to 1.9 g/cm^3. Based on this curve, the density of the part in Figure 5 varied from 1.67 to 1.79 g/cm^3.

A second way for displaying this data is in binary format (Figure 6). Using binary format, upper and lower threshold limits are selected for the CT number. Any pixels having CT numbers within these boundaries appear black, while any pixels having values above or below the boundaries are white. In this way, the density patterns can be observed as the upper boundary value is changed. Also, binary displays are quite useful in that they provide one means of enhancing an image to facilitate stereological analysis of the image.

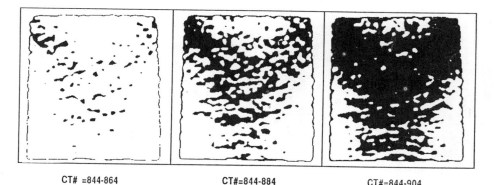

CT# =844-864   CT#=844-884   CT#=844-904

Figure 6.   Binary CT images of green, die compacted SiC cylindrical plug.

CONCLUSIONS

Medical CT scanners can provide a means of non-destructively characterizing industrial samples as long as close attention is payed to the fact that these samples differ in composition and density from the objects for which the medical scanners are designed. Very good results can be obtained by compensating for these differences. For example, cupping artifacts can be eliminated in two ways: 1) The sample can be immersed in a medium, such as aqueous KI solution or sand, that pre-hardens the beam and moves the cupping away from the sample. The optimal medium should have a CT number approximately 2/3 of the CT number of the sample in absolute Hounsfield units. 2) The beam correction algorithms can be recalibrated using a phantom that attenuates X-rays in a manner similar to the sample.

Image enhancement and quantitative analysis on off-line computers increase the ability to relate material structure and properties to manufacturing processes. This could allow CT to replace more costly, destructive techniques. A single part can be characterized for density and compositional variations, and the presence of voids and inclusions, at each stage of the manufacturing process.

REFERENCES

1.   W. D. Friedman, R. D. Harris, P. Engler, P. K. Hunt and M. Srinivasan, Characterization of Green Ceramics with X-ray Tomography and Ultrasonics, Proc. Non-Destructive Testing of High Performance Ceramics at Boston, MA, Aug. 25-27, 1987, Amer. Ceramics Soc., ed. by A. Vary and J. Snyder, pp. 128-131.

2.   P. K. Hunt, P. Engler and W. D. Friedman, Industrial Applications of X-ray Computed Tomograph, in: "Advances in X-ray Analysis", vol. 31, C. S. Barrett et al., ed., Plenum, New York, (1988), pp. 99-105.

3.   S. L. Wellington and H.J. Vinegar, X-ray Computerized Tomography, J. Petroleum Tech., August:885 (1987).

4.   P. K. Hunt, P. Engler and C. Bajsarowicz, Computed Tomography as a Core Analysis Tool: Instrument Evaluation, and Image Improvement Techniques, J. Petroleum Tech., in press.

LM-ACT FOR IMAGING RAM DEVICES

IN X-RAY DIFFRACTION TOPOGRAPHS

Warren T. Beard* and Ronald W. Armstrong+

*Laboratory for Physical Sciences, College Park, MD   20

+Department of Mechanical Engineering, University of
Maryland, College Park, MD   20742

INTRODUCTION

Analysis of semiconductor material and associated integrated circuits (IC) is imperative for ensuring quality products. Currently,
routine circuit testing is dominated by measurement of the optical and
electrical material/device properties through final device performance
and parametric testing.

Characterization of the crystal microstructure still is not considered a routine process test. Structural characterization usually is
based on double-crystal rocking curves[1], x-ray topography[2], or a combination of these techniques. Rocking curves provide excellent angular
characterization of the crystal quality in small areas. Topography
reveals lattice strains and misorientations over large areas.

Recently Qadri, Ma, and Peckerar[3] have used topographic results to
correlate lattice strain and device performance in silicon integrated
circuits. Barnett, Brown, and Tanner[4] have used topography for characterization of Liquid Encapsulated Czochralski (LEC) GaAs material. In
this report we present details of a new topographic technique for high
resolution reflection imaging, entitled Line Modified-Asymmetric Crystal
Topography (LM-ACT). For this reflection method, no particular sample
preparation is required. This nondestructive technique can be used after
any step in the process and the wafer returned for further processing.

LINE MODIFIED-ASYMMETRIC CRYSTAL TOPOGRAPHY

The Asymmetric Crystal Topography (ACT) method is known to give fine
angular resolution in diffraction images of varying crystal orientations
spread over large areas[5]. Improved x-ray diffraction topography (XRDT)
images have been produced with our topography system which employs a
horizontal line x-ray source. This technique has been described in an
earlier paper[6] as Line Modified-Asymmetric Crystal Topography. In this
new arrangement, the horizontal line source is perpendicular to the vertical rotation axes of both the monochromating and specimen crystals.
The purpose of the changed design was to minimize the deleterious effect

of large vertical divergence on spatial resolution of intensity variations in topographic images. The significance of controlling the vertical divergence for high resolution was given emphasis by Newkirk[7]. A similar x-ray system was reported by Bonse[8].

Figure 1 shows the new LM-ACT system on a stereographic projection basis[9] relating to the imaging of structural details within a proprietary Random Access Memory (RAM) device, produced on a silicon substrate. Coupled stereographic projections of the monochromating crystal and specimen crystal are shown with the monochromating crystal on the right. A schematic view of the XRDT experiment is shown below the coupled stereographs. The coupled stereographs show along the equatorial plane that the vertical divergence at the specimen is controlled by the source height and the source-to-specimen total distance.

In the LM-ACT representation (Fig. 1), the horizontal line source is 0.04×8.0 mm^2. The silicon monochromating crystal is asymmetrically cut 11^0 off the (111) toward the [010]. Using the (111) Bragg reflection ($\Theta_B = 14.22^0$, Cu Kα_1) rotated about the [$\bar{1}$01] axis, the x-ray incidence angle relative to the crystal surface is 25.22^0 and the reflection angle relative to the surface is 3.22^0. A compression ratio of (sin 3.22^0/sin 25.22^0) = 0.132 produces a reflected beam from the monochromator which is collimated and intensified, and which contains both Kα_1 and Kα_2 wavelength components. In Fig. 1 this is the collimated intense monochromatic beam (CIMB). The beam height is limited by the monochromating crystal which is approximately 3 cm.

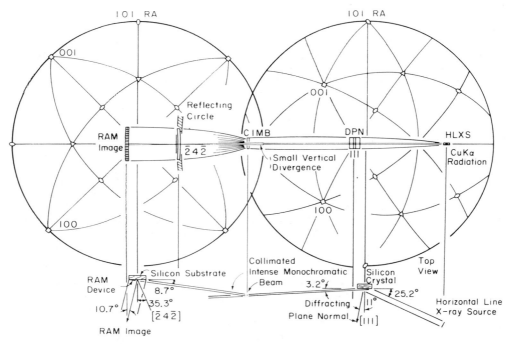

Fig. 1.   Coupled stereographic projections for silicon monochromating crystal and silicon RAM device studied by line modified-asymmetric crystal topography (LM-ACT) method.

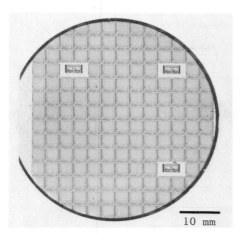
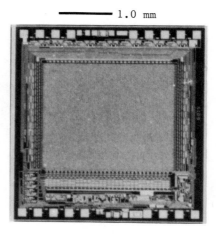

1.0 mm

10 mm

Fig. 2.   Optical photographs of the RAM device.   (a) Total wafer; (b)
           single unit.

The (010) face of a silicon-based RAM device was set for Bragg
reflection from the inclined ($\overline{2}4\overline{2}$) plane ($\Theta_B$ = 44.0°, Cu Kα_1) when the
wafer was rotated 8.7° about the [$\overline{1}$01] axis. The zero reference for the
rotation is taken as the position where the incident x-ray beam is
parallel to the crystal surface. With the [$\overline{2}4\overline{2}$] diffracting plane normal
positioned within the reflecting circle at an angle of ($\pi/2$)-Θ_B from the
CIMB, the RAM device image emerges from the specimen surface at 10.7°
away from the device surface normal. The near orthogonal exit beam
direction allows for capturing the XRDT image close to the device surface. This is important to minimize the effect of the emergent spread in
intensity from each point on the crystal surface. The vertical
divergence from a point on the crystal specimen looking towards the x-ray
source is ($\Delta\beta$=0.04mm/406.4mm) 9.84×10^{-5} radians (20.3 arcsec).

THE MICROSTRUCTURE OF THE RAM DEVICE

Figure 2a,b shows low magnification optical photographs of the RAM
device surface structure. The authors determined by Laue back-reflection
orientation analysis that the silicon substrate was cut approximately 3°
off the (010) toward the [$\overline{1}$01]. The individual device elements are about
3600 μm square.

LM-ACT TOPOGRAPHY EXPERIMENTS

Figure 3a,b,c shows several Bragg reflection "spots". Each print
shows the actual nuclear plate exposures from our experiment. Two
general points should be noted: 1) with the film-to-specimen distance on
the average of 3 mm, both Kα_1 and Kα_2 images are separately resolved via
the diffraction process; 2) the white mark on each topograph is a wire
shadow used as a fiduciary mark. Fig. 3a is the ($24\overline{2}$) reflection with
the specimen rotated 90° counterclockwise relative to Figs. 1 and 2.
Fig. 3b is the ($\overline{2}4\overline{2}$) reflection as in the geometry shown in Fig. 1. Fig.
3c is the same ($\overline{2}4\overline{2}$) reflection as Fig. 3b, however, in Fig. 3c the

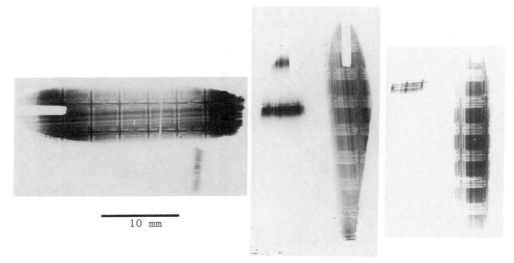

10 mm

Fig. 3.   Bragg "spots" in LM-ACT experiments. (a) (24$\bar{2}$) reflection; (b)
(2$\bar{4}$2) reflection; (c) ($\bar{2}$4$\bar{2}$) reflection with reduced vertical
divergence.

monochromator-to-specimen distance was doubled from that used for Fig.
3b. This was done to verify that vertical resolution was determined by
the angular divergence of the x-ray source as seen by a point on the specimen crystal. As mentioned earlier, Fig. 1 indicates that this condition is met across the equatorial plane of the coupled stereographic
projections.

Tanner and Bowen have provided the relation[10]:

$$\text{Resolution Limit} = (h \ d_2)/[d_1 + D + d_2(1-h/H)], \qquad (1)$$

where h is the source height, D is the source-to-monochromator distance,
d_1 is the monochromator-to-specimen distance, d_2 is the specimen-to-film
distance, and H is the monochromating crystal height. Even at the low
magnification (approximately 2X) of the diffraction images, better resolution can be seen for Fig. 3c which had a significantly larger value of
d_1. Although at $d_2 = 3$ mm, the $K\alpha_1$ and $K\alpha_2$ images are separated as a
result of the horizontal divergence of the beam, the small vertical
divergence still maintains high resolution in the XRDT image. For the
highest resolution XRDT images, equation (1) shows that the distance d_2
should be minimized. When d_2 is minimized, the $K\alpha_1$ and $K\alpha_2$ images are
largely superimposed on the film plate with an overlap of only a fraction
of a micron.

COMPARATIVE LM-ACT AND ACT IMAGES

Figure 4a,b shows a direct comparison of LM-ACT and ACT images.
Both topographs were exposed with all distances in equation (1) being
equal. However, the ACT topograph uses the x-ray source in a vertical
orientation which gives h=8.0 mm. A relatively small value of d_2
(approximately 1 mm) was employed. As for normal operation of the ACT
system, D=320 mm and d_1=115 mm.

$$\frac{\rho_s}{\rho_p} = \frac{3\pi^2}{16} \frac{1}{\xi \mu} \frac{\overline{C^2}}{\overline{C}^2} \quad [1-\cot\Theta_B \tan\chi]\sin(\Theta_B+\chi), \qquad (2)$$

where μ is the linear absorption coefficient, $(-\chi)$ is the angle between the diffracting plane normal and the surface normal, C is the x-ray polarization, and ξ is the extinction distance. The form of ξ is given by

$$\xi = \frac{mc^2}{e^2} \frac{\pi V}{C \lambda |F|} \quad [|\sin(\Theta_B-\chi)| \sin(\Theta_B+\chi)]^{1/2}, \qquad (3)$$

where $(e^2/mc^2) = 2.82 \times 10^{-13}$ cm, V is the cell volume, λ is the x-ray wavelength, and F is the structure factor. Reflection from the $(\overline{2}4\overline{2})$ silicon substrate gives $\xi = 16.12$ μm and $(\rho_s/\rho_p) = 4.04$.

For a relatively perfect crystal, ξ is of the same order as the width of an individual dislocation line or the size of the mosaic blocks employed to model the (secondary) intensity enhancement. Bragg and Azaroff[12] have reported block sizes of approximately 15 μm for rather perfect single crystal silicon. Based on the smallest (horizontal) dimension of enhanced intensity in Fig. 5a of 5 μm, we assume a mosiac block size ten times smaller (0.5 μm) is reasonable. In standard alloying treatments dislocation loops on the order of 0.05 μm might

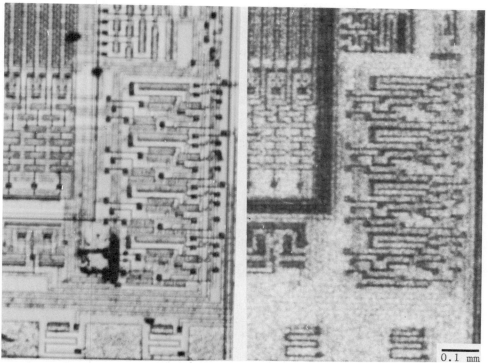

0.1 mm

Fig. 5.   Enlargement (approximately 100X) of (a) optical photograph and (b) LM-ACT topograph of the corner region of a RAM device. Both images are of the same RAM unit.

—————————— 1.0 mm

Fig. 4.   Line modified-ACT (a) and normal ACT (b) images obtained under
           otherwise identical conditions.

The fine details captured in the LM-ACT image from the ($\overline{2}4\overline{2}$) Si
substrate are obvious. However, note the clear horizontal resolution of
the ACT image despite the image smear produced by the large vertical
divergence. This large vertical divergence is traced to the large vertical x-ray source height. We see in Fig. 4b that the fine angular resolution of the ACT system carries over to produce good resolution along
the horizontal axis. In the LM-ACT system the Bragg condition controls
the horizontal resolution, which is excellent even with small d_2 where
$K\alpha_1$ and $K\alpha_2$ images are combined.

The small vertical divergence which we have employed in LM-ACT
ensures the vertical resolution of details seen in Fig. 4a. The somewhat
diffuse dark bands running at an included angle to the vertical axis in
Fig. 4a are the result of variations in the intensity of the x-ray
source. These variations have been proven via pinhole photographs
obtained of the x-ray source.

HIGH RESOLUTION LM-ACT IMAGING

Figure 5a,b shows the match of the LM-ACT diffraction contrast image
and the optical image of features observed in one corner of one RAM
device. Both images are shown at larger than 100X magnification. Almost
all of the details observed in Fig. 5b are the results of enhanced x-ray
intensity. Localized lattice strains produced by alloying treatments are
very likely the major cause of the intensity-based diffraction contrast.

An estimate of the possible intensity enhancement is obtained from
the ratio of integrated intensities for an ideal mosaic structure
(exhibiting secondary extinction), ρ_s, and that for a perfect crystal
subject only to primary extinction, ρ_p, as given by[11]:

reasonably be expected. Thus, the enhanced diffraction image contrast is attributed to lattice strains associated with the diffusion/alloying processes involved in fabricating the RAM device.

SUMMARY

Excellent spatial resolution of microstructural details is obtained for a RAM device with the LM-ACT technique. The method should prove useful for nondestructively monitoring device fabrication and performance. The authors are currently upgrading the LM-ACT apparatus with a high precision goniometer and a specially-cut 75 mm diameter monochromating crystal. With these modifications, the LM-ACT system should be capable of giving XRDT images of the entire surface of 75 mm wafers at a resolution of 0.1 μm.

ACKNOWLEDGMENTS

The authors wish to thank Dr. X. Jie Zhang for his excellent photographic work. Also we wish to thank T.S. Ananthanarayanan of Brimrose Corporation for supplying the silicon RAM wafer. R.W. Armstrong has been supported by the Office of Naval Research, Contract N00014-86-K-0286.

REFERENCES

1.   X. Chu and B.K. Tanner, "Double Crystal X-ray Rocking Curves of Multiple Layer Structures," Semicond. Sci. Technol. 2: 765 (1987).

2.   G.A. Rozgonyi and D. Miller, "X-ray Characterization of Stresses and Defects in Thin Films and Substrates," in: Methods and Phenomena 2-Characterization of Epitaxial Semiconductor Films, H. Kressel, ed., Elsevier Scientific Publishing Co. (1976), p. 185.

3.   S.B. Qadri, D. Ma, and M. Peckerar, "Double-Crystal X-ray Topographic Determination of Local Strain in Metal-Oxide Semiconductor Device Structures," Appl. Phys. Lett. 51: 1827 (1987).

4.   S.J. Barnett, G.T. Brown, and B.K. Tanner, "The Distribution of Lattice Strain and Tilt in LEC Semi-Insulating GaAs Measured by Double-Crystal X-ray Topography," in: Inst. Phys. Conf. Ser. No. 87: Section 9 (1987), p. 615.

5.   W.J. Boettinger, H.E. Burdette, M. Kuriyama, and R.E. Green, Jr., "Asymmetric Crystal Topographic Camera," Rev. Sci. Instrum. 47: 906 (1976).

6.   W.T. Beard, T.S. Ananthanarayanan, and R.W. Armstrong, "Line-Modified Asymmetric Crystal Topography (ACT) Technique," in: 16th Symposium on Nondestructive Evaluation, Nondestructive Testing Information Analysis Center, Southwest Research Institute (1987), p. 23.

7.   J.B. Newkirk, "The Observation of Dislocations and Other Imperfections by X-ray Extinction Contrast," Trans. Metallurgical Soc. AIME 215: 431 (1959).

8.   U. Bonse, "X-ray Picture of the Field of Lattice Distortions around Single Dislocations," in: Direct Observations of Imperfections in Crystals," J.B. Newkirk and J.H. Wernick, eds., John Wiley and Sons, New York (1961), p. 431.

9.   R.W. Armstrong, "X-ray Diffraction Topography Description via the
      Stereographic Projection," in: Applications of X-ray Topographic
      Methods to Materials Science," S. Weissmann, F. Balibar, and J.-F.
      Petroff, eds., Plenum Press, N.Y. (1984), p. 33.

10.   B.K. Tanner and D.K. Bowen, private communication, March, 1988.

11.   R.W. Armstrong, "Laboratory Techniques for X-ray Reflection
      Topography," in: Characterization of Crystal Growth Defects by X-ray
      Methods, B.K. Tanner and D.K. Bowen, eds., Plenum Press, N.Y.
      (1980), p. 349.

12.   R.H. Bragg and L.V. Azaroff, "Direct Study of Imperfections in
      Nearly Perfect Crystals," in: Direct Observations of Imperfections in
      Crystals, J.B. Newkirk and J.H. Wernick, eds., John Wiley and Sons,
      New York (1961), p. 415.

AUTHOR INDEX

Abuhasan, A., 471
Adl, T. P., 303
Ahonen, A., 59
Alam, S. O., 545
Alfthan, C. V., 59
Andermann, G., 261
Andersen, C. A., 489
Aoki, S., 141
Arai, T., 21, 83, 131
Armstrong, E. E., 651
Armstrong, R. W., 659
Artz, B. E., 121

Bamberger, M., 293
Barrett, C. S., 609
Beard, D. W., 531
Beard, W. T., 659
Beiter, T. A., 561
Benson, D. K., 609
Berneike, W., 105
Bessières, J., 285
Borgonovi, G. M., 397
Briden, F. E., 437
Bullock, D. C., 269
Butler, B. D., 389

Cantrell, J. S., 561
Carpenter, D. A., 115
Castillo, G., 323
Caussin, P., 531
Chandra, D., 609
Chatfield, E., 593
Chaudhuri, J., 279
Chen, M.-M., 323
Cohen, J. B., 341
Correale, S. T., 617
Couture, R. A., 233
Crowder, C. E., 507
Crystal, K. R., 311

D'Amico, K. L., 641
Davis, M. E., 507
Deckman, H. W., 641
Dunsmuir, J. H., 641

Edmonds, J. W., 545

Ellingson, W. A., 629
Engler, P., 651

Flannery, B. P., 641
Friedman, W. D., 651
Fujiwara, F., 261
Fukushima, S., 155
Futernick, R., 149

Garces, J. M., 507
Gazzara, C. P., 397
Giauque, R. D., 149
Gilfrich, J. V., 1
Goehner, R. P., 311, 577
Gohshi, Y., 141, 155, 167
Gonsui, S., 131
Gorman, G. L., 323
Gruner, S. M., 641

Harada, H., 365
Harbison, J. P., 279
Harding, A. R., 31, 39, 221
Hart, M., 481
Hayakawa, S., 141
Hegedüs, F., 251
Heizmann, J. J., 285, 415, 423
Hietala, M., 49
Higashi, Y., 155
Hirose, Y., 443, 451
Hirvonen, M., 59
Holcombe, C. E., 115
Holomany, M., 539
Holynska, B., 45
Hom, T., 545
Howard, J. K., 261
Howard, S. A., 523
Huang, T. C., 261, 269, 331
Hunt, P. K., 651

Ihara, I., 459
Iida, A., 141, 167
Ikeda, S., 197
Ingham, M. N., 227
Istone, W. K., 601
Iwamoto, K., 131